# Exemplary STEM Programs: Designs for Success

# Exemplary STEM Programs: Designs for Success

Edited by Robert E. Yager and Herbert Brunkhorst

NSTA press
National Science Teachers Association
Arlington, Virginia

NSTA press®
National Science Teachers Association

Claire Reinburg, Director
Wendy Rubin, Managing Editor
J. Andrew Cooke, Senior Editor
Amanda O'Brien, Associate Editor
Amy America, Book Acquisitions Coordinator

ART AND DESIGN
Will Thomas Jr., Director

PRINTING AND PRODUCTION
Catherine Lorrain, Director

NATIONAL SCIENCE TEACHERS ASSOCIATION
David L. Evans, Executive Director
David Beacom, Publisher

1840 Wilson Blvd., Arlington, VA 22201
*www.nsta.org/store*
For customer service inquiries, please call 800-277-5300.


17 16 15 14 4 3 2 1



**Library of Congress Cataloging-in-Publication Data**
Yager, Robert Eugene, 1930-
Exemplary STEM programs : designs for success / by Robert E. Yager and Herbert Brunkhorst.
pages cm
"National Science Teachers Association."
Includes index.
ISBN 978-1-941316-03-0
1. Science--Study and teaching--United States. 2. Technology--Study and teaching--United States. 3. Engineering--Study and teaching--United States. 4. Mathematics--Study and teaching--United States. 5. Curriculum planning--United States. 6. Instructional systems--United States--Design. I. Brunkhorst, Herbert, 1944- II. National Science Teachers Association. III. Title. IV. Title: Exemplary science, technology, engineering, and mathematics programs.
Q181.Y25 2014
507.1'073--dc23

2014022582

# Contents

# Foreword

## Exemplary STEM Programs: Designs for Success

*Robert E. Yager*
*University of Iowa*
*Iowa City, IA*

As 2015 emerges, the reforms of science teaching in every respect center on STEM efforts (Science, Technology, Engineering, and Mathematics). It defines the reform efforts in the *Next Generation Science Standards* (*NGSS*) and all the National Science Teachers Association (NSTA) publications and projects. But, the reforms of the late 1960s and early 1970s relate to current efforts and should serve as reminders of the real changes needed that could provide exciting outcomes in the present day. We should try to match those championed by Jim Rutherford's Project 2061 (AAAS 1989). Rutherford envisioned that major educational changes would take 70+ years (2061 the year Halley's Comet would again be seen from Earth). It is important to note that most changes envisioned and tried will *not* take place overnight—not even in a year or two; but, they could take less time with such exciting plans and actions being taken and tried across the United States.

The recent Common Core efforts indicate where we are with science education reforms. The Common Core was initially designed to provide a consistent and clear understanding of what concepts students would be expected to learn in mathematics and the English language. Unfortunately, science was not a part of this Common Core. Mathematics and the English language are important subjects where students can memorize content and then be tested for remembering! But, did such "courses" really indicate any evidence for real student "learning"?

The new efforts with STEM reform were recently described by Rodger Bybee (2010, 2013) and Harold Pratt (2013). These efforts encourage success with the national *NGSS* reforms and indicate specific efforts to define STEM reforms.

Continuing efforts should be encouraged as ways of using STEM results that indicate "More Success" with school science. Bybee reviewed the changes proposed originally by Project 2061 (AAAS 1989). He was a past student of both Jim Rutherford and the late Paul DeHart Hurd. Project 2061 was a project concerning student understandings and ways of thinking that are essential for all citizens in a world shaped by science and technology. The purpose of education is to prepare people to lead personally fulfilling and responsible lives. Therefore, science education should include all four facets, namely, science, technology, engineering, and mathematics. It should help students develop the understandings they need to become persons able to think for themselves and to solve the problems they encounter and define. We offer this monograph to provide exemplary teaching of current STEM reform efforts. But, the reforms outlined must be related to successes with actually "doing" science (also including science education itself)!

## *NGSS* as a Model for Achieving Success

Specific goals framing the National Science Education Standards (NSES; NRC 1996) were recommended and unanimously approved as ways to educate students who:

1. experience the richness and excitement of knowing about and understanding the natural world (doing science);
2. use appropriate scientific processes and principles in making personal decisions (solving personal problems);
3. engage intelligently in public discourse and debate about matters of scientific and technological concern (solving societal problems); and
4. increase their economic productivity through the use of the knowledge, understandings, and skills of the scientifically literate person in their careers (career choices). (Harms and Yager 1981).

The *NGSS* support the following eight science features as goals for improving student learning and mastery of the *practices* of science. They are: (1) Asking questions; (2) Developing and using models; (3) Planning and carrying out investigations; (4) Analyzing and interpreting data; (5) Using mathematics and computational thinking; (6) Constructing explanations; (7) Engaging in argument from evidence; and (8) Obtaining, evaluating, and communicating information (Hazzard 2014). The *NGSS* also urges most of all that the major focus is on crosscutting concepts and core ideas. These practices and foci all indicate aspects of personally "doing" science successfully, which result from student problem solving.

## Learning From Earlier Reform Efforts

The ideas guided by the reform authors of the 1960s and 1970s were these: (1) Learning should be fun; (2) Subject matter should be relevant; (3) Science should be easy; and (4) All students can and should experience the actual "doing" of science. But, these four essential features were not typically achieved!

One example from the 1960s and 1970s was the use of 14 process skills and named Science—A Process Approach (SAPA 1965). Central to SAPA was having students develop their own thinking and reasoning skills. This National Science Foundation (NSF) reform program was adopted fully and used in many classrooms, including those in Cedar Rapids, Iowa. It was suggested as a new focus to replace or add to concepts comprising typical curricula.

Unified Science and Mathematics for Elementary Schools was another project (USMES; NSF 1973). It was a major focus on problem solving that emphasized student involvement. Some of the major examples include: (1) Pedestrian Crossings, (2) Lunch Line Management; (3) Burglar Alarm Designs; (4) Electromagnet Devices; (5) Consumer Product Testing; and (6) Soft Drink

Consumption. These were all issues used by students for solving their own problems rather than the focus being defined as topics, concepts, or ideas selected and taught by teachers.

Another example of reform that illustrates STEM was the Man-Made World (Truxal and Piel 1971). The project was conceived when it became clear that fewer and fewer students were electing to take chemistry and physics courses (often offered only as college preparation). It was alarming to hear that the United States was entering an "age of technology" and yet science did not fit with standards or textbook-defined curricula. The invention of the computer promised changes and yet it took over 30 years for the changes needed for it to happen. The "Man-Made World" was a focus on engineering as identifying one aspect of the STEM acronym. But it resulted with little success in many schools. It was not seen as a course needed or important for students in most high schools.

## Problems With Defining Science

Recently a dozen former students who were new science teachers were asked to check with a typical samplings of students in their schools to indicate whether or not they felt they were really "doing" science in their schools. Over 90% out of the 300+ involved students answered that they *did* science in class in their courses in elementary, middle, and high schools. But, further questioning resulted in 90% of the students reporting they were actually only doing assignments, working in laboratories, reading from textbooks, and taking tests that measured their remembering. They did not associate science as something to be done outside of what was presented by the teacher or included in textbooks. It related to nothing outside of the classroom.

Most researchers accept the definition of science to be the "exploration of the natural universe and seeking explanations of the objects and events encountered." The information gathered from the sample of students did not reveal *any* evidence of students personally "doing" science itself or have any understanding for use of science outside of classrooms. We all need to realize that "actually doing" science does not include dependence on textbooks, typical laboratories, or doing directed experiments and then expecting students to recall what was told to them. If science is unrelated to students' own lives and their own perceived problems, there is often no "real" learning of what science is! We need to focus on student understanding and use of science as known and accepted by science researchers.

## Current STEM Efforts

The 20 examples of STEM teaching in this monograph were selected from over 100 submitted draft chapters. The authors were told that all new NSTA Press books must connect with and reference the *Next Generation Science Standards* (*NGSS*) as well as *A Framework for K–12 Science Education*. From the STEM draft chapters received, many showed no evidence of successful learning by students. This was a major and common problem with many of the authors. Most indicated that they had little data that showed the actual "doing of science" by students or how their STEM efforts illustrated their effectiveness with student learning.

The goal of this monograph was to provide 24 chapters that followed the outline, with major emphasis concerning success with student learning. The outline used considered the following features:

**Setting** (institution, industry, university, or organization) of author(s)

a. Needs to introduce the author(s)
b. Where they worked
c. Where the project(s) took place
d. How (if more than one author) how they came to collaborate

**Overview** of the specific features of the STEM program

a. How it fits with student goals for school science?
b. Who are the Do-ers?
c. Who are the target learners?

**Major Features** of the instructional program needed for success with STEM as Reform

a. History of the reforms addressed
b. How does it exemplify STEM features?
c. Teachers as partners in learning (not merely students as receivers!)

**Evidence for Success** with students (at least one-third of the 25+ page draft)

a. Types of information collected from students
b. Varied users of the program
c. Outside evaluation/observers
d. "Voices" of instructors/students targeted

**Next Steps**

a. What yet to try?
b. Proposed use of the data indicating success
c. Ties to other major reform efforts
d. Questions raised regarding use of the STEM exemplary program by others

As was predicted by Jim Rutherford and others, it takes 70+ years and longer for educational change to happen significantly in most schools. It should be remembered that for change to happen, it must include evidence for successes with student learning. Even though change in public education happens at a very slow rate; we must continue to learn and adjust the speed to address current science reforms. The importance of errors, interpretations, and use of inaccurate information—even by famous scientists—is demonstrated through some examples of erroneous assumptions. The statements indicate that even the most noted scientists make errors. Examples include:

1. "There is no likelihood man can ever tap the power of an atom," (Robert Millikan, Nobel Prize in Physics, 1923);

2. "I think there is a world market for maybe five computers," (Thomas Watson, Chairman of IBM, 1943);

3. "Heavier-than-air flying machines are impossible," (Lord Kelvin, President, Royal Society, 1895); and

4. "Everything that can be invented has been invented," (Charles H. Duell, Commissioner, U.S. Office of Patents, 1899). (Rinkwords 2014)

Even experts may have wrong ideas regarding the efforts to promote STEM reforms. As many states are becoming more involved with STEM efforts, we need to engage the most successful teachers in understanding the meaning of STEM as reform and how it should affect changes in teaching, learning, and student preparation for the future. We all need to have a real understanding of science itself (exploration of nature and explanations provided for the objects and events encountered). In addition, technology and engineering must be included in school programs as well as urging success with reading and use of mathematics. If art and music were taught as science is taught, the concentration would be on how you would use a musical instrument or how to prepare a specific art piece before you were allowed to touch the instrument or a specific painted picture. We seem to teach information collected by scientists and never consider how it was produced or how it might be used.

If we are to accomplish the real "doing" of science for all students, we must provide real evidence of its success with student learning. The focus on reports of success should not be dependent on student memories or ideas identified by teachers. STEM has different meanings and use for many people—just as "doing" science has a different meaning for many people. Doing science is not just reading textbooks, working in laboratories, and taking tests by memorizing. "Doing" real science begins with personal problems and proposing specific results. It means collecting evidence for the explanations provided for different students. "Doing science" should help with successful learning for all students and teachers and it should be useful throughout the whole lifetimes of both students and teachers.

Be prepared to learn from 24 successes of STEM examples included in this monograph. Evidence for success is the most important outcome that shows science and student learning. The focus should be on students and encouraging them to do more with explorations and attempts to explain objects and events they encounter. It is important to change while continuing efforts that result in meaningful learning. It is important to focus on questions that are personal, local, and current and invite collaboration. Enjoy the experiences of some 50 or more authors in the pages that follow!

## References

American Association for the Advancement of Science (AAAS). 1989. *Science for all Americans.* Washington, DC: AAAS.

Bybee, R. W. 2010. Advancing STEM education: A 2020 vision. *Technology & Engineering Teachers* 70 (10): 30–35.

Bybee, R. W. 2013. *Translating the* NGSS *for classroom instruction.* Arlington, VA: NSTA Press.

Harms, N. C., and R. E. Yager. 1981. *What research says to the science teacher, vol. 3.* Arlington, VA: NSTA Press.

Hazzard, E. 2014. A new take on student lab reports. *The Science Teacher* 81 (3): 57–61.

NGSS Lead States. 2013. *Next Generation Science Standards: For states by states*. Washington, DC: National Academies Press. *www.nextgenscience.org/next-generation-science-standards*

National Research Council (NRC). 1996. *National science education standards.* Washington, DC: National Academies Press.

National Research Council (NRC). 2011a. *A framework for K–12 science education: Practices, crosscutting concepts, and core ideas*. Washington, DC: National Academies Press.

National Science Foundation (NSF). 1973. *Unified science and mathematics for elementary schools: Teacher's resource book*. Newton, MA: Education Development Center.

Pratt, H. 2013. *The NSTA reader's guide to* A framework for K–12 science education: Practices, crosscutting concepts, and core ideas. Arlington, VA: NSTA Press.

Rinkworks. Things People Said—Bad Predictions. *www.rinkworks.com/said/predictions*

Science—A Process Approach (SAPA). 1965. The process instrument. Commission on Science Education. Washington, DC.

Truxal, J. G., and E. J. Piel. 1971. *The man-made world: Engineering concepts curriculum project.* New York: Polytechnic Institute of Brooklyn.

# Acknowledgments

## Members of the National Advisory Board for the Exemplary Science Series


**Lloyd H. Barrow**
Professor, Science Education
University of Missouri
Columbia, MO 65211

**Stephen A. Henderson**
President, Briarwood Enterprises, LLC
Richmond, KY 40475

**Herb Brunkhorst**
Professor Emeritus
California State University, San Bernardino
San Bernardino, CA 92407

**Bobby Jeanpierre**
Associate Professor Science Education
University of Central Florida
Orlando, FL 32816

**Rodger W. Bybee**
Executive Director (Retired)
Biological Sciences Curriculum Study (BSCS)
Golden, CO 80401

**Page Keeley**
Author and Consultant, Science Education
NSELA Region A Director
Jefferson, ME 04348

**Janet Carlson**
Executive Director
Center to Support Excellence in Teaching
Stanford Graduate School of Education
Stanford, CA 94305

**LeRoy Lee**
Director
Wisconsin Science Network
DeForest, WI 53532

**Charlene M. Czerniak**
Professor
The University of Toledo
Toledo, OH 43615

**Lisa Martin-Hansen**
Professor and Chair of Science Education
California State University Long Beach
Long Beach, CA 90807

**Pradeep (Max) Dass**
Director, Center for Science Teaching and Learning
Northern Arizona University
Flagstaff, AZ 86011

**LaMoine L. Motz**
Managing Partner
The Motz Consulting Group, LLC
White Lake, MI 48386

**Linda Froschauer**
Consultant, Science FUNdamentals
Westport, CT 06880

**Brian Newberry**
Associate Professor
California State University, San Bernardino
San Bernardino, CA 92407

**Edward P. Ortleb**
Consultant Author
St. Louis, MO 63139

**Carolyn F. Randolph**
Science Education Consultant
Blythewood, SC 29016

**Pat Shane**
Associate Director and Professor (retired)
Center for Mathematics & Science Education
University of North Carolina at Chapel Hill
Chapel Hill, NC27514

**Gerald Skoog**
Co-Director
Center for the Integration of STEM Education and Research
College of Education
Texas Tech University
Lubbock, TX 79409

**Cathy Spencer**
Mathematics Educator
Coordinator for the MA in Education, Mathematics and Science Education Option
California State University San Bernardino
San Bernardino, CA 92407

**Jennifer Wilhelm**
Associate Professor
Science and Mathematics Education
Chair of STEM Education Department
University of Kentucky
Lexington, KY 40506

**Mary Ann Mullinnix**
Assistant Editor
University of Iowa
Iowa City, IA 52242

# About the Editors

## Robert E. Yager

**Robert E. Yager**—an active contributor to the development of the *National Science Education Standards*—has devoted his life to teaching, writing, and advocating on behalf of science education worldwide. Having started his career as a high school science teacher, he has been a professor of science education at the University of Iowa since 1956. He has also served as president of seven national organizations, including NSTA, and has been involved in teacher education in Japan, Korea, Taiwan, Indonesia, Turkey, Egypt, and several European countries. Among his many publications are several NSTA books, including *Focus on Excellence* and two issues of *What Research Says to the Science Teacher.* He has authored over 700 research and policy publications as well as having served as editor for 10 volumes of NSTA's Exemplary Science Programs (ESP). Yager earned a bachelor's degree in biology from the University of Northern Iowa and master's and doctoral degrees in plant physiology from the University of Iowa.

## Herbert Brunkhorst

**Herbert Brunkhorst** has been a science educator for the past 47 years. He started his career as a secondary science teacher and has been a professor of science education and biology, and chair of the department of science, mathematics and technology education for 16 years at California State University, San Bernardino. He is past-president of the Association for the Education of Teachers of Science (now ASTE) and has served as the preservice director on the board of NSTA. He has several publications and numerous presentations in science education nationally and in Germany, Japan, China, Australia, and Greece. He is a fellow of the AAAS and an associate lifetime member of the National Academy of Sciences. Brunkhorst earned a bachelor's degree in biology from Coe College and master's and doctoral degrees in science education and plant physiology from the University of Iowa.

# STEM Starters

## An Effective Model for Elementary Teachers and Students

*Ann Robinson, Debbie Dailey, Gail Hughes, and Tony Hall*
*University of Arkansas at Little Rock*
*Little Rock, AR*

*Alicia Cotabish*
*University of Central Arkansas*
*Conway, AR*

### Setting

In a midsize rural school district near the capitol city of Arkansas, a fourth-grade classroom is full of students eagerly engaged in building electrical circuits to power a fictitious city. With brightly colored neon goggles and lab aprons, groups of students tackle the city's electrical design. One group is constructing a baseball park, and another is diligently wiring a replica of a fast food eatery. Other groups are building model storefronts and businesses, which all require electrical circuitry. The classroom teacher and a peer coach, an expert in science, circulate the room, monitoring student progress. The classroom teacher pauses at one group and asks, "Why do you suppose the park will not light up?" One boy jumps in, "Because there is a broken connection." The teacher promptly asks the group to review their circuitry diagram in front of them before proceeding. Next, the peer coach chimes in to elaborate, "If any part of the circuit is opened, or broken, the current will not flow." She then asks the students, "What happens when the current will not flow?" Students respond that the device will not receive power and the lights will not work. The peer coach asks for the students' attention as she reviews the basic rules of electrical

STEM Starters students explore circuitry.

circuits on the interactive whiteboard. The class discussion continues as the peer coach provides further explanation of the behavior of electricity and then tag-teams with the classroom teacher to deliver this well-orchestrated science lesson.

Combined with problem-based science curriculum, expert peer coaching, and intensive teacher professional development focused on science, this classroom is one of 60 from two rural communities in central Arkansas that participated in STEM Starters. A team of researchers collaborated with classroom teachers, gifted and talented facilitators, and building principals to create a culture of inquiry, and to engage in lively instruction that "hooks" children early through the excitement of science and innovation.

Grounded in scientifically based research studies, STEM Starters project components, goals, objectives, and activities focus on increased science learning for all students in grades 2–5, and increased content knowledge and process skills in the STEM disciplines for their elementary teachers. The model was developed with funding from the U.S. Department of Education and validated through randomized field studies of teacher and student outcomes.

## Overview of STEM Starters

To advance the national STEM agenda, the National Science Board (NSB 2010) recommended students be provided early opportunities to engage in inquiry-based learning with real-world problem solving and that their teachers be supported with research-based STEM preparation and professional development programs. The NSB emphasized early exposure to STEM opportunities to develop and nurture science interest in young innovators. In *A Framework for K–12 Science Education*, the National Research Council (NRC 2012) cautioned that omitting science at *any* grade level, including the early grades, potentially impacts students' conceptual learning and places additional demands on teachers in higher grades.

STEM Starters was guided by the principle that sustained and embedded teacher professional development, coupled with the implementation of an inquiry-based science curriculum, can positively influence student achievement. Brandwein (1995) and more recently, Worth (2010) stressed the importance of both the teacher and the curriculum in developing the science talent in young students. Brandwein emphasized that the greatest barriers to developing science talent included inadequately prepared teachers and an outdated curriculum that neglects the needs and interests of the child. Bitan-Friedlander, Drefus, and Milgrom (2004) suggested that the lack of follow-up support received by teachers constituted a major barrier in implementing a new innovation. Bitan-Friedlander and colleagues recommended that professional development be lengthy enough to internalize the innovation and extend into the *real-world* of the classroom. Finally, Johnson, Kahle, and Fargo (2007) reported increases in science achievement among students when their teachers participated in a sustained and embedded professional development program focused on standards-based practices, including inquiry-based instruction. With this theoretical framework guiding the intervention, STEM Starters focused on two major components: (a) teacher professional development and (b) the implementation of standards-based science curriculum in classrooms.

## Major Features

The major features of STEM Starters included a two-year job-embedded professional development component and support and resources for implementing a standards-based science curriculum. The STEM Starters intervention is presented in Figure 1.1.

**Figure 1.1.** STEM Starters Intervention Model

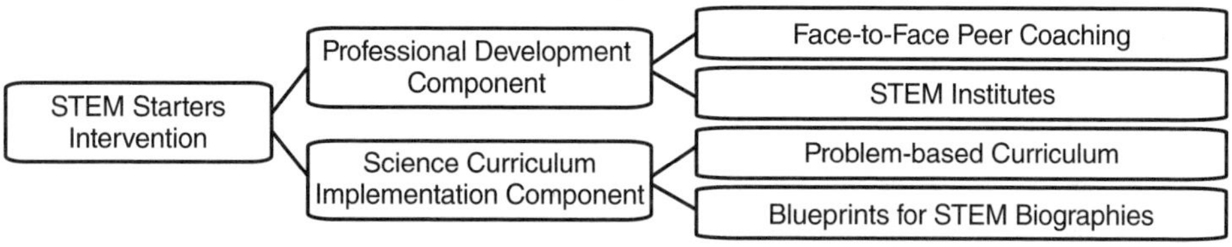


## Professional Development Component

To provide teachers with sufficient professional development to change practice, STEM Starters combined summer institutes and peer coaching sustained over a minimum of two years. Multiple research studies support the efficacy of increased professional development contact time on teacher instructional practices (Gerard, Varma, Corliss, and Linn 2011; Lumpe, Czerniak, Haney, and Beltyukova 2012; Sandholtz and Ringstaff 2011) and ultimately on student achievement (Desimone 2009; Guskey 1986; Johnson, Kahle, and Fargo 2007; Shymansky et al. 2010). Specifically, researchers recommended teachers receive 80+ hours of professional development to implement an inquiry-based curriculum effectively (Cotabish, Dailey, Hughes, and Robinson 2011; Corcoran, McVay, and Riordan 2003; Roehrig et al. 2011; Supovitz and Turner 2000).

### *STEM Institutes*

STEM Starters teachers participated in weeklong summer institutes across two summers. The institutes focused on science content and pedagogy, specific curriculum units aligned with science standards, technological applications, and differentiation of instruction. As recommended by Duschl, Schweingruber, and Shouse (2007), Haney and Lumpe (1995), and Penuel, Gallagher, and Moorthy, (2011), the summer institutes provided explicit instructions necessary for the implementation of a new program. For example, teachers actively took the role of students during the curriculum units. They participated in multiple activities such as using the Taba Model for Concept Development to brainstorm, categorize, and make generalizations about the overarching concepts of the specific units. They also participated in laboratory investigations such as determining the effect of temperature on the evaporation rate of water. In addition, they were engaged with real-world problems and scenarios and were guided to generate evidence in support of a conclusion. Through the implementation of the curriculum units, an expert science instructor modeled effective science pedagogy and checked for teachers' understanding of science content. The science instructor emphasized overarching concepts, higher-order thinking skills, inquiry-based instruction, experimental design, and the use of technology as recommended by the National Research Council (2012) and now embedded in the *Next Generation Science Standards* (*NGSS*; NGSS Lead States 2013).

During the STEM Starters summer institutes, an emphasis was placed on the use of technology in the classroom to enhance the learning processes. Teachers were exposed to multiple web-based resources that aligned with their specific units. These resources provided content information for the student and for the teacher as well as offering multiple activities and games to motivate student learning. During the second year of summer institutes, STEM Starters offered iPad training to teachers. Teachers were provided with information on iPad applications to enhance their teaching experience and their students' learning.

## *Peer Coaching*

A unique feature of STEM Starters is the use of one-to-one peer coaching to deliver embedded professional development in science. The configuration of pairing a generalist teacher and a science expert is uncommon in most elementary schools; however, results indicate that the STEM Starters approach to increasing teacher science content knowledge and student achievement in science works. With the need for increased hands-on STEM education in the elementary grades (NSB 2010), and the lack of science expertise among elementary educators (Fulp 2002), the marriage between peer coaching and STEM makes sense.

In a peer coaching intervention, the relationship forged between the teacher and the peer coach is crucial; it must be one of mutual trust and shared purpose (Caccia 1996). The National Foundation for the Improvement of Education (NFIE 1999) provided a list of qualities necessary for an effective peer coach including: demonstrating commitment to lifelong learning, being flexible and open-minded, being viewed as an expert in pedagogy and content area, exhibiting confidence in teaching, collaborating well with others, providing positive and productive critiques, maintaining confidentiality, and being approachable and patient.

Tschannen-Moran and Tschannen-Moran (2011) further elaborated on the characteristics of an effective coach. They suggested coaches focus on five concerns to mentor teachers effectively, including concerns for consciousness, connection, competence, contribution, and creativity. Tschannen-Moran and Tschannen-Moran described a *concern for consciousness* as a nonjudgmental awareness of what is going on in the teacher's classroom. Specifically, when making suggestions for improvement, an effective coach will focus on the positives that happen in the classroom. By emphasizing what a teacher does well and how it connects to student progress, the teacher is more likely to increase the frequency of the positive aspects of his/her teaching. Tschannen-Moran and Tschannen-Moran maintained "By discovering and developing their strengths, teachers can transform their weaknesses without having to tackle them head on" (p. 16).

In addition, Tschannen-Moran and Tschannen-Moran (2011) stated a *concern for connection* allows the coach to build a trusting relationship with the teacher; thereby, opening necessary channels of communication. In this instance, a teacher is more likely to share his/her fears and frustrations with the coach, then the pair of professionals can work together to make improvements. The authors suggested a coach demonstrate a *concern for competence* by appreciating teachers' current level of expertise, thus, allowing them to focus and build on their strengths. Tschannen-Moran and Tschannen-Moran stated a *concern for contribution* involves teachers having opportunities to voice concerns in a nonjudgmental format that confirms professional equality between the coach and the teacher. Finally, Tschannen-Moran and Tschannen-Moran said "For true learning to take place, coaching must also unleash creativity" (p. 15). The authors described the *concern for creativity* as a desire to instill in teachers a motivation and ethic for continuous improvement. Through a desire for improvement, teachers will search continuously for creative and new avenues to increase their effectiveness.

Peer coaching in the STEM Starters model is used to extend professional development beyond the summer institute to the classroom, enabling teachers to practice their newly learned skills in familiar surroundings with support from their peer coach. The positive effects of peer coaching in multiple domains have been reported. For example, Slater and Simmons (2001) found teachers increased knowledge, skills, and confidence due to participation in a peer coaching program and Showers (1984) reported increased achievement scores among students whose teachers had participated in peer coaching.

In addition to the general literature on peer coaching, there is an emerging literature on coaching in the STEM disciplines. In a peer coaching study with middle school science teachers, Appleton (2008) found teachers reported benefits from the support provided by the peer coach. In the intervention reported by Appleton, the peer coach provided teacher support through modeling instruction, facilitating classroom discussion, reflecting on the previous lesson, and collaboratively planning the upcoming science lesson. The teachers indicated their science instructional practices had improved due to the support provided by the peer coach. In addition, the teachers felt the presence of the peer coach increased their confidence in leading students in exploratory and self-discovery activities. The STEM Starters studies reported in a later section of this chapter add to the science-specific literature on coaching.

## *STEM Starters' Peer Coaching Intervention*

STEM Starters provided peer coaching on a weekly basis to the participating teachers. During the developmental stages of the model and in the studies published on the model, the peer coach was a former secondary chemistry and physics teacher as well as a gifted and talented facilitator. During the school year, the peer coach was in each classroom two to three times per month, providing approximately 60 hours of professional development over two years. Initially, the peer coach acted as an instructional leader and modeled effective science teaching for the teachers and the students. Eventually, the role of the peer coach transitioned to an instructional facilitator, where she assisted the teacher with instruction and monitored and encouraged student involvement. The peer coach also acted as a materials facilitator by ensuring all necessary science materials were available in the schools and ready for use in the individual classrooms. The peer coach maintained continuous contact with the teachers through e-mail and personal visits, ensuring that their needs were being met.

The primary objective of the STEM Starters peer coach was not to evaluate the teachers but rather to encourage and support them in implementing a new science program. In the early months of the intervention, teachers were hesitant about the program. They were nervous about the role of the peer coach, the extra time needed for implementation, and about their own lack of science content knowledge. Once they realized the peer coach was not in the school to pass judgment but to help them, they welcomed her into their classrooms. In addition, the reaction of the students to the program was very positive, which in turn encouraged the teachers. The professional development intervention is summarized in Table 1.1.

**Table 1.1.** STEM Starters Teacher Professional Development Across Two Years

| | Summer Institute | Peer Coaching |
|---|---|---|
| **Year 1** | • 30 hours (out-of-school)<br>• curriculum units<br>• inquiry-based strategies<br>• differentiation for high-ability learners<br>• *talent-spotting* of students from underrepresented groups | • 30 hours (in-school)<br>• implementation of curriculum units<br>• model teaching<br>• instructional facilitator<br>• materials facilitator<br>• science content expert |
| **Year 2** | • 30 hours (out-of-school)<br>• science content development<br>• inquiry-based strategies<br>• classroom management | • 30 hours (in-school)<br>• instructional facilitator<br>• materials facilitator<br>• science content expert |

## Science Curriculum Implementation Component

Given the commitment to inquiry-based pedagogy, particularly with a problem-based learning approach, STEM Starters researchers reviewed and selected rigorous curriculum that had been validated with low-income students. Due to the low average-income demographic of the districts in the initial field trials, both quality and cost of the materials were considerations. In addition, the units selected are of sufficient length that they provide an in-depth inquiry experience. They are self-contained and therefore can be *post-holed* into the elementary curriculum schedule that generally focuses heavily on literacy and numeracy rather than on science and engineering.

### *William and Mary Science Curriculum Units*

The William and Mary science curriculum units implemented through STEM Starters situated science learning in the context of a real-world problem. Each unit introduced students to advanced content, engaged students in problem solving and critical thinking, and was focused on specific overarching concepts that were integrated throughout the unit, including change (Grades 2 and 3) and systems (Grades 4 and 5). To increase student understanding, students were asked to brainstorm examples and non-examples, categorize, and make generalizations about each overarching concept.

The Grade 2 and 3 units were inquiry-based learning units focused on exposing young students to science concepts and scientific processes. These units engaged students in creative and critical-thinking opportunities through investigations and problem solving (Bracken et al. 2008). Each unit provided real-world scenarios where students were to use their newly acquired knowledge and skills to solve a meaningful problem. For example, in *What's the Matter?* students were challenged to solve the mystery of the missing water. The unit was introduced through the following scenario: "The principal of the school approached the class and asked for help. She had recently brought to her office a bowl of ocean water. After a couple of days, her water disappeared." Through the re-creation of the scene and using a need-to-know-board, the students were able to conclude that the water evaporated. Through this process, students developed an understanding of the water cycle and the phase changes of matter.

The Grade 4 and 5 units used by STEM Starters were problem-based science units that involved students in real-world problem solving. One unit, *Electricity City,* involved students in designing a model city, complete with electrical lighting. Once the city was completed, students were faced with a problem. A storm ravaged the city and damaged the electrical circuitry of their buildings. The students worked together to find the location of the break and reconnect the power to light the buildings. Through this unit, students developed an understanding of the properties of electricity and the behavior of electrical circuits. In *Acid, Acid Everywhere*, students were exposed to the problems of a chemical spill. They had to work together to solve environmental problems (effects of acid on living things) and logistical problems (rerouting traffic, cleanup). Through this unit, students gained an understanding of the differences between acids and bases and their effects on the environment.

### *Blueprints for Biography – STEM Starters Series*

To address the concerns and needs of elementary educators to integrate science through literacy, *Blueprints for Biography* (Robinson 2009) was expanded to include a STEM Series. *Blueprints for Biography* are teacher curriculum guides with high-level discussion questions, creative and critical-thinking activities, a persuasive writing component, and rich primary resources. STEM *Blueprints* focus on eminent scientists and inventors for whom exemplary children's biographies exist in trade book form. STEM Starters *Blueprints* currently exist for the following: George Washington Carver, Galileo Galilei, Thomas Edison, Marie Curie, Alexander Graham Bell, Michael Faraday, Louis Pasteur, and Albert Einstein. Each guide engages students in four activities: persuasive writing, point-of-view analysis, portrait study, and primary source analysis. The guide concludes with a classic experiment for students to carry out. For example, in the Thomas Edison *Blueprint*, students assume his persona to write a letter as if they were Thomas to persuade others about the benefits of electricity in the 1880s. In the primary source application, students are asked to analyze an audio clip of Edison speaking on the importance of technology. The portrait study is of a young Thomas sitting at a table working on the phonograph and finally, the classic experiment engages students in creating a lightbulb. Table 1.2 (p. 8) highlights the curriculum units used by STEM Starters, the biographies linked to them, and the alignment with the *Next Generation Science Standards*.

## Evidence for Success

The efficacy of STEM Starters has been reported in a series of studies on teacher implementation and on teacher and student outcomes (Cotabish, Dailey, Hughes, and Robinson 2011; Cotabish, Dailey, Robinson, and Hughes 2013; Dailey 2013; Dailey, Cotabish, and Robinson, 2013; Dietz and Robinson 2013; Robinson, Dailey, Hughes, and Cotabish 2013; Robinson, Dailey, Hughes, and Cotabish, forthcoming).

**Table 1.2.** Curriculum Units With *Next Generation Science Standards* by Grade Level

| Grade Level | Problem-Based Learning Units | Blueprints for Biography |
|---|---|---|
| 2 | Weather Reporter: *2-LS2; 2-ESS1; K-2-ETSI<br>Budding Botanist: *2-ESS2; K-2-ETSI | George Washington Carver<br>Louis Pasteur |
| 3 | What's the Matter: *2-PS1; 5-PS1; 3-5-ETSI<br>Dig It: An Earth and Space Unit: *3-ESS2; 3-LS4; 4-ESS3; 3-5-ETSI | Albert Einstein<br>Galileo Galilei |
| 4 | Electricity City: *4-PS3; 3-5-ETSI<br>Invitation to Invent: *4-PS3; 3-5-ETSI | Thomas Edison<br>Michael Faraday |
| 5 | Acid, Acid Everywhere: *5PS1; 5LS2; 5ESS3; 3-5-ETSI<br>Nuclear Energy: *5PSI; 5ESS3; 3-5-ETSI | Alexander Graham Bell<br>Marie Curie |

*Note: See chapter appendix for *Next Generation Science Standards* (*NGSS*).

## *Participants*

To examine the effects of the intervention, two school districts were selected to participate in the STEM Starters intervention based on their geographical proximity to the researchers and because their demographics were representative of the state. From the school districts, 70 teachers from grades 2–5 were randomly assigned to experimental or control conditions. In school district one, 9% of students were culturally diverse and 40% participated in the free-and-reduced-lunch program. In school district two, 30% of the population was culturally diverse and 69% of the students participated in the free-and-reduced-lunch program. All teacher participants were female with the exception of two males. Years in teaching ranged from 0 to 34, with an average of 12.8 years ($SD$ = 7.48).

Students assigned to experimental teachers were designated as experimental students, and those assigned to control teachers were designated as comparison students ($N_{year1} = 1750$; $N_{year2} = 1711$). Throughout the life of the intervention or until students reached grade 6, school districts agreed to keep experimental students with experimental teachers and comparison students with control teachers. The number of experimental and comparison students in this study are summarized in Table 1.3.

**Table 1.3.** Number of Experimental and Comparison Students by Grade Level and Year

| Grade Level | Experimental Students | | Comparison Students | | Total |
|---|---|---|---|---|---|
| | Year 1 | Year 2 | Year 1 | Year 2 | |
| 2 | 197 | 206 | 220 | 216 | 839 |
| 3 | 206 | 220 | 256 | 182 | 864 |
| 4 | 194 | 273 | 235 | 203 | 905 |
| 5 | 221 | 206 | 221 | 205 | 853 |
| Total | 818 | 905 | 932 | 806 | 3461 |

### Teacher Effects

To analyze the effects of the intervention across two years, researchers examined teachers' science teaching perceptions and their ability to design an experiment. Both experimental and control teachers were administered the instruments prior to the intervention, at the end of year 1 and at the end of year 2 (conclusion of the intervention).

To examine teachers' science teaching perceptions, the Perceptual Assessment of Science Teaching and Learning (PASTeL) questionnaire was used. The PASTeL consists of 50 items arranged in two scales, the Teaching Scale and the Student Learning Scale (Bracken et al. 2008). Using a four-point Likert-type scale, the PASTeL assesses how teachers perceive their ability to teach, as well as their students' ability to learn, science. A higher score indicates a greater degree of confidence in science teaching and/or student science learning. Sample items are presented below.

> *Teacher Scale: "I am confident in my foundational knowledge of the science content I teach." "I guide students to generalize concepts across multiple science lessons." "I am comfortable explaining experimental and control variables to my students."*
>
> *Student Scale: "My students correctly use scientific terminology presented in class." "When called upon, students in my class can explain how to conduct a scientific investigation." "My students discuss science concepts from multiple perspectives."* (Bracken et al. 2008)

To examine the ability of teachers to design an experiment, researchers used the Adapted Fowler Test (also known as the "Diet Cola Test"). The Adapted Fowler Test is an open-ended assessment of one's ability to design a simulated controlled experiment in response to a scientific question (Fowler 1990). For example, participants are asked to design an experiment to answer the question, "Are bees attracted to diet cola?" Responses describing the participant's proposed design are scored across five criteria: (a) generates a prediction, (b) lists materials needed, (c) lists experiment steps/arranges steps in sequential order, (d) plans the data collection, and (e) states plan for interpreting data for making predictions. Ratings range from *no evidence* (0) to *strong evidence* (3) with two additional points possible on one criterion resulting in 17 points possible for this instrument.

When comparing the experimental teachers to the control teachers, experimental teachers demonstrated statistically significant improvement in perceptions of science teaching and learning and in their ability to design an experiment from baseline to year 1 and year 2 (Cotabish, Dailey, Hughes, and Robinson 2011; Dailey, Cotabish, Robinson, and Hughes 2011; 2012). A comparison of percent gains demonstrated by experimental and control teachers across two years is presented in Table 1.4. The inferential statistical data and results are provided in Table 1.5 (p. 10).

**Table 1.4.** Teacher Percent Gains Across Two Years

| | Experimental Teacher Gains on a Pre-/Post-Assessment | Comparison Teachers Gains on a Pre-/Post-Assessment |
|---|---|---|
| **Science Process Skills** | 40% | 1% |
| **Science Teaching Perceptions** | 46% | 9% |

**Table 1.5.** Summary of Teacher Questionnaire/Assessment Results

| Instrument | Statistical Test | Results |
|---|---|---|
| **Pastel** | **Year 1** | **Year 1** |
| | One-way MANCOVA comparing adjusted posttests between experimental and control groups. Post Hoc: Bonferroni adjusted alpha level of .025 | Hotelling's $T= 5.41$, $p = 0.008$, $n_p^2 = .18$; Post Hoc: Teacher Scale [$F(1, 50) = 6.12$, $p = 0.017$, $n_p^2 = 0.11$; $M_E = 2.97$, $M_C = 2.56$]; Student Learning Scale [$F(1, 50) = 10.98$, $p = 0.002$, $n_p^2 = 0.18$, $M_E = 2.89$, $M_C = 2.39$] (Dailey, Cotabish, Robinson, and Hughes 2011). |
| | **Year 2** | **Year 2** |
| | One-way MANCOVA comparing adjusted posttests between experimental and control groups. Post Hoc: Bonferroni adjusted alpha level of .025. | Wilks' Lambda = 14.53, $p < 0.000$, $n_p^2 = .38$; Post Hoc: Teacher Scale [$F(1, 48) = 29.64$, $p < 0\ .000$, $n_p^2 = 0.38$; $M_E = 3.21$, $M_C = 2.63$]; Student Learning Scale [$F(1, 48) = 14.59$, $p = 0.000$, $n_p^2 = 0.23$, $M_E = 2.96$, $M_C = 2.59$] (Dailey, Cotabish, Robinson, and Hughes 2012). |
| **Fowler** | **Year 1** | **Year 1** |
| | One-way ANCOVA comparing adjusted posttests between experimental and control groups | $F(1, 49) = 64.16$, $p < 0.001$, $= 0.57$]; $M_E = 13.56$, $M_C = 4.61$ (Cotabish, Robinson, Hughes, and Dailey 2011). |
| | **Year 2** | **Year 2** |
| | One-way ANCOVA comparing adjusted posttests between experimental and control groups. | $F(1, 51) = 16.22$, $p < 0.001$, $= 0.24$; $M_E = 10.68$, $M_C = 6.52$ (Cotabish, Dailey, Hughes, and Robinson 2011). |

## *Teacher Voices*

In the report from the external evaluator of the STEM Starters project (VanTassel-Baska 2010), teachers praised the efforts of the intervention, particularly noting the effectiveness of peer coaching in the model. Specifically, teachers made comments such as, the peer coach "is a gem, she is in touch with teachers and science, just being out of the classroom herself. The students also relate well to her." Another teacher referred to the peer coach as a "Human Zanax." The external evaluator reported that teachers stated the peer coach "has been wonderful by helping with experiments, providing support, modeling lessons, being positive, providing websites and other resources, and engaging [students] in open communication weekly." Although the teachers acknowledged difficulties in implementing a challenging science program (time constraints, background knowledge, inquiry-based instruction), overall they praised STEM Starters and its effect on student enthusiasm, engagement, and higher-order thinking.

## *Student Effects*

To analyze the effects of the intervention on students, researchers examined pre- and posttest results from curriculum-embedded assessments, including measures of science content knowledge, science concept knowledge, and experimental design. Both experimental and comparison students were administered the assessments prior to the intervention, at the end of year 1 and at the end of year 2 (conclusion of the intervention).

Specifically, pre-post embedded curriculum-based assessments were used to capture student learning gains in content knowledge. Open-ended concept mapping was used in Grades 2–3 to measure student understanding of science content topics; whereas, short-answer questions were used in Grades 4–5 to assess student understanding of the science content. Short-answer questions were used to assess student understanding of science concepts in all grades. Specifically, the curriculum-based concept assessments measured student understanding of overarching concepts that unify STEM disciplines (i.e., systems, change, patterns, cause and effect). The same instrument that measured teachers' ability to design an experiment (Adapted Fowler) was also used with the students. Although the instrument was designed with students in mind, the test is appropriate across a variety of age-levels due to the high ceiling of the instrument (VanTassel-Baska 2011).

When comparing experimental students with comparison students, experimental students demonstrated statistically significant increases from pre- to posttest on measures of science content knowledge, science concept knowledge, and experimental design after one year of intervention (Cotabish, Dailey, Hughes, and Robinson 2013). Researchers noted that the preliminary results after two years of intervention reveal similar findings (Robinson, Dailey, Hughes, and Cotabish, forthcoming). A comparison of percent gains demonstrated by experimental and comparison students across two years is presented in Table 1.6. The inferential statistical data and results are provided in Table 1.7 (p. 12).

**Table 1.6.** Student Percent Gains Across Two Years

| | Year 1 Gains | | Year 2 Gains | |
|---|---|---|---|---|
| **Science Content Knowledge** | Experimental | 32% | Experimental | 45% |
| | Comparison | 7% | Comparison | 10% |
| **Science Concept Knowledge** | Experimental | 27% | Experimental | 29% |
| | Comparison | 3% | Comparison | 9% |
| **Science Process Skills** | Experimental | 20% | Experimental | 21% |
| | Comparison | 11% | Comparison | 9% |

**Table 1.7.** Summary of Student Assessment Results

| Instrument | Statistical Test | Results |
|---|---|---|
| Content Assessment | Year 1 | Year 1 |
| | Hierarchical Linear Modeling comparing three models of fit. | Teacher participation in the treatment program accounted for 12% of the variability in students' posttest scores (*N* = 964) (Cotabish, Dailey, Robinson, and Hughes 2013). |
| | Year 2 | Year 2 |
| | Hierarchical Linear Modeling comparing three models of fit. | Teacher participation in the treatment program accounted for 22% of the variability in students' posttest scores (*N* = 1170) (Robinson, Dailey, Hughes, and Cotabish, forthcoming). |
| Concept Assessment | Year 1 | Year 1 |
| | Hierarchical Linear Modeling comparing three models of fit. | Teacher participation in the treatment program accounted for 14% of the variability in students' posttest scores (*N* = 739) (Cotabish, Dailey, Robinson, and Hughes 2013). |
| | Year 2 | Year 2 |
| | Hierarchical Linear Modeling comparing three models of fit. | Teacher participation in the treatment program accounted for 12% of the variability in students' posttest scores (*N* = 1213) (Robinson, Dailey, Hughes, and Cotabish, forthcoming). |
| Fowler | Year 1 | Year 1 |
| | Hierarchical Linear Modeling comparing three models of fit. | Teacher participation in the treatment program accounted for 7.5% of the variability in students' posttest scores (*N* = 713) (Cotabish, Dailey, Robinson, and Hughes 2013). |
| | Year 2 | Year 2 |
| | Hierarchical Linear Modeling comparing three models of fit. | Teacher participation in the treatment program accounted for 9.0% of the variability in students' posttest scores (*N* = 1184) (Robinson, Dailey, Hughes, and Cotabish, forthcoming). |

### *Student Work Examples*

To further illustrate the differences in student learning attributable to the STEM Starters intervention, we provide examples of student responses to the science content assessment for the Grade 3 unit on matter. Students were to create a concept map on everything they know about matter. In Figure 1.2, the student example from the pretest revealed an understanding of concept maps, but clearly the student did not have a content understanding of matter. Although this concept map is an endearing statement of the importance of family to this child, no understanding of matter as science is evident. In contrast, by examining the posttest, we find a dramatic increase in content knowledge. In fact, the student demonstrated an understanding of a complex concept by acknowledging that liquid will change by adding or taking away heat. Figure 1.3 (p. 14) displays the elaborated learning due to the child's participation in STEM Starters.

STEM Starters teacher guides a second grade student in a bubble activity.

**Figure 1.2.** Grade 3 Student Pretest for Science Content Knowledge of Matter

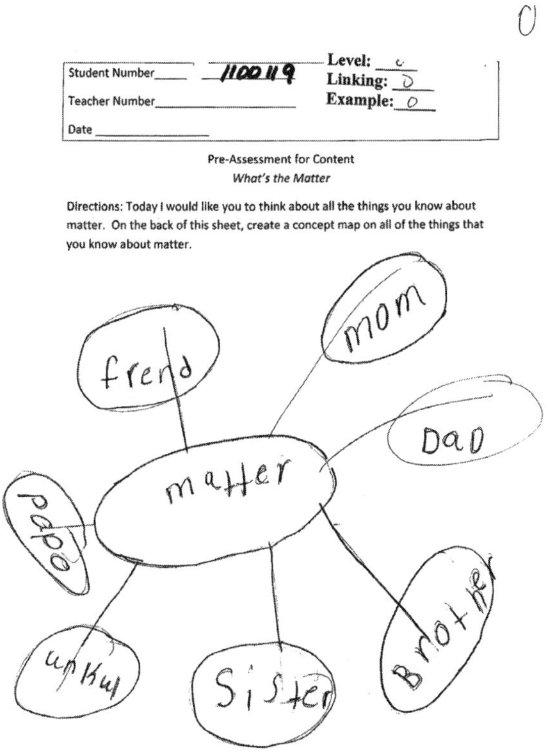

**Figure 1.3.** Grade 3 Student Posttest for Science Content Knowledge of Matter

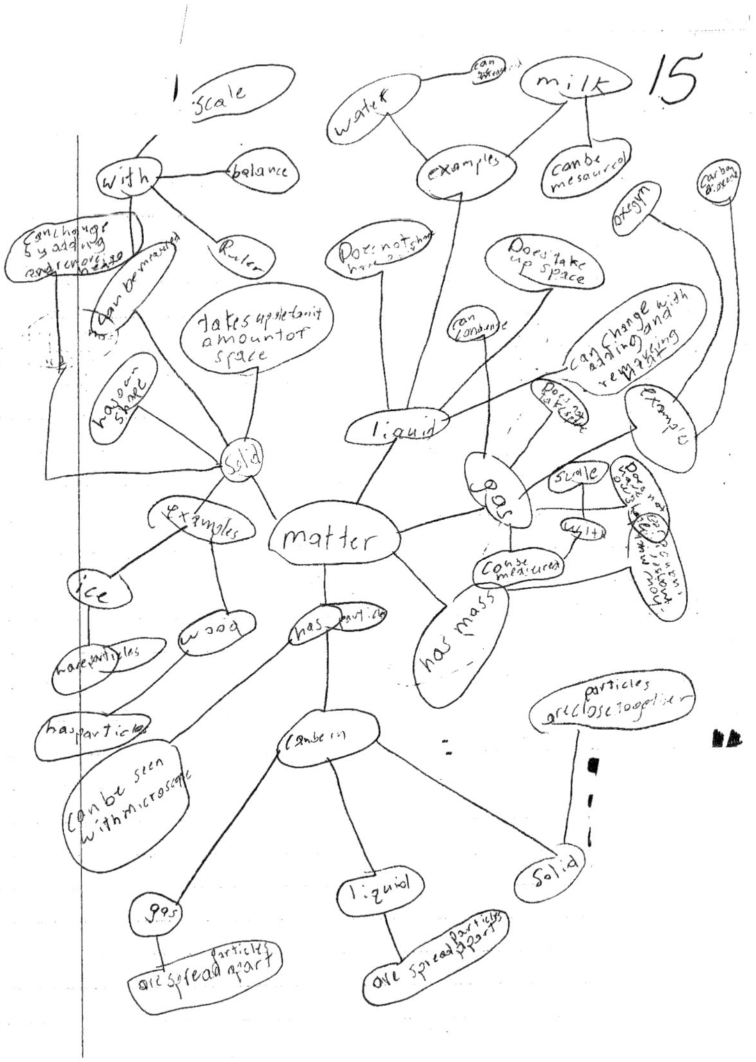


## Next Steps

The rigorous randomized field test of STEM Starters and its effectiveness with teachers and students leads to several next steps. First, additional scholarly publications are under development and review to increase dissemination of information about the program model and its effectiveness. For example, what are the effects when the external peer coaching supports are withdrawn? Do teacher leaders emerge in individual school buildings? Do districts modify staffing patterns, schedules, and teaching assignments to sustain the coaching? Second, are there ways to deliver coaching through real-time technologies? In what ways can face-to-face coaching interventions be translated effectively and efficiently to virtual environments? Beyond the opportunity for further research studies, a dissemination plan for the STEM *Blueprints* should be implemented. The materials were highly rated by teachers (Deitz and Robinson 2013). The teachers' suggestions for revisions need to be incorporated and the guides made easily accessible to a larger audience of educators. As part of reform efforts, links to the *Common Core State Standards* and the *Next Generation Science Standards* should be made explicit in the STEM *Blueprints*. Finally, given the efficacy of the intervention as measured by the gold standard of a randomized control trial (rct), STEM Starters researchers and staff should pursue a scale-up plan to increase access to STEM Starters nationally.

## References

Appleton, K. 2008. Developing science pedagogical content knowledge through mentoring elementary teachers. *Journal of Science Teacher Education* 19: 523–545.

Bitan-Friedlander, N., A. Dreyfus, and Z. Milgrom. 2004. Types of "teachers in training": The reactions of primary school science teachers when confronted with the task of implementing an innovation. *Teaching and Teacher Education* 20: 607–619.

Bracken, B. A., K. A. Holt, M. L. Lee, C. J. McCormick, C. L. Reintjes, J. I. Robbins, and T. L. Stambaugh. 2008. *Perceptual assessment of science teaching and learning preliminary examiner's manual*. Williamsburg, VA: The College of William and Mary.

Brandwein, P. F. 1995. *Science talent in the young expressed within ecologies of achievement* (RBDM 9510). Storrs, CT: The National Research Center on the Gifted and Talented, University of Connecticut.

Caccia, P. 1996. Linguistic coaching: Helping beginning teachers defeat discouragement. *Educational Leadership* 46 (8): 37–47.

Corcoran, T. B., S. McVay, and K. Riordan. 2003. *Getting it right: The MISE approach to professional development*. Philadelphia, PA: Consortium for Policy Research in Education.

Cotabish, A., D. Dailey, A. Hughes, and A. Robinson. 2011. The effects of a STEM professional development intervention on elementary teachers' science process skills. *Research in the Schools* 18 (2): 16–25.

Cotabish, A., D. Dailey, A. Robinson, and A. Hughes. 2013. The effects of a STEM intervention on elementary students' science knowledge and skills. *School Science and Mathematics* 113 (5): 215–226.

Dailey, D. 2013. The effects of a stem professional development intervention on elementary teachers. PhD dissertation, University of Arkansas at Little Rock.

Dailey, D., A. Cotabish, and A. Robinson. 2013. A model for STEM talent development: Peer coaching in the elementary classroom. *TEMPO* XXXIII (1): 15–20.

Dailey, D., A. Cotabish, A. Robinson, and G. Hughes. 2011. Interim effects of implementing a STEM initiative on elementary teacher perceptions and concerns about science teaching and learning. Paper presented at the annual meeting of the American Education Research Association, New Orleans, LA.

Dailey, D., A. Cotabish, A. Robinson, and G. Hughes. 2012. The effects of implementing a STEM initiative on elementary teacher perceptions and concerns about science teaching and learning. Paper presented at the annual meeting of the American Education Research Association, Vancouver, BC.

Desimone, L. M. 2009. Improving impact studies of teachers' professional development: Toward better conceptualizations and measures. *Educational Researcher* 38: 181–199.

Dietz, D., and A. Robinson. 2013. *Gifted education teachers' perceptions on implementation of Blueprints for Biography in a STEM intervention*. Manuscript submitted for publication.

Duschl, R., H. A. Schweingruber, and A. Shouse. 2007. *Taking science to school: Learning and teaching science in grades K–8*. Washington, DC: National Academies Press.

Fowler, M. 1990. The diet cola test. *Science Scope* 13 (4): 32–34.

Fulp, S. L. 2002. *Status of elementary school science teaching.* Chapel Hill, NC: Horizon Research. *http://2000survey.horizonresearch.com/reports/elem_science/elem_science.pdf*

Gerard, L. F., K. Varma, S. B. Corliss, and M. C. Linn. 2011. Professional development for technology-enhanced inquiry science. *Review of Educational Research* 81: 408–448.

Guskey, T. R. 1986. Staff development and the process of teacher change. *Educational Researcher* 15 (5): 5–12.

Haney, J. J., and A. T. Lumpe. 1995. A teacher professional development framework guided by reform policies, teachers' needs, and research. *Journal of Science Teacher Education* 6 (4): 187–196.

Johnson, C. C., J. B. Kahle, and J. D. Fargo. 2007. A study of the effect of sustained, whole-school professional development on student achievement in science. *Journal of Research in Science Teaching* 44: 775–786.

Lumpe, A., C. Czerniak, J. Haney, and S. Beltyukova. 2012. Beliefs about teaching science: The relationship between elementary teachers' participation in professional development and student achievement. *International Journal of Science Education* 34: 153–166.

National Foundation for the Improvement of Education (NFIE). 1999. *Creating a teacher mentoring program. www.neafoundation.org/downloads/NEA-Creating_Teacher_Mentoring.pdf*

National Research Council (NRC). 2012. *A framework for K–12 science education: Practices, crosscutting concepts, and core ideas.* Washington, DC: National Academies Press.

National Science Board (NSB). 2010. *Preparing the next generation of STEM innovators: Identifying and developing our nation's human capital* (NSB-10-33). *www.nsf.gov/nsb/publications/2010/nsb1033*

NGSS Lead States. 2013. *Next Generation Science Standards: For states by states.* Washington, DC: National Academies Press. *www.nextgenscience.org/next-generation-science-standards.* Appendix A: Conceptual shifts.

Penuel, W. R., L. P. Gallagher, and S. Moorthy. 2011. Preparing teachers to design sequences of instruction in earth systems science: A comparison of three professional development programs. *American Educational Research Journal* 48: 996–1025.

Robinson, A. 2009. *Blueprints for Biography: STEM Series.* Little Rock, AR: Jodie Mahony Center, University of Arkansas at Little Rock.

Robinson, A., D. Dailey, G. Hughes, and A. Cotabish. 2013. *The effects of a STEM intervention on gifted elementary students; science knowledge and skills.* Manuscript under review.

Robinson, A., D. Dailey, G. Hughes, and A. Cotabish. Forthcoming. *The two-year effects of a STEM intervention on elementary students' science knowledge and skills.*

Roehrig, G. H., M. Dubosarsky, A. Mason, S. Carlson, and B. Murphy. 2011. We look more, listen more, notice more: Impact of sustained professional development on Head Start teachers' inquiry-based and culturally-relevant science teaching practices. *Journal of Science Education and Technology* 20 (5): 566–578.

Sandholtz, J. H., and C. Ringstaff. 2011. Reversing the downward spiral of science instruction in K–2 classrooms. *Journal of Science Teacher Education* 22: 513–533.

Showers, B. 1984. *Peer coaching: A strategy for facilitating transfer of training* (Report No. ED 271 849). Eugene, OR: Center for Educational Policy and Management, College of Education, University of Oregon.

Shymansky, J. A., T. L. Wang, L. A. Annetta, L. D. Yore, and S. A. Everett. 2010. How much professional development is needed to effect positive gains in K–6 student achievement on high stakes science tests? *International Journal of Science and Mathematics Education* 10: 1–19.

Slater, C. L., and D. L. Simmons. 2001. The design and implementation of a peer coaching program. *American Secondary Education* 29 (3): 67–76.

Supovitz, J. A., and H. M. Turner. 2000. The effects of professional development on science teaching practices and classroom culture. *Journal of Research in Science Teaching* 37: 963–980.

Tschannen-Moran, B., and M. Tschannen-Moran. 2011. The coach and the evaluator. *Educational Leadership* 69 (2): 10–16.

VanTassel-Baska, J. 2010. Evaluation report for STEM Starters Javits Project at the University of Arkansas at Little Rock. Unpublished report.

VanTassel-Baska, J. 2011. Implementing innovative curriculum and instructional practices in classrooms and schools: Using research-based models of effectiveness. In *Content-based curriculum for high-ability learners*, ed. J. VanTassel-Baska and C. A. Little, 437–465. Waco, TX: Prufrock Press.

Worth, K. 2010. Science in Early Childhood Classrooms: Content and Process. *http://ecrp.uiuc.edu/beyond/seed/worth.html.*

**Table 1.8.** ***Next Generation Science Standards*** **Aligned With the Curriculum Used in the Stem Starters Intervention**

| Standard | |
|---|---|
| **2-PS1-1** | Plan and conduct an investigation to describe and classify different kinds of materials by their observable properties. |
| **2-PS1-2** | Analyze data obtained from testing different materials to determine which materials have the properties that are best suited for an intended purpose. |
| **2-PS1-3** | Make observations to construct an evidence-based account of how an object made of a small set of pieces can be disassembled and made into a new object. |
| **2-PS1-4** | Construct an argument with evidence that some changes caused by heating or cooling can be reversed and some cannot. |
| **2-LS2-1** | Plan and conduct an investigation to determine if plants need sunlight and water to grow. |
| **2-LS2-2** | Develop a simple model that mimics the function of an animal in dispersing seeds or pollinating plants. |
| **2-ESS1-1** | Use information from several sources to provide evidence that Earth events can occur quickly or slowly. |
| **2-ESS2-1** | Compare multiple solutions designed to slow or prevent wind or water from changing the shape of the land. |
| **2-ESS2-2** | Develop a model to represent the shapes and kinds of land and bodies of water in an area. |
| **2-ESS2-3** | Obtain information to identify where water is found on Earth and that it can be solid or liquid. |
| **K-2-ETS1-1** | Ask questions, make observations, and gather information about a situation people want to change to define a simple problem that can be solved through the development of a new or improved object or tool. |
| **K-2-ETS1-2** | Develop a simple sketch, drawing, or physical model to illustrate how the shape of an object helps it function as needed to solve a given problem. |
| **K-2-ETS1-3** | Analyze data from tests of two objects designed to solve the same problem to compare the strengths and weaknesses of how each performs. |
| **3-ESS2-1** | Represent data in tables and graphical displays to describe typical weather conditions expected during a particular season. |

| | |
|---|---|
| **3-ESS2-2** | Obtain and combine information to describe climates in different regions of the world. |
| **3-LS4-1** | Analyze and interpret data from fossils to provide evidence of the organisms and the environments in which they lived long ago. |
| **3-LS4-2** | Use evidence to construct an explanation for how the variations in characteristics among individuals of the same species may provide advantages in surviving, finding mates, and reproducing. |
| **3-LS4-3** | Construct an argument with evidence that in a particular habitat some organisms can survive well, some survive less well, and some cannot survive at all. |
| **3-LS4-4** | Make a claim about the merit of a solution to a problem caused when the environment changes and the types of plants and animals that live there may change. |
| **4-PS3-1** | Use evidence to construct an explanation relating the speed of an object to the energy of that object. |
| **4-PS3-2** | Make observations to provide evidence that energy can be transferred from place to place by sound, light, heat, and electric currents |
| **4-PS3-3** | Ask questions and predict outcomes about the changes in energy that occur when objects collide. |
| **4-PS3-4** | Apply scientific ideas to design, test, and refine a device that converts energy from one form to another. |
| **4-ESS3-1** | Generate and compare multiple solutions to reduce the impacts of natural Earth processes on humans. |
| **4-ESS3-2** | Generate and compare multiple solutions to reduce the impacts of natural Earth processes on humans. |
| **5-PS1-1** | Develop a model to describe that matter is made of particles too small to be seen. |
| **5-PS1-2** | Measure and graph quantities to provide evidence that regardless of the type of change that occurs when heating, cooling, or mixing substances, the total weight of matter is conserved. |
| **5-PS1-3** | Make observations and measurements to identify materials based on their properties. |
| **5-PS1-4** | Conduct an investigation to determine whether the mixing of two or more substances results in new substances. |
| **5-LS2-1** | Develop a model to describe the movement of matter among plants, animals, decomposers, and the environment. |
| **5-ESS3-1** | Obtain and combine information about ways individual communities use science ideas to protect the Earth's resources and environment. |
| **3-5-ETS1-1** | Define a simple design problem reflecting a need or a want that includes specified criteria for success and constraints on materials, time, or cost. |
| **3-5-ETS1-2** | Generate and compare multiple possible solutions to a problem based on how well each is likely to meet the criteria and constraints of the problem. |
| **3-5-ETS1-3** | Plan and carry out fair tests in which variables are controlled and failure points are considered to identify aspects of a model or prototype that can be improved. |

(NGSS Lead States 2013)

# Science in Our Backyard

## How a School Is Turning Its Grounds Into a Living Lab

*Jyoti Gopal and Ella Pastor*
*Riverdale Country School*
*New York, NY*

### Setting

Riverdale Country School is an independent school located in the Bronx, New York City. The majority of our student body lives in Manhattan in close and personal contact with technology, concrete and very little natural space. The bus ride to and from the lower school often can take as much as an hour for our students, most of whom are armed with music players, tablets, video games, and other gizmos that keep them entertained but disconnected from their surroundings.

As soon as the bus pulls up in the school driveway and our students tumble out with their backpacks, our goal is to keep them as tuned in to their surroundings and as disconnected from electronic entertainment as possible.

For many years, however, our extensive grounds were rarely used for anything except an outdoor play space. About 15 years ago, the three kindergarten teachers, under the leadership of then early childhood educational administrator, Jonina Herter, started discussing the potential of our outdoor campus as a learning tool. They built raised beds to plant herbs and vegetables and created a winding rocky bed for capturing water flow during heavy rains. Abutting our campus, on the banks of the Hudson River, is a woodlands park managed by the Parks Department. Weekly, the children went on nature walks to observe and record changes in this space through the seasons. About five years ago, after the success of our original planting beds, the school built 10 additional beds on a different part of the campus. These would now be used by grades preK–5 as a tool for learning about soil, water, garden ecology, as well as for growing food that would be eaten by the students and faculty.

Our goal was a simple one: We wanted our students to spend more time in the natural world and to have the opportunity to literally get their hands dirty. Our premise was equally simple: The more time they spent with nature, developing an understanding of its science, the more they would appreciate, respect, conserve, and most importantly, nurture it.

## What Does STEM Mean to Us?

Rodger Bybee (2010) identified several challenges that STEM faces in school curricula. Two of these challenges, context-based education and curricular integration, have been particularly problematic for many schools given the traditional nature of how disciplines are taught.

As elementary school educators, nothing is more important to us than making learning meaningful and contextual. Young children are naturally driven to explore the world around them and derive meaning from what they see and experience. We created a curriculum that deliberately harnessed student curiosity and used our campus as an extension of the classroom. Children learn by doing, and young children especially so. Even more important, young children learn by "repetitive doing." By this, we do not mean doing the same thing over and over again, although that certainly is an important aspect of learning for young children. Rather, we define "repetitive doing" as providing children the opportunity to learn the same concepts and skills in many different ways many times throughout the year. Integrating science, math, technology, engineering as well as music, art, and social studies is critical for providing these repetitive but layered opportunities. This approach allows for different types of learners to access the same information through a number of avenues. Similarly, in the lower school environmental education program, learning is context-based and much effort is put into integrating various disciplines into our work. In conjunction with classroom teachers, areas of study outdoors overlap and blend into the more traditional disciplines taught within the four walls of a classroom. Using the outdoor campus and linking it with indoor experiences provides an ongoing continuity of learning experiences throughout the elementary years. Additionally, it enables learners to see the links and connections between a science or math concept and the real world. This deliberate practice also allows for children to make personal connections to their learning by not only allowing for individual constructions of knowledge but also the opportunity to act on this knowledge to build further learning experiences and to initiate "change-making" behavior.

### *The Big Picture*

Ultimately the goal of education should be to create critical thinkers who can be the next-generation change makers and problem solvers of our society. Too often, however, we lose sight of this and focus instead on preparing children for the next step of their schooling. Elementary school teachers worry that their students must be prepared for middle school. Middle school teachers worry about high school and high school teachers about college. The recent push for STEM comes from the recognition that education has become removed from the challenges of real life. Rather, if we concentrate on preparing our students for life, focusing on the big understandings and essential questions that will impact their own lives as they grow into adulthood, school would become a much more meaningful place as well as a powerful learning environment.

At Riverdale, one way we start our students on this journey is to encourage children to think critically about their connection to the "place" in which they live. This place is as small as our classrooms, where they are a community of learners and citizens, each with a responsibility to self as well as to others. "Place" is also our campus and the adjacent woods where we ask our students to explore questions about the Earth on which we live and their relationship to it as well as the living beings that live alongside us. As the children get older, the concept of "place" stretches further to include a more global study of the planet, including the effects of climate and human intervention on our changing planet.

The curriculum we have developed is one that has expanded and grown over the past few years, fueled by collaborative planning between the early learning team (preK–1st grade), the environmental educator, and other classroom teachers.

In developing such a curriculum, our goals were manifold:

1. We wanted to target some big understandings and essential questions about the natural world, emphasizing the interdependence of all living things.
2. We wanted our students to have the opportunity for a hands-on, yearlong, and repetitive in-depth study that would encompass scientific practices and a number of crosscutting concepts outlined in the current science framework.
3. We wanted to create interdisciplinary connections that would encompass language arts, math, social studies, art, music, movement, and technology.
4. We wanted to weave the curriculum into the daily life of children so they could revisit the same practices and concepts throughout the year, as well as year after year, through a range of activities and lessons.
5. We wanted to encourage children to reflect on their personal relationship to the natural world and the impact, both positive and negative, that their own behavior could potentially have on their environment.
6. Finally, through this reflection and study, we wanted to develop a sense of caring, which translates into sustainable practices.

## *Investigations in Kindergarten*

Using Grant Wiggins and Jay McTighe's Backward Design template, we outlined the framework for our yearlong exploration and decided to focus on the theme of decomposition for our study. Decomposition is not a pretty process but it is an essential one and, at its core, a truly beautiful part of the life cycle, enabling a dying organism to return its nutrients to the earth for reuse.

The study of decomposition allows us to explore a variety of crosscutting concepts as well as a number of core ideas that *A Framework for K–12 Science Education* has outlined. The inquiry is rooted in scientific practices but with opportunities for expression and understanding through multiple disciplines. It provides various experiences for investigating both natural and managed decomposition. It also emphasizes classroom practices that seek to create in each student a sense of environmental stewardship and personal responsibility for their behavior. It therefore easily meets the goals of the *National Science Education Standards* as it allows our students to "experience the richness and excitement of knowing about and understanding our natural world" as well as gives them opportunities to "use appropriate scientific processes and principles in making personal decisions."

The following are the areas of investigation through which the study of decomposition is woven into our kindergarten classrooms and each area is explored in further detail in this chapter.

- Classroom Routines
- Red Wiggler Study
- Garden Work
- Seed Study
- Nature Walks and Log Hotels
- Organic/Inorganic Litter Study
- Bottle Cap Recycling
- Pumpkin Study

## *Classroom Routines*

In September, as the children enter kindergarten, they become familiar with the rhythms and routines of our classroom. One of the first things they learn is that there are four receptacles for their waste: a paper recycling bin for paper; a comingles bin for glass, metal, and recyclable plastic; a food scraps bucket for certain types of food and paper towels from the sink; and a garbage can for what can't be put in the first three. Our goal, we tell them, is to keep the garbage can as empty as possible.

Jobs are assigned on a weekly basis and the two environmental stewards are responsible for taking the food scraps bucket out to our garden and emptying the contents into the outdoor composter. Most children know that food "goes bad" or "rots" and we tell them that putting it into the composter is a way of helping the food rot faster and recycling it back into the earth. They begin to use the words *compost, composter, recycle,* and *environmental steward* as part of their daily language. A rap about composting sung by some wiggly worm puppets becomes our class song and is a favorite Youtube video at lunch! Periodically, we remove decomposing matter from the composter and investigate it at various stages.

Different stages of compost

Children are also assigned the job of litter patrol, checking the playground and garden for litter and cleaning up if necessary. Through these classroom routines, children begin to take ownership and gain awareness of their own behavior with regards to waste. They begin to classify and sort the various types of litter and start constructing their own beginning understanding of why some items go in the composter and some not.

## *Our Class Pets, the Red Wigglers*

We soon introduce the children to our class pets, *Eisinia fetida,* the red wiggler worms. Our worm bin is brought into the classroom in the mornings so that children can dig into the bedding and investigate the worms and their home. The children help with cutting fruit and vegetable leftovers from our snacks and lunches and feeding it to the worms. As they observe the worm bin day after day, they see the food they put in slowly get eaten up as worms swarm over the various delicacies. They tear newspaper strips to refresh the bedding and moisten them down for the worm's comfort. They notice that what looks like soil in the worm bin increases and learn that the worms are recycling the food by eating it and that what looks like soil is actually worm poop, or as they learn to call it, castings. This always goes down well!

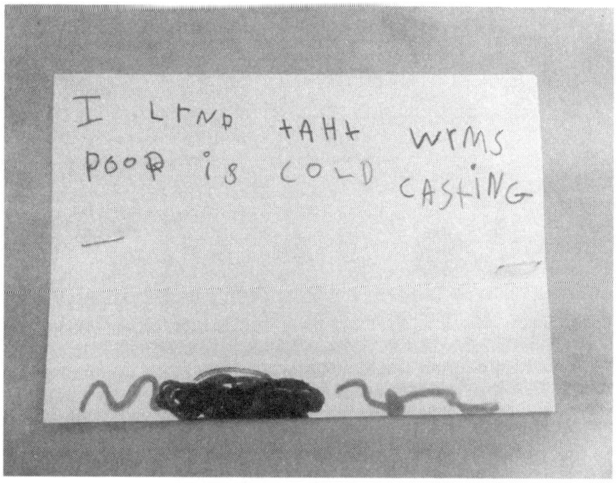

A kindergartner writes what he knows

Our students learn to distinguish adult worms from "teenage worms" and baby worms. Even our most squeamish kindergartners are helping with worm tallies by January! In addition to seeing worms in the worm bin at different stages of growth, they also come across other living organisms, like snails, nematodes, and mites, and we introduce the word *decomposers*, explaining that all these organisms are helping to break down the food we are putting in the bin.

Later in the year, once the children have had plenty of opportunities to investigate the worms on their own, we present a formal lesson on worm anatomy and a worm's role in enriching the soil. The children are now able to use their experience with the worms to express what they have observed and use this, along with the formal information of the lesson, to build their conceptual understanding. We give them the accurate scientific terminology and they use it with great enjoyment and accuracy! Using magnifying glasses and powerful microscopes, they take a closer look and identify the different body parts as well as observe their movements. During these formal observations, students are asked to figure out answers to questions such as "Do worms have eyes?" "Do worms like water?" "Do worms like light?" and "Do they move forward or backward or both?"

By this time in the school year, many of our students already know the answers to these questions based on their earlier informal experiences with the red wigglers in our worm bins. In fact, these experiences enable them to either design experiments or provide us with evidence that will prove their answers correct. For children who did not spend as much time with the worms, these answers are not as evident and formal experiences like these provide them with the opportunity to start learning more. In either case, because the children will continue to care for the worms and harvest their castings, they will continue to have plenty of opportunities to become familiar with and build their knowledge of these marvelous creatures.

## *Garden Work*

Concurrently, the children are becoming familiar with our outdoor garden beds where they see herbs, pumpkins, and other plants that they can smell, touch, pick, and taste. They weed and

Harvesting compost from one of our composters

mulch and later in the year, harvest compost from the outdoor composter and the castings from the worm bin to mix into the beds and indoor plantings.

We show them some good places to dig for earthworms. Comparisons with our red wigglers' anatomy are inevitable. We point out a log under which they may also find earthworms in addition to some other fascinating creatures. Observing the various residents underneath logs is a good lead-in to our study of decomposers.

Daffodil bulbs are planted in the fall for later spring observation. Indoors, we look at paperwhite bulbs comparing them to the daffodil bulbs we planted, and then we grow them indoors in shallow containers where our students can observe the life cycle, measuring the height of the shoots, monitoring the time it takes for the bulbs to bloom, and observing their decay. The children also dissect the bulb at the end of the life cycle to see the internal anatomy and once done, the children place the remains in our outdoor composter.

In early spring, they see the beginnings of the daffodil shoots pop through the soil from the bulbs we planted in the fall. We start informal observations to compare them with the paperwhites—will they be the same as or different from the paperwhites? What are the similarities? Where are the differences?

### *Seed Study*

Our seed study starts in the fall, examining the acorns, horse chestnuts, and pinecones that abound on our campus, and continues into spring when we explore a variety of seeds such as lima beans, various kinds of bulbs, and wheat berry seeds. In the fall especially, we engage in a lot of collecting and counting, and the seeds we collect are used for various math and art activities. In the spring, our emphasis is on the comparisons of various kinds of seeds and their anatomy, as well as observation of plant growth and the conditions needed for a plant to survive. The children help us design an experiment to see what conditions a lima bean plant will need for optimal growth.

Some of the conditions designed by our students to investigate seed growth

Students are always surprised by the plant that grows in the dark and amazed to see its pale coloring—and even more amazed at how quickly it turns green once left out in sunlight. It's a great demonstration of how important sunlight is to the health of a plant.

The children also notice that all seeds do not sprout at the same time, despite having been planted at the same time and the question they raise is "why?" This question arises from the very personal concern that the seed they planted has not begun to sprout but it gives rise to a much more generic discussion about reasons, both environmental and innate, that may have prevented growth. Some of the reasons the children came up with:

- "The seed got too much water—that's why there's fungi on it."
- "The baby plant is not ready to come out."
- "The seed does not have a baby plant inside."

One activity that the children love is to pretend to be a lima bean seed: They all crouch on the ground curled up and slowly they act out, as we "water" them and the "sun shines down," the process of the seed coat breaking and the cotyledon slowly springing out. Acting out the entire life cycle of the bean is another way for our students to access this information. We will always have that one precocious kindergartner who will inevitably state that he (or she) is not planted in good soil (or did not get enough water) and so cannot grow and will remain curled up on the ground and what a lesson that is!

Looking at a log decomposing in the woods

## *Nature Walks and Log Hotels*

We start our nature walks in September, in the woodlands adjacent to our campus, and one of our favorite walks is the one we go on early in the year with our fourth-grade buddy class. The fourth graders share with their kindergarten buddies what they have been studying about log hotels. Immediately, the kindergartners start comparing what they see in the woods with what is under the log in our garden. They also start noticing the different fungi on the log and the words "rotting" and "decomposers" come up.

In the classroom, we set up a temporary log hotel for observation and exploration. Students start to identify and name the different creatures they see, looking for invertebrates, arachnids, insects, and gastropods.

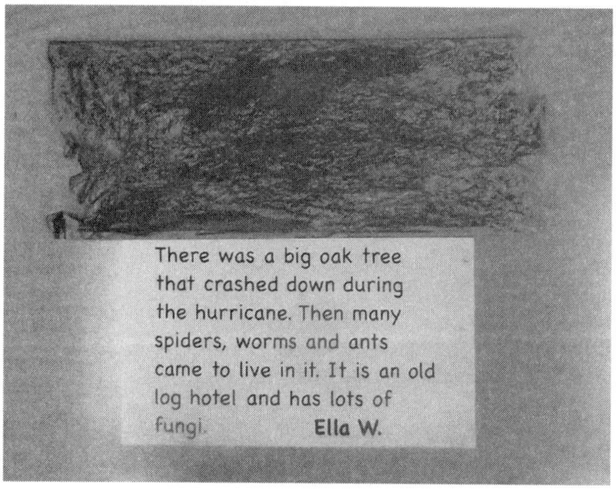

There was a big oak tree
that crashed down during
the hurricane. Then many
spiders, worms and ants
came to live in it. It is an old
log hotel and has lots of
fungi.
Ella W.


Through art, children share their understanding

We read books on log hotels and discuss the life cycle of a tree. The children begin to learn that even a tree that has fallen down and is dead continues to be a source of life for other living things.

The topic of decomposition comes up again as we discuss how the living organisms that live in this tree eventually break the tree down back into the soil. During our weekly nature walks, children see the evidence of this decomposition as they touch logs that crumble.

The FBI song (Fungi, Bacteria, and Invertebrates) is a favorite tune that the children belt out. Created and performed by the Banana Slug String Band, it is an alternative tool for learning about nature's recyclers.

Through the seasons, nature walks continue to be an excellent avenue for children to observe decomposition in its natural state. As we trudge through fall leaves and investigate fallen trees, the evidence of decomposition is all around us. In the spring, as we walk and see signs of new life all around, from budding seeds underneath log hotels to new buds on trees, we encourage our students to notice that there are very few leaves left lying around and that indeed through winter, many of the same leaves that created a soft cushiony layer to walk on during the fall have almost disappeared. A few skeletal leaves remain, evidence of the decomposition process.

## *Organic and Inorganic Litter Study*

Around mid-October, we add on another layer of our decomposition curriculum: our Organic and Inorganic Litter Study. All of the trash from the day's lunch—apple cores, banana peels, plastic bags, juice boxes, napkins, and some added leaves and twigs from the playground—is put on a tray.

We play "What's the Rule?" and as we place the litter into two different trays, the children try to figure out what rule we are using to separate the litter. It's always fun to listen to how they start narrowing it down, starting off with a color attribute if we put two things down of the same color, to "things we can eat/things we cannot eat," "made in a factory/grown in the garden," "things we can put in the food scraps bucket/things we cannot," and "food for the worms/ not food for the worms." At the end of the exercise, we introduce the words *organic* and *inorganic* and we clarify the difference.

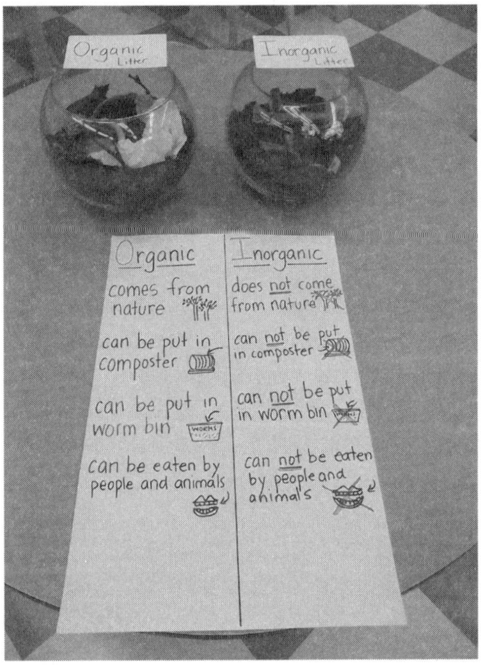

Differentiating between organic and inorganic litter

We bring out two glass bowls filled with soil—our mini-Earths. We fill the two bowls with the two types of litter. The children write (or dictate) their predictions about what they think will happen to the litter. Some examples of predictions include:

- "They both will stay the same."
- "They both will sink into the soil."
- "The organic litter will get rotten."
- "The inorganic litter will get smaller."

For the rest of the year, the children observe the organic and the inorganic litter, confirming or refuting what was predicted. We model the use of the words *decomposing* and *biodegradable* to define what the children are observing. Inevitably, we keep adding to the mini-Earth with the organic litter as the litter decomposes. The mini-Earth with the inorganic litter, however, gets full very quickly and for the rest of the year remains exactly the same. It's a small but powerful visual.

A journal entry that records an observation

At times, children will ask to add something else to the organic litter bin, usually something we have tasted as a class and they will want to see what happens to it. Once a student surmised that if we added some "organic stuff" to the bowl with the inorganic litter, it might help the inorganic litter to decompose, so we tried it.

These experiences, child-initiated and born out of their curiosity, allowed for our students to be the architects of their own learning and thus vested in the outcome. There is always a healthy debate about what they think will happen and predictions are made. Sometimes, we ask them to record or dictate their predictions into their science journals, sometimes we do not. Always, we ask them to tell us "Why?" and then explain the reasoning behind their predictions. As they hear each other's explanations, some incorrect and some right on the mark, they become each other's teachers and we merely the facilitators.

## *Bottle Cap Recycling*

New York City recyclers are not equipped to deal with certain types of plastics and, unfortunately, bottle caps have to be thrown in the garbage. Often these caps end up in rivers and oceans. Birds and other marine creatures mistake them for food with tragic results. The magnitude of this pollution problem is devastating to our oceans and wildlife.

In partnership with community schools, preschool through 12th grade, Aveda, a company that makes hair and skin products, has built a recycling program for plastic bottle caps and our lower school has enrolled in this program. Starting in September, the children bring in bottle caps from home and throughout the year they sort, count, and collect; each full collection box is mailed to Aveda. Originally initiated by the kindergarten team, this bottle cap recycling program caught on with the older grades and before we knew it, children were popping into our classrooms at all times of the school day to bring in caps from home or from their classrooms. The caps are also used for math (estimation, counting, sorting, and patterns) and art.

We put some caps into the mini-Earth with the inorganic litter so that children could keep track of changes in the bottle caps, and with each passing month, the children noted that the caps

Counting bottle caps by tens for Aveda

not only do not decompose but in fact stay exactly the same and as we added more inorganic litter, crowd the space. They can literally see the difference they are making every time we ship off a box of 2,000 caps (yes, we count them!) to Aveda for recycling instead of throwing them in our garbage and sending them off to landfills.

## *Pumpkin Study*

Our Halloween pumpkin study is another layer that we add to our decomposition curriculum. After Halloween, the pumpkin is placed in the dirt outside our classroom doors. Every day as they run out to recess, the children check on the pumpkin to see the state it is in. Photographs are taken and kept in their science journals. The children both speak and write about the changes they see. We compare the changes we see in the pumpkin to what we have seen happening to the organic litter and to the food in the worm bin. The word *decomposing* becomes a regular addition to our daily vocabulary. During this observation, we go back to the list of predictions they had made before we put the pumpkin out and we check off those that we see are coming true. We note what observations did not pan out—such as "It is going to get bigger"—and what may still happen: "It will turn into soil and make more pumpkins." We also note that many of the predictions are part of one long process called decomposition.

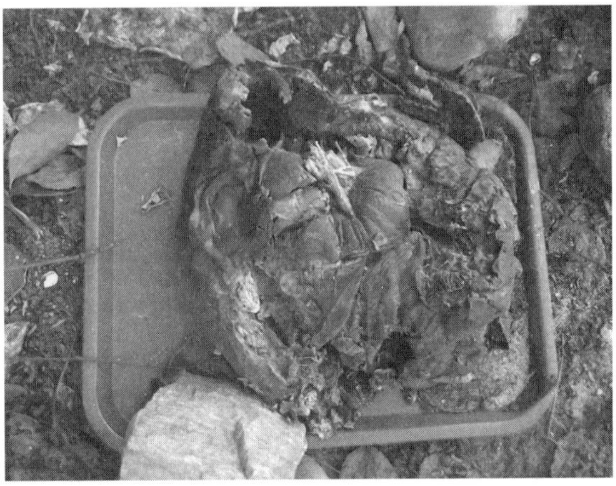
Our decomposing pumpkin

The observation of the pumpkin becomes child driven, as each day, at least one child comes running to a teacher to update us on the status of the pumpkin, or to remind us to take a picture of the pumpkin to add to our timeline.

On cold winter days, we bring the decomposing pumpkin inside so children can take a closer look at some of the decomposers at work. Fungi, of different variety, usually abound. A centipede or a beetle will hurriedly crawl out. The children are immediately assailed with the smell of decomposition as well as the very real sight of what rotting food looks like. Yet, despite the few "gross" or "eew" comments, the children continue to exhibit scientific behavior in the way they investigate the pumpkin, using magnifying glasses from the science center to take a closer look.

Our students are always amazed at how long the stem takes to decompose—way after the rest of the pumpkin is already part of the soil, the stem hangs around, stubbornly refusing to go anywhere. When it finally does, one spring day, it is a cause for celebration and an opportunity to count how many days it had taken for the stem to finally become a part of the soil.

Students writing down their pumpkin observations

## Making Connections

Once, as we were preparing to add some red pepper leftovers to our "organic" mini-Earth, one of our kindergartners said, "Ms. Gopal, I think the stem of the pepper is going to take longer to decompose than the soft part." We pointed out to the class that Savannah had made a hypothesis and asked her for her reasoning. She immediately replied, "Do you remember how long it took the stem of our pumpkin to decompose?" Here was scientific thinking based on extrapolation from previous data!

Along with such anecdotal evidence, we also use journal entries and representational drawings to encourage children to document and share their understandings or observations. One day, we tasted Brussels sprouts (as part of our weekly taste tests connected to our letter study). We placed the leftover chopped Brussels sprouts from a taste test into the organic mini-Earth and decided to add a whole one as well. We asked our students to predict if one or the other would decompose faster and why. They recorded their predictions into their science journals and in the process we were able to see that most of the children immediately picked up on one of two attributes to explain their reasoning: either the size of the sprouts or the consistency of the sprouts. A couple of children were not sure what would happen; they would need more experiences with decomposition and more time to make the connection.

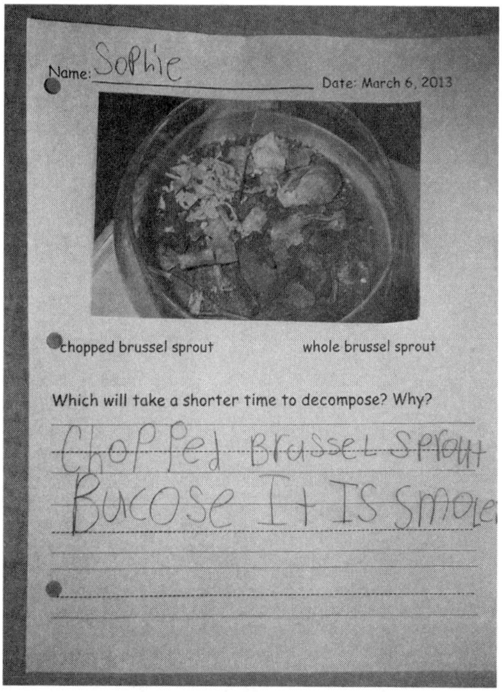


A prediction on decomposition rates

We also use Voice Thread for documentation and assessment. Voice Thread is a web-based interactive collaboration and sharing tool that enables users to add images or videos that other

Making a life-size Giant Gippsland worm out of modeling clay

users can then comment on by either audio or text. It is especially effective for young learners whose expressive language exceeds their writing abilities. "I See, I Think, I Wonder," a Visible Thinking strategy tool developed by Harvard's Project Zero, is another excellent tool to help determine what connections and conceptual understandings our students are developing.

Art projects set up throughout the year (mixed media collages, tempera painting, oil pastel murals, botanical sketches and work with clay) provide opportunities for artistic representations of the children's experiences and knowledge. Making a life-size replica of the giant Gippsland worm out of modeling clay brings it to life, as does becoming a giant Gippsland worm when making a line move to Music.

## Stewards of Our Environment

Through this yearlong study, the children start to learn that they can make an impact on their environment, in both positive and negative ways. They start to make the connection between the kind of litter they throw away and the possible impact it may have on their surroundings. They learn that they can make a difference, whether in recycling bottle caps, composting, or being on litter patrol. They are more aware of using containers that can be reused or recycled. They learn that even the smallest creature has a deeply important role to play in the health of our planet. They begin to take a closer look at the natural world, appreciating the colors and beauty of the smallest leaf or seed. Some of our students internalize this, others develop a beginning understanding, and still others may need more experiences to continue them on this journey.

Regardless, it is a journey that is well worth it at this early developmental stage and which sets them toward a path of both scientific inquiry and respect for the environment as they move up in the lower school.

## Investigations in Grades 1–5

Delighted that our kindergartners were delving deep, we also wanted to continue building their knowledge and interest as they go through the lower school. It is very much a work in progress and only recently have we begun to knit the curriculum vertically, but our school has come far from the first beds built.

In each grade, the overarching goal remains to build on the concepts and process skills from the year before in order to deepen our students' understanding of as well as their personal connection to the natural world. Through all of the explorations, experiments, and ongoing observations, the big ideas are always kept in mind. They form the backbone of the program and are used in varying ways. For example, the fact that each organism is dependent on its environment to get the things it needs to survive is emphasized. This is true in studying plants, animals, or in constructing something to become part of the environment.

Changes in the environment affect the viability of one species, which affects the whole balance of creatures and plants in the ecosystem. This is a theme that is repeated throughout the curriculum, but explored in greater detail as the children get older. Another example of an ongoing theme is the subject of soil science. An overriding goal for the program is to make children familiar and comfortable with the natural world, learn about how it works, and develop a deep understanding and commitment to follow sustainable practices.

Interdisciplinary projects that allow for real-world applications remain a priority, and the STEM initiative that our school has embarked on (we call it STEAM, with the A standing for the Arts) encourages the integration of the various disciplines. Technology, defined broadly by the *Framework* as any modification of the natural world that is made to fulfill a human need or want, is an integral part of our investigations and explorations. Whether the tools are shovels or magnifying glasses, tablet apps like Leaf Snap or Nature Tap, computer microscopes like the MiScope, or more sophisticated computer applications to help with robot design or construction projects, technology is used as a vehicle for learning. It is also important for our students to know not only how to use these tools but when so that they are maximally effective.

Laurie Bartels, our STEAM integrator, is key to the hands-on explorations and investigations conducted in the older grades. She skillfully integrates technology with the various disciplines and has enabled our students to not only experience but effectively use technology to articulate questions and discover the answers.

We have chosen to outline the areas of investigation that the older grades pursue to give a better sense of the deliberate overlap and spiraling in our curricula, as well as the linkages with the kindergarten curriculum. While some grades may focus on one area more than another, depending on the interests and passions of their teachers, the following are the broad areas of investigation that our students experience as they go through the lower school:

- Observation and Exploration
- Gardening, Composting, and Soil Study
- Dendrology
- Hydrology

## *Observation and Exploration*

The pre-kindergarten consists of one small group of four-year-olds and the curriculum is designed around observation and exploration. As in kindergarten, once each week the children go for "nature walks," either on campus or in the adjacent woodlands. Bending over the ground, they explore beneath the rocks and under leaf litter. Their wonder abounds, and they can be heard saying "Look, I found so many bugs, even worms." It is at this point that the children are introduced to the richness of the natural world—decomposers of the forest and the variety of colors, trees, seeds, and flowers. These walks never cover much distance with the youngest ones, but then how far do you have to go to discover something interesting in the woods? We often collect specific things on our walks, like pinecones, which the children count and label and put on display in our environmental lab museum. Questions about the things they collect are inevitable and can be used to generate further investigations.

The first graders are not that different from the younger ones in wanting to dig and explore in the woods. Finding living things that are not immediately visible often brings squeals of delight. One day when we were on a nature walk, we recorded video of one of those moments when the children turned over a log in the woods.

| | |
|---|---|
| *Teacher:* | What do you see? |
| *Student:* | I see a worm. Is it a red wiggler? |
| *Teacher:* | Does it look like a red wiggler? |
| *Another student:* | And a slug. Ooh gross! It's tickling me. |
| *Another student:* | I found a worm too. Which is the front? |
| *Teacher:* | Which end is leading and which is following? |
| *Teacher:* | It's interesting that all these critters we're finding live in the same place. |
| *Student:* | I found one hidden in the leaves. |
| *Teacher:* | What do you call that? |
| *Student:* | Camouflage. |

This sort of conversation that accompanies their discoveries enables us to facilitate richer discussions (and investigations) about a variety of topics such as: Do red wigglers live in the woods, how are they similar to or different from other earthworms, what is the purpose of camouflage and what other living things use it?

On a beautiful spring day, one class went for a "meditation walk" during which they found a quiet spot to sit still and listen to the sounds in the woodlands. It was not easy for the children to quiet down and concentrate on listening. When they finally did, they were able to identify the soft sound of a breeze blowing through the leaves, in addition to the birdcalls and the sound of a train passing by. It was a magical moment for us.

Change over time becomes a focus of these walks as our students observe seasonal changes, decomposition in its natural state, and various life cycles. There are opportunities to experience how change can occur quickly (as in when a sudden storm causes trees to fall or branches to break) or can take a long period of time as in a slowly decomposing log near a trail.

In our gardens, the children closely study insects and other critters, observing, identifying, and drawing what they see. We learn about what insects need to survive, note their similarities and differences and come to a realization that living things rely on their environment and that changes in the environment, either natural or made by people, can have an effect on their survival. This leads to a more in-depth study of animal life cycles, as exemplified by the study and observation of ants working together in a community, caterpillars turning into butterflies, and a yearlong observation of walking sticks and earthworms. In studying worms, we observe them in their natural habitat, note the conditions they like, and create an environment for them in the classroom where we can observe them. The study of butterflies offers a good opportunity to see a metamorphosis where the babies evolve into their adult counterparts. Ants are studied in conjunction with a classroom study of community and the children have an opportunity to observe the interactions and jobs of different members of the community and how each ant contributes to the survival of the whole colony.

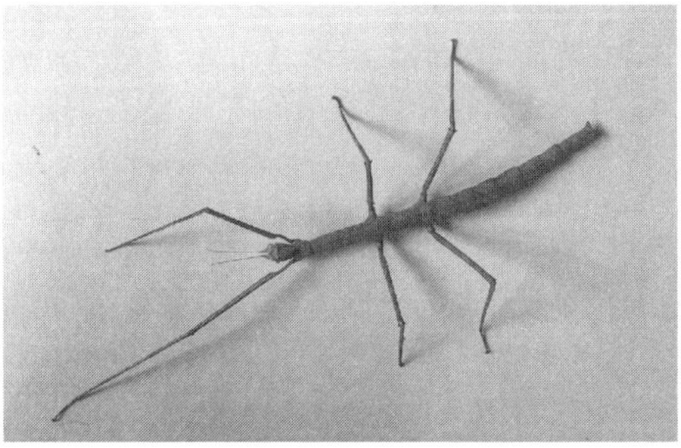

A walking stick ready for close-up observations

## *Gardening, Composting, and Soil Study*

It seems such a simple thing and yet many children who grow up in an urban environment are so removed from nature's processes that it is almost magical to them. Yet to truly understand the importance of soil to food production and sustainable practice, children must experience firsthand the life cycle of the food on their plates. The year starts off with our students harvesting some of the fall produce and engaging in the process of weeding, mulching, and putting the beds to sleep for the winter. Compost from our composters is harvested in the spring and added to the soil when the land begins to awaken again and the cycle is begun anew with seeds being planted in our environmental lab during the colder months and later transplanted into the beds once the frost has passed.

The vermicomposting program is firmly entrenched in kindergarten. This year we started a first-grade worm bin, which was requested by the children who missed the worms and the composting snack activity they experienced in kindergarten. We hope to extend it further into the older grades to more deeply explore the science of composting (temperature monitoring, moisture retention, decomposition rates, and so on) as well as the technology of vermicomposting (designing and building bigger worm bins that can handle more waste).

Spreading compost

Our youngest children learn about how things grow and discover firsthand by planting seeds they find and observing growth and sprouting under different conditions. They plants snap peas in early spring and are amazed to find that they have begun growing when they return from spring break. Kindergartners grow strawberries, beans, lettuce, and scallions as do the first graders. The children enjoy visiting their plants, drawing pictures of how much they are growing, learning songs that teach about the parts of a plant, and watching them flower and then produce fruits and vegetables.

"Garden recess" is a time when an environmental educator joins in the children's recess and solicits their help in tending the garden. There is usually a small group of dedicated gardeners ready and willing to explore and discover, participate in growing plants, and compost the garden waste. They learn about the usefulness of tools and explore their uses in the garden. With wheelbarrows and shovels, rakes and hoes, they get "up close and personal" with the growing process.

Throughout the experience, attention is always given to the importance of the soil, water, and sunlight in making living things grow. The adding of compost from our own composters encourages a discussion about the quality of the soil needed to grow good produce, and the children are always excited to see that what was originally fruit and vegetable scraps has now become a rich, dark, crumbly material that can be added back to the earth. A chant that we often sing is "sun, soil, water, and air—it's everything we eat and everything we wear," written by the Banana Slug String Band.

Second graders continue the exploration of seeds with a study of seeds found in common vegetables. They predict the number and color of seeds in each, open them up, observe, count, compare, and then graph the results. The class decides to open an avocado, remove its seed, and start to root it in the classroom. Students discover that it takes time and patience for a seed to grow. They observe and journal their observations as the roots appear and finally a green stalk. One student writes: "The avocado looks orange, white, and brown. It feels soft and smooth. It smells like fruit." The process of germination takes much more time than the children predicted.

Then they move on to germinating seeds in the environmental lab for planting in the garden. The students start lettuce and scallion seeds in early March and revisit the lab periodically to check on their plants, which are growing with the help of grow lights.

In early April, students transplant their seedlings and observe them as they grow. This year the lettuce grew steadily but the scallions failed to thrive after being transplanted. This formed the basis for a conversation about the different nature of plants and their abilities to withstand changes in their environment.

While learning about the spice trade and Columbian trade routes, the fourth grade does a lot of tasting of food from various places, as well as work in the garden to produce food representative of different regions of the world. It is always fascinating to the children that Italy did not have tomatoes until they were brought from the New World to Europe, after Columbus's voyages. A Native American bed with the "Three Sisters" planted (corn, beans, and squash) invites children to explore why a community chose to plant these particular crops in this way and discover for themselves the interdependent relationship between these vegetables.

This year, the fourth graders became deeply involved in planning trellises and tipis to support their various plantings, collecting fallen branches from our woodlands and constructing the support structures using tape measures, saws, and twines. All this work in the garden naturally integrates with social studies, science, math, and history, and makes our students' learning immediate as well as more meaningful.

Fifth graders are involved in learning about healthy eating and spend part of their health classes working in the garden, learning about healthy soil, its contribution to the health of plants and therefore our bodies, and planting cover crops before putting the garden beds to sleep for the winter. They learn about the role of nutrients in the soil, they test the soil, plant cover crops to replenish the nutrients used up by the plants, and in the springtime they do the hard work of turning over the soil in all the garden plots in preparation for planting.

Fifth graders turn the soil in preparation for spring planting

Last year, a group of fifth-grade students independently developed a garden area in one of the fields during recess time and populated it with bulbs transplanted from different parts of the campus. Another year, after having researched the habitats of local butterflies, our students cleared a piece of land to create a butterfly garden with buddleia, parsley, and lemon balm to attract the local specimen. As it turned out, the spot they chose did not get as much sunlight as was needed and the butterfly garden did not flourish. It was a rich learning experience for them and one they would not soon forget.

Building the stage

For another project, the fifth-grade teachers decided that an outdoor theater platform would encourage further interaction with the outdoors, so an interested group of children learned to use tools, design, measure, cut, and build a wood stage that was heavily used for the remainder of the year.

## *Dendrology*

In the younger grades, trees are a focus of study throughout the year, whether during our nature walks or around our campus. The children learn about tree anatomy and a tree's life cycle, the similarities and differences between the different trees, and that the tree is an ecosystem and continues to give life even after it dies or is felled by a storm. Several grades adopt trees to observe through the seasons and the amount of information the students collect and record gets more sophisticated as they get older.

In the fall, our students also comb the campus for seeds of all sorts and compare and contrast them. They look at them under a wireless microscope, whose image is projected onto an interactive whiteboard. Studies about trees begin with a conversation about the shape, color, and texture of seeds and leaves; similarities and differences in bark; size and girth of the trunk; and (as they get older) more detailed discussions about and experimentation with chlorophyll and energy. Tree rings are investigated as a source of information about the tree's age and history.

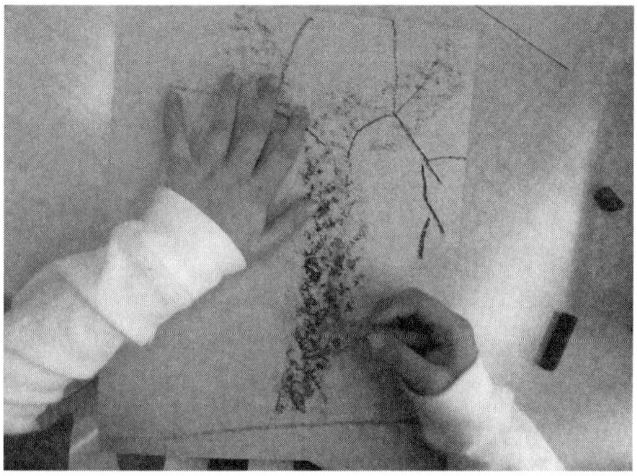

A kindergartner sketches a winter tree

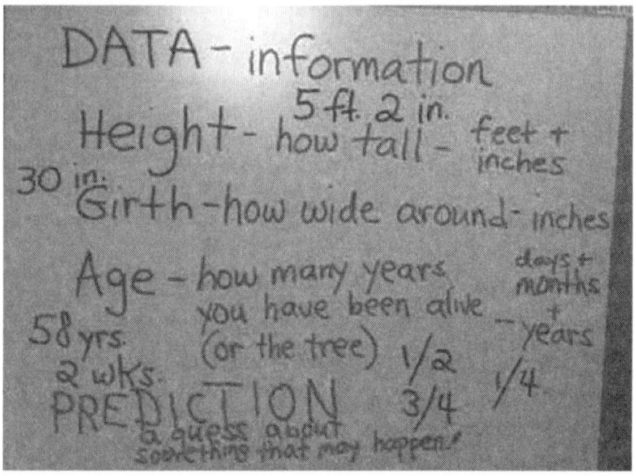


2nd graders get more detailed information about a tree

After researching and collecting information on trees, one class worked in small collaborative groups to write scripts, with the goal of producing a video to share the information from its research.

Another class compared the height of a giant sequoia (247 feet) in California with that of a horse chestnut tree on our campus. To get a sense of 247 feet, the children unfurled three 100-foot tape measures in a straight line. Then they went over to the horse chestnut tree and did some educated guessing and approximation (using rulers or their own height) to figure out how tall it might be.

Yet another class was posed the following question by Ms. Bartels in winter: *It is January. It is winter. I collected a number of leaves from trees (they were STILL on the trees) during the second weekend of January. What color do you think the leaves will be?*

The children made predictions (she had holly leaves and pine needles that were hidden in her bag) and then examined the leaves for color, shape, and texture. This led to a conversation about how the differently shaped leaves would capture snowfall, which then led to questions about why some trees have thin leaves while others have broader leaves and why some trees remain green in the winter whereas others do not. It was a wonderful way to introduce our students to the concept of adaptations and survival strategies.

Art is an important medium through which trees are explored in the elementary years and it is a common sight to see different grades sitting on a hill sketching the trees on our campus during the different seasons. Songs are also sung about the parts of a tree and their functions.

## *Hydrology*

The study of water has recently begun to resonate on our campus. With the Hudson River our constant neighbor, it seems only natural that we use our sense of "place" to anchor our study of water.

Our younger students—after a visit to the Little Red Lighthouse (officially named Jeffrey's Hook Light) situated by the Hudson and watching a video of the journey from the river's source—were inspired to create their own river outside in our playground. The work they did was truly amazing to see, as they dug a "source" for the river, created channels of flow that would lead out to the "ocean" and even created a "tributary" like the one we saw on one of our nature walks by the Hudson. When we flooded the "source" with water, there was much excitement as students watched the water flow through. They also tried to make the water flow faster by removing debris and smoothing the channels. There was no doubt in our minds that they were learning the beginnings of not only how a river worked but also a good deal of physics. The winding rocky bed on our playground that occasionally captures water flow during heavy rains is another source of water play.

Creating a bridge over the "river"

The indoor experiments with snow, ice, water, and the explorations with puddles and ice blocks outside help to build student knowledge about the properties of water, as well as the water cycle. Children of all ages participate in these activities. Once, when the kindergarten was designing an experiment to prove whether xylem does indeed carry water up a plant, the children determined that we would have to cover the vases of colored water in which the celery stalks were sitting in order to make sure that the decreasing water level was caused by the xylem taking the water up rather than evaporation. Anecdotes like these are indications to us of sophisticated scientific thinking and are a delight to witness.

A rain chain built by third graders

Older grades explore water in its many forms, its many uses, its distribution throughout the world, its importance for life, and the issue of water pollution. Throughout all the investigations, the necessity of clean water for life to thrive on Earth is stressed. The water cycle is explored in greater depth than earlier, and students go about their explorations in a variety of formats.

In one project, children placed rain gauges in different areas of the campus, collecting and charting the data. They predicted how their results would compare to local statistics on average rainfall in a particular time of the year. A comparison with local statistics was made and the children then relocated their rain gauges, making more predictions, and charting and comparing the data. This project led to a discussion of downspouts, gutters, and rain chains. The students then designed and created rain chains from recycled bottles and

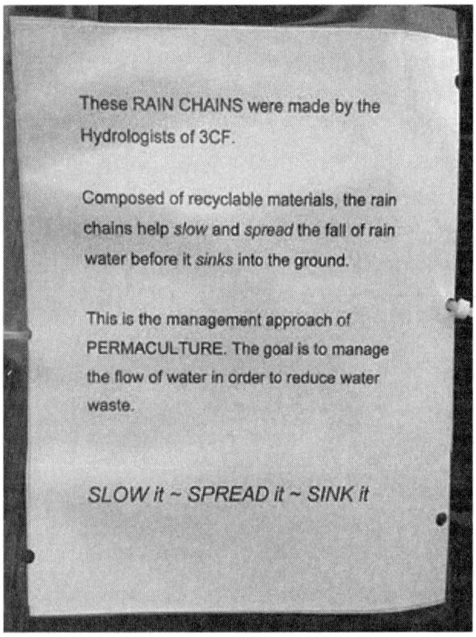

These RAIN CHAINS were made by the Hydrologists of 3CF.
Composed of recyclable materials, the rain chains help *slow* and *spread* the fall of rain water before it *sinks* into the ground.
This is the management approach of PERMACULTURE. The goal is to manage the flow of water in order to reduce water waste.
*SLOW it ~ SPREAD it ~ SINK it*


A sign next to the rain chains

yogurt containers. They hung the rain chains in places where puddles often formed to see if the chains would solve the problem. They did!

Another solution that the children came up with was to suggest we get a rain barrel to alleviate the pileup of water near one of the buildings. We will try it out this year. This would also provide a water source for a part of the campus without much water access.

In another interesting project, older children explored the World Monitoring Challenge and proceeded to do tests on local saltwater areas and compare the results, as well as tests of tap water. In these explorations, they measured temperature, turbidity, Ph and dissolved oxygen in water and researched the effects of the factors.

Such projects enable children to make the links between their academic studies and their real-world applications. They are also helpful in our assessment process as we note the role students take, their level of participation or input, and what this tells us about the depth of their understanding.

Trips are an integral part of the program. As part of learning about the water cycle, a trip was made to the Grass Island Waste Treatment Plant in Connecticut, where the children got a more thorough understanding of where the water we use goes. They also visited the Bruce Museum's exhibit of extreme habitats, where they learned about the characteristics of places with very little water as well as those with an overabundance of water.

One third grader wrote this poem about water:

"When water falls in the sink
It makes the sound of a clink
When water drops
The ground soaks it up
When water evaporates from a stream
It rises like steam
Rain can drizzle
Rain can pour
Rain makes puddles
On the floor
Rain, ocean, river
There is so much more
I'm sure, I'm sure."

Water is a big part of fourth-grade study as well. They explore the world's oceans, their inhabitants, their importance to the health of the Earth, as well as ways in which people can participate in movements devoted to ocean health. Rachael Miller, founder and executive director of the Rozalia Project and Atlantic Cup intern Laura Migliaccio visited our campus and our students were treated to some eye-opening bits of garbage, specifically garbage that winds up in the Earth's waters and pollutes both the water and the marine life, all of which then impacts the human-animal food web.

## Who Makes It All Happen?

It should be noted that the participants in our program not only consist of the students and teachers directly related, but numerous staff members who are involved during the summer, as well as during the school year. Upper school students serve as interns helping to oversee the garden beds in July and August as well as planting a new crop of vegetables for use in the fall. In the fall, herbs are picked and dried and distributed to many faculty members as a Thanksgiving treat. Dave Buske, the chef who prepares the school meals, is often seen in the morning wandering through the herb box with scissors, smiling.

The school administration, the parents, the teachers, and the people who help us build compost bins, water our garden, and maintain our school, are also crucial to the success of our work. A strong parent committee dedicated to promoting sustainable practices at the school made a large contribution as well as a forward thinking headmaster and a supportive engineering, gardening, and grounds staff. It is the wide base of support that is helping us in our efforts.

## Links With Other Reform Efforts

In addition to aligning our science curriculum with *A Framework for K–12 Science Education*, Riverdale Country School seeks to engage and partner with other communities and organizations with similar goals.

Our school is a charter member of the Green Schools Alliance. Our school is dedicated to being a green institution and lowering its carbon footprint.

The NYC Compost Project, under the aegis of the NY Department of Sanitation, has designated us as a community compost site, and our goal is to become a demonstration site so that we can go into outreach.

We are partnered with Aveda in their nationwide bottle cap recycling efforts and have enthusiastic participants in every grade of the lower school.

Our goal to make our campus a living lab closely aligns with the goals of the National Wildlife Federation, which encourages the development of school campuses as wildlife habitats. Our campus has been certified by the NWF as a wildlife habitat.

We are also active members of NAAEE, the North American Association of Environmental Educators, which offers workshops, training, and a chance to network with other environmentalists.

Another resource for integrating environmental studies into the curriculum is CELF (The Children's Environmental Literacy Foundation). Their workshops promote active, project-based work that yields real-life information about school campuses that students can then act on.

## What Next?

After losing two fabric greenhouses to storms last winter, we are upgrading to a polycarbonate greenhouse, which is fastened to a concrete foundation. We hope to be able to explore plants in a different way with the help of this controlled environment. We will also try to extend the growing season, which is a very short time before the children leave for the summer. We've enlisted the help of a Wave Hill staff member (Wave Hill is a neighboring botanical garden) to consult with us regarding its use and care.

In consultation with BSCS, an organization whose mission is to transform the teaching of science so that we have scientifically literate citizens, our school is engaged in articulating a vertical science curriculum that encompasses the preK–12th grades. The committee engaged in this work is comprised of a representative from each grade at the lower school, the health teacher, the STEAM integrators, the environmental educator, and all the middle school and upper school science teachers. Using *A Framework for K–12 Science Education* as our compass, we have begun outlining strands of study as well as identifying crosscutting concepts and scientific and engineering practices that we can focus on in our teaching and learning.

Upper school students are now involved in working in our gardens during the summer. We are planning to involve the Upper School Student Sustainability Club by having them come to the lower school during activity periods to work alongside the younger ones. We've also thought about possible participation in local green markets during the summer with the help of the older students.

Our school has also formed ties with the Monteverde Friends School in Costa Rica and one kindergarten classroom at Riverdale is linking with a kindergarten class in Monteverde to explore our two habitats, investigating the similarities and differences in plant and animal life, seasonal changes, and the way nature recycles. It will be an exciting opportunity for our students to learn with other children engaged in the same investigations.

Formal assessment modules for our curricula are determined grade by grade and need overhauling, a process we hope will be facilitated by the work with BSCS. However, we do think it is crucial to allow for formative assessment as a method of not only determining what students are learning but as a tool to inform teachers for further lesson planning and investigations. This is especially crucial to keep in mind given that different students learn at different paces according to their interests and abilities. Even summative assessments should be used for the same reasons.

With regard to current assessments, we rely on informal observations of students and their conversations, student writing, journals, observation logs, and project work. Independent projects that our students take on (whether it is the creating of the butterfly garden by fourth graders or the building of a worm bin by first graders) are also clear indicators of an engagement with and commitment to our environment.

## Conclusions

Studies have shown that a close engagement with and observation of the natural world enhances the acquisition of not only science-process skills and content knowledge but also improves performance in math, social studies, and reading. Equally important, the more time children spend with nature, the healthier they are both physically and socio-emotionally.

Organizations as diverse as the National Education Association, the National Science Teachers Association, the No Child Left Inside Coalition, the Outdoor Industry Federation and various healthcare organizations agree on this. Yet policy makers and educational administrators across the nation continue to ignore research results, and the amount of time children now spend outdoors has declined significantly. Today, many schools are being forced to cut down on not only environmental programs but even on the time that students get to spend outside. Our children are increasingly becoming more and more dissociated from the natural world and ultimately, this will be to the detriment of our future on this planet.

Visionaries in any field will recognize that without a healthy sustainable planet to live on, our efforts for job creation, economic growth, business development, and energy expansion are less than useless. A stable economy cannot exist without a healthy environment. The way STEM is articulated in schools will mean the difference between students who merely use more technology and students who develop the thinking skills needed to use technology effectively to build a sustainable future. Our students will be the future leaders, and what could be more vital for their education and for that matter, sustainability, than an appreciation and a deep understanding of the natural world and the way in which we are interconnected?

## References

Archie, M. 2003. *Advancing education through environmental literacy.* Alexandria, VA: Association for Supervision and Curriculum Development.

Charles, C., R. Louv, and S. St. Antonie. *2010. C&NN report: Progress, initiatives, studies, tools, networks and innovation*. Minneapolis, MN: Children and Nature Network.

The National Environmental Education and Training Foundation. 2000. *Environment-based education: Creating high performance schools and students.* Washington, DC: National Environmental Education and Training Foundation.

National Research Council (NRC). 1996. *National science education standards.* Washington, DC: National Academies Press.

National Research Council (NRC). 2012. *A framework for K–12 science education: Practices, crosscutting concepts, and core ideas.* Washington, DC: National Academies Press.

# Promoting STEM Practices for All Students

*Karen E. Johnson*
*STEM Magnet Lab School*
*Northglenn, CO*

## Setting

The STEM Magnet Lab School is located in a suburb of the Denver area. This school has approximately 450 students, grades K–8. STEM Lab is a public magnet school in which students are selected through a lottery process, based largely on where they live within the school district. There are no testing requirements to enroll in the school; students need only a passion for learning science, technology, engineering, and mathematics. Upon completion of the first year of the opening of the school, more than 400 families were on the wait list to enroll in year two. After the second year of opening, more than 400 families continued to be on the waitlist.

The demographics of STEM Lab consist of approximately 67% of the student population being Caucasian, 18% Hispanic, 6% Asian, 4% Black, and 2% American Indian. Special education students are approximately 5% of the student body and 9% are English Language Learners. The identified gifted and talented student population is approximately 13%. In addition, approximately 24% of the student body participates in the free-and-reduced-lunch program.

As a K–8th-grade school, a variety of programs are implemented. Extended day kindergarten; electives classes such as art, music, and physical education; and an entrepreneur course are to be a focus for all students in the new school year. A middle school philosophy is implemented at the grades 6–8 level, which includes five core classes in language arts, social studies, math, science, and engineering. Middle school students have an elective wheel of courses which consist of art, music, physical education, health, foreign language, and entrepreneur. There are two classes each with approximately 27 students in each class for each grade level.

An effective model of STEM exists in this public K–8 school. The vision for this particular school focuses on an integration of authentic, real-world issues that nest in a problem-based learning approach. Students are immersed in inquiry-based instruction where they are pushed to apply their content learning to engineering practices. Technology and mathematics are tools for student learning throughout the entire process. This vision does not only apply to middle school students or to high ability students! It applies to *every* student within the building on a daily basis.

Teachers at STEM Lab believe that content should be relevant to society in the whole world. There is also a schoolwide belief that students learn best when working in cooperative groups and in collaborative environments. Technology is used as a tool that connects students to the outside world. Teachers use curriculum resources in a variety of ways, which includes district-mandated curriculum in addition to teacher-created materials. Students determine the direction that the curriculum may take.

## Essential Features of STEM: Merging Science and Engineering Practices

### *Classroom Scenario: Weather Around the World*

Students are at computers, quickly uploading current weather data, evaluating the data, and modifying their presentations. In a few minutes, they will be presenting to a panel of adults—not just any adults, but scientists, meteorologists, high school teachers, parent experts, and business partners. This is the culminating day of a four-week-long problem-based learning task that sixth-grade students have completed.

Based on the *Next Generation Science Standards* (NGSS Lead States 2013) and supported by *A Framework for Science Education* (NRC 2012), students were immersed in science, technology, engineering, and mathematics on a daily basis throughout the Weather Around the World unit of instruction. Scientific skills were put to the test as students learned how to predict the weather and create a three-day weather forecast for a particular location. Teams consisted of groups of students in which each team member chose an expert role: meteorologist, oceanographer, topographer, climatologist, and historian. Each of these perspectives provided important information to the team regarding weather predictions for a location. The classroom teacher focused on teaching science content related to weather instruments, collecting weather data, and using data to analyze weather patterns such as fronts and pressure systems. Classroom instruction used an inquiry-based approach. It also included a variety of outside resources such as field excursions and webinars, which included the participation of meteorologists and scientists. It also included real-time weather data using the internet.

### *Classroom Scenario: Volcano Alert*

Robots moved around a model volcano. Students placed sensors at various locations to monitor potential volcanic eruptions. As students worked in teams to program robots and create a final presentation of their ideas, excitement was in the air! An outside visitor observed high levels of student engagement, excitement, innovative ideas, creativity, and most of all, persistence. As robots stopped in their tracks, failed to release sensors, or fell down the side of the volcano, student learning prevailed.

This scenario represents one of the final days of the Volcano Alert problem-based learning task that engaged sixth-grade students in an authentic problem that scientists and engineers face: How do scientists and engineers monitor and predict volcanic eruptions? How are evacuation plans developed and carried out in areas located with this potentially deadly force of nature? Students learn science content related to volcanic eruptions using an inquiry-based approach to study topics such as plate tectonics, types of volcanoes and eruptions, the flow of volcanic materials such as lava, pyroclastic flow and lahars, and calculating evacuation times based on mathematical models. Engineering practices were incorporated throughout the task as students studied and designed sensors that monitor potential volcanic eruptions.

Bybee (2011) has described science and engineering practices that provide for effective STEM instruction in all classrooms. Additionally, the National Research Council (2012), and the *Next Generation Science Standards* (NGSS Lead States 2013) described effective science and engineering practices. In this school, science and engineering practices are integrated throughout instructional units. Opportunities for asking questions and defining problems, developing and using models, planning and carrying out investigations, using mathematics and computational thinking, constructing explanations and designing solutions, engaging in argumentation, and communicating information to others, are all at the forefront as classroom practices. A summary of classroom examples is included as Table 3.1.

**Table 3.1.** Examples of Science and Engineering Practices

| Science and Engineering Practices (NRC 2012) | Weather Around the World | Volcano Alert |
|---|---|---|
| **Asking Questions and Defining Problems** | Students generate questions related to predicting the weather from different perspectives. | Students identify a problem related to volcanic eruptions and determine criteria affecting evacuation plans of a volcanic area. |
| **Developing and Using Models** | Students use computer simulations and models to predict weather patterns. | Students create models of a volcano to test a proposed idea. |
| **Planning and Carrying Out Investigations**<br><br>**Analyzing and Interpreting Data** | Students create and investigate weather topics such as the seasons, air pressure, and heat transfer. | Students develop and investigate earth processes that influence volcanic activity such as flow rates of materials and eruption patterns and determine the efficiency of sensor design ideas. |
| **Using Mathematics and Computational Thinking** | Students use math models to depict weather data and variables related to real-time weather data. | Students calculate velocity of lahars, lava flow and pyroclastic flows and use this data to determine evacuation times of residents in the area. |
| **Constructing Explanations and Designing Solutions** | Students construct explanations related to weather forecasting and determine natural disaster predictions related to weather. | Students design sensors that measure various aspects of volcanic eruptions. |
| **Engaging in Argument From Evidence** | Students propose weather forecasts and defend their explanations. | Students compare evacuation ideas to determine which is the most feasible. |
| **Obtaining, Evaluating, and Communicating Information** | Students present their ideas to a panel of adults. | Students present their solutions and evacuation plans to an adult panel. |

Student questioning is at the forefront of each of these classroom scenarios. A classroom teacher not only needs to design lessons based on the standards but also provide opportunities for student questioning within the context of the task. For example, within the Weather Around the World unit, students were provided a few guiding questions related to the "role" they elected to play. They were provided the opportunity to generate their own questions to be answered while completing the task. Some of the questions that students generate are researchable and students must explore and find accurate information from members in the scientific community. Other questions require experimentation in order for students to collect data and develop scientific understanding of the problem.

An important aspect of each of these scenarios is the use of authentic problem-based scenarios for students to solve. The engineering practices include a scenario that is written clearly for students and includes realistic constraints related to the task. For example, students are challenged to create an evacuation plan, including a sensor system that can potentially save lives of residents living in a potential volcanic region.

Throughout each of these scenarios, students use a variety of models to develop understanding of scientific and engineering practices and content. For example, students practice predicting weather patterns by using current weather data, computer simulations, and taking a field excursion to NCAR (National Center for Atmospheric Research) and NOAA (National Oceanic and Atmospheric Administration). Students create and use models that indicate how water cycles exist across the Earth, demonstrate evidence of ocean currents, and determine ocean water temperatures. In the volcano unit, students create volcano scale models, simulate volcanic eruptions using models, and create models of sensing equipment. Through the development of models, students are able to create and design, which support their ideas with realistic visions.

Effective STEM integration includes the opportunity for students to plan and carry out investigations related to science and engineering. For example, students conduct scientific investigations related to weather topics such as air pressure, density, unequal heating of the Sun, and seasons. Within a volcano study, students investigate volcanic eruptions and velocity and viscosity of various volcanic materials such as pyroclastic flow, lava, and lahars. They also are expected to determine evacuation times needed for various cities living in areas of a volcano. A variety of "hands-on" investigations helps build background knowledge for student learning and application to engineering practices. With the opportunity to explore and investigate, students have the background information to apply the content to higher-level thinking, which is a necessity during effective problem-based learning experiences.

Throughout investigations, students analyze and interpret data. Data may include student generated data from investigations or real-time data from various web resources. For example, throughout the Weather Around the World unit, students collect real-time weather data, construct graphs, and analyze patterns to predict weather patterns and forecast the weather. While studying volcanoes, students calculate and graph the velocity of lahars, lava flow, and pyroclastic flow to determine the evacuation time needed for residents.

An additional application of mathematics includes opportunities for students to create and design their own video games and simulations. Students begin by creating their own version of "Frogger," "Space Invaders," and "Pac-Man" games. By the end of the school year, students apply their computational thinking skills to create a science or math simulation. For example, students created simulations depicting a forest fire, a tsunami, a volcanic eruption, and a hurricane. These experiences displayed STEM integration at its best.

To build on scientific and engineering practices, students must explain their ideas and design solutions to problems they encounter. For example, students needed an understanding of the sensoring equipment used for monitoring volcanic eruptions. Once they began learning about current equipment, they easily began designing their own solutions. Students are engaged in designing, testing their designs, and redesigning to find appropriate solutions. Constraints are addressed, collaboratively integrated into one solution, and critiqued by classmates.

Students defend their explanations to others and collaborate with other teams to determine the best explanation of weather events. Students have opportunities to grapple and defend their understandings of the evidence. Students revise their ideas based on the ideas and questions generated by others. For example, when predicting the weather, students must defend their weather forecast to their peers. This practice not only heightens the learning of the team, but provides opportunities for peers to think critically about content and concepts as well.

Students are immersed in problems that required them to research information and evaluate the evidence for accuracy. For example, as students develop background information for a particular problem they are researching, it is imperative to practice the skills in the National Educational Technology Standards (ISTE 2000). Students must be able to find relevant information, evaluate that evidence for accuracy and its relevance to the problem they are addressing, and pull the information together to communicate their understandings to others. While students are creating weather forecasts, they need to learn about the area of the world they were focusing on and use that information to effectively communicate their solutions. In addition, students also evaluate their own solutions and communicate their designed and redesigned solutions to an audience.

## STEM Features and Characteristics of Best Practice

Although models of STEM practices vary greatly among schools, districts, and within the walls of classrooms, a number of characteristics exist that create unique environments for STEM learning. The application of engineering, with a focus on interdisciplinary problem-based learning, inquiry-based instruction, and using technology as a tool for incorporating 21st-century skills, is at the forefront of instruction at the school. All staff members are committed to this vision and strive to include these practices into daily planning.

### *Application of Engineering*

During the first year of school opening, the staff focus on engineering practices and developing strong problem-based learning units of instruction. At the elementary level, resources such as "Engineering Is Elementary" (Museum of Science) and LEGO engineering curriculum units were used. At the middle school level, Adventure Engineering units and LEGO curriculum units were implemented. As the year progressed, teachers began to see the connections between science content and engineering practices and began to develop purposeful interdisciplinary units of study. For example, elementary teachers began with a focus on engineering careers while implementing engineering curriculum units. But as teachers became more confident with the implementation of engineering, they were able to make meaningful connections to the science content for the particular grade level. Engineering careers were brought to life through community partnerships with outside resources. Field trips to various facilities in the community were also incorporated that helped teachers bring engineering concepts to the real lives of their students. Teachers began making connections between science and engineering content and focused on helping students make these connections as well.

Within the upper grades, teachers began developing more authentic engineering tasks that complemented the science content. For example, while students were learning about space exploration, they were challenged to design a space probe that could provide data regarding a particular object in the solar system. Students were challenged with constraints, researched current technologies used to study space, proposed design ideas, built a model, and created scale drawings of their designs. This project began as a small task but developed into a larger problem-based learning experience for students. The integration of science, engineering, technology, and mathematics became a reality and necessity for planning and instruction.

## *Problem-Based Learning (PBL)*

Staff also developed a variety of problem-based learning units of instruction. Problem-based learning (PBL) encompasses an authentic problem in which students work to develop a solution. In our school, PBL looks different at various grade levels but may include the following key structural features: (a) standards based, (b) a focus on big ideas, (c) multiple solutions are possible, (d) effective uses of technology, (e) use of partnerships and field excursions, and (f) a presentation to an authentic audience.

All PBL units are based on the standards for a particular grade level. Teachers began the development of their first problem-based learning experience by focusing on the science standards for a grade level. For example, second-grade students were immersed in studying insects and raised questions regarding pine beetles and their effects on trees in the mountain community. After much research and investigation, in and outside of the school walls, student teams developed a solution, along with cost analysis, to the pine beetle problem and presented them to an adult panel. As this example illustrates, a variety of content standards were addressed, and multiple solutions were possible and encouraged. Students used a number of resources in the form of field trips and guest speakers. This ultimately led to a purposeful, student-motivated presentation to an adult audience.

PBL experiences may evolve from social studies, mathematics, or literacy content. In one math classroom, the teacher posed a problem regarding the entrance to the school building. There were three sections of rock beds that looked very uninviting for guests entering the building. Students decided to redesign the area by measuring the space, researching appropriate drought-tolerant plants, and creating a landscape design plan within a budget. Students presented their plans to the staff and a design was chosen for implementation in the spring. Students not only identified a community problem but became the experts in determining a solution as well.

Teacher immersion into a PBL experience was the initiation into the development of a schoolwide model for PBL. A PBL template was provided to teachers to facilitate planning.

As teachers began to delve into the implementation of a PBL for their students, components emerged. Some teachers began with a global issue related to their standards and others began with a community issue. Some teachers used guiding questions to facilitate student research and others used a student inquiry-based approach. Some teachers used cooperative learning roles. As teachers developed PBL experiences, the experiences became diverse, based on the students and teacher involvement.

### *Technology as a Tool*

Technology has been used as a tool throughout all classes and grade levels within this school on a daily basis. Technology began as the primary tool used for presentation to a panel of adults as the culminating aspect of a PBL experience, but it has evolved. The incorporation of Web 2.0 tools has been at the forefront of instruction throughout the entire process. Technology was used to plan during a writing lesson or research project, through tools such as WordArt or the creation of graphic organizers and mind maps.

Technology was used to research and investigate topics of study. Technology was also used to assist students in organizing their PBL experience through the use of wikis and web pages. Technology has also been used to present to audience members in a variety of creative ways. Although schools have made advances in technologies, equipment alone does not create an effective STEM environment. More important is the integration of the National Educational Technology Standards (ISTE 2000), where students are immersed in collaboration, creativity, and innovation on a daily basis.

### *Incorporating Business Partnerships*

One of the strongest aspects of this school is the inclusion of community partnerships and the role they play with both teachers and students. The school vision began with input from various business partnerships and the communication among them continues to expand. Partnerships included businesses, educational facilities, corporations, and government agencies. Within each PBL experience, outside resources became a part of the school environment and learning. Partners were included as guest speakers, consultants regarding content, experts in the field, and as panel members. At the completion of each experience, students shared their findings and solutions to a panel of adults. The panel provided feedback and asked questions of students during this time.

## Evidence of Student Learning

How do students develop understandings related to STEM practices? The practice of engaging students in authentic real-world applications of engineering began with a state math/science partnership grant in 2005 (Mooney 2008). The major goals of this project were to (a) enhance science teacher content knowledge and teacher quality, and (b) improve science content knowledge of middle school students. Student learning was assessed through the administration of student pre- and post-content assessments, unit assessments that measured critical thinking and problem solving, and student science and engineering attitude questionnaires. Teachers participated in ongoing professional development in summer academies intended to increase content knowledge while integrating engineering practices into middle school classrooms. The middle school science and engineering teacher at this STEM school had participated, as a researcher and instructor, in this project. This experience led to the formation of an integrated engineering focus at the school.

### *Pre- and Post-Student Content Knowledge Assessments*

Middle school students took districtwide content tests prior to and at the completion of each unit. The assessment included a multiple-choice section, a constructed-response section, and a scientific and engineering practices section. Assessments were scored using a rubric developed by the district science team and administered to all sixth-grade students across the district. Complete data sets indicated an increase in student content knowledge in all areas. In particular, student content knowledge increased by 79%, compared to an increase of 53% by non-STEM students, especially in the area related to predicting the weather, which was the primary focus for the Weather Around the World unit. Other areas of growth include characteristics of oceans (an increase of 52% for STEM students as compared to 41% for non-STEM students) as it relates to weather and water standards. Standards related to the atmosphere showed an increase of 74% by STEM students compared to 47% for non-STEM students. Table 3.2 provides a summary of district science assessments for Weather Around the World and additional student assessment results.

**Table 3.2.** Summary of District Science Assessments: Weather Around the World PBL

| Standards Topics | Pre-Assessment | | Water Assessment | | Weather Assessment | |
|---|---|---|---|---|---|---|
| | Non-STEM | STEM | Non-STEM | STEM | Non-STEM | STEM |
| Atmosphere | 11% | 24 % | 58% | 98% | — | — |
| States of Matter | 47% | 47% | 68% | 91% | — | — |
| Oceans | 14% | 39% | 55% | 91% | — | — |
| Water | 16% | 80% | 80% | 72% | — | — |
| Weather Concepts | 13% | 26% | — | — | 80% | 39% |
| Weather Prediction | 10% | 11% | — | — | 63% | 90% |

Students were also given a district common assessment instrument related to the study of Earth processes. This assessment included multiple-choice items, constructed response items, a performance assessment and scientific practices assessment questions. Table 3.3 provides a summary of district science assessments for Volcano Alert. The PBL indicates that content knowledge of STEM sixth graders increased by 40% as compared to 11% for non-STEM students in the area of natural events related to volcanos. In addition, science inquiry skills and engineering practices were measured with an increase of 70% for STEM students and 22% increase for non-STEM students in using patterns. Furthermore, STEM students increased by 61% in the use of technology in science and engineering, as compared to 45% for non-STEM students. Table 3.4 (p.

52) provides a summary of district science assessments of scientific and engineering practices. It displays the breakdown of skills regarding the assessments administered.

**Table 3.3.** Summary of District Science Assessments: Volcano Alert PBL

| Standards Topics | Pre-Assessment | | Plate Tectonics Assessment | |
|---|---|---|---|---|
| | Non-STEM | STEM | Non-STEM | STEM |
| Earth Changes | 26% | 61% | 67% | 85% |
| Minerals/Rocks/Soil | 17% | 28% | 75% | 95% |
| Natural Events | 9% | 39% | 20% | 79% |

## *Post-Curriculum Unit Assessments*

Upon completion of each curriculum unit, students were administered the end-of-unit exams with higher-level thinking questions. The exams were designed to evaluate the critical thinking, data analysis, and problem-solving skills of students. Assessments were piloted with middle school students across the state and inter rater reliability was established among student samples. Results of students who completed the Volcano Alert unit indicated that 87% of students displayed a proficient or advanced rating in critical thinking, data analysis, and problem-solving skills.

The Weather Around the World end-of-unit assessment is reported as Table 3.2. Students made significant gains in content related to the atmosphere, states of matter, oceans, and predicting the weather. In addition, scientific practices increased in the areas of using evidence, formulating explanations, use of technology, and predicting patterns, as compared to non-STEM students.

## *Pre- and Post-Student Attitude Survey*

Students also completed a pre/post survey associated with the implementation of the engineering based Volcano Alert unit. This survey was previously administered to students across the state who had participated in the Volcano Alert unit, in addition to other engineering-based curriculum units related to the Math Science Partnership mentioned previously (Mooney 2008). The Likert-type survey asked questions related to student anxiety toward science, value of science and engineering, beliefs regarding if engineering being perceived as "fun," and the characteristics of engineering such as "creativity" and "brilliant." On average, the data show an increase in student understanding of what engineering is as well as a more positive attitude about engineering. A desire to do engineering increased, and self-confidence in both science and engineering improved.

A rubric (Appendix A) was developed and used to measure content and presentation of final solutions to the problem-based learning objectives. Adult panel members were given the rubric and scored the team during the final exercise. Students were scored on a scale of Advanced, Proficient, Partially Proficient, and Unsatisfactory. Results for the weather presentation indicate that students displayed content knowledge at a deeper level when presenting to a panel. For example, during the weather project, 96% of the student teams scored an average of "proficient" or above. During the volcano project, 89% of the student teams scored "proficient" or above.

**Table 3.4.** Summary of District Science Assessments: Scientific and Engineering Practices

| Standards Topics | Pre-Assessment Non-STEM | | Plate Tectonics Assessment | | Water Assessment | | Weather Assessment | | Pre-Scientific Practices | Post-Scientific Practices |
|---|---|---|---|---|---|---|---|---|---|---|
| | | | STEM | Non-STEM | STEM | Non-STEM | STEM | Non-STEM | | |
| Science Evidence | 10% | 26% | — | — | — | — | 71% | 73% | 88% | 90% |
| Hypothesis | 6% | 2% | — | — | 56% | 65% | — | — | 57% | 65% |
| Models | 10% | 50% | 16% | 73% | — | — | — | — | 50% | 87% |
| Science Explanations | 20% | 21% | 43% | 80% | 62% | 67% | 63% | 63% | — | — |
| Variables | — | — | — | — | — | — | 32% | 86% | 40% | 77% |
| Predicting Patterns | 19% | 26% | 41% | 85% | — | — | 53% | 96% | — | — |
| Technology | 20% | 24% | | | | | 65 % | 84% | 63% | 85% |

Although using a rubric may result in scores that are biased, panelists reflected upon the high level of content understanding and presentation skills that students displayed. For example, one high school Earth science teacher commented that the teams of sixth-grade students were more advanced in their understanding of predicting the weather, compared to the 10th-grade student projects that he taught.

Upon the completion of the volcano unit, student reports included a robot that could move around a volcano model, place sensors at various locations, and show the evacuation plan for area residents. Results showed that 89% of the students were proficient or advanced. These results show the importance of providing opportunities for students to communicate and share their knowledge. More importantly, students report that the challenge of presenting to adults holds them accountable for their learning as well. The student comments after their presentation experience referenced the feelings that they encountered. One student commented:

> *Presenting to adults is different than writing answers on a test. I was nervous when it came time for the panel to ask us questions. I was afraid that I wouldn't know the answer. Now that it is over, I have more confidence in answering questions. I learned that I do know the answers and can explain them, even to an adult. I really like presenting.*

In addition to student learning, students also reported that they are more accountable with regards to spelling, grammar, and punctuation, when sharing information to a panel. One student reported:

> *I feel more pressure presenting my work to adults. I want to be sure that my spelling is correct and that they can read my presentation. The first time I presented, I learned that the color of my font did not show up very well and was hard to read. Now, I check my presentation. I can see that fancy fonts are not the best way to show others what you know.*

Students also reported the importance of using and understanding how to manipulate various technologies and programs when sharing information. Students needed to know how to use Web 2.0 tools in effective ways to display their information. For example, throughout the "Weather Around the World" unit, teams used PowerPoint, embedded videos and images, created videos, and used Google Earth to interpret "live" weather data. Teams needed to be prepared to manipulate each of the tools and technologies they used. Some teams used interactive whiteboard technologies and incorporated polling of the panel members to check for understanding. A student reported:

> *We really had to practice using the computer and finding all of the links and programs that we used in our presentation. I was nervous that our video link might not work, but we learned to have a back-up plan, just in case. I was glad we did, because our first video link did not work but I had the video open and we were able to continue our presentation. We learned that it is important to open everything we want to use, so it is easier to find in an emergency.*

Student learning was assessed in a variety of ways. Teachers used formative assessments as well as summative assessments throughout their instruction. Teachers have come to realize that assessment does not always require a written test, but should be offered through providing multiple opportunities for student learning, resulting in a wealth of knowledge.

How do teachers develop understanding related to STEM practices? The practice of engaging students in authentic real-world applications of engineering began with teacher professional development focusing on problem-based learning. Building professional development included immersing teachers in a common PBL experience. Field excursions to local partnerships kicked off the learning. Teachers learned content that they could apply in their classrooms and developed basic understandings related to developing their own PBL experiences for students. Teachers collaborated and had time to plan together, brainstorming possibilities for their students. The administrator facilitated teacher learning and encouraged risk taking. Teachers had the luxury of dreaming of the endless possibilities, because support was provided along the way.

Throughout the inaugural year, teachers participated in various conferences based on the needs of each teacher. For example, a second-grade teacher was interested in learning deeper science content. She was supported in attending the National Science Teachers Association conference with a fellow middle school teacher from the school. This opportunity was instrumental in changing a teacher's learning about science and engineering practices. She immediately returned to her classroom and began implementing her new skills and ideas. Her identity as an elementary teacher has transformed her into a model STEM teacher.

## Conclusions

Based on the first two years of implementing STEM as a model school, a number of areas are considered. These include:

- Authentic and real-world problems can provide students with purposeful and meaningful learning experiences that boost student achievement.
- Problem-based learning allows students to develop and analyze solutions in a manner that builds student confidence in their learning and provides a forum in which students become empowered by their decisions and solutions.
- Engineering practices are integrated with science practices and content.
- Technology is used as a tool for all aspects of STEM integration and across all content areas.
- Providing students with opportunities for communicating scientific and engineering practices to adults allows for deeper understanding of content and confidence in learning.
- Business partnerships can add to student learning through field excursions, guest speakers, and panelists. They provide effective learning for all parties including teachers, students, parents, and school administrators.
- Teacher preparation must incorporate engineering and science practices in addition to implementation of effective STEM-integrated units and problem-based learning experiences.

## Next Steps

The current eighth-grade students are the first class to experience three years of this STEM program. These students are making decisions regarding the high school they are going to attend. One option is a STEM-focused high school that has worked closely with this school to develop an effective high school program for STEM students. Approximately 90% of the STEM students are planning on attending this high school. This is encouraging, as one of the major goals of STEM education is to promote STEM careers. Following these students will be imperative as we focus on STEM education for all students.

As this school continues to refine practices related to integrating engineering and science practices and content, a number of goals remain. Professional development will be an ongoing process as problem-based learning expands into additional curricular areas such as social studies, language arts, and the arts. Engineering practices will also be a focus in order to encompass all grade levels and content across the K–8 spectrum. Technology is ever-changing and the school community will need to build on those changes and incorporate them into classroom practice. Business partnerships will need to be fostered and creative approaches to the effective use of partnerships will be imperative to fully use resources to their potential. An additional area of focus will be the development of an entrepreneur course and professional development that spans all grade levels.

With each of these goals, a strong research agenda needs to be put into place to clearly define various aspects of STEM practices that are effective and what is not effective. Research related to parent beliefs regarding STEM and attitudes toward the existing practices will be important. In addition, research with regards to outside stakeholders such as the district administration, community partnerships, and the implications for schools across the district is imperative. District support is needed to facilitate research in these areas. A longitudinal study needs to be put into place as students ultimately will be spending eight years in a STEM-focused setting before entering high school. Following these students can potentially lead to major improvements in STEM learning for all.

## References

Bybee, R. W. 2011. Scientific and engineering practices in K–12 classrooms: Understanding *A framework for K–12 science education. The Science Teacher* 78 (9): 34–40.

International Society for Technology in Education (ISTE). 2000. *National educational technology standards. www.iste.org/standards*

Mooney, M. 2008. Increasing teacher and student science content knowledge in a Denver metropolitan area. Denver, CO: Colorado Department of Education.

Museum of Science, Boston. 2013. Engineering is elementary. *www.eie.org*

NGSS Lead States. 2013. *Next Generation Science Standards: For states by states.* Washington, DC: National Academies Press. *www.nextgenscience.org/next-generation-science-standards*

National Research Council (NRC). 2012. *A framework for K–12 science education: Practices, crosscutting concepts, and core ideas.* Washington, DC: National Academies Press.

**Table 3.5.** Volcano Alert: Final Version of Rubric

| | 4 (Advanced) | 3 (Proficient) | 2 (Partially Proficient) | 1 (Unsatisfactory) |
|---|---|---|---|---|
| **The Big Idea or Purpose of Project** | The scenario is clearly explained and includes what the engineering team was asked to accomplish.<br>Roles and responsibilities for each team member are described.<br>Team Name is clearly labeled. | The scenario is explained but may not include what the engineering team was asked to accomplish.<br>Roles and responsibilities for each team member are described.<br>Team Name is clearly labeled. | The scenario is vaguely explained.<br>Roles and responsibilities are not addressed.<br>Team Name is labeled. | The scenario is not addressed. |
| **Define the Problem** | Defines the problem or goal of the project in detail. Constraints are clearly addressed. | The problem is clearly stated but lacks some detail. Constraints are addressed but may lack details. | The problem is somewhat stated but lacks details and/or does not address constraints. | No problem is stated or the problem does not relate to the task. |
| **Summarize Gathered Information** | A thorough summary of information/research gathered throughout the unit is described. For each activity, an explanation including what the team did and a description of the results are included. The team also described what was learned for each activity and how this assisted in forming a solution.<br>Activities include: (volcano locations, volcanic types, hazards, volcanic eruptions, viscosity/velocity, modeling volcanoes, sensors used and evacuation plans). | Information is summarized in an organized way. Explanations include either what the group did or how the group accomplished the task.<br>A description of the results is clear but may lack detail regarding how activities assisted the group.<br>The team described between 5– 7 activities while researching/investigating the problem. | Information is summarized but lacks detail on what the group did and how the group accomplished the task.<br>Descriptions of results are described but do not include an explanation of how the results assisted the group.<br>The team described 3–4 activities while researching the topic. | Information is not summarized or explained. The team has little research/ investigation to assist in understanding the topic. |

**Table 3.5** *(continued)*

| | 4 (Advanced) | 3 (Proficient) | 2 (Partially Proficient) | 1 (Unsatisfactory) |
|---|---|---|---|---|
| **Quality of Content** | Product shows 90–100% understanding of the concepts and correctness of ideas.<br>Product describes at least four types of sensors used to study volcanoes and explains the location of sensors on the volcano.<br>Product clearly describes an evacuation plan for the towns affected by a volcanic eruption.<br>Robots and programming are used to clearly explain the tasks of placing sensors on the volcano and evacuating the towns. | Product shows 80–89% understanding of the concepts and correctness of ideas.<br>Product describes at least three types of sensors used to study volcanoes and explains the location of sensors on the volcano but may lack details.<br>Product describes an evacuation plan for the towns affected by a volcanic eruption.<br>Robots are used to explain the tasks of placing sensors on the volcano and evacuating the towns but some programming details are lacking. | Product shows 70–79% understanding of the concepts and correctness of ideas.<br>Product describes one or two types of sensors used to study volcanoes or explains the location of sensors on the volcano but lacks details.<br>Product describes an evacuation plan for the towns affected by a volcanic eruption but lacks details.<br>Robots are used to explain the tasks of placing sensors on the volcano or evacuating the towns. | Product shows 69% or less understanding of the concepts and correctness of ideas.<br>Product does not describe types of sensors used to study volcanoes or the explanation of the location of sensors on the volcano.<br>Product does not describe an evacuation plan for the towns affected by a volcanic eruption.<br>Robots are not used to explain the tasks. |
| **Analysis of Solutions** | An explanation of why and how the different activities helped your group determine a final solution for the problem is included. | An explanation of why activities helped the group is described. | Explanations of activities are described but lack detail regarding how the activities helped the group. | No explanation for how the final solution is described. |

**Table 3.5** *(continued)*

| | 4 (Advanced) | 3 (Proficient) | 2 (Partially Proficient) | 1 (Unsatisfactory) |
|---|---|---|---|---|
| **Final Design Solution** | A final appropriate solution and details are described. Early warning systems and evacuation times are described in detail and why the team chose the final solution is included.<br>A comparison of why this solution is better than others is included. | A final solution and characteristics are described. Details are lacking regarding early warning systems and evacuation times.<br>A description of why this solution is the best is included.<br>A comparison to other solutions is included but may lack details. | A final solution is described but details are lacking.<br>An explanation of why the solution is the best is included.<br>A comparison to other possible solutions is missing or incomplete. | A final solution is described but an explanation or comparison is not included. |
| **Media Elements** | Product contains the use of video, audio, images, AND text including citation information. All are 90–100% effectively used. | Product is missing the use of one media element (video, audio, images, OR text). All sources include citation information. They are 80–89% effectively used. | Product is missing the use of two media elements (video, audio, images, OR text) and/or citation information is missing. They are 70–79% effectively used. | Product contains 69% or less of the required elements. |
| **Organization of Material** | Content is presented in an orderly manner with perfect chronology. Exceptional explanation of ideas. | Content is presented in an orderly manner. Additional explanation of ideas may or may not be present. | Content is presented in a somewhat orderly. manner. | Content is presented haphazardly. |

## Table 3.5 *(continued)*

| | 4 (Advanced) | 3 (Proficient) | 2 (Partially Proficient) | 1 (Unsatisfactory) |
|---|---|---|---|---|
| **Presentation and Appearance** | The product is cleanly constructed and very pleasing to the eye 90–100% of the time. Product is highly organized. Creativity is at or above expectations. | The product is clean looking and pleasing to the eye 80–89% of the time. There are 2–4 minor mistakes are visible. Good use of creativity. | The product is clean-looking and pleasing to the eye only 70–79% of the time. There are 5–7 mistakes visible. Organization is lacking. Creativity was lacking but an attempt was made. | The product is clean-looking 69% or less of the time. More than seven visible mistakes. Little or no creativity and organization. |
| **Spelling and Grammar** | The grammar and spelling of the project are at or near flawless levels. | There are 2–3 minor grammar or spelling mistakes in this project. | There are 4–6 grammar or spelling mistakes in this project. | There are 7 or more grammatical or spelling mistakes. |
| **Oral Presentation** | Speaks clearly and audibly. Maintains eye contact with audience throughout the presentation. Each group member has a role. | Speaks clearly and audibly with eye contact being maintained most of the time. Each group member has a role. | Speaks clearly and audibly with eye contact being maintained some of the time. Two or three group members present the information for the group. | Difficult to hear and understand. Minimal eye contact is maintained. One person completes the presentation. |

# STEMRAYS

## After-School STEM Research Clubs

*Morton M. Sternheim*
*University of Massachusetts Amherst*
*Amherst, MA*

*Allan Feldman*
*University of South Florida*
*Tampa, FL*

## Setting

STEMRAYS was an after-school environmental science research program for students in grades 4–8. It was funded by a three-year grant from the National Science Foundation's (NSF's) Academies for Young Scientists program and subsequently by funds from the Massachusetts Board of Higher Education for an additional year. In each of the three years about 200 students met at their schools once a week for 30 weeks. The teachers leading the groups were usually classroom teachers from that school. Higher education STEM faculty researchers served as leaders of research teams that included 3–5 teachers and their respective 8–12 participating students. The teams worked on problems related to the research of their faculty mentor and reported their results at a June science conference. Partners included nonprofit organizations, governmental agencies, and a variety of businesses, which provided opportunities for field trips and contacts with experts.

## Overview of the Program

In 2006, NSF announced support for a program of Academies for Young Scientists (AYS) that would create, implement, evaluate, and disseminate effective models to attract K–8 students to prepare them for and retain them in science, technology, engineering, and mathematics (STEM) disciplines, leading to an increase in the pool of students continuing in STEM coursework in high school and considering careers in STEM fields (NSF 06-560).

NSF stipulated in the request for proposals (RFP) that models for the projects should be built on sustainable partnerships among formal and informal education providers, business and industry, and colleges of education. It also required that the projects take place during out-of-school time (OST), and that they be synergistic both with in-school curricula and with the special attributes of the setting. In addition, the projects were to be structured so that they provide highly motivational experiences for the students while preparing them for further study of the STEM disciplines. NSF recognized in the RFP that this would require professional development for the educators providing the OST experiences. Finally, NSF saw the AYS program as a way to explore a variety of models in urban, rural, and suburban settings with diverse student populations. The expectation was that evaluation and research on the projects would "inform NSF and the broader educational community of what works and what does not, for whom, in what settings" (NSF 06-560).

NSF funded a total of 16 three-year projects under this program in 2006; no NSFAYS funding was offered in any other year. Each of the 16 projects had a different model. STEMRAYS and some other projects were located at schools, while others were at informal educational institutions such as museums and aquariums. Some, like STEMRAYS, met after school, some on Saturdays, and some during the summer. Staffing varied widely too. STEMRAYS clubs were staffed by regular classroom teachers at the schools, while others were run by college students or museum personnel.

In STEMRAYS, higher education faculty researchers (not graduate students) trained the teachers (who in most cases had little science background), hosted campus visits, and provided ongoing support at monthly meetings, and by e-mail and telephone. The faculty was drawn from a university, a liberal arts college, and a community college. Their graduate students sometimes helped as well. The clubs were engaged in long-term research projects related to the faculty members' own research. For example, some clubs studied environmental arsenic contamination associated with pressure-treated wood, in bodies of water, in foods, and in fabrics. Other clubs made a variety of observations and measurements of water quality in local streams. Bird populations and behavior was another focus. There were also clubs that focused on engineering design problems.

Some clubs had the students work together on joint explorations throughout the year. Others, once the basics had been covered, encouraged individual or small-group projects within the general framework. In two of the years STEMRAYS also offered one-week summer camps. The 30-week academic year program concluded with a June Science Conference where the students presented their research findings with posters, PowerPoint presentations, and other materials. Although science fairs were not an explicit part of STEMRAYS, some students went on with their projects to the regional fairs. One won the statewide competition and several did well in the region.

Each year STEMRAYS included children in grades 4–8 attending about 20 different elementary and middle schools. The original NSF-funded program included schools in old New England mill towns as well as in rural areas, and the state funding added two college towns with considerable ethnic and income diversity. Neither the students nor most of the teachers had previously had an authentic research experience.

## Major Features

The STEMRAYS model was based on extensive research about the role of apprenticeships in training scientists. It was built around after-school clubs that conduct original research on a particular research thread. Participants were selected randomly if too many applied. In general the clubs had approximately equal numbers of boys and girls, and minority participation mirrored the school populations. Each year there were approximately 20 clubs with typically

8–12 students in each club. The clubs met weekly for two hours at their own school. The leader was usually an experienced teacher from that school, which was a major factor in STEMRAYS' success. This arrangement has many advantages:

- Teachers have an understanding of children, teaching, and classroom management, which college students or other leaders may lack (Rahm, Martel-Reny, and Moore 2005). They are also aware of the diverse cultural backgrounds of their students and their implications (Caplan and Calfee 1998).
- The children often have positive impressions of the teachers from direct contact or by their reputation and are attracted to working with them.
- The teachers know the building personnel and facilities, making it easier to access resources.
- It avoids the problems of transportation and of schools with different dismissal times faced by programs located at central sites.
- An added benefit is the increase in the teachers' knowledge of the nature of science and of scientific research that may carry over into their classroom teaching (Feldman and Pirog 2011).

Teacher turnover was very low, avoiding the financial and programmatic costs incurred in some programs with less experienced staff. Several teachers continued through the entire four-year program; most others participated for more than one year.

## History and Rationale

There is a long history in the United States of promoting inquiry in science education. For more than 50 years the federal government has invested in the development of inquiry-oriented curricula and the training of teachers to teach using inquiry methods. Both U.S. and international reform documents have made strong cases for using an inquiry approach (e.g., NRC 2000; OECD 2003). More recently, *A Framework for K–12 Science Education* (NRC 2012) and the *Next Generation Science Standards* (NGSS; NGSS Lead States 2013) have called for students to learn the practices of science. These include

1. Asking questions;
2. Developing and using models;
3. Planning and carrying out investigations;
4. Analyzing and interpreting data;
5. Using mathematics and computational thinking;
6. Constructing explanations;
7. Engaging in argument from evidence; and
8. Obtaining, evaluating, and communicating information. (NRC 2012, p. 42)

One way to accomplish this is to engage teachers and students in authentic science activities in partnership with practicing scientists. This is at the heart of the STEMRAYS model.

The STEMRAYS model is supported by two sets of literature. One is on the role of out-of-school-time (OST) learning in STEM education. The other is concerned with the engagement of school students in authentic science activities referred to as *research apprenticeships*. We begin with a look at OST science.

## Out-of-School-Time (OST) Science

Out-of-school-time (OST) programs were first institutionalized in the United States during the 19th century (Halpern 2002). These programs, which were primarily affiliated with settlement houses and boys' clubs, were implemented to care and protect children while also providing opportunities for learning and play. They were also designed to enculturate immigrant children into mainstream American culture (Halpern 2002). As schools have become more accountable for the academic successes and failures of their students, OST programs have shifted their mission to focus on remediation for high-stakes examinations (Bhanpuri, Naftzger, Margolin, and Kaufman 2005; Mass Insight Education 2002). Unfortunately, there is a conflict between the goal of preparing students for high-stakes tests and the goal of providing them with opportunities that increase their motivation and enthusiasm to pursue STEM careers (House of Representatives 2005).

A study of OST programs in Massachusetts found that remedial activities resulted in higher test scores, but enrichment activities such as sports and the arts are more likely to result in a greater enthusiasm among students that "often transfers to the more formal environment of school" (Mass Insight Education 2002, p. 2). Also, children are more attracted to programs that are youth-centered (McLaughlin 2000; Rahm, Martel-Reny, and Moore 2005) and are seen as fun or play (e.g., sports or arts) rather than those that are curriculum-centered. Recently there has been a growing interest in developing OST activities that are more like the ways that people engage in personal interests such as hobbies and sports (Azevedo 2006). These programs may be designed around themes or topics such as environmental awareness, engineering, or astronomy. They may also be structured based on the focus of their host institution, like a museum, zoo, or aquarium. The Afterschool Alliance (Afterschool Alliance 2011) identified 19 STEM afterschool programs, many of which began as part of the NSF program that funded STEMRAYS. An analysis of the evaluation reports from these programs found that overall they had positive effects on students, including improved attitudes toward STEM fields and careers, increased STEM knowledge and skills, and a higher likelihood of high school graduation and the pursuit of a STEM career (p. 2).

We end this section by noting that OST STEM activities can be the sites for learning the science and engineering practices described in *A Framework for K–12 Science Education* (NRC 2012) and the *NGSS* (NGSS Lead States 2013). Bell and his colleagues argued that informal settings, in contrast with formal settings like schools, are not bound by constraints such as those imposed by high-stakes examinations, 50-minute class periods, and large class size. Therefore, in OST programs, students can

> *... experience excitement, interest, and motivation to learn about phenomena in the natural and physical world ... [and] think about themselves as science learners and develop an identity as someone who knows about, uses, and sometimes contributes to science.* (Bell, Lewenstein, Shouse, and Feder 2009, p. 4)

The STEMRAYS model made this happen by providing opportunities for students and teachers to partner with practicing scientists in research apprenticeships

## Research Apprenticeships

In the STEMRAYS model, teachers and students collaborate with practicing scientists in authentic science activities. This type of experience is called a research apprenticeship. Before looking more closely at research apprenticeships, we first want to take a brief look at what we mean by authentic science.

### *Authentic Science*

Authentic science activities are those that are similar to what scientists actually do. These include "asking questions, planning and conducting investigations, drawing conclusions, revising theories, and communicating results" (Lee and Songer 2003, p. 923). For this to happen, the activities should be student-oriented and open-ended, and be concerned with topics investigated by practicing scientists (Braund and Reiss 2006). While there are many school science activities that mimic the work of scientists, they differ both cognitively and epistemologically from what scientists do (Chinn and Malhotra 2002). By partnering with practicing scientists, the STEMRAYS model was designed so that teachers and students would be legitimate, although peripheral, participants (Lave and Wenger 1991) in actual scientific research studies. Therefore, they would engage in science that is authentic rather than school science.

### *Apprenticeships in Science*

The STEMRAYS model is derived from research on how people participate in scientific activities and on the resulting growth in expertise on how to do scientific research (Barab and Hay 2001; Bleicher 1996; Charney et al. 2007; Crawford 2012; Etkina, Matilsky, and Lawrence 2003; Richmond and Kurth 1999; Sadler, Burgin, McKinney, and Ponjuan 2009). STEMRAYS co-PI Allan Feldman recently studied the effects on teachers participating in an NSF-funded study of the natural remediation of acid mine drainage (Feldman, Divoll, and Rogan-Klyve 2009; 2013). He found that the process of learning to do research is a developmental continuum that progresses from novice researcher and proficient technician to knowledge producer. Novice researchers have little or no experience with scientific research. When novice researchers are allowed to be legitimate peripheral participants in a research group, they are seen as members who can develop the skills to help maintain the laboratory and collect data, but are not expected to contribute much if anything to the analysis of data or the creation of new knowledge. But through their participation, with guidance from more knowledgeable researchers, they can gain the skills necessary to become skilled practitioners in their fields. That is, as proficient technicians they can develop a researchable question, design an appropriate study, and collect and analyze data. With additional experience and training, they can become knowledge producers who formulate their own research questions, develop new research methods, and add to the literature.

The process of learning to be a scientific researcher is best understood using an apprenticeship model. An important characteristic of apprenticeships is the indistinguishable nature of learning and the practice of work (Lave and Wenger 1991). This is quite different from how students are taught in formal settings in which they learn skills in isolation from their use. In an apprenticeship model, successful teaching is the ability to partition tasks into appropriate sizes that are useful for the developmental trajectory of the apprentice. The apprentice demonstrates learning by performing tasks in a way that is analogous to the expert (Lave and Wenger 1991). At the beginning of the STEMRAYS program, students and teachers are novice researchers. However, teachers can develop a deep and sustained participation in the research and can move quickly along the continuum toward proficient technician. In an apprenticeship situation, a person who is more knowledgeable and skilled can be the mentor for someone less knowledgeable and skilled. Therefore, a teacher who is at the level of proficient technician can be a successful apprenticeship mentor to students who are novice researchers.

The STEMRAYS model is also supported by research by Bell and his associates on the development of everyday expertise (Banks et al. 2007). Recently they identified six strands of science learning (NRC 2009), all of which are included in the STEMRAYS model. STEMRAYS provides experiences for students that excite, interest, and motivate them to learn about the natural world (Strand 1). Students in conjunction with the teachers and scientist mentors "generate, understand and use concepts, explanations, arguments, models, and facts related to science" (Strand 2). They manipulate, test, explore, observe, and make sense of the natural world (Strand 3). All STEMRAYS participants reflect on the nature of science and how they learn about phenomena (Strand 4). Clearly they also participate in scientific practices using scientific language and tools (Strand 5). Finally, through the STEMRAYS experience they come to identify themselves as authentic participants in science (Strand 6).

## Evidence for Success

The STEMRAYS evaluation and research components focused on its goals, which were included in the original proposal to NSF. These were aligned with the overall goals and objectives of the NSF Academies for Young Scientists initiative and included:

1. Teacher Goals
   a. Increase teacher understanding of the process of doing scientific research.
   b. Improve the ability of teachers to engage and motivate students in scientific research.
   c. Expose teachers to new and varied instructional strategies.
   d. Increase teachers' knowledge of science content.
2. Student Goals
   a. Stimulate interest among students in grades 4–8 in the pursuit of science careers.

b. Provide challenging educational experiences in science to these students seamlessly through grades 4–8.
c. Increase students' conceptual understanding and appreciation of the role that the sciences play in the world.
d. Increase students' knowledge of science content.

Evidence that sought to determine whether STEMRAYS achieved these goals was gathered as part of the STEMRAYS evaluation and of its research component. The evaluation was done by the SageFox Consulting Group; Allan Feldman, STEMRAYS co-PI and the co-author of this paper, directed the research component. Overall the findings from the evaluation and research study indicated that STEMRAYS had a major impact on the participating teachers, students, and schools. This is summarized in SageFox's year 4 report:

Results of evaluation activities across Years 1–4 of the STEMRAYS program are characterized by overwhelmingly consistent positive feedback from the different stakeholders. The evaluation has demonstrated the strong leadership and coordinating role of the UMass STEMRAYS staff and the positive and successful collaboration with participating teachers. As reported since Year 1, the STEMRAYS team has been successful with the recruitment of teachers and coordination of activities and program support across varied educational institutions of local schools and higher educations. (SageFox 2010).

Data sources included teacher interviews; focus groups and surveys; observations of research group meetings, workshops, club meetings and the June Science Conference; student surveys; focus groups; pre- and posttests; and parent and administrator surveys.

## Teacher Impact

We noted above the work of Feldman, comparing learning to doing scientific research with apprenticeships. Feldman and Pirog (2011) completed a case study of the arsenic group and found results consistent with that model. They studied how the professor mentored the teachers and how they in turn mentored the children. In less than one academic year, the elementary school teachers with little formal science education gained the expertise needed to facilitate the children's legitimate participation in authentic STEM research. The children gained the methodological and intellectual proficiency needed to contribute useful data and findings to the scientist's research program.

Interviews with the teachers, analysis of the session reports, and observations of the clubs indicated a large and rapid growth in all the teachers' abilities to understand the processes of science and to engage students in scientific research. In addition, for the majority of teachers this was the first time that they used instructional methods that prepared the students to engage in authentic research activities (Feldman and Pirog 2011). Teacher surveys, based on an instrument developed by Kardash (2000), showed significant gains in their research skills over time (see Table 4.1, p. 68).

**Table 4.1.** Comparison of Year 1 Teachers' Research Skills to Those of Year 2 Teachers at the Start of Year 2

| **Item: Do what extent do you think you are able to ...**<br>**1 = not at all, 2 = very little, 3 = somewhat, 4 = quite a bit, 5 = a great deal** | ***t*-Test** |
|---|---|
| Understand contemporary concepts in your STEMRAYS research area. | $p < 0.05$ |
| Make use of the primary scientific research literature in your STEMRAYS research area. | $p < 0.2$ |
| Identify a specific question for investigation in your STEMRAYS research area. | $p < 0.05$ |
| Formulate a research hypothesis based on a specific question. | $p < 0.01$ |
| Design an experiment or theoretical test of the hypothesis. | $p < 0.01$ |
| Understand the importance of "controls" in research. | $p < 0.01$ |
| Make observations and collect data. | $p < 0.001$ |
| Analyze numerical data. | $p < 0.01$ |
| Interpret data by relating results to the original hypothesis. | $p < 0.001$ |
| Reformulate the original hypothesis (as appropriate) | $p < 0.001$ |
| Relate results to the "bigger picture" in your STEMRAYS research area. | $p < 0.001$ |
| Orally communicate the results of your research. | $p < 0.01$ |
| Write a scientific report of your research. | $p < 0.001$ |
| Think independently about your STEMRAYS research area. | $p < 0.001$ |

(Feldman and Pirog 2011, p. 500).

Table 4.2 reports findings of a survey administered to teachers by SageFox that included questions about themselves, their students, and their schools. The figures in the table represent the percentage of teachers who responded "Yes" to the impact of STEMRAYS on the different items. The survey results indicate, with some variance across the four years, increases in the teachers' interest in reading about science, the changes in the way they will teach and relate to their students, and interest in engaging in scientific research. The data show a decrease in the teachers' interest in pursuing additional professional development.

**Table 4.2.** Teacher Survey on Impact on Teachers, Students, and Schools

| Impact | Year 4 | Year 3 | Year 2 | Year 1 |
|---|---|---|---|---|
| **Impacts on the teacher** | | | | |
| Increased interest in reading about science | 92% | 63% | 64% | 57% |
| Changes in the way I will teach my classes this coming year | 75% | 38% | 43% | 14% |
| Changes in the way I related to my students | 50% | 13% | 43% | 29% |
| Increased interest in pursuing future professional development activities | 67% | 63% | 79% | 71% |
| Increased interest in doing my own scientific research | 58% | 75% | 50% | 43% |
| **Impacts on the students** | | | | |
| Developed new ways of thinking about science | 100% | 100% | 71% | 71% |
| Increased interest in science | 92% | 75% | 79% | 71% |
| More students interested in joining the program than were originally accepted | 33% | 68% | 64% | 43% |
| Better performance in school | 33% | 13% | 14% | 0% |
| **Impact on the institution** | | | | |
| Heightened institutional awareness of STEMRAYS | 92% | 75% | 71% | 71% |
| Heightened institutional interest in science activities | 67% | 50% | 57% | 57% |
| More support for after-school activities | 33% | 13% | 50% | 29% |

The SageFox focus group interviews with the teachers focused primarily on the impact of STEMRAYS on their students. However, the teachers also commented on their experiences. For example, one teacher reported that, "I have been doing this for four years now and it has brought really good science opportunity to our school … I've grown a lot as a teacher, a scientist and now as an engineer" (SageFox 2010, p. 5). Another teacher told SageFox about the importance to her of working with a practicing scientist or engineer:

> *I love having the mentor scientist or engineer to work with, the whole idea of doing authentic science is very powerful—having someone who is really doing this kind of work out there and being a model for us is amazing, for me and the kids.* (SageFox 2010, p. 5)

One of the mentors reported favorably about the ways in which the teachers collaborated with each other and the mentors:

> *It's so great to see this group of teachers working together as a community of mutually supporting colleagues. They help each other with problems and questions, very powerful—thank goodness for email. Seeing the enthusiasm in the classroom and the professionalism of the teachers—gives*

*me the confidence that incredible things are happening. I see 100% engagement when I visit.* (SageFox 2010, p. 6)

Although no data were collected to determine or measure teachers' content learning as a result of their participation in STEMRAYS, there are indications that this did in fact occur. First, a majority of the teachers were elementary school teachers with little formal background in science. Many had no more than two science courses while in college, which for some was over 20 years ago. That said, Feldman and Pirog (2011) were able to show that the teachers were able to facilitate their students' authentic scientific research. This required not only an increase in their knowledge of how to do science (as seen in Table 4.1), but also the content knowledge needed to understand the research.

## Student Impact

SageFox's evaluation of STEMRAYS also sought to uncover effects related to the project's student goals (SageFox 2009; 2010). Increasing the students' interest in science and engineering, student goal (a), was evaluated using data from surveys administered to teachers, students, parents, and administrators. Overall, SageFox found that STEMRAYS increased the students' interest. Teacher survey data (see Table 4.2) indicate that they believed that participation in STEMRAYS increased student interest in science. The large positive results are in spite of teachers noting that many of the students entered the program with an initial strong interest in science. Teachers' open-ended responses support the numerical results. For example, one teacher told SageFox:

*For these kids this is their "sport" and it is their identity at school and at home. I have kids that came in over vacation break to work on their projects and I said to them "you really love this!" and they all said "oh yeah this is what we want to do—be scientists." This gives them a really significant peer group, it really functions like a club, they have inside jokes with each other and they are all really close.*

Students also indicated that participation in STEMRAYS increased their interest in science. In open-ended responses students told of their new interest in doing science at home and in school, and of the possibility of engaging in scientific careers. One student wrote "When I first started I had no idea what engineering was, now I want to get my master's degree in engineering." In addition, administrators and parents were clear about the increase in interest in science that they saw in the students. For example, parents reported the following to SageFox about their children: "She has started to develop her own love of science when prior to this her interests were mainly in language arts" and "STEMRAYS has opened his eyes to the possibilities of science. He had no interest before starting but is very enthusiastic and really wants to participate next year." Overall, SageFox found "Parents consistently noted an increased interest and understanding of important environmental questions" and that "There were numerous comments about students' increased interest in studying scientific disciplines in college and pursuing careers in science" (SageFox 2009, p. 5).

Student goal (b), providing challenging science education opportunities, was evaluated using observations of club activities, teachers' session reports of club activities, and parent and administrator surveys. SageFox found that STEMRAYS provided the students with challenging experiences in science, primarily in the form of "new and engaging teaching and learning opportunities, and ... hands-on science experiences that relate to the real world" (SageFox 2010, p. 3). Their observations found that teachers did consistently engage the students in challenging science activities. At the beginning of the academic year the activities focused on preparing students to engage in science research by introducing them to the content and context of the research theme and by teaching them research skills. The focus then shifted to the development of research questions. During the final part of the year students and teachers collected and analyzed data, and prepared their reports and posters for the Annual Research Conference (Pirog and Feldman 2009). Parent and administrator surveys also noted the value of these hands-on experiences (SageFox 2009). In addition, SageFox found that these opportunities and science experiences were categorized by 91% of the teachers in year 3 as "authentic science" (SageFox 2009).

Student goal (c), increasing student conceptual understanding and appreciation of the role that the sciences play in the world, and goal (d), increasing content knowledge, were evaluated using teacher, parent, and administrator surveys, which indicated substantial increases. All the data indicate that the students gained the understanding and knowledge specified in goals (c) and (d). Table 4.2 shows that 100% of teachers in Years 3 and 4 responded that their students developed new ways of thinking about science. Parents reported that their children became engaged in scientific research activities. Their comments include

- I believe that my child's involvement with STEMRAYS provided him with a deeper understanding of stating a hypothesis, running his experiment, collecting data and finding information to support his hypothesis or not.
- [It provided my child with] an understanding of the scientific method—process, data collection, variables.
- [My child learned] how to do research (form hypothesis, test hypothesis, issues of error and control).

They also reported that their children increased their knowledge about science and engineering, and about renewable energy and consumption; they also increased their critical-thinking skills.

At the end of Year 4, students were surveyed by SageFox about the most important things they learned in STEMRAYS. In addition to learning about science concepts related to their science club topics, student responses reflected the important themes of positive learning and experiences related to teamwork, collaboration, and relationship-building. Another important theme related generally to improved confidence in their independent thinking and problem-solving skills. Typical comments from students:

- If you work together you get things done quicker.
- That working together can be the difference between a successful project or not.
- I learned how to pick a topic and form a conclusion.
- That I have the ability to do something other than sports.
- How to set up experiments, the club really made me want to do more experiments at home.

- If you work hard you can accomplish anything. When I first started I had no idea what engineering was, now I want to get my master's degree in engineering.

In addition to the SageFox evaluation, in the 2008–09 academic year Feldman examined the effects of students' participation in STEMRAYS on their learning of content. In collaboration with the teachers taking part in STEMRAYS that year, he developed a pre- and posttest for measuring changes in content knowledge for the four club themes: Birds, Sustainability, Global Environmental Change, and Engineering. The pre- and posttest data for students in each of the themes demonstrated that they learned the content selected by the teachers ($t < 0.001$ for all themes).

## School Impact

Administrators were surveyed about the nature of after-school programs at their school, the benefits of STEMRAYS to students and teachers, and their partnership experience with STEMRAYS. The overall response from surveyed administrators was very positive in terms of the benefits for teachers and students in increasing scientific knowledge, new and engaging teaching and learning opportunities, and more hands-on science experiences that relate to the real world. Administrators all responded that STEMRAYS supported the mainstream curriculum and science content standards well. Their comments highlighted that STEMRAYS reinforces core standards, provides additional hands-on learning experiences beyond the regular curriculum, gives additional real-world science experiences, and further engages student interest in science. In addition, one administrator commented on the positive experience for students with special learning needs and attention issues who had the opportunity to work collaboratively in groups over a long period of time. Administrators cited benefits to teachers: the opportunity to advance their own scientific knowledge and opportunities to learn project-based approaches and engage more deeply with students over longer periods of time. Teacher feedback in Table 4.2 and in a focus group also reported a strong STEMRAYS school impact. One teacher noted

*Our school desperately wants this to continue happening and are looking at ways to keep funding it. It has a very high profile in our building and a huge effect on our school since we [our engineering design group] have helped soundproof the cafeteria as one project and improve milk carton recycling as another.*

## Next Steps

Like most NSF grants, the STEMRAYS award funded the development and testing of the model, but provided support for a limited period. An exploratory effort to extend and broaden the program in cooperation with 21st Century Community Learning Centers had some success and may lead to a large-scale expansion. The basic model of connecting academic researchers with existing after-school programs is very attractive wherever these researchers are available for a modest stipend or as volunteers. Since most communities are in reasonable proximity to a college or university, the model potentially has widespread, low-cost applicability. We are also currently exploring possible state-funded opportunities to partner with other OST programs in improving their STEM offerings. These include locally funded after-school programs and youth programs such as Girl Scouts, 4H, and Girls, Inc. Feldman has recently had success replicating the model on a small scale in Florida with high school students studying the growth of algae for biofuels in collaboration with an engineering professor from the University of South Florida (Chapman 2013; Chapman et al. 2013).

## 21st Century Community Learning Centers

We noted above that we have begun to work with the 21st Century program. It is a very large program and offers the potential for a large-scale application of the STEMRAYS model.

In Massachusetts, this program serves approximately 19,500 students in 41 districts at 184 sites; 59% of the students are from minority groups. It has had extensive evaluations (Massachusetts Department of Education 2009) and was examined in depth by the Massachusetts After School Research Study (MARS 2005). A highlight of the evaluation report was the finding that participating students made significant subject area gains, and that limited English proficient and special education students made the greatest gains. The MARS study had two major goals: to identify those program characteristics that are most closely related to high quality implementation, and to explore the links between program quality and youth outcomes. Some key points in their report:

- Staff made a difference in program quality, and staff that have a strong educational background and appropriate training are key to program quality.
- Most programs in the MARS sample had very low staff-to-child ratios, typically between 1:7 and 1:9. We found clear links between low ratios and high quality, as has previous research in the field.
- Most of the activities we observed tended to be short-term in nature and seemed only minimally related to program goals. Most did not have a clear connection to larger themes, curricula, or projects and did not require higher level or critical thinking.

STEMRAYS recognizes these points. We recruit regular classroom teachers to be the after-school leaders, and provide initial training plus ongoing content and pedagogy support. The clubs normally have at most 12 students, although occasionally teachers opt to admit more than 12; leaders may have an assistant if they have nine or more students. The long-term research projects provide students with opportunities to develop a deep understanding and to pose meaningful questions that they can investigate. Accordingly, the 21st Century sites that become part of STEMRAYS will have much stronger programs than before.

## Out-of-School-Time STEM Conference

Much of what we learned in STEMRAYS is consistent with the conclusions drawn from other NSFAYS programs. A conference was held in June 2010 to synthesize what has been learned from the NSFAYS and other OST programs and what research is needed (Bevan et al. 2010). The key findings included:

- There is no one model of OST STEM. Most programs are designed to reflect local resources, needs, and communities and therefore program goals, design, and outcomes vary widely. This diversity enriches the science learning landscape, providing multiple points of entry for many different learners. This diversity has implications for funding, research, assessment, and scaling up. (STEMRAYS was very different from most of the other programs.)
- OST STEM programs offer important developmentally supportive environments for children to develop their interest in, affinity with, understanding and pursuit of STEM. The low-stakes nature of these programs can provide all students with learning opportunities and activities that legitimate them as productive science learners. (STEMRAYS allowed children with a wide range of skills and abilities to be successful.)
- Most OST STEM programs are not extensions of school STEM. They draw on different resources, have different time frames, and therefore have different potentials and outcomes. They must be researched and evaluated differently. (These programs cannot be judged, for example, by the results of the statewide STEM tests.)
- There is a need to better understand the complex connections between school and OST experiences in STEM and how they contribute to lifelong engagement with science, including career pursuits. This work will require longitudinal and ethnographic studies. (We have attempted without success to obtain funding for such critical studies. Very little work has been done on research on the long-term impact of OST programs.)
- There is a need to better understand the ways in which OST STEM programs can provide important professional development sites for classroom teachers, both preservice and inservice. (STEMRAYS demonstrated the efficacy of its model in in-service professional development.)
- Funding for OST programs should provide sufficient time and resources for these complex spaces to develop partnerships, programs, and evaluation strategies that can test innovations, investigate sustainability, and support strategic scale-up efforts. (Again, funding for such long-term studies is difficult to obtain.)

## Conclusions

There are two conflicting pressures today. On the one hand, numerous reports stress the importance of STEM education to the future of our country, and note the mediocre results American students achieve in international comparisons. On the other hand are the pressures to improve test scores in math and English, driving many schools and teachers to minimize the time devoted to other subjects, including science. To the extent that high-stakes science (and technology) testing is being introduced, much of the science teaching becomes driven by the required factual content, depriving students of the opportunity to be engaged learners of the ways of science. Out-of-school-time science can provide students with rich experiences that have the potential to increase their interest in STEM courses and careers. However, there is generally a low priority for funding the staffing of these programs, the professional development of their teachers, and the research into their efficacy and long-term impact.

## References

Afterschool Alliance. 2011. *STEM learning in afterschool: Analysis of impact and outcomes*. Washington, DC: Afterschool Alliance.

Azevedo, F. 2006. Personal excursions: Investigating the dynamics of student engagement. *International Journal of Computers for Mathematical Learning* 11 (1): 57–98.

Banks, J. A., K. H. Au, A. F. Ball, P. Bell, E. W. Gordon, K. D. Gutiérrez, and M. Zhou. 2007. *Learning in and out of school in diverse environments: Life-long, life-wide, and life-deep*. Seattle, WA: The LIFE Center.

Barab, S. A., and K. E. Hay. 2001. Doing science at the elbows of experts: Issues related to the science apprenticeship camp. *Journal of Research in Science Teaching* 38 (1): 70–102.

Bell, P., B. Lewenstein, A. W. Shouse, and M. A. Feder, eds. 2009. *Learning science in informal environments: People, places and pursuits*. Washington, DC: National Academies Press.

Bevan, B., V. Michalchik, R. Bhanot, N. Rauch, J. Remold, R. Semper, and P. Shields. 2010. *Out-of-school time STEM: Building experience, building bridges*. San Francisco: Exploratorium.

Bhanpuri, H., N. Naftzger, J. Margolin, S. Kaufman. 2005. Ensuring equity, access, and quality in 21st century community learning centers. *Policy Issues* 19: 1–24.

Bleicher, R. E. 1996. High school students learning science in university research laboratories. *Journal of Research in Science Teaching* 33 (10): 1115–1133.

Braund, M., and M. Reiss. 2006. Towards a more authentic science curriculum: The contribution of out-of-school learning. *International Journal of Science Education* 28 (12): 1373–1388.

Caplan, J., and C. S. Calfee. 1998. Strengthening connections between schools and after-school programs. *www.ncrel.org/21stcclc/connect/index.html*

Chapman, A. 2013. An investigation of the effects of authentic science experiences among urban high school students. PhD diss., University of South Florida.

Chapman, A., A. Feldman, T. Halfhide, S. Ergas, F. Alshehri, D. Ozalp, and V. Vernaza- Hernández. 2013. An investigation of an algal biofuel project as an authentic science experience for urban high school students. Paper presented at the Annual Meeting of the Association for Science Teacher Education, Charleston, SC.

Charney, J., C. E. Hmelo-Silver, W. Sofer, L. Neigeborn, S. Coletta, and M. Nemeroff. 2007. Cognitive apprenticeship in science through immersion in laboratory practices. *International Journal of Science Education* 29 (2): 195–213.

Chinn, C. A., and B. A. Malhotra. 2002. Epistemologically authentic inquiry in schools: A theoretical framework for evaluating inquiry tasks. *Science Education* 86 (2): 175–218.

Crawford, B. A. 2012. Moving the essence of inquiry into the classroom: Engaging teachers and students in authentic science. In *Issues and challenges in science education research: Moving forward*, ed. K. C. D. Tan and M. Kim, 25–42. New York: Springer.

Etkina, E., T. Matilsky, and M. Lawrence. 2003. Pushing to the edge: Rutgers astrophysics institute motivates talented high school students. *Journal of Research in Science Teaching* 40 (10): 958–985.

Feldman, A., K. Divoll, and A. Rogan-Klyve. 2009. Research education of new scientists: Implications for science teacher education. *Journal of Research in Science Teaching* 46 (4): 442–459.

Feldman, A., K. Divoll, and A. Rogan-Klyve. 2013. Becoming researchers: The participation of undergraduate and graduate students in scientific research groups. *Science Education*: 97 (2): 218–243.

Feldman, A., and K. Pirog. 2011. Authentic science research in elementary school after-school science clubs. *The Journal of Science Education and Technology* 20: 494–507.

Halpern, R. 2002. A different kind of child-development institution: The history of after-school programs for low-income children. *Teachers College Record* 104 (2): 178–211.

House of Representatives. 2005. *House Report 109–272.*

Kardash, C. A. 2000. Evaluation of an undergraduate research experience: Perceptions of undergraduate interns and their faculty mentors. *Journal of Educational Psychology* 92 (1): 191–201.

Lave, J., and E. Wenger. 1991. *Situated learning: Legitimate peripheral participation*. New York: Cambridge University Press.

Lederman, N. G., F. Abd-El-Khalick, R. L. Bell, and R. S. Schwartz. 2002. Views of nature of science questionnaire: Toward valid and meaningful assessment of learners' conceptions of nature of science. *Journal of Research in Science Teaching* 39 (6): 497–521.

Lee, H. S., and N. B. Songer. 2003. Making authentic science accessible to students. *International Journal of Science Education* 25 (8): 923–948.

Mass Insight Education. 2002. *Schools alone are not enough: How after-school and summer programs help raise student achievement*. Boston, MA: Mass Insight Education.

Massachusetts After-school Research Study (MARS). 2005. *http://supportunitedway.org/asset/massachusetts-after-school-research-study-mars*.

Massachusetts Department of Education. 2009. Massachusetts 21CCLC Year-End Reports. *www.doe.mass.edu/21cclc/reports.html*

McLaughlin, M. W. 2000. *Community counts: How youth organizations matter for youth development*. Washington, DC: Public Education Network.

National Research Council (NRC). 2000. *Inquiry and the national science education standards.* Washington, DC: National Academies Press.

National Research Council (NRC). 2009. *Learning science in informal environments: People, places, and pursuits*. Washington, DC: National Academies Press.

National Research Council (NRC). 2012. *A framework for k–12 science education: Practices, crosscutting concepts, and core ideas*: National Academies Press.

NGSS Lead States. 2013. *Next Generation Science Standards: For states by states*. Washington, DC: National Academies Press. *www.nextgenscience.org/next-generation-science-standards*

Organisation for Economic Co-operation and Development (OECD). 2003. *The PISA 2003 assessment framework: Mathematics, reading, science and problem solving knowledge and skills. www.pisa.oecd.org/dataoecd/46/14/33694881.pdf*

Pirog, K., and A. Feldman. 2009. From science teacher to scientist. Paper presented at the Annual Meeting of the Association for Science Teacher Education, Hartford, CT.

Rahm, J., M. P. Martel-Reny, and J. C. Moore. 2005. The role of afterschool and community science programs in the lives of urban youth. *School Science and Mathematics* 105 (6): 283–291.

Richmond, G., and L. A. Kurth. 1999. Moving from outside to inside: High school students' use of apprenticeships as vehicles for entering the culture and practice of science. *Journal of Research in Science Teaching* 36 (6): 677–697.

Sadler, T. D., S. Burgin, L. McKinney, and L. Ponjuan. 2009. Learning science through research apprenticeships: A critical review of the literature. *Journal of Research in Science Teaching* 47 (3): 235–256.

SageFox. 2009. *STEMRAYS 2008–9 (Year 3) Evaluation Report*. Amherst, MA: SageFox Consulting.

SageFox. 2010. *STEMRAYS 2009–10 (Year 4) Evaluation Report*. Amherst, MA: SageFox Consulting.

Schwartz, R. S., N. G. Lederman, and B. A. Crawford. 2004. Developing views of nature of science in an authentic context: An explicit approach to bridging the gap between nature of science and scientific inquiry. *Science Education* 88 (4): 610–645.

# STEM Education in the Middle School Classroom

*Margie Hawkins*
*Winfree Bryant Middle School*
*Lebanon, TN*

*Leadership tomorrow depends on how we educate our students today, especially in math, science, technology, and engineering.*
—President Barack Obama, January 6, 2010

## Setting

STEM lessons were first incorporated into my sixth-grade classroom in the Lebanon Special School District in 2007. Lebanon is a small community in Middle Tennessee, about 30 miles west of Nashville. The district has 3,692 students, with 66% economically disadvantaged, 8.2% English Language Learners and 12.16% Special Education students. With such a large percentage of economically disadvantaged students, very few of the students, even the brightest ones, ever gave much thought to careers in the STEM field. When career choices were discussed, they seemed to envision those jobs being meant for the rich kids in the big cities—not a middle/lower class child in rural Tennessee. These students needed something to help them realize the world of opportunities in STEM careers was within their reach. STEM education was a way to help them discover their talents in science, technology, engineering, and math, so that they would be confident enough in their skills to pursue careers in any field they chose and never feel like any doors were closed to them. The effect that STEM education has had on these student's attitudes and achievement is astounding.

## The Effect of STEM Education on Student Attitudes and Achievement

For decades STEM education has been mostly reserved for accelerated learners, the advanced students. One of the most surprising aspects of teaching science with math, technology, and engineering integrations, is the skyrocketing achievement of struggling learners! All students enjoy learning this way. It is fun, exciting, relevant, and engaging. Real-life explorations requiring them to gather and analyze data; to create models; to make observations; to build, test, redesign, and redefine their ideas; all in order to discover a scientific concept or hidden truth … it is riddle solving at its finest! Yes, all students love to learn this way, but the struggling learners especially seem to flourish. Students have relayed the following feelings about learning through STEM lessons: "It is so much easier to understand the science when you can see the answer in the math," "When we graph the data on the computer, I can really see what is happening," "Now I see why we need to know how to do math!" or, during the engineering design lessons, "I love that you don't expect us to get it perfect the first time, that it is ok for us to try over and over." But, the all-time best student response to STEM education has to be, "This looked like it was going to be so hard; all that math and stuff, along with the science—but it is so much easier to understand this way!" Struggling learners started experiencing a much deeper a level of understanding when learning science with math, technology, and engineering integrations, and it really boosted their self-confidence.

That increased self-confidence paid off in many ways. The interwoven nature of their lessons and the constant application of math skills in science class made them perform better in math class too. Imagine the feeling of victory for them when they started getting higher grades in both science and math. Even better than getting higher grades, they actually understood the concepts better. They became a force to be reckoned with! These struggling students, who used to sit quietly during group discussions, now could not wait to give their input. After a STEM lesson, they felt like they so completely understood the concept that even the most reticent students would now be the first to raise their hands to answer questions. They eventually got to the point that they would argue their position and point out data that supported it—these formerly struggling students now felt, and acted, like real scientists! The empowering effect of STEM education on struggling learners was probably the most wonderful surprise encountered on this journey.

At Winfree Bryant Middle School, all subgroups of sixth-grade science students made remarkable gains in the 2013 TCAP tests; however, the gains of the lowest-performing students were the highest. One student moved from Below Basic all the way to a fraction of a point from Advanced in one year's time, and he made similar gains in math also. When congratulated and asked how he did it, he simply said, "It's easier to 'get' when you make us think harder about it with all that math and stuff." That is what STEM education does for students: It requires them to dig deeper into what is happening and why, which makes it easier for all students to "get" the concept.

This note sent to me from a student just three weeks into the school year says it all: "Mrs. Hawkins, Thank you for teaching science in the hands-on way that you do. For letting us actually DO science. You are the science teacher I have waited my entire life for—one who realizes that a textbook can only take you so far!"

Somehow, learning science by *doing* science made all students retain the knowledge longer, and understand it to a much deeper level. Experiencing science along with the math, technology, and engineering that you must use to really *do* science, made the learning experience so much more effective. So many students would relay how science and math used to be their "most hated" subjects and now they are their favorites. Students are learning in a way that makes sense to them, and they want to learn more and more because they see the knowledge as useful now—something they will use to solve problems and mysteries of the world—not just something they have to memorize for a test.

## Development of Marketable Skills

STEM education develops many marketable skills in students. This way of learning trains students to be creative problem-solvers, effective communicators, self-directed learners, collaborative workers, and efficient processors of information. These are exactly the skills needed in the innovative workforce we are trying to develop.

The first year implementing STEM lessons can be difficult for students and teachers alike. The students really struggle with having to figure things out for themselves. They are used to the teacher telling them how science works and expecting them to remember, or memorize, what they were told. For a teacher new to the discovery learning process of STEM lessons, it is really hard not to tell their students what should happen and what to watch for, especially when they are making mistakes in their procedures and drawing wrong conclusions. However, if students are working on teams, they have their peers to collaborate with, so the teacher just can simply question in a way to kick-start a discussion in the direction needed. Questioning is one of the most powerful tools a teacher has. In STEM lessons especially, teachers learn how to question in a way that leads students to rethink things, to recheck data and technology calibrations, to examine their procedures, and to look for every possible avenue for error. It is important to even question the right conclusions when they share them, just to see if they are confident enough in their position to stand behind it. We know it is much easier for students to remember something they have experienced than it is to remember something they have only been told. It is even easier to remember something they have had to figure out for themselves! When students learn a science concept by doing an experiment, using technology, collecting and analyzing data, while always practicing that "try and try again, examine and question everything" spirit of the engineering design process, they understand it more completely than ever before. After they have had to explain and support their thinking with evidence, they are ready to apply that knowledge to any applicable situation. Today's students will develop into top-notch problem solvers and collaborative, innovative workers because of the learning processes they are experiencing in the STEM classroom.

Anyone who has ever spent much time with a 12-year-old knows how much they love to debate. In sixth grade they begin to develop their own ideas about things, not just blindly adopting the beliefs of the adults in their lives. Being able to take a stance and prove that they are right is so empowering to them—they will argue with anyone, anytime, about anything! Needless to say, they absolutely love the aspect of STEM lessons where they have to take a position and support it with evidence from their data or observations. They are so confident of their thinking at that point because by then they have gone through the collaborative refining of those findings and the evidences with their peers. They especially love to find an error in another student's reasoning during that process. As long as there are controls in place to keep the discussions on a respectful tone, this can be the most effective part of their learning experience. This time of collaborative argumentation helps the students refine their understanding of the concept by questioning and re-thinking their findings. It is the time when, as a group, they are able to produce the most valid evidence-based reasoning. This time to collaborate with their peers is the part of the lesson where formerly unrecognized misconceptions surface. The best part for the students is to be able to clear any misconceptions up for themselves during the productive discourse happening in their group, before they have to present their findings to the class.

Most scientists will agree that one of the most engaging aspects of their careers is the process of critique and argumentation. Having a chance to share their ideas and hear the ideas of others, to collectively identify flaws in procedures, conclusions, or data collection. Students enjoy this process just as much as the professionals do, if not more. In fact some students, ones who thrive on persuasive argumentation, have expressed that they think scientists have to be even more persuasive than lawyers and they are now considering careers in science or engineering because of that aspect.

STEM education gives our students an opportunity to experience the joy of new discoveries, the invigoration of truly productive collaboration, the confidence of having evidence to support your position, and the victory of finding a way to solve a problem that has eluded others, all of the rewards that a STEM career has to offer. They find this exciting and fun! Many of the future innovators in our world will have followed the path to a STEM career because of the experiences they had and the skills they developed in STEM education classes early in their academic life.

## STEM and Today's Learners

Today's learners have changed. The middle school students of 2013 are quite different from the students encountered in 2001. Today's students expect to be entertained at all times. They have grown up with constant entertainment. Anywhere you encounter children you will find a video playing, or a video game being played, music playing or being played, a live show, YouTube, social media, video chatting, Skyping, and tweeting. They have grown up with constant entertainment all around them. Because of this change in our audience, our teaching styles have had to change and STEM education fits this style change perfectly. Students find it exciting to have a question posed that requires them to conduct research, do labs, collect data with instruments, plot and graph the data on the computer, and make observations to find the answer—so much more entertaining than the teacher giving them the answer while they take notes to take home and memorize!

Today's learners are multitaskers. They have grown up in a society of multimedia, multisensory, and multi-stimulation. They function quite well juggling several things at once and, in fact, seem to feel a bit bored or unproductive when they are not engaged in more than one activity at a time. Most of our highest achieving students are quite skilled at listening to music while they do their homework, with their smart phone or tablet nearby so they can be tweeting, chatting with friends, or updating their social media site. There are times and tasks that require targeted focus and they seem to be able to do that when needed, but otherwise, they are tending to several things at once. Students today seem to do this quite successfully and with amazing ease. The structure of STEM education requires that students be monitoring and considering input from several different sources at the same time. They must evaluate data and observations from different viewpoints. The multitasking brain that has developed in today's students is one of the assets that will make them better innovators, scientists, and engineers. STEM education trains students to develop this skill to its fullest potential.

Students of today expect relevance. Even though they do not know it by that term, they want their learning to be meaningful and real. If they are given a learning task that relates to their lives, they will be as highly engaged as you could ever imagine. In the overview of life that I have been afforded because of my multitude of years on this planet, I believe this has come about because knowledge is so easy to acquire now. Our students have become discerning consumers of knowledge. Back when I attended school, the only sources of academic knowledge my classmates and I had were our teachers and our textbooks. We grew up in a small town with no public library and we did not have a library at school. We would ask for books instead of toys for Christmas. We would read any print material we could get our hands on. Certainly a stark contrast to the information overload students encounter today! But that was in B.G. (the time before Google). Today's students have all of the information mankind possesses at their fingertips at all times on their smartphones. With all of this knowledge at their avail, they have become more discerning, only interested in acquiring the knowledge that pertains to their lives. They do not want to waste their time gathering knowledge that is not relevant to them. In response to that change, teachers have begun to design real-life lessons that develop the concept knowledge and skills required in their state standards, while students progress through a relevant, real-life problem to be solved or question to be answered. STEM education is the ideal venue for relevant learning experiences.

## Vive La Différence!

We all know that there are as many different teaching styles as there are different learning styles. Depending on the teaching assignment, each teacher's implementation of STEM education will vary tremendously. In the past two years, while serving as the steering committee chair and middle school strand leader for the NSTA STEM Forum & Expo, the master of ceremonies for the TN STEM Leadership Academy, and master teacher on TN Tech's STEM Around Us program, and training elementary Math and Science teachers to develop challenge-based STEM lessons using the Legacy Cycle, I have met hundreds of teachers and STEM experts. In those interactions with educators from around the country, I have realized that there are a multitude of different and truly delightful approaches to STEM education being put into place around the country.

There is a popular saying, "Everything in moderation, including moderation." That attitude could be applied to education by saying, "Everything in moderation, including standardization." We need a certain amount of standardization in education, especially with the mobility of families these days. We are no longer preparing students for careers only in their local communities, or even their home state. Today's students must be prepared to be competitive in careers anywhere on this planet! With the development and implementation of the *Next Generation Science Standards* and *Common Core Standards, Mathematics and Language Arts*, we are moving toward a standardization that has been long overdue. We now have a cohesive set of national standards that include concept knowledge, but are also focused on student practices—how students gather and apply the knowledge they have gained. The *Next Generation Science Standards* mesh perfectly with STEM education! We also need nationally standardized assessments if we are going to have valid data to identify best practices to achieve our national goals. However, in the classroom, teachers need the freedom to achieve those goals in their own way. Just as students learn best when the material is presented in their particular learning style, teachers are also most effective when they can facilitate that learning using their own teaching style.

Just as teachers have different teaching styles, they also have different teaching assignments. These variations in the school system, schedule, and grade level in which the STEM program is being implemented, will mandate different approaches. Early elementary teachers usually teach all subjects, making total integration easy and completely manageable; however, their main focus is not on the science but rather on the reading and math skills, which it should be at that level. So lower-elementary teachers would structure their STEM lessons integrating some science into their reading and math lessons. Some upper-elementary teachers are departmentalized and some are self-contained. The same holds true with middle school teachers. Their STEM program structures would also vary, but would probably have a stronger science component and they may or may not be able to collaborate with colleagues to implement integrations across all subject areas. However, high school teachers are usually specialized to one subject/department, so at that level they would probably find it easier to implement all components of a STEM program in their own classroom. If you do not teach all subjects, buy-in and commitment from your colleagues will be a deciding factor on how to implement STEM integrations. Obviously, your STEM program can be more powerful if all concept area teachers collaborate on the integrations and implement them in every subject area. This idea has been in practice in the Metro Nashville STEM Schools for a few years now. All teachers on the same grade level plan together and come up with one Big Question for the students to answer, but work on standards-based explorations of the topic in every subject area during the unit to gather knowledge needed to answer the Big Question. By the end of the unit, students have used all the knowledge they have acquired to answer that overriding question and design a presentation to share their answers. It seems to be very engaging for the students, but it takes a huge amount of common planning time to accomplish this and that is not possible in most school schedules. However, even if total integration of all subjects is not feasible, it is still possible for the science teacher to teach STEM lessons in the classroom, without the involvement of others. Just having students using technology to gather, analyze, and graph data and communicating their ideas either verbally or in writing will result in them understanding science concepts at a much deeper level. All these variations in the possibilities of implementation mean that STEM education programs will never look the same in all settings, and vive la différence!

In the same vein, a teacher's approach to STEM education will probably change a bit every year. An effective teacher is always looking at what works and what doesn't and finding new ways to address new challenges that every new school year and every new class brings with them. This is a good thing. It keeps teaching fresh and customized to each set of students they teach. When STEM integrations first started in Lebanon, I was teaching all subjects in a self-contained sixth-grade classroom. It was incredibly easy to design learning experiences that encompassed all subject areas and standards. It was possible to structure my daily schedule any way needed, since I was not relegated to conform to anyone else's schedule, except for lunch and related arts timing. Using the Science Café style of integrating, the students thrived! The second year of science and math integration, 75% of these students score Advanced in math and 80% scored Advanced in science. It was magical!

Then they restructured our school system and we departmentalized. Teachers were assigned to teach four classes a day, 55-minute classes, and had to end at each bell no matter where we were in the lesson. This made Science Café integrations impossible. My TVAAS scores went down from a level 5 to a level 3 and it was devastating! My principal was able to shine some light on what had happened when he stated, "I have watched you teach. I would be in your room during language time and you would be talking about what the students had learned in social studies, or I would be in there during science and you would be using the math skills they learned that day—now that you aren't teaching them all subjects, you can't make those connections for them." That was a life changing revelation! The next year I asked my colleagues to keep me informed of what skills and concepts they were teaching in the other subject areas so those connections could be pulled into science lessons and my TVAAS scores went back up to a level 5 that next year.

With growth comes change, and as the district grew and built a new middle school, our schools were reconfigured again. We are now on block scheduling, with science teachers teaching six different science classes, and only seeing students every other day—2½ blocks a week, for 90-minute blocks. So again, it is important to find new ways to implement STEM lessons. With block scheduling and 90-minute blocks, teachers have plenty of time for integration and full implementation of STEM lessons. However, they must complete two full lessons each time a class meets to stay with district pacing. STEM lessons are usually completed in the science classroom, but when a class runs short on time, or the lab takes longer than expected, science teachers can count on their colleagues to help out. In such a case students will do the lab and collect the data in science class, they take that data to math class, where the math teacher will have them plot, graph, and analyze it, and then they will do a literacy task for language class explaining their findings, persuading action, sometimes writing a narrative of the experience and a description of their observations or procedures. The connections teachers are making for their students, the prior knowledge and experience the students take from class to class, all result in a deeper understanding in all subject areas, so all teachers are excited to make those connections for the students any way they can.

## Challenge-Based Learning

This school year has brought another exciting discovery in STEM lesson design. In working on a team with Dr. Bruce Howard, Dr. Martha Howard, and Dr. Sally Pardue, Director or the Oakley STEM Center at Tennessee Tech University, on the STEM Around Us program this past 18 months, I realized that the challenge-based learning experiences provided in the Legacy Cycle lesson design by Dr. Tony Petrosino are the perfect strategy to solve the time dilemma of block scheduling. Teachers can structure real-life learning challenges that encompass several standards at a time, and the connections will give the students a better understanding of the big picture and the interconnectedness of science concepts while accomplishing all STEM components in a timely manner, with the added bonus of the excitement of the "challenge." The Legacy Cycle uses challenges as anchors for learning. The challenges are designed to create an increased depth of knowledge in a specific subject. The combination of well-designed challenges and meaningful learning activities provides a rich environment for both the students and the instructor. Students who have experienced Legacy Cycles this year are enjoying science and learning more than ever. Because the challenges encompass several standards, they are also a little ahead of district pacing at this point.

In the first challenge, the Fuel Cell Challenge, the students were told they were part of a design team at an automotive plant. The Board of Directors had been discussing the issue of solar-powered vehicles for general use. Some felt that a completely solar-powered vehicle was not possible, and that some other (or added) power source and fuel were needed. For strategic planning over the next year, the board needed to know just what was possible. They had heard a lot of talk about "hybrid" systems. The students were directed to run several experiments on solar-powered and hydrogen fuel cell cars and compile a recommendation for the board on which direction to take, whether to develop a completely solar-powered vehicle or if a hybrid system would be better. Students were required to cite evidence from their experimentation that supported their recommendation.

Students researching pros and cons of solar vs. hybrid cars

During the second Legacy Cycle, The TN River Crisis Challenge, the classroom became an Emergency Response Center deciding how to respond to earthquakes, landslides, flooding, breeches in the dams, a flooded nuclear plant, and evacuations. In this challenge, an earthquake threatens to break some of the dams located along the Tennessee River. Students are placed on teams, monitoring real-time data online, to develop an action plan. Teams interact live with a mission commander from the Challenger STEM Learning Center as they monitor changing online data. They worked on teams, deciding how to respond to the various disasters in order to minimize loss of life and property. This was a real-life disaster response adventure of data analysis and problem solving at its best! The geology team had to monitor data coming in from inclinometers that registered ground movement and they had to plot the sectors of movement on maps. If the ground movement was along the edge of the water they had to report to the hydro team, which would do math calculations to determine how much water displacement would occur from the amount of erosion happening. When the water displacement got critical, the hydro team had to report to the engineering team, which would decide when to open and close the flood gates at the dam to release water and the Emergency Response team had to make critical decisions about evacuations. It was amazing to watch, and so true-to-life that the students

Students saving TN from the TN River Crisis!

felt like it was really happening! The students said they now understand why communication skills are so important—they said one of the hardest jobs was relaying the information accurately and in a way that the other teams could understand what the crisis was. Here are some quotes from the students involved in this challenge:

- "It was the most stressful thing I have ever done, but it was also the most coolest, funniest thing I have ever done as well!"
- "This mission was both stressful and exciting. I loved sending reports and releasing the floodgates. But, when I found out the dam I was assigned to monitor broke down, and I had to tell the Rapid Response Team to evacuate Chickamauga it got stressful—but I LOVED IT!"
- "It was such an amazing and challenging experience! It was amazing to be able to make decisions that saved people from the flooding. But, it was challenging because we had to figure out what to do very quickly!"

- "It was difficult because there were complex details to every choice we had to make."
- "As safety manager for the hydro team, I learned how stressful it is to be a leader! I now have an idea of how much work these people put into this job. I feel like I will be more successful in my career when I grow older due to the fact that I had this experience!"
- "It was very stressful, but it was really enjoyable!"
- "This was a once in a lifetime experience for me! It was fun and stressful all at the same time!"
- "It makes you very stressed, in a good way. You have to make split-second decisions!"
- "Being one of the communicators on the Rapid Response Team was difficult because I had to answer questions coming in from other teams: I had to type really fast and make sure I was giving correct information. Although it was a difficult job, it was really fun! I may want to do this as my career in the future."
- "I loved how everyone involved treated it like a real case scenario. It was like getting our own jobs and getting started in our careers. We entered a whole new world and it felt real! I sure hope we get to do something like this again!!"
- "This was the most awesome experience I have ever had! It was fun working together and communicating with other teams, giving them information they needed!"
- "I loved the sudden events and the need for quick problem solving! I also like that there is no clearly right or wrong answers!"
- "It was a fun and challenging project. It certainly helped me with my graphing skills and helped me to get a feel of what people do in those types of jobs."

And from our Superintendent of Schools, who stopped by to observe the mission: "I enjoyed seeing our students working collaboratively on such an engaging project! Good stuff!"

This program was only possible because UTC Challenger Center needed to test their new beta version of the program; I met their representative at a conference and I offered my class for testing—otherwise it would have cost $500 to bring this program in, which would have been way out of my $100-a-year classroom budget. Networking at conferences and looking for these types of resources for our students outside of the schoolhouse is critical! All of my best lending resources have come from connections I have made at conferences and other professional development events. The hydro fuel cell cars were loaned to us by the University of Tennessee TN-SCORE, and I have also borrowed equipment from the Oakley STEM Center at TN Technological University.

The moral of this story is that no matter what grade level, configuration, teaching assignment, or teaching style a teacher has, whether they go for the gusto with Legacy Cycles or they just work a few math, ELA, and technology skills into the science lesson, whether they live close to a University to borrow from or they have to fund their own projects, some level of STEM education *can* always be implemented, and *should* be implemented, for the academic success of their students. Depending on the individual circumstances, the STEM programs and integrations will take many different forms and this is as it should be. All academic programs should be customized to the population and system implementing them.

STEM education is more like real life than the traditional way of teaching, which makes it more relevant to students. It gives students a feel for what it might be like to have a STEM career. And, if we want to get them excited about the possibilities the future holds for them, there is no better way to do that than to let students experience the thrill of STEM challenges with true integration that will afford them greater success in all subject areas and will enable them to become tomorrows' innovators.

## The Need for STEM Education

Since President Obama's "Educate to Innovate" in 2009, we have seen the development of a nationwide campaign to strengthen STEM education in hopes of moving American students from the middle to the top of the pack in science and math achievement over the next decade. In today's world, the four disciplines of science, technology, engineering, and math have become one in the workplace—working in concert to answer contemporary questions and solve problems. Our students need to see and experience them as one. A former professor of engineering at Yale University, Dr. Ainissa Ramirez, spoke at the 2013 NSTA Stem Forum & Expo about the need for "smashing silos." She related that life does not come at you in neat little pockets of math, science, or technology; in real life they are all intertwined and they all come at you at once. Dr. Ramirez says we can no longer keep our subject areas isolated in their own little silos. Our students will need to be able to apply the skills of all disciplines in unison, to solve real-world problems, in order to be competitive in the workforce of the future.

Ultimately, it always begins with education," said Gail Hardinge, executive director of the STEM Education Alliance (SEA). "We must expose students to careers within an appropriate educational environment if we expect them to be interested in becoming the engineers and scientists of the future." If we structure the educational experience to mirror real-life career experiences, our students will not only see the relevance in their education, but will move into the workforce better prepared for what they will experience in their careers. However, several hurdles are being encountered by teachers all across our country as they try to develop STEM education programs in their classrooms and school districts, hurdles that will need to be addressed if we want to achieve our president's goal of moving American students from the middle to the top of the pack in science and math achievement.

## The Struggle of Middle School Teachers to Facilitate STEM Education

In my professional work outside of the classroom when discussing the state of STEM education with hundreds of teachers, administrators, and STEM experts from all over the country, everyone seems to agree that this is the path modern education should be taking and students love to learn this way—so why haven't we seen STEM programs developed in more of our schools? There are still many hurdles to be cleared before STEM education can be successfully implemented across the country. There is a popular saying that, "There is no problem, other than death, that enough time and money cannot fix." Those are the two main hurdles to be cleared in STEM education, time and money.

## *Commitment Hurdles*

Science education has always taken the back seat to language and math in elementary education. Most elementary students do not have formal science class until fourth or fifth grade. While being able to read, write, and "cipher" are basic skills that should be the main focus of early education years, by middle school one would expect that science would be on common ground with those other disciplines. Sadly, it is not. Most of the middle school science teachers interviewed report that their students get much less time for science instruction than they do for math and language. In a lot of schools middle school students get only half as much time for science and social studies lessons as they do for math and language. In our modern world, where technology reigns supreme and science and engineering jobs are growing at a higher rate than any other career field, this seems like a practice that needs to change.

A lot of science teachers I spoke with voiced frustration at being expected to teach math in science class when they get so much less time to teach their science standards than the math teacher does. Now, with the implementation of Common Core Curriculum standards, science teachers are being told they have to teach literacy skills in their science classes too. We all know this is a more effective way of teaching and learning and that student understanding of science concepts would obviously be deeper if students were really using math and writing skills in the science lessons, but in order to teach science standards, while reinforcing math skills and having students performing literacy tasks to report their evidence-based reasoning, science teachers say they need their fair share of the instructional day.

Science teachers would love to embrace STEM education, but it is difficult to do without the commitment of priority from their school districts that they will be given time in the daily schedule to implement STEM education effectively.

Another time commitment hurdle mentioned by both teachers and STEM high school students was the need for STEM businesses to get involved in STEM education. A panel of high school students following the STEM career path spoke at the wrap-up session of the 2013 NSTA STEM Forum & Expo. When asked about what was needed in STEM education, the students noted that one of the most motivating factors for them was the opportunity to shadow someone in a STEM career field. It would be equally motivating to middle school students to have STEM professionals come into their classroom and share what a day in their field consists of, what classes students should take in high school and college to get into that career, show some of the technology used in their field, and answer questions about the challenges and rewards of STEM careers. The STEM professional could even donate outdated equipment to the schools so students could have more hand-on experiences in STEM education. It could be a wonderful partnership. We need to open that door a little wider and welcome our STEM professionals into our schools.

## *Funding Hurdles*

Every project has funding constraints. STEM education has them too; funding for curriculum development, funding for equipment, funding for technology, funding for consumables. All of these funding challenges, in classrooms engaged in STEM education, are coming at classroom teachers. We all know that public school teachers do not make enough money to be spending hundreds and in some cases thousands of dollars each year so their students can engage in STEM education, yet they do. We see a lot of government money being earmarked to support STEM education; however, it seems to be absorbed by administration programs and little of it ever seems to reach the classrooms. The teachers themselves are left to supply money for equipment and supplies for their STEM programs out of their family's budget. For STEM education to be implemented nationwide, this must change. Teachers are no longer just buying pencils, crayons, and backpacks for their students. Teachers are now purchasing computers and digital probes, microscopes, glassware, and all consumables needed for their STEM classrooms from their family coffers and their families are making sacrifices so they can do this. This is not right, but it is the only way teachers can think of to provide their students with the kind of education they will need in the workforce of the future. There are grants available for some of the equipment, but not for consumables. Teachers, however, spend their evenings correcting papers and preparing things for the next day's lesson and have very little time to write grants. Grants are not really an option for providing what students need in the classroom right now. Our state and federal governments need to make monetary commitments to STEM education that will actually reach the classroom. If we want our future workforce to be technologically literate, students need to learn on that technology in the schools. If we know students learn better by *doing* science, then all science teachers should have a sufficient budget to buy the supplies for their labs each year. Most of the middle school science teachers interviewed have reported having 100–200 students and getting $100 a year for their classroom, that is not even enough to cover the classroom supplies to start the year or the consumable supplies for even one lab. It is not surprising that most of those teachers do not do hands-on labs because they can't afford it. They do a demo in front of the class, and they give the students the data to record. The students get to see science, but they do not get a chance to do science. As our government is deciding how to spend our tax money, it seems like the education of the people who will be running our country and providing the innovations to keep our economy sound in the future should be a higher priority.

All that being said, the outlook is bright! Our nation has wisely seen the need for STEM education and is moving in the right direction. Our teachers, students, parents, and STEM professionals are all on board. Universities are establishing more and more lending programs to get needed equipment out to the schools. States are establishing critical networking opportunities with STEM hubs. School districts are funding more professional development and networking opportunities for their teachers. We all know this is critical to our future, to our survival on this planet, for our competiveness in the world market, and our economic security. But, most of all, we know in our hearts that the youth in our country are some of the most brilliant on the planet. They are looking to us to help them develop into the creative problem solvers, effective communicators, self-directed learners, collaborative-workers, and efficient processors of information that will make them competitive in the future workforce so that our country can become the greatest innovation nation in the world. We can't let them down.

## References

NGSS Lead States. 2013. *Next Generation Science Standards: For states by states*. Washington, DC: National Academies Press. *www.nextgenscience.org/next-generation-science-standards*.

Petrosino, T. 2014. Legacy cycle. *www.edb.utexas.edu/visionawards/petrosino*

Ramirez, A. 2013. Smashing silos! *Edutopia*. June 24. *www.edutopia.org/blog/smashing-silos-ainissa-ramirez*

Zagursky, E. 2012. STEM education alliance. *Ideation*. May 29. *www.wm.edu/research/ideation/ professions/ stem-education-alliance6478.php*

# Introducing STEM to Middle School Students

## A World of Excitement and Inquiry

*Matthew Cieslik*
*Rosa International Middle School*
*Cherry Hill, NJ*

### Setting

Rosa International Middle School, located in Cherry Hill, New Jersey, is a public school providing education to students in grades 6–8. Rosa is an IB (International Baccalaureate) World School authorized in the Middle Years Program. Cherry Hill is a suburban community outside of Philadelphia. A majority of students score in the proficient and advanced proficient categories as determined by the NJ ASK standardized test, which is given to all middle school students in the state. The subjects tested in middle schools in New Jersey include science, language arts, and mathematics, although science is not assessed until eighth grade. In 2008 Rosa was named a national school of character and in 2009 the school was a recipient of the National Blue Ribbon Award for excellence in schools.

Beginning with the 2010–2011 school year, the school was forced to eliminate a team of teachers as a result of budget cuts. At the time of the cuts Rosa students attended three "specials" classes, which in other districts may be known as exploratory courses. Since the learning communities at Rosa would be increasing from 90–100 students to 125 or more students, the administration faced a crisis in regards to the specials courses. A team of 120 students, for example, would require each of the three specials courses (art, technology, and music) to teach 40 students at once, in facilities that were less than adequate for this number of students. As a result it was decided that Rosa would add a new exploratory course, thereby limiting the number of students in any class to approximately 30. The fourth specials course was added to the technology, art, and music classes that were already in place at Rosa. After some research, which included discussions with Rosa faculty, it was decided that the new course would be known as STEM (Science, Technology, Engineering, and Mathematics).

### Overview of the Program

While the decision to add a fourth exploratory course was initially based on practical considerations, the vision of the STEM course allowed students an opportunity to research, reflect, and discover aspects of the natural world. Goal 1 of *the National Science Education Standards* (*NSES*) called for students to "experience the richness and excitement of knowing about and understanding the natural world" (NRC 1996). Engaging in basic and (sometimes) fun and challenging engineering activities allows students to encounter math, science, technology, and engineering in a practical manner while emphasizing the higher level thinking skills needed to succeed in the process of design and construction while at the same time deemphasizing other thinking skills, such as recall of textbook information.

The emphasis on bringing excitement to STEM disciplines takes place in an environment that merges learning and play. While STEM activities are challenging, they also allow students to compete with one another and themselves while working to successfully complete basic engineering challenges that require and develop skills such as imagination, analysis, and creativity. Conducting activities in a manner that generates excitement has benefits that include increased motivation. Possible reasons for excitement and enjoyment may be collaboration, hands-on learning, and the permission to satisfy curiosity.

Rosa's STEM program is designed to not only enrich students' educational experiences with additional math and science content but also to provide experiences that place the content of these disciplines in the background. The STEM program places its primary focus on scientific processes by immersing students in an authentic environment where skills such as problem solving, critical thinking, and creativity are valued as life skills, as well as for skills that will allow students to achieve academically.

Our STEM program is a movement to shift the emphasis of instruction from memorization of information toward the other end of the learning spectrum where children are encouraged to become independent learners who use their own original ideas. In Rosa's STEM program students are constantly encouraged to make decisions regarding the processes that they will take to solve a problem and to make decisions, including what to build, how to build it, and what materials will be most effective.

The STEM program at Rosa attempts to prepare students by developing what Dr. Robert Sternberg of Yale University describes as "successful intelligence." Dr. Sternberg, known for his work in the field of intelligence, states that people who are successful possess high levels of analytical, practical, and or creative intelligence (Sternberg 1997). In recent years, as emphasis on standardized testing has increased, educators have placed greater emphasis on analytical intelligence. Some would argue that this emphasis has come at the expense of the development of practical and creative intelligences. The STEM program at Rosa does not ignore analytical intelligence. Rather, it attempts to develop analytical intelligence while simultaneously building practical and creative intelligences.

## *Student-Centered Learning Environment*

Many science and math courses have traditionally used strategies that would typically be found in a teacher-centered classroom. In some science courses, note-taking is the most used learning activity, while students in some math classes typically spend a great deal of time observing teachers solving problems and then use the teacher-modeled solutions to solve similar problems. There are certainly benefits to this type of teaching and learning. Benefits would include the ability of the teacher to ensure that the entire curriculum is covered. Also, teacher-centered classrooms tend to be more orderly. Finally, once students learn how this "system" works, they often continue to have academic success as evidence by strong test scores and good grades. The winning formula of a traditional classroom is for a student to sit quietly, take great notes, complete the homework, and study for the test. If students follow these steps, they are likely being successful if our definition of success is limited to grades and test scores. When a student does not do well in a traditional class, the teacher will often point to deficiencies in one of these steps as the reason.

One of the outcomes that we hope to accomplish in STEM is student enjoyment. A unique feature of the STEM program that helps toward this end is the lack of a mandated curriculum, along with the fact that there is not a formal assessment at the conclusion of the course or unit. This allows the teacher to focus on a few "big ideas" at the expense of the smaller details. This format is also conducive to taking advantage of teachable moments. If students show interest in a topic that deviates from the planned curriculum, there will be no testing repercussions for allowing or even encouraging students to stray from the typical planned course outline.

Challenges can also generate fun! During one STEM activity, students adjust a ramp until a toy car stops at a specific distance. Students then use these data to mathematically determine how the slope should be adjusted to stop at another distance specified in the activity. The objective is to have the car stop at that point with the fewest possible attempts. Sometimes rewards are given but in most cases the reward is the opportunity to share the math, engineering, and problem-solving strategies that lead to success with the members of the class.

Unlike a traditional STEM-type course that might rely more heavily on note taking, STEM students are not given information such as definitions and formulas upfront. Rather, STEM activities begin with a two-to-three-minute video, followed by a short discussion that introduces a topic and explains why and how this topic is relevant to the lives of middle school students. Students do not take notes or listen to a content lecture. Instead, the instructor's introduction for the day's activity is limited to metacognitive considerations such as various approaches that can be taken or skills that may become valuable in the attempt to complete the activity successfully. In some cases the benefits and pitfalls of specific strategies may be discussed—especially if certain strategies present safety concerns.

Even though class has only been in session for 5–10 minutes, at this point students will begin diagnosing and understanding the challenge. Once students have a potential solution, they will ask for approval before attempting to put their solution into practice. However, it is worth noting that student approval is for safety reasons only and students are not initially told if their ideas will or should lead to success. As the students begin the design and building process, discoveries are made and new ideas hatched. Thus begins a process, a cycle that begins with the problem, leads to an attempt at solving the problem, and ultimately leads to newer and better ideas that build on initial attempts. In the course of completing this cycle, students discover and make connections to math and science that enhance solutions. Depending on the student, the teacher may make suggestions or guide students in a particular direction.

### *Emphasis on Creativity*

The value of creativity as a skill is evident in the "new" Bloom's Taxonomy (see Figure 6.1) which now places creativity at the apex of the pyramid. The connection between creativity and STEM is strong, as students are asked in every class to provide original ideas and products. While the task of producing original work and ideas is challenging, it is the belief of the creator of the STEM program that being creative will serve students well after their time spent in STEM class is completed.

**Figure 6.1.** Bloom's New Taxonomy

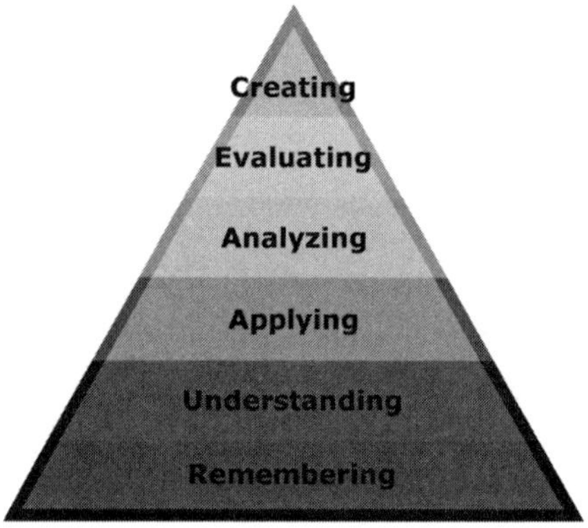


Class discussions in STEM also focus on the value of being a creative thinker. In the scope of these conversations, the value of our creativity includes the potential benefit of an individual's creativity to society. Included in conversations of this nature are examples where those who have come before us used their own creative skills to further the current body of knowledge in a particular field of study.

The world of math and science are filled with stories of discoveries that were made despite mistakes—or even because of mistakes. However, in some more traditional science and math classes we often assess students on one attempt, usually in the form of a pen-or-pencil test that may only measure how well students are able to retain information. This type of learning essentially limits students to use of the aforementioned analytical intelligence. If students are able to recall information or patterns for solving problems, they are labeled as successful. In many cases this is not reflective of real life, where people do make mistakes, and steps are taken to ensure that ample opportunities are given for working through mistakes before a final product is unveiled. The STEM course attempts to mimic the latter setting by offering learning environments where students are encouraged to work with the understanding that many problems can indeed have many correct solutions, but along the road to discovering the best solution we may encounter a number of mistakes. A goal of Rosa's STEM program is for students to realize that we can learn a great deal from our mistakes and that our mistakes can actually assist us through the process of arriving at a better solution.

The success of our STEM program can be attributed to a cooperative effort of the Rosa administration, the school district's central administration, PTA, and parents. When the STEM program was in the early stages of development, our searches lead us to believe that similar programs to the STEM program that we envisioned at Rosa are not common. The administration at Rosa and the Central Administrators for the district trusted that the program would be good for kids, even though there was no curriculum map or standards from a similar program that would be followed. The administration also trusted that there were people in the building and in the district who could essentially invent a program that is based on and reflects the strengths and needs of the students at Rosa.

## Evidence for Success

As the STEM program proceeded through its first year, it received a great deal of positive feedback, from both students and parents. However, this feedback was informal and it was soon recognized that in order to gauge the effectiveness of the program more concrete data would be needed.

A colleague, Gretchen Seibert, provided a tangible piece of data. This teacher-researcher happened to be looking for a topic to study for graduate school research. While teaching next door to the STEM classroom, the teacher-researcher became curious regarding the sounds of cheering and animated discussions that often permeated the classroom walls. When she approached the STEM teacher about using the STEM program as a potential study, both teachers agreed that a study related to the newly created STEM program could provide valuable information. Collectively, it was decided that the research could center on the role of creativity, critical thinking, and problem solving in STEM. This study was conducted with the assumption that these skills can be developed. This study used a series of Likert scale statements that asked students to comment on their own creativity.

The 2011–12 academic year offered the opportunity for a thorough study to be completed. Data were collected in the form of an online student survey that was completed by many of the STEM students on the final day of the course. The survey was completed in school.

The student survey was completed by 147 of the students who attended STEM during the first marking period. Of these students, 88 were males and 59 were females. The discrepancy between males and females could be explained by the fact that some students attend a music class called choral workshop. These students attend choral workshop every eight days and so during the course of STEM these students are expected to occasionally miss class. Of the 147 students who completed the survey, 56 students were sixth graders, 45 were seventh graders, and 46 students were eighth graders.

**Skill #1, Critical Thinking:** Students were asked to respond to the following statement (see data in Figure 6.2, p. 96):

**Critical thinking is a skill.** If Critical thinking means you make decisions based on information (instead of simply guessing), do you think your critical-thinking skills have improved during the STEM course?

**Figure 6.2.** Student Responses to Question About Critical-Thinking Skills

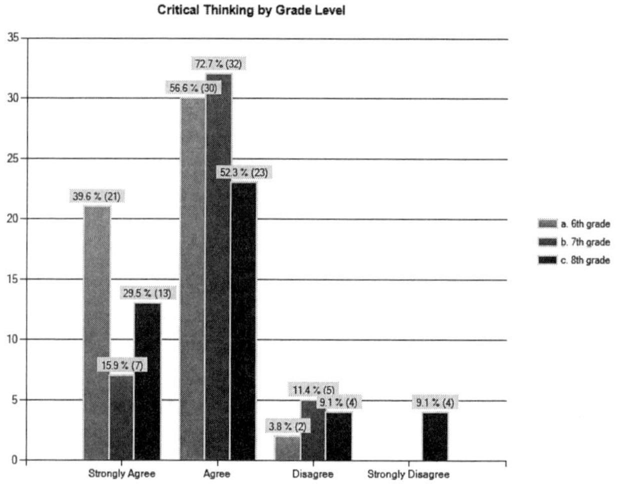


- 89.4% of students surveyed perceived that their critical-thinking skills improved as a result of the STEM course.
- Under the "strongly agree" category there is a great difference between males and females.
- When broken down by grade level, no sixth graders placed themselves in the "strongly disagree" category while 18.2% of eighth graders placed themselves in the "disagree" or "strongly disagree" categories.

Students were asked to provide examples of critical thinking or when they believed they had used critical thinking during the STEM activities. The responses provide a glimpse of the types of activities that students were asked to complete. Some are:

- "We made different shaped boats and had to throw out tons of them because they could not hold a single marble without sinking. We also had to calculate the slope of a ramp. It did not go as planned. Finally, we are making kites and were having a bit of trouble. We think it may be ready for flight. There are more tests but I want to go fly the kite."
- "I learned new ways to make things with stuff that is trash. We test out the first thing, and find out what went wrong and add on to what didn't work."
- "I had to make a kite out of given materials … not a kit. This tested my skills because I truly had to think."
- "Making a mini-golf course and when we made things equal on the fulcrum."
- "When we did a toilet paper rope, I did not think the rope would hold a lot of weight, (which it didn't). So after it broke, we were thinking about how we could have done it better."

**Skill #2, Problem Solving:** Students were asked to respond to the following statement (see data in Figure 6.3):

**Problem-solving is a skill.** Problem solving refers to the process or steps that you use to come up with a solution. Do you think that your problem-solving skills have improved during the STEM course?

**Figure 6.3.** Student Responses to Question About Problem-Solving Skills

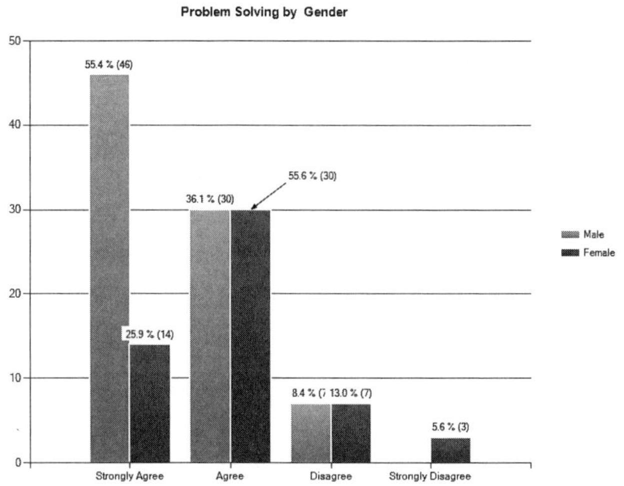


- 87.6% of those who responded placed themselves in the "strongly agree" or "agree" category for improved problem solving.
- 55.4% of males strongly agree that their problem solving has improved.
- No males placed themselves in the "strongly disagree" category.
- 92.2% of sixth graders placed themselves in the "strongly agree" or "agree" categories for improved problem solving.

Students were asked to provide examples of problem solving and when they believed they used problem solving during the STEM activities. The responses again provide a glimpse of the types of activities that students were asked to complete.

- "During almost all of the STEM projects, what students were doing had not worked the first time. During STEM we have learned that if we do not do it right the first time, we should always get back up and try again. One example is when we made parachutes. If it didn't fly the first time, then we cut it again and kept trying until it floated beautifully".
- "When your bridge fails—you try to find multiple ways to light your bulb when the one you have isn't working—when your sailboat sinks."

- "You will almost never get your first trial correct. Practice makes perfect. That's what I have learned."
- "My problem solving skills have improved. In STEM you experience tons of trial and error, which in the end, you end up solving your problem."

**Skill #3, Creativity:** Students were asked to respond to the following statement (see data in Figure 6.4):

**Creativity is a skill.** Creativity is the ability to move beyond the regular way of doing something to come up with a unique way of doing it. Do you think you became more creative during the STEM course?

**Figure 6.4.** Student Responses to Question About Creativity Skills

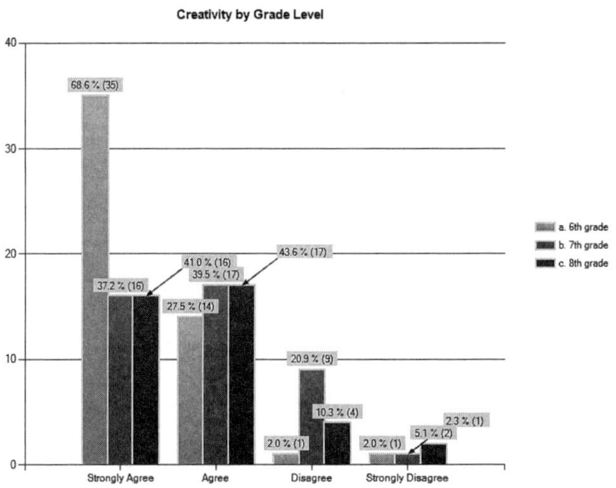


- 86.5% of the students perceived that they became more creative during the STEM course.
- 50.4% of the students strongly agreed that they became more creative. This is interesting because this was the only skill that we targeted for the study in year 1 of the program.
- 68.6% of all sixth graders placed themselves in the "strongly agree" category.

Students were asked to provide examples of creativity or when they believed they utilized creativity during the STEM activities. Again the responses provide a glimpse of the types of activities that students were asked to complete.

- "In every experiment because of the lack of things we can use we have to be very creative."
- "With the kites if you don't have creativity, you won't be able to make them because we don't get kits."

- "It may not be the best way of doing things, but the creative answer is always more exciting to test."
- "I would use different kinds of materials and would keep on working if I mess up. And, I would, add on, instead of removing it."

**Use of STEM skills inside and/or outside of school:** While completing the online survey, students were asked to strongly agree, agree, strongly disagree, or disagree with the following statement (see data in Figure 6.5):

Participating in the STEM class has strengthened my use of the following skills in other classes and/or outside of school: critical thinking, problem solving, curiosity, imagination, creativity, risk taking.

## **Figure 6.5.** Student Responses to Question About STEM Skills Inside and Outside School

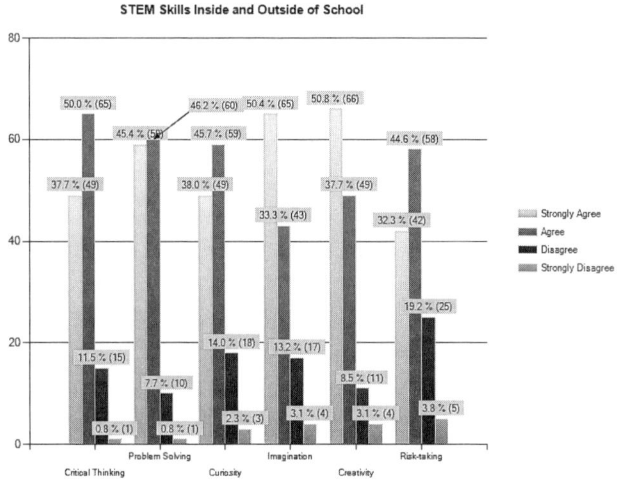


- 88% of students surveyed strongly agreed or agreed that participating in STEM has strengthened their critical-thinking skills inside and/or outside of school.
- 91% of students surveyed strongly agreed or agreed that participating in STEM has strengthened their problem-solving skills inside and/or outside of school.
- 84% of students surveyed strongly agreed or agreed that participating in STEM has strengthened their curiosity inside and/or outside of school.
- 84% of students surveyed strongly agreed or agreed that participating in STEM has strengthened their imaginations inside and/or outside of school.
- 88% of students surveyed strongly agreed or agreed that participating in STEM strengthened their creative skills inside and/or outside of school.

- 67% of students surveyed strongly agreed or agreed that participating in STEM strengthened their risk-taking skills inside and/or outside of school.

Voices of the participants involved in the STEM Program were also recorded as follows:

- *Instructor:* To the outside observer, STEM can appear chaotic but the best thinking, real thinking, creative thinking in STEM, occurs when students do not feel restricted by directions and limited by materials. The STEM class at Rosa encouraged a relaxed atmosphere. After the first week of class students recognize that this is not a grades-driven class. Once this transformation occurs, many students thrive in such an environment where they are encouraged to focus on the scientific processes, to come up with original ideas and to put these ideas into practice, without being penalized if the ideas are unsuccessful (Matt Cieslik, creator and teacher of STEM at Rosa).

The following quotes were taken from the end of the marking period reflections given by students. While one voice certainly does not represent an entire group, these comments were chosen because they are representative of the stated thoughts and feelings of many students.

- "I like in STEM that we got to build things that were fun to build with a group. STEM made me problem solve and think hard on how to complete the task." (eighth-grade boy)
- "I love that we don't just get told information. You put your thinking to the test." (eighth-grade girl)
- My favorite thing about STEM is the fun problems and activities we have to solve. I very much enjoy the combination of my two favorite classes—math and science. Every activity was unique and fun and was always a challenge for me." (eighth-grade girl)
- "I like that you get to use your ideas instead of following instructions." (seventh-grade boy)
- "STEM is a class that really isn't a class. Basically, STEM is a learning experience where you don't realize you are learning but you are." (seventh-grade girl)
- "I like being creative and exploring new ways of accomplishing tasks. STEM has given me the chance to do this more than before." (seventh-grade boy)
- "My first year of STEM was great and I can't wait for next year when the activities will probably get even more challenging, which will make STEM even more fun." (seventh-grade boy)
- "I like everything about STEM. Science, Technology, and Math are 3 of my favorite subjects. I really wish engineering was a class because I'd use it and it would be fun." (seventh-grade boy)
- "I like the mixture of science and engineering and math and technology in STEM." (seventh-grade girl)
- "When I first heard of STEM I thought we have to do math problems. But it turned out we didn't have to unless it helped us do better on the project." (seventh-grade boy)
- "The thing I like about STEM is that you can be creative and try new things. We can do a lot of stuff and take a lot of risks." (seventh-grade boy)

- "Doing these activities helped me think harder and made me take my time doing things when I get frustrated." (seventh-grade girl)
- "I liked that I had the freedom to explore several ways to approach each task. I also liked that I wasn't graded on success but instead, on my trials … I liked how we were allowed to express our individuality in each problem and how there is more than one way to complete each task." (seventh-grade boy)
- "What I liked most was the creative freedom. We were allowed to go out on our own to try to correctly and effectively find a solution to a problem. There were very few set guidelines or rules to follow. We were just allowed to express our own thoughts and individuality creatively. This is one of the very few classes where this is possible." (seventh-grade girl)
- "I like how STEM gives us a break from the very linear school day that is all about facts. STEM is very non-linear. STEM is like a wide, expanding valley, in which you are free to create anything you want (as long as you follow the rules!). STEM shows you that failing is a good thing. It shows that you can learn from your failures, and grow from them." (seventh-grade boy)

## Next Steps

The most obvious goal of Rosa's STEM program is to allow students to experience integration of science and math and to facilitate activities that reinforce integration. To this end more work must be done to strengthen the math-science connections. In some lessons the desired degree of integration is already in place, but in others the connection between the two disciplines could be stronger. Conversations between the STEM teacher and math and science teachers are necessary to ensure that progress continues.

The work being done in Rosa's STEM program needs to become more visible. It is possible that many science and math teachers understand the need for integration but lack the resources to introduce integration into their own classrooms. STEM can act as a conduit for the manner in which science and math can be taught. Rosa's STEM program can model activities for teachers who are interested in learning about integration as well as the other features of the program. While the exact lessons may not lend themselves to every classroom, the model itself may serve as a useful resource.

Several voices were tabulated to get a glimpse for what was yet to be tried. These included:

- In two short years STEM has risen from an idea to a class that is attended by each of the 840 students who attend Rosa International Middle School. Although there are some data that the format of the program may be beneficial to students, there are a number of questions, which continue to be investigated. Research must begin to focus on the effectiveness of the program in reaching academic proficiency. For example, while creativity is a valuable skill in and of itself, it remains to be seen if development of creativity leads to increased comprehension and improved skills for future work.
- Research on the emphasis of fun can also be conducted. Most would agree that if students can have fun during a class, it is a good thing. However, does fun in the STEM classroom

improve students' general academic performance? Continuing with the emphasis on skills, the teacher has had many conversations where students attribute their abilities to recall something that they have created or studied in STEM on the fact that students were encouraged to be creative and innovative. But, do students also possess long-term recall of the content that accompanied the memorable experiences?

- It has been stated that the creator/instructor of the STEM program believes that the skills that have been targeted in the research are valuable in life as well as in the classroom. If activities are designed to facilitate the development of these skills and if students use and develop these skills in STEM, then will students possess these skills as adolescents or even adults? In other words, did the experience in the STEM course cause long-term growth? Also, did students utilize these skills after they have left STEM?
- The first year of the STEM program coincided with a rise in standardized test scores in both math and science. The relationship between the STEM program and the test scores needs to be explored. If there is a link between standardized test scores and STEM, would the impact be greater or less if STEM were not limited to 23 class periods?
- Edward Canzanese, principal of Rosa, has stated that he would be interested to learn if the STEM program has any crossover influence in the way our students specifically approach math and science. If there is an increase in student achievement in either of those subjects, are there any relations between STEM and that success?
- Mr. Canzanese adds that if he were to be really critical of our program, it would be that it is delivered in relative isolation. If we could make improvements, it would be to tie it into both the math and science curricula. Before that could happen, we need to tie those two curricula together. Could the STEM program be the catalyst that leads to more science-math connections?
- Finally, Mr. Canzanese would love to see our teachers engage in more collaborative, self-directed, problem solving in their classrooms. Could the STEM program be the catalyst to make that happen? He is careful to note that we need more than the anecdotal evidence to move forward.

## Specific Ties to Other Reform Efforts

The STEM class at Rosa is designed to be a student-centered learning environment. The role of the teacher has changed from the traditional teacher-centered role where lectures are given and students are expected to follow directions. The STEM teacher is less vocal and less visible. The role of the teacher in STEM is comparable to being a coach: offering encouragement, making suggestions, recognizing efforts, and recapping what has been learned. In a course that values students having fun, learning by discovery, and the value of mistakes, the teacher has found more success when serving in this less authoritative role. The perception that the teacher is more effective in this less traditional role can possibly be attributed to the freedom that students have been given to think and create.

A typical STEM class will find the teacher moving from student to student as solutions and ideas are hatched. The instructor and students both gauge how much assistance is necessary to challenge the student while at the same time not pushing the student to the point where they are likely to give up in defeat. The conversations that take place between the teacher and each student are unique and tailored to the needs of that student. In essence, he/she encompasses the art of teaching in a student-centered classroom.

It is the hope of the creators of Rosa's STEM program that the student-centered classroom can lead to changed perceptions regarding STEM disciplines. Many traditional STEM classes follow a teacher-centered model where outcomes to activities are explicit and possibly announced beforehand. Exposure to this traditional model may reinforce an inaccurate perception that STEM-related disciplines must rely heavily on memorization and application of existing strategies to new problems. The STEM course at Rosa has attempted to take a contrary approach with the idea that solutions to problems are yet to be discovered.

## References

Anderson, L. W., D. R. Krathwohl, et al., eds. 2001. *A taxonomy for learning, teaching, and assessing: A revision of Bloom's Taxonomy of Educational Objectives*. Boston: Allyn and Bacon.

Bloom's New Taxonomy. *ww2.odu.edu/educ/roverbau/Bloom/blooms_taxonomy.htm*

Bronson, P., and A. Merryman. *Newsweek*. 2010. The creativity crisis in America. July 10.

Bybee, R. 2010. Advancing STEM education: A 2020 vision. *Technology and Engineering Teacher* 70 (1): 30–35.

Grigorenko, E., L. Jarvin, and R. Sternberg. 2009. *Teaching for intelligence, creativity, and success*. Thousand Oaks, CA: Corwin Press.

Louv, R. 2005. *Last child in the woods*. New York: Workman.

Millar, G. W. 2002. *The Torrance kids at mid-life: Selected case studies of creative behavior*. CT: Ablex.

National Research Council (NRC). 1996. *National science education standards*. Washington DC: National Academies Press.

Pink, D. H. 2005. *A whole new mind*. New York: Riverhead.

Pink, D. 2009. *Drive: The surprising truth about what motivates us*. New York: Penguin.

Seibert, G. 2011. STEM's effect on essential skills in middle school students. Unpublished Manuscript.

Sternberg, R. 1997. *Successful intelligence: How practical and creative intelligence determine success in life*. New York: First Plume Printing.

# STEM Challenges and Academic Successes

*Cheryl Frye, Melinda Jodoin, Terri Ladd, Shelly Muñoz, and Lisa Waller*
*Menifee Union School District*
*Menifee, CA*

*Taylor Predmore*
*University of California-Davis*
*Murrieta, CA*

*Robert Voelkel, Ed.D., N.B.C.T.*
*Menifee Union School District*
*Menifee, CA*

## Setting

Menifee Valley Middle School is a public school that is part of the Menifee Union School District in Riverside County, California. We are a union district with nine elementary schools and three middle schools. Our site enrollment averages around 1,000 students in a rural area, and we are on a modified traditional calendar. Our classes consist of sixth- through eighth-grade students. Our ethnicity is as follows: 46.9% Hispanic, 40.7% White, 4% African American, 3% Asian, 3% Filipino, and 3% Other. Our school demographics include 14.7% English Language Learners, 8.9% students with disabilities, and 43% of our students are socioeconomically disadvantaged. Student mobility is low.

Our school is in the sixth year of our Professional Learning Community model as defined by DuFour and Eaker (1998, 2008). Our staff has made great strides in improving student achievement as is evidenced by our state standardized test results. We have an overall student increase of 145 points over a six-year period; the highest gains in our district by 42 points. Our academic culture has been slowly moving in a positive direction, where students are proud of succeeding and strive to do their best.

Yet we continue to have a significant population of students who do not feel school is connected to the real world and do not want to apply themselves. We have implemented an intervention period where students who need additional support and time or refuse to complete assignments are given curriculum support in the needed content areas, but we want students to be engaged in learning and feel that what they are learning is applicable to their lives. A group of teachers with varied backgrounds felt strongly that a STEM (Science, Technology, Engineering and

Mathematics) curriculum would satisfy these needs. We applied and received a grant through the California Postsecondary Education Commission's Improving Teacher Quality State Grants Program to implement STEM elective classes for seventh- and eighth-grade students. We began with two seventh-grade classes and one eighth-grade class in the 2011–2012 school year and expanded to two seventh-grade and two eighth-grade classes in 2012–2013.

We focused on our English Learners and felt they would benefit from a STEM elective through in-depth and quality conversations, writing, and problem solving with inquiry-based projects where all subjects are integrated. Our district has 205 long-term English Learners (five years in American schools and maintain CELDT level 3 or below), and their significant gaps in academic vocabulary and background knowledge have the potential to be overcome by using language, writing, and oral skills with peers in a collaborative environment and high-interest STEM projects.

Our STEM team discovered that our school climate, as measured by student surveys, was in need of a variety of rigorous academic electives to ignite passions and encourage self-motivation before the students enter high school. Our school's academic climate is slowly rising, but we felt a STEM elective would propel our school and students to the top in readiness for a global economy.

## Overview

Our STEM program aligns with the national goals for STEM instruction by addressing many different facets placed fourth by the National Research Council on "Successful K–12 STEM Education: Identifying Effective Approaches in Science, Technology, Engineering and Mathematics" report (NRC 2011). The first goal addressed "expanding the number of students who ultimately earn advanced degrees and careers in STEM fields and broaden participation of women and minorities in those fields." By implementing a rigorous and integrated STEM program at our school, we are exposing all students to STEM careers, including engineering, food science, and computer programming. We specifically address this goal by requiring students to research STEM careers that would be relevant to the project in which they are engaged. This research is compiled within the project portfolio and ultimately shared with younger students during our tri-annual symposium. The portfolio consists of student work throughout the project, such as student research, scaled drawings, problem-solving notes, justifications for changes made, presentation notes, and reflections. The tri-annual symposium is where students have the opportunity to share their knowledge with everyone on campus. Two or three students present on the stage to an audience. The exposure to the vast number and types of STEM careers allows our students to explore advanced degrees in the future.

The next goal we addressed was to "increase STEM literacy for all students, including those who do not pursue STEM-related careers or additional study in STEM disciplines" (NRC 2011). One of the steps taken to meet our literacy challenge is for each student to write every day. Our students write journal entries about every 10 minutes in STEM classes, describing what they accomplished, the changes they made, and most importantly, justifying why they made each change. We began having students write more frequently because the students were unable to remember what they accomplished and why they did something day after day. Students were not used to thinking about their thinking, even though we had taught lessons and modeled metacognition and its role in critical thinking and problem solving. Our English Learners found this extremely challenging at the beginning of the process, but we assured them each day that we needed them to write even if words were spelled wrong or the grammar was incorrect. Once they saw we were not going to critique this part of their work, they became more expressive (examples) in their writing. Students became aware of their metacognition and were provided time to write these thought processes down in a journal. Over time, their writing became exceedingly complex, including growth in sentence structure and academic vocabulary because of peer review (initiated by the English Learner) and by practice. Student growth was measured by a rubric. The amount of discussion among students also played a role in improving their writing.

Our second step undertaken to meet the literacy goal consists of students reading technical and informational text and applying the knowledge to a different situation or problem. Students are required to research background information for each unit. We taught lessons regarding validity and reliability of websites, so students are educated about using the internet before research begins. Students used their background information to justify changes made in their project, therefore applying information in new and innovative ways. Toward the end of their projects, students researched 5–10 careers that may be involved in the project they just completed. They researched college requirements, future needs, salary ranges, and job duties. Again, students were reading informational text, which supports the Common Core claim of reading informational text closely and analytically (C1.1).

Our STEM program began five years ago when two teachers attended a STEM summer institute at a local state university. The summer institute consisted of lessons where educators were given specific objectives and then carried out the project. The last few days of the summer institute we observed instructors facilitating lessons with students from nearby public schools. We found this to be an extremely important motivator for us to implement our classrooms plans on a larger scale. We realized we had been conducting STEM-like lessons sporadically in our classroom, but now we had the impetus to move forward. Our backgrounds in scientific industries before we changed careers to teaching had us conflicted because we knew what students would need for careers in science, technology, engineering, and mathematics such as problem solving, conflict resolution, creativity and critical thinking, but we are constrained by state standards and set-pacing guides.

The two teachers began attending every professional development regarding STEM education that they could, networking and attempting to gain funding for the program they had created. Two years after our initial professional development, we applied for a grant through the California Postsecondary Education Commission (CPEC) and the California Department of Education (CDE) through the Improving Teaching Quality (ITQ) Program. At this point, we recruited teachers for our STEM team. Our team consists of teachers from each grade level and science, technology, and AVID. This combination of teachers and experiences proved to be vital in implementing our goals and objectives. The administration was crucial in organizing our classes to allow for us to teach STEM along with our other classes. We began with two seventh-grade classes and one eighth-grade class. Our grant was funded for two years and was a unique experience because it was composed of two components: one for student materials and the other for professional development. The professional development was unique in that it allowed us to choose the professional development we attended. As a team we felt we were weakest in the "engineering" component of STEM, so we attended the International Technology and Engineering Educators Association (ITEEA) conference. This was a pivotal conference for our STEM team because we realized that we had indeed been teaching the engineering design process through scientific experimentation. We felt the engineering design process was the "glue" that was going to meld all the subjects together. We wrote two units of our STEM curriculum that week at the conference. We were extremely excited and motivated. In hindsight, the grant offered us an advantage by allowing us to attend the conference together where we could attend different workshops, then come together, share information, and then create.

Our team met before the first year of implementation once a month during our Professional Leadership Community (PLC) for 20 minutes. After the first few months of implementation, most of our communication was through email, running into each other on campus, and during conferences.

Our goal regarding learners follows that of the National Research Council (2011) in that we intend to "broaden the participation of women and minorities in the STEM fields." We went from 90 students to 120 students enrolled in classes during year two. The administration was able to increase the STEM classes by one eighth-grade class in year two.

One of our goals was to open the STEM class to all students who expressed an interest; therefore, we had English Learners, GATE (Gifted and Talented Education), IEP (Individualized Education Plan), RSP (Resource Specialist Program) and Special Education students enrolled in the STEM classes. This was an important aspect for us because many of the STEM class examples and designs were within charter or magnet schools where students are not as heterogeneous as a public school classroom. Our first experiences with the students were enlightening. We had the advantage of having our STEM students in our regular subject classrooms as well, so we could see the difference in their behavior and motivation between the classes. As previously mentioned, our STEM elective classes include English Learners where the in-depth conversations, writing about experiences, and experimentation while they are kinesthetically involved in a real-world problem or project would improve their understanding and achievement. The integration that would be provided in the STEM electives would also enhance transfer of knowledge across the curriculum.

Students complete three to four projects per year. Our students are involved in many different projects, depending upon their grade level, since we are aligning them with *Common Core State Standards* along with the *Next Generation Science Standards* (*NGSS*). Two examples of our projects are building truss bridges and designing a new student backpack. We adapted ideas from many different sources to create rigorous, authentic, and integrated projects for our students.

Building truss bridges began with the students researching the different types of bridges, examining the direction of the forces placed on those bridges, and experimenting with digital programs where they could test their design, all aligned with the *NGSS*. Students then followed directions on building a truss bridge from manila folders (*http://bridgecontest.org/resources/vtutorial*). Students followed architectural design plans by using exact measurements and following technical directions to create the pieces of the bridge. Students used the *Next Generation* Science Engineering Design Standards along with the Standards for Mathematical Practice, including making sense of problems and persevering when solving them, using appropriate tools strategically, and attending to precision. Throughout the process, student writing included analysis and summarization of different media reports on bridge building, structural integrity and engineering practices. Once students completed their truss bridge, it was tested for structural integrity. Their final project was an oral presentation on the different types of bridges, the engineering challenges of building a bridge, and how they would redesign their bridge and the supporting justifications. Students would build their redesign and test.

Another example of one of our projects was designing a new student backpack. Students created a survey that they passed out to our student population asking what components students liked about their current backpack and what they wanted changed. Students researched the history of the backpack and why certain components were chosen over others. Students then designed a backpack based upon their survey results. They drew a scaled drawing, made a paper pattern out of butcher paper, and used this pattern to create their backpack out of fabric. We taught students basic sewing skills. Students either sewed by hand or used a sewing machine in class.

As we collected and analyzed data throughout year one, we realized that all students need to be exposed to the STEM program because we observed academic growth in our STEM students compared to those from the general population.

## Major Features

The reforms first addressed were how we would use STEM as a conduit for integration. We needed to align our STEM curriculum with our state standards for language arts, math, science, and technology. We felt this was important because the only way to quantitatively measure our STEM students' growth and affect was through state standardized testing. Our thought process regarding this was that we wanted to ensure that our STEM program would continue indefinitely. But we knew we would need student achievement data to keep the program and teacher resources. We finally used sixth-, seventh-, and eighth-grade standards in all subjects because our STEM curriculum was encompassing so many standards. We discussed the possible reasons why our whole school population repeatedly scored much higher on their state standardized testing in science versus all the other subjects. Based on research, we know students have a higher level of transfer when applying skills for a given purpose or goal. The eighth-grade science test requires students to apply their math skills to answer questions of speed, buoyancy, and density, yet those students would score surprisingly low on basic math skills but scored very high on the science portion. Our conclusions were that our science students continually use their math skills for finding answers and evidence during experimentation. Therefore, we wanted to enhance the transfer of knowledge through the STEM classes.

By using the engineering design standards in the *NGSS* and the engineering design process as the catalyst for each STEM unit we created, we easily were able to integrate the other subjects. We were committed to integrating all subjects into each unit because we repeatedly were exposed to STEM programs at various conferences (state, regional, and national) where projects were focused on one subject one week, then another subject the next week, and so on. We became very frustrated because this structure is not how real-world problems and challenges are met. We felt very strongly that a "true" STEM education would require the continuous and concurrent use of science, technology, engineering, and mathematics on a daily basis. The students need to use the integrated subjects in realistic ways and experience how they work in conjunction, not as separate entities. For example, when students were working on designing a fitness trail for our community park, they needed to consider the population age and fitness level. They needed to integrate how different fitness levels were challenged with specific equipment. Students used their research on aerobic and anaerobic exercise to determine the order of the equipment and the fitness level it would satisfy. Students were also challenged with the topography of the area, including elevation. Students calculated elevation and determined that at higher elevations exercise would be more difficult.

The next goal that we needed to address was student learning. Since the focus has been on standardized test scores, Depth of Knowledge 1 and 2, such as memorization, has been the most important requirement of most students. We use productive talk moves in the science curriculum on a daily basis and as students enter eighth grade, the application of knowledge becomes the focus. This transition from rote to conceptual learning is a very difficult one. The National Science Resources Center (NSRC) suggests "learning and teaching through guided inquiry, to replace rote memorization of textbook facts" as a critical element of science education reform. We knew this transition would be difficult as the STEM classes were implemented. The first students in STEM found the class very difficult, and in the first few weeks every student knew the STEM class was "really hard." This was a completely new way of thinking for students in that they were required to use creativity and problem solving along with justifying their decisions.

Student learning was the easiest goal because, after the initial increase in difficulty and frustrations with the high level of thinking required, we saw students begin to thrive in the STEM environment, specifically students that did not find the traditional classroom setting successful. We found our students resisted solving their own problems in the beginning because they continually were asking us if what they were doing was "right" or confirming with us whether or not this is what you do or asking us if what they were doing would work. We had discussed, as a team, how we would react to these questions because we ran into them repeatedly in science classes. We decided we would respond by asking students what they thought, and reiterating repeatedly what would they have to do to determine if something would work—they need to test it. Students would be very frustrated with these responses. Our team would determine if students had reached too high a frustration level, and we continually referred to our notion that these were skills the students would need in the future. They were unlearning years of teacher-centered learning, so we stuck with it. Eventually, the students began to depend on themselves and their peers through collaboration.

Another goal observed in our professional development experiences was that of integrating creativity. Students are given general parameters, but each problem requires creativity to solve. For example, when students are building wind turbines from recycled materials. The problem the students must solve is to lift a bucket tied to three meters of string off the ground with wind power. Students need to not only be creative with materials but with their turbine structure. Luckily, our team was highly creative and from previous industry experience was already using creativity in teaching science, math, technology and AVID classes, so this was a natural integration for us. Gardner (2006) describes creative people as those who take risks without fear of failure while seeking the unknown or challenging the status quo. We particularly embrace this definition because we found our students' "fear of failure" to be the most debilitating to the creative process. Cullen, Harris, and Hill (2012) state that "creative thinking thrives in environments that offer individual freedom, alternative thinking, safety in risk-taking, and collaboration and teamwork. Our STEM environment specifically provided for each of these skills, but it took time for our students to adapt to this new type of environment. Their fear of failure stopped them at many steps throughout each project, in particular when students were required to commit to a decision they made and progress toward the end product. They needed a lot of reassurance in the form of letting them know they can go back and change their decisions, as long as they justified their decision-making process. "Students with high self-efficacy are more persistent in their learning in the face of difficulties. They interpret failure not as a personal failing but as a single poor performance that can be overcome with hard work" (Cullen, Harris, and Hill 2012). "A student's sense of self-efficacy depends on his or her type of goal orientation: toward learning goals or toward performance goals" (Dweck and Leggett 1988).

Our STEM program exemplifies many different STEM features. We implemented project-based learning with integrated content across STEM subjects. Project-based learning is a student-centered instructional approach used to promote active and deep learning by involving students in investigating real-world issues in a collaborative environment (Krajcik and Czerniak 2014). Since we had observed many STEM programs that did not successfully integrate subject matter, we felt this goal needed to be a top priority for maximum exposure of real-world problems for our students.

The first project our seventh-grade students engaged in was the Fitness Trail Project. The foundation for this project was provided through the Maryland Public Television through Math by Design (*www.mathbydesign.com*). We adapted and expanded this project to reach our goals of providing complete subject integration and rigorous expectations for our students.

Students must find a solution for a given problem and design a fitness trail for a specific area with varied physical fitness levels and ages. Students followed the Science and Engineering Practices (SEP) ideas which include: (1) asking and defining problems; (2) developing and using models; (3) planning and carrying out investigations; (4) analyzing and interpreting data; (5) using mathematical and computational thinking; (6) constructing explanations and designing solutions; (7) engaging in arguments from evidence; and (8) obtaining, evaluating, and communicating information.

On days 1–3 students were asked to brainstorm 10 questions to address the following questions: What kind of problem am I facing?, What do I need to do?, and What are my limitations? Students were also given a copy of a timesheet for them to use.

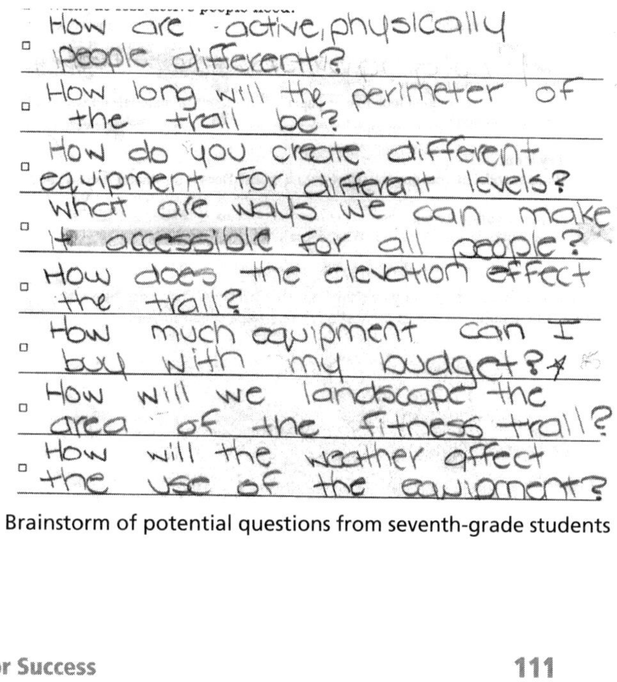
- How are active, physically people different?
- How long will the perimeter of the trail be?
- How do you create different equipment for different levels?
- What are ways we can make it accessible for all people?
- How does the elevation effect the trail?
- How much equipment can I buy with my budget?*
- How will we landscape the area of the fitness trail?
- How will the weather affect the use of the equipment?

Brainstorm of potential questions from seventh-grade students

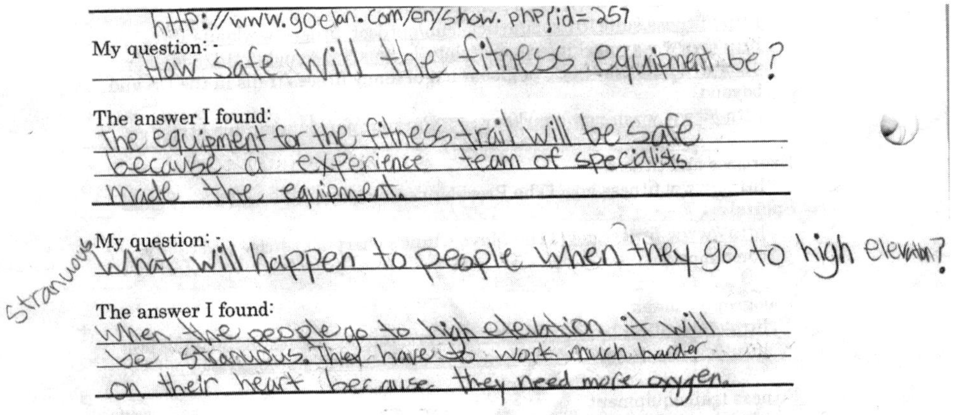

http://www.goelan.com/en/show.php?id=257

My question: How safe will the fitness equipment be?

The answer I found: The equipment for the fitness trail will be safe because a experience team of specialists made the equipment.

My question: What will happen to people when they go to high elevation?

strenuous

The answer I found: When the people go to high elevation it will be stranuous. They have to work much harder on their heart because they need more oxygen.

Student-generated research questions of a seventh-grade English learner

On day 4 students were introduced to topography. Students were given topographic maps, vocabulary definitions, and an activity to apply their new knowledge of topography. Questions that the students began to consider were about how the elevation would affect people exercising.

Days 5–9 were spent at the computers researching their 10 questions. At the beginning of each class period, students were reminded to fill out their timesheets, and students shared interesting websites, frustrations, and successes. Day 10 was dedicated to teaching students how to make a spreadsheet to organize their equipment and costs. Day 11 was spent teaching students how to draw to scale. They needed to transfer the small postage-size topography map to a larger grid. Students were required to use mathematical conversions, such as miles to feet. Students spent five days completing this process.

Students were asked to research the health and fitness components of the equipment they were researching. Students had difficulty narrowing down websites and information. Students were then provided with an outline, questions, and specific websites to assist with this challenge. The outline focused on target heart rates and the difference between cardiovascular, fat burning, aerobic, and anaerobic exercise. Students were also asked to justify why they chose each piece of equipment over another and communicate through writing. The project required 12–15 pieces of fitness equipment. Some students worked on the health and fitness components, while others continued drawing to scale. Students worked and earned points individually to this point. On day 17, there were six students still drawing to scale.

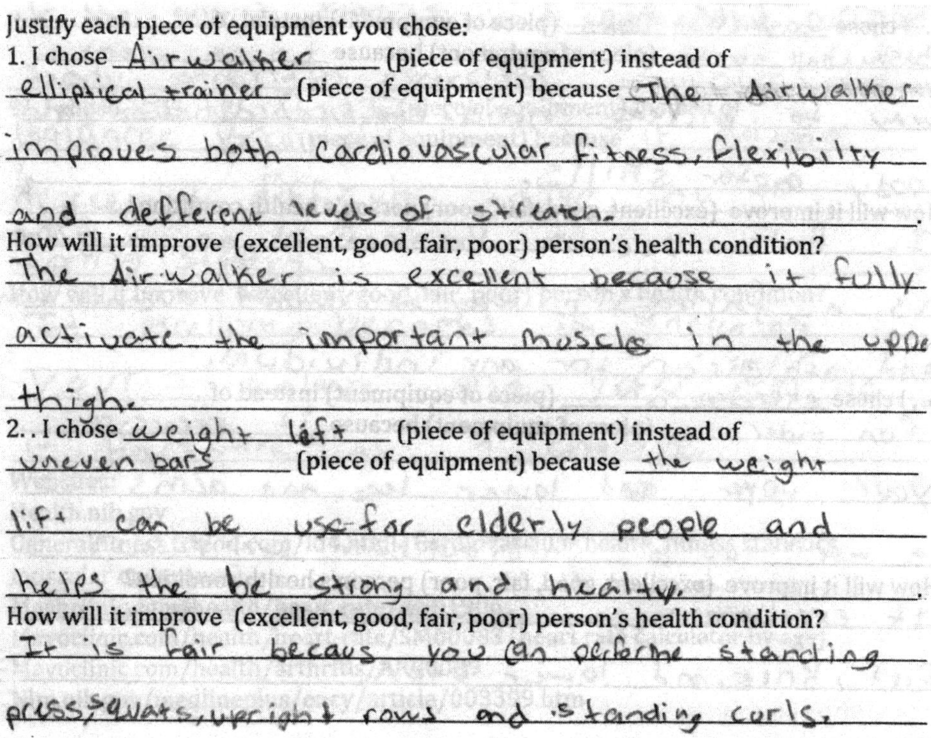

Justify each piece of equipment you chose:
1. I chose Airwalker (piece of equipment) instead of elliptical trainer (piece of equipment) because The Airwalker improves both cardiovascular fitness, flexibilty and defferent levels of streatch.
How will it improve (excellent, good, fair, poor) person's health condition? The Airwalker is excellent becaose it fully activate the important muscle in the upper thigh.
2. . I chose weight lift (piece of equipment) instead of uneven bars (piece of equipment) because the weight lift can be use for elderly people and heles the be strong and healty.
How will it improve (excellent, good, fair, poor) person's health condition? It is fair becaus you can perform standing press, squats, upright rows and standing curls.

Student justification from a seventh-grade English learner (early intermediate writing)

Days 18–20 were spent either drawing the topography map on the larger grid paper or researching exercise equipment. Students earned points for completing the health and fitness component and answering their original 10 questions at this point. On day 21 students were given large foam boards to transfer their topographic maps. Students needed to create an accurate grid to draw the contour lines. This became a challenge since none of them had been shown how to draw straight lines on a large space. We spent time demonstrating how to calculate how many millimeters apart each square needed to

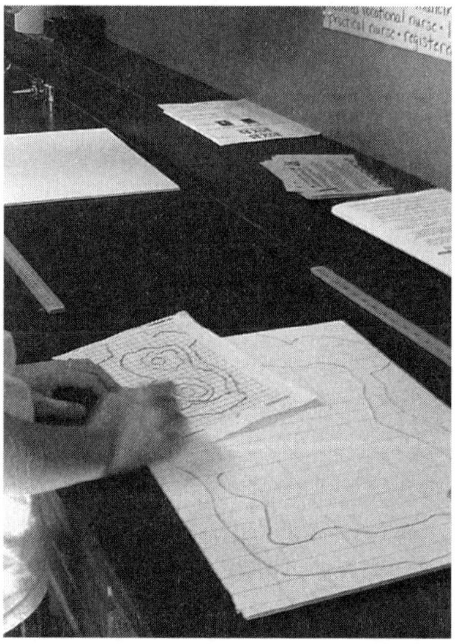

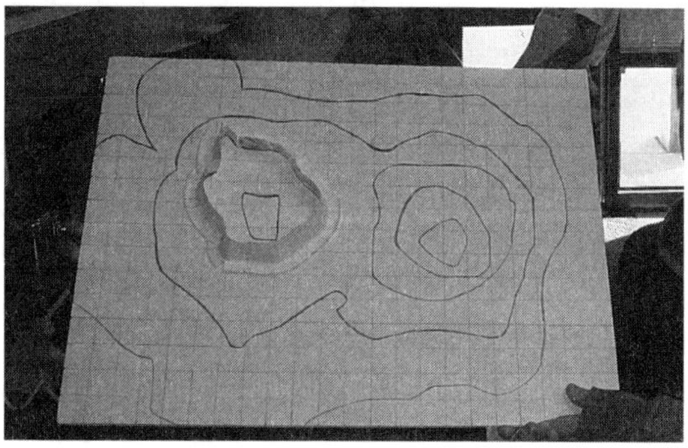

be for the number of squares needed in the space given. Students calculated the number of millimeters and we showed the class how to make "tick" marks on each side and then line up the meter stick to connect these "tick" marks.

Days 22–30 were spent transferring their topographic map to the foam board grid. Once students had completed this step, they began calculating the scale they wanted to use to build up the contour elevations. Students were given the contour line scale in feet, so they needed to find a realistic scale for their foam board model. Students used manila folders to create height for the elevations. Small cuts were made at the bottom of each folder so it could be folded around the contour lines. Clay was used to lay over the contours so the students had an accurate topographic model for their fitness trail. Students painted the clay after drying.

Days 31–39 were spent completing various tasks depending on each student's motivation and focus. Some groups were gluing contour elevations and painting, some groups were researching equipment measurements, and others began making their equipment. Students had to calculate the scale for their exercise equipment. Students made their equipment out of floral wire and flat, wooden stir sticks.

Day 40 was spent giving students the project rubric and requirements for the oral presentation and persuasive essay. For days 41–55, students continued working on their models, building equipment, and working on media presentations and persuasive essays. Students practiced their oral presentations in front of their peers before presenting to the school. Student presentations were given symposium style. The school population rotated through each student station using a checklist that the fitness trail students had to address in their presentation. The symposium occurred over three days in the multipurpose room.

The fitness trail project concluded with students answering reflection questions that required them to analyze their actions and skills. Questions included: What did you learn from this project? How would you do things differently next time? How can you use the skills from this project in other classes and your future? Name three things you enjoyed about this project and why. How will you use the skills you learned in other classes and your futures?

We chose two groups that had exceptional media presentations supported by excellent research that eventually proposed a fitness trail for a nearby park at our city council meeting.

Another STEM feature that our students engage in is authentic assessment and communication of STEM skills through various media. Our student assessment consists of daily conversations with each student as to where they are in the project and an overview of their journal entries to ensure that students are answering the "why" and justifying their decisions. We are very conscientious about focusing on the process, not specifically on the end product. We want our STEM students to be creative, use experiences and background knowledge to make decisions, problem solve on a daily basis, and justify choices, because these processes mimic most STEM careers. We realize that solely producing a product can be completed at home with adult assistance, but we expect our students to not only produce a product in class, but be entrenched in the learning process, which requires frequent reflection about their thinking (metacognition).

Another component of student assessment is through communication of learning to peers, younger students, and adults. Students have an option as to how they present their background knowledge, research, original design, decisions made, and final product design. Most students choose PowerPoint or Prezi, but are required to produce a script before they begin practicing their oral presentation skills. We are introducing the presentation skill of PechaKucha, which is a different way to communicate findings or a product to a group of people. Students have pictures for 20 slides and each slide is 20 seconds long, so the total presentation is six minutes. We are finding this very useful for our students because they tend to want to read verbatim off the slides in a regular presentation, but in the PechaKucha there are no words on the slide, so students feel there is a purpose to practicing their oral presentation. By presenting at the end of each project, students become more confident and show growth in their oral presentation skills. Since communicating evidence or data is crucial in all STEM careers, we are reinforcing skills necessary in the workplace along with integrating language arts standards.

## Evidence for Success

We collected data from the STEM students and the general population in both seventh and eighth grade. We had a total of 52 seventh-grade STEM students and 25 eighth-grade STEM students. We had 199 seventh-grade general population students and 118 eighth-grade general population students. The class consisted of various students: Gifted and Talented (GATE), English Learners, students with Individualized Education Plans (IEP), and students with Behavioral Support Plans (BSP). STEM is a yearlong class that meets one period each day. Student surveys were given on the first day of the STEM class and pre- and posttests were given in the students' core math and science classes in August 2011 and June 2012. The California Standards Test (CST) was administered in April 2012. Trimester grades were collected for math, science, and language arts. Our data was first organized and analyzed by the STEM teachers, but we needed further evaluation to ensure data reliability and validity.

### *Results*

We analyzed attendance data for the school year, comparing our STEM students to the general population for period 2. There was no statistically significant difference between the two populations for attendance.

One of our goals was to analyze how much growth our STEM students showed on the California Standards Test (CST) in mathematics, language arts, and science compared to the general population. We compared scores from the 2010–2011 school year and those of 2011–2012 for math and language arts. The CST for science is only administered in eighth grade, so we looked at the number of students that earned proficient and advanced scores, which is considered mastery of the subject. We looked at student CST scores because they are standardized and how states determine students are reaching proficiency. We needed to show that our STEM students were showing growth to continue our STEM program. Since we integrated language arts and math completely, we used these scores to determine student growth. We recognize that the CST does not measure our goal of improving critical thinking and problem solving at this point. It also does not measure the increased opportunities and experiences the STEM students engaged in throughout the year. Our data was analyzed using $p = 0.129$ and if alpha was larger than $p$, we determined it was significant. We used a 0.05 significance level for all calculations.

Our 2011 sixth-grade students who were then enrolled in STEM in 2012 as seventh-grade students demonstrated growth in language arts. The percentage of advanced students remained the same at 45.1%, while proficient students increased 5.88%, basic decreased 1.96% and below basic decreased 3.92%. Our school goal is to decrease the number of basic, below basic, and far below basic by moving students up at least one band.

Our 2011 sixth-grade students who were then enrolled in STEM in 2012 as seventh-grade students demonstrated growth in mathematics. The percentage of advanced students increased 7.92%, while proficient students decreased by 3.92%, basic decreased by 3.92%, below basic decreased by 1.96%, and far below basic increased by 1.96%.

Our 2011 seventh-grade students who were enrolled in STEM in 2012 as eighth-grade students demonstrated growth in language arts. The percentage of advanced students increased 17.39%, while proficient students decreased by 4.35%, basic remained the same, and below basic increased by 13.04%. We took those students that were enrolled in STEM as seventh-grade students and again as eighth-grade students for our 2011–2012 student data. As stated previously, we lose students to leadership and ASB classes during the eighth-grade year.

We have positive results in our data in the following areas: seventh-grade comparable (2011 LA vs. state; 2011 math vs. state; 2012 LA vs. state; 2012 math vs. state), eighth-grade comparable (2011 math vs. state; 2012 LA vs. state), seventh-grade STEM vs. non-STEM (2011 sixth-grade LA; 2012 seventh-grade LA; 2012 seventh-grade math), eighth-grade STEM vs. non-STEM (no positive results), seventh-grade girl comparable (no positive results), seventh-grade girl vs. state (2011 sixth-grade LA; 2012 eighth-grade LA), eighth-grade girl comparable (2012 eighth-grade science), eighth-grade girl vs. state (2012 eighth-grade LA), seventh-grade English Learner comparable (no positive results), and eighth-grade English Learner comparable (2011 seventh-grade LA).

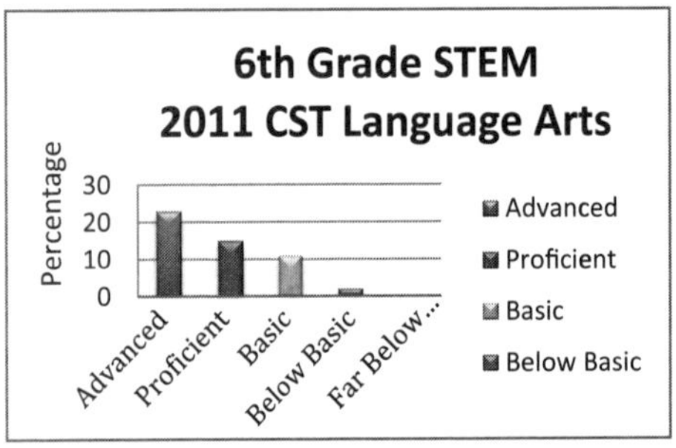


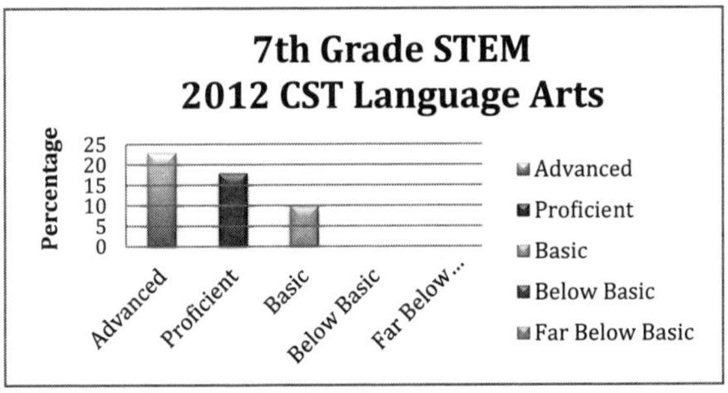


The above two graphs are showing growth from non-STEM students as sixth graders to STEM students in seventh grade in language arts.

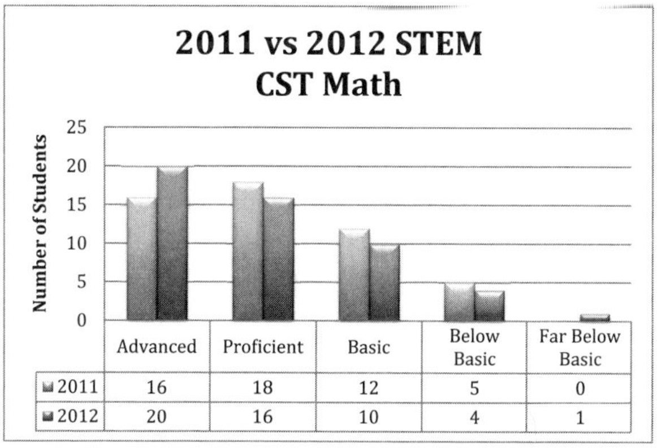


We compared CST Mathematics data from 2011 to 2012. Our STEM students were 34% in advanced/proficient band in 2011 and 36% in 2012.

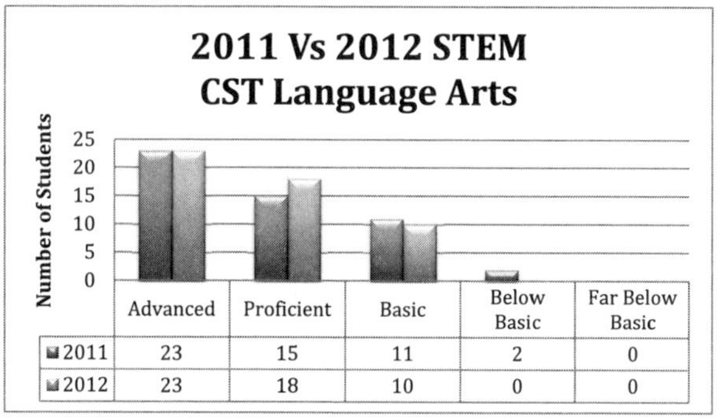


We compared CST Language Arts data from 2011 to 2012. Our STEM students were 38% in advanced/proficient band in 2011 and 41% in 2012.

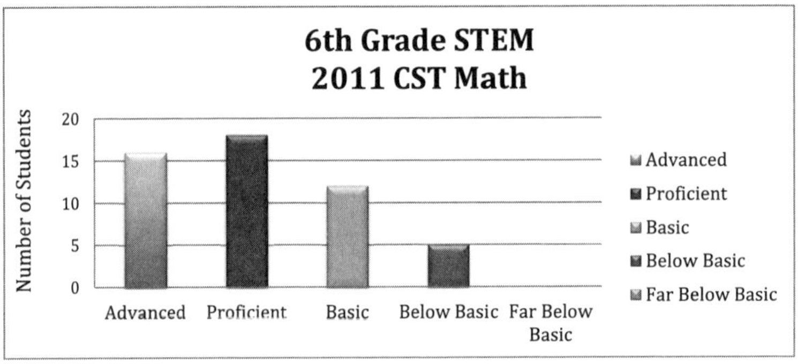


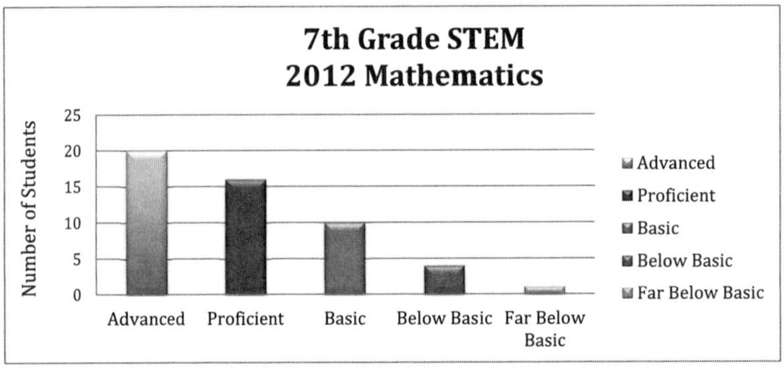


The above two graphs are showing growth from non-STEM students as sixth graders to STEM students in seventh grade in mathematics.

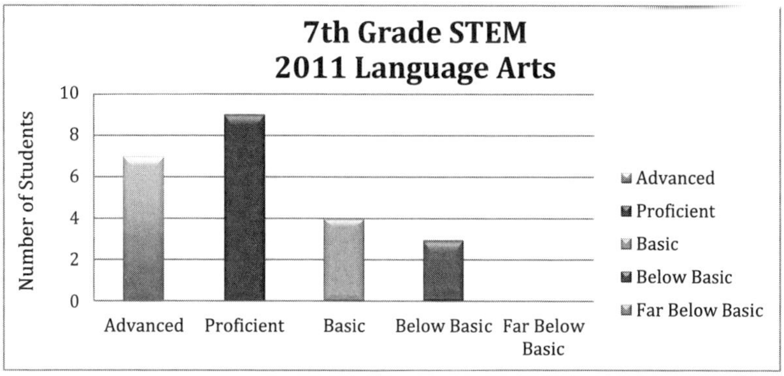


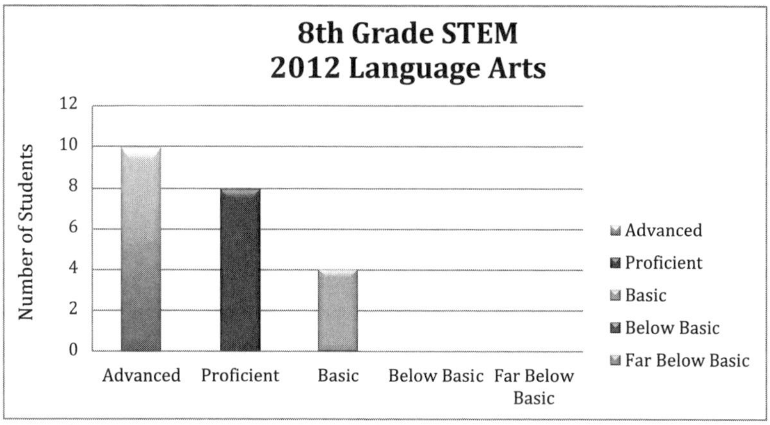


The above two graphs are showing growth from non-STEM students as seventh graders to STEM students in eighth grade in language arts.

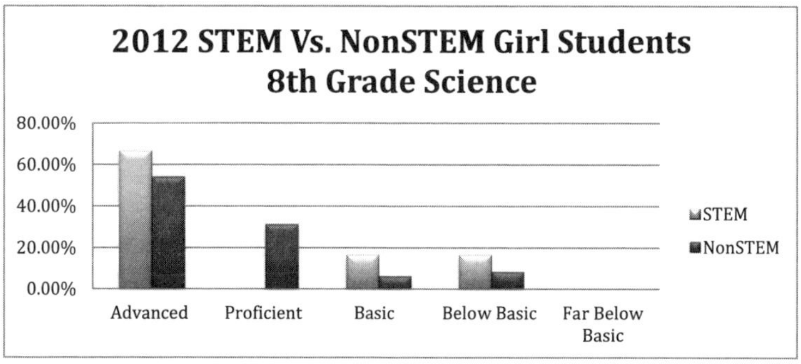


We looked at the girls that were enrolled in STEM for both seventh- and eighth-grade CST Science scores compared to non-STEM girls. There is a noticeable increase in the science advanced band, but there is a concerning gap in the proficient band with basic and below basic bands higher than that of the non-STEM girls. The STEM bar is zero at the proficient band.

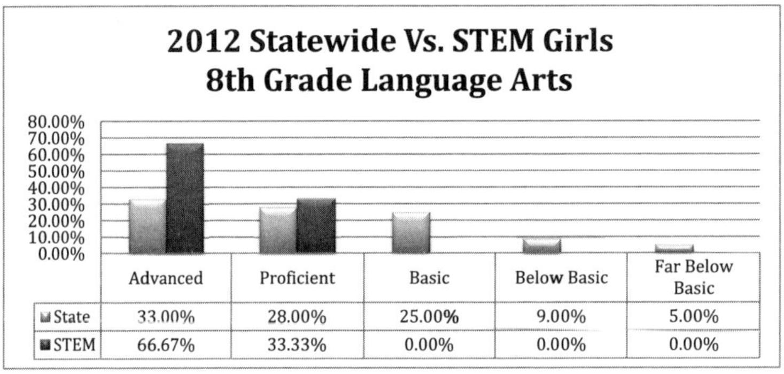


| | Advanced | Proficient | Basic | Below Basic | Far Below Basic |
|---|---|---|---|---|---|
| State | 33.00% | 28.00% | 25.00% | 9.00% | 5.00% |
| STEM | 66.67% | 33.33% | 0.00% | 0.00% | 0.00% |

We compared our STEM girls (girls that were enrolled in STEM as both seventh- and eighth-grade students) to that of state scores in Language Arts. We observed a percent increase result with 100% of the girls in STEM scoring in the advanced/proficient band while 61% of the state scored in the advanced/proficient band. We saw that three students increased their band and that none decreased in band. None of the girl students who were in STEM both as seventh- and eighth-grade students were in the basic, below basic, or far below basic. We also looked at the growth from seventh to eighth grade and observed that every student increased their score. The scores increased at least 22 points with the highest score increase of 110 points. We analyzed girls, specifically, who were enrolled in STEM both as seventh- and eighth-grade students. We saw a 16.67% increase in the advanced band, a 16.66% increase in the proficient band, a 16.67% decrease in the basic and below basic band. We also analyzed growth from seventh to eighth grade and observed a 22–110 point increase in their language arts scores.

We analyzed our STEM student compared to California student data. Our STEM students had a 16.1% increase in the advanced band compared to state scores, a 2.29% increase in proficient, a decrease of 4.39% in the basic band and a decrease of 10% in the below basic band.

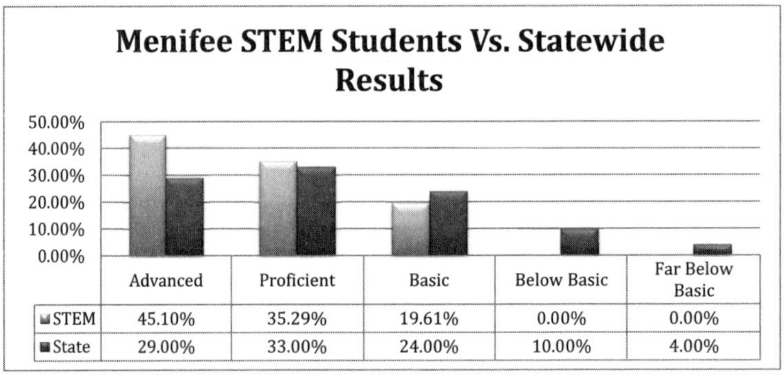


| | Advanced | Proficient | Basic | Below Basic | Far Below Basic |
|---|---|---|---|---|---|
| STEM | 45.10% | 35.29% | 19.61% | 0.00% | 0.00% |
| State | 29.00% | 33.00% | 24.00% | 10.00% | 4.00% |

## *CST Data 2011–2012*

We also analyzed the CST data for our English Learners in STEM: 33% of our English Learners demonstrated growth in math and 67% growth in language arts. Our additional English Learner data was found to be insignificant due to the low number of English Learners enrolled in STEM due to the double block of language arts that is required.

We analyzed data for our eighth-grade students taking the CST in science: 92% of our STEM students were in the proficient/advanced band and 70.3% of the general population was in the proficient/advanced band.

The goals that were targeted for students and instructors include providing an integrated STEM program to all students enhanced by professional development geared toward a highly rigorous and purposeful STEM curriculum.

Our students fill out an index card the first day of school answering the question: "What do you hope to learn in STEM this year?" The only background students have about STEM is the four-minute assembly presentation we have the previous year, so their understanding is limited at this point. Student answers included:

- "I want to get the skills to be able to become better in math and science. I hope that STEM will help in different subjects. I want the knowledge in STEM to become better at something different."
- "I want to get knowledge out of STEM that will help me in science, technology, engineering and math."
- "I want to learn great math skills to help me in the subject. I also want to start liking science to stay focused and entertained. I will be using technology in my future job of a music producer. If I need to know other things for my future job I would like to know."
- "I want to use my creativity to help me find out new things in technology by trial and mistake. I also think STEM will expand my creativity and problem-solving skills. I want to advance in math and science, which I sometimes have trouble with. Last, I wish to work on something to help me choose a career."
- "I want to learn about STEM because I want to go to college."
- "One skills I would like to get out of STEM is engineering because I want to be a military engineer."
- "The first thing I want to get out of STEM is to start building something and finish it."
- "I would like to change my feeling about math." These statements were taken from seventh-grade students.

At the end of the year, we hand the students their original index card from the beginning of the year and ask them to write if they learned what they hoped from the STEM class. Eighth-grade responses included:

- "Yes, I learned how to present my projects and the steps of designing projects"
- "This year in STEM I learned a lot of information about designing things. I learned why certain designs worked and why some didn't. I also learned how to apply the engineering design process to almost every problem I have. Finally, I learned how to efficiently modify my projects to fix any problem I have."
- "I accomplished everything I wanted and I practiced and challenged myself at what I couldn't do to be able to."

- "Yes, because my presenting things in front of people. Building things was what I want to be better at and I did it! This class did help me in math and my other classes."

Students are required to reflect upon their learning for each project, along with an essay (persuasive, narrative), which reinforces the integration of writing and language arts. One of the six questions we ask students to respond to in their reflections is: "What skills did you use during this project? Explain how or why." Seventh-grade answers included:

- "The skills I used the most during this project were skills I will always apply in my life. I only hope these skills will become better and stronger in the future. The skills I learned to use and make stronger were researching, creativity, and applying math skills to projects. I used these skills to find things that will work better than what I first thought to do or use. I used creativity to build, paint, and design the fitness equipment."
- "Some skills I used were reading skills, math skills, and science skills. We had to read the internet for information. We also had to use our math and science skills to find our scale and building mountains."
- "I think the skills I used most was math. We used math the most because we had to find out the scale. Also, we had to find out the cost of labor and everything else."
- "The skills that I used the most during this project were math because for mostly everything you had to multiply numbers and add them together, to get the scale, to see how much grass was. Also, adding up the cost of everything and figuring out how much recycled tires was."

At the end of year one, the five STEM instructors met and discussed our successes, challenges, and how we would improve our STEM program. We agreed on many aspects of the program that allowed us to improve student and teacher learning. We realized we are very good at discussing lessons on a daily basis during a few moments during passing period, e-mail, and after school. We were allowed to take one day to meet and look over the first student portfolios together. This time was invaluable, but we were unable to repeat it due to budget issues. We discussed our expectations based on the rubric we supplied the students at the beginning of the project. We collected all the portfolios for all grade levels and each blindly graded the portfolios using the rubric. After we each graded them, we looked to make sure that at least three of us had graded within the same grade band. This allowed us to establish a baseline for each letter grade for subsequent projects.

We determined that we really enjoyed the STEM format and most admitted that this was what we had hoped teaching would be when we started. We are truly facilitators in the STEM classroom. We thoroughly enjoy listening to the students solve problems and critically think with each other. The conversations we hear are at a high level because there is a purpose behind each project. Students feel each project is authentic; therefore they are committed to it. Every day, students come into the STEM classroom and immediately get to work. We each have specific students that have convinced us that we need to teach all students as we teach our STEM students.

A qualitative example is a student who was disengaged from school since seventh grade. After a couple of days, we saw this student was very capable, but was not willing to do any work.

Once this student realized the steps needed to complete this project, the student became actively engaged. After a few weeks, our team discussed using the STEM class to help motivate this student in the other classes the student was failing. We conferenced with the student, stressing the many abilities and potential we observed. We discussed with the student the future opportunities that were available as long as all the classes were passed. This student went from a 0.167 GPA to a 3.5 GPA. This student became more outgoing, worked hard, and self-signed up for tutoring every day at lunch as to not fall behind. We continue to observe students who do not thrive in the traditional classroom setting be successful in the STEM program. We continually revisit these situations because we believe we are reaching students that have become disengaged from school due to not making connections between school and real careers.

With five instructors as part of the STEM team, we provide checks and balances for each other, ensuring we meet our goals and objectives of the program. We each research, analyze, implement, modify, and discuss new projects and ensure we are integrating all subjects on a daily basis.

## Next Steps

Our next steps consist of continuing to improve our curriculum, attending professional development, sharing our experiences with other educators, and increasing the number of STEM classes. Recently, our district has come to us wanting to become a STEM district. We intend to read new research regarding STEM curriculum and integrate it with what we are observing in our classrooms. We need to continue with our professional development because it has been the foundation we built our program upon, and we have been sharing our experiences with teachers and administrators at state, regional, and national conferences. We intend to increase the number of students we reach through the STEM program with the end goal being every student enrolled in a STEM class.

## References

Cullen, R., M. Harris, and R. R. Hill. 2012. *The learner-centered curriculum: Design and implementation*. Hoboken, NJ: Jossey-Bass.

DuFour, R., R. DuFour, and R. Eaker. 2008. *Revisiting professional learning communities at work: New insights for improving schools*. Bloomington, IN: Solution Tree.

DuFour, R., and R. Eaker. 1998. *Professional learning communities at work: Best practices for enhancing student achievement*. Alexandria, VA: Association for Supervision and Curriculum Development.

Dweck, C., and E. L. Leggett. 1988. A social-cognitive approach to motivation and personality. *Psychological Review* 95: 256–272.

Gardner, H. E. 2006. *Multiple intelligences*. New York: Basic Books.

Krajcik, J. S. and C. M. Czerniak. 2014. *Teaching science in elementary and middle school: A project-based approach*. New York: Routledge

National Research Council (NRC). 2011. *Successful K–12 STEM education: Identifying effective approaches in science, technology, engineering, and mathematics*. Washington, DC: National Academies Press.

NGSS Lead States. 2013. *Next Generation Science Standards: For states by states*. Washington, DC: National Academies Press. *www.nextgenscience.org/next-generation-science-standards*

Yager, R. E. 2012. *Exemplary science for building interest in STEM careers*. Arlington, VA: NSTA Press.

# Middle School Engineering Education

*Angelette M. Brown*
*Columbia Heights Public Schools*
*Columbia Heights, MN*

*Gillian H. Roehrig*
*University of Minnesota*
*St. Paul, MN*

*Tamara J. Moore*
*Purdue University*
*West Lafayette, IN*

## Setting

Columbia Heights is a small community located just north of Minneapolis, Minnesota. The school district, which includes a total of 3,000 students across its three elementary schools, one middle school, and one high school, has undergone many changes over the last 20 years. Its once almost all-white student body has changed to reflect a more diverse student population with 75% of students qualifying for free-and-reduced lunch, the highest percentage of any school district in the Twin Cities metro area. The focus of this chapter is Columbia Academy, the district middle school, where the student population is 33% Black/Somali students, 33% White, 25% Hispanic, 6% Asian, and 3% American Indian; 23% of students at Columbia Academy are English Language Learners.

Corresponding with changes in the student population, the district began to notice a drop in student engagement and learning, as evidenced by decreases in academic scores, especially in the areas of mathematics and reading. Hand-in-hand with declining student achievement was the loss of students to neighboring school districts. Open-enrollment policies in Minnesota have forced districts to evolve and provide programs aimed at retaining students, and additionally draw in students from neighboring districts. For Columbia Heights, the competition included two neighboring school districts that already offered International Baccalaureate programs. The district administration wanted a program that would set them apart and get families to *want* to come to their schools. The challenge for the district was to find a way to update the curriculum to re-engage students, as well as providing the district an edge in the education market. Paralleling national education reforms and documents (NRC 2012), we chose to center on STEM, with a focus on integrating engineering into our middle school course offerings.

## Focus on STEM

National reform documents (NRC 2007; President's Council of Advisors on Science and Technology 2010) call for the United States to prepare more students with a strong STEM background in order to be competitive in a global society, claiming that economic growth and national security are at risk. Another common theme within these documents is the importance of educating scientifically and technically literate students who possess 21st-century skills including problem solving and critical thinking. However, while providing compelling rationales for the improvement of K–12 STEM education and the development of STEM schools, these documents provide minimal direction for achieving these goals.

Ironically, with such intense national focus on STEM there is a lack of clarity around the meaning of STEM. Bybee (2010) notes that "for most, it means only science and mathematics, even though the products of technology and engineering have so greatly influenced everyday life" (p. 996). He goes on to argue that "a true STEM education should increase students' understanding of how things work and improve their use of technologies" (p. 996). Proponents of K–12 engineering also argue that engineering provides a real-world context for learning science and mathematics, and question the isolation of STEM subjects in schools, noting that "in the real world, engineering is not performed in isolation—it inevitably involves science, technology, and mathematics" (National Academy of Engineering 2009, pp. 164–165). Researchers argue the following benefits for the inclusion of engineering into K–12 classrooms (Brophy, Klein, Portsmore, and Rogers 2008; Hirsch et al. 2007; Koszalka, Wu, and Davidson 2007):

- Engineering provides a real-world context for learning mathematics and science.
- Engineering design tasks provide a context for developing problem-solving skills.
- Engineering design tasks are complex and as such promote the development of communication skills and teamwork.

Indeed, engineering design is a prominent feature of *A Framework for K–12 Science Education* (NRC 2012). The framework authors argue that,

> *the major goal of engineering is to solve problems that arise from a specific human need or desire. To do this, engineers rely on their knowledge of science and mathematics as well as their understanding of the engineering design process. (p. 27)*

While no single engineering design cycle exists, it is agreed that the design process is "both iterative and systematic" (NRC 2012, p. 46) and there are common characteristic steps that include:

- identifying the problem and defining specifications and constraints;,
- generating ideas for how to solve the problem,
- testing of potential solutions through the building and testing of models and prototypes,
- analyzing how well the various solutions meet the given specifications and constraints, and
- evaluating what is needed to improve the leading design or devise a better one.

The specific approach we selected for the middle school engineering program and the writing of engineering challenges and activities is illustrated in Figure 8.1. In the following sections, we highlight the use of this engineering design cycle in our middle school STEM program and provide successful examples implemented at Columbia Academy.

**Figure 8.1.** An Engineering Design Cycle for Middle School STEM Courses

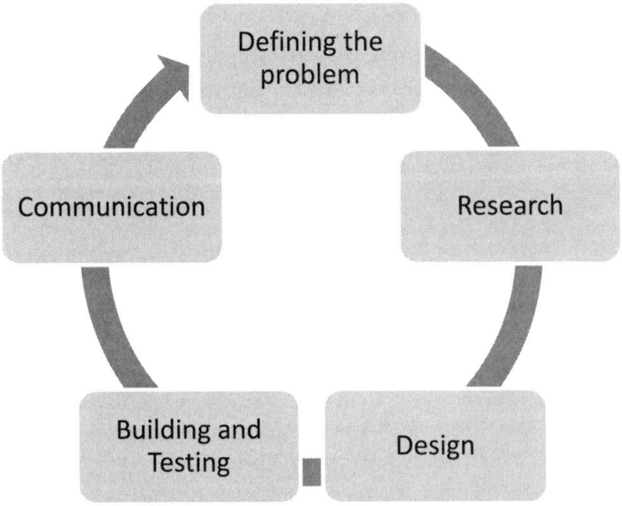


## An Approach to Middle School STEM Curriculum

Quality STEM curricula should have a motivating and engaging context. Students need to have personally meaningful contexts that provide access to the activity (Brophy, Klein, Portsmore, and Rogers 2008; Carlson and Sullivan 2004; Frykholm and Glasson 2005). For example, connecting to current events—when an earthquake occurs or an airplane pilot demonstrates supersonic flight, we are provided with the perfect opportunity to show our students the relevance of what they are learning. Engineering is a way of thinking used to solve problems for a purpose. In engineering approaches to STEM instruction, relevance and purpose is incorporated through the use of a client. In our STEM classes, student teams operate as engineering companies with a client in need of the student's problem-solving expertize.

To be both meaningful and worthy of inclusion in a middle school STEM course, the content must drive the learning objectives of the activity (Fortus et al. 2004; Harris and Felix 2010; Mehalik, Doppelt, and Schunn 2008). Planning for our STEM activities begins with determining the science content standards we need to teach. After selecting the specific science standards and learning goals, we look for a local or global news story that highlights the science content. And with this context in mind, we develop a background story for a client who needs a product or process designed for a specified purpose. Within the story the client provides constraints, such as due dates and budget, that student teams must follow. Finally, we review the context and design challenge for specific mathematics standards and skills that are needed to complete the engineering design challenge, focusing on the application of mathematics in an authentic setting. The final layer of the story is technology, which can be as simple as the use of hot glue guns up to CAD programs and Vernier sensors.

## *STEP 1: Defining the Problem*

Step 1 of the design cycle has several learning objectives:

- Students will understand how to be an effective team member.
- Students will be able to identify and choose information needed to solve the problem.
- Students will understand and apply vocabulary and concepts needed to solve the problem.
- Students will manage their time by laying out the first three steps of the engineering process in a calendar.

To give the students a feel of authenticity, we start each new class by hiring the students as engineers for our company, Academy Engineering. We go over a contract with them that covers the job description, a company Code of Conduct, and safety protocols. After each student has signed the contract, they are placed into company teams. Within their teams students review the team jobs and agree on who will cover each position (see Figure 8.2). The students have an opportunity to do three of the four team jobs over a semester course. Throughout the curriculum we use the design activities as a learning opportunity for building teaming skills. At the end of a challenge students are often able to identify how poor teaming negatively affected their work or will feel a great sense of pride that they were able to overcome their personal problems to go on and successfully complete the challenge.

**Figure 8.2.** Teams of Students Are Composed of Four People, Each With a Different Job.

**Team Jobs**

**MANAGER:** Make sure everyone has a job. Watch the clock. Make sure the team stays on schedule. Assist with building. Assist with marketing. Act as accountant whenever that student is absent.

**ACCOUNTANT:** Purchase supplies, manage the budget sheet, and calculate discounts, and assist with building.

**ASSEMBLER:** Get supplies from the metrology station, double check measurements, make sure the building is done correctly.

**DESIGNER:** Complete detail design for client, make changes to detail design as needed. Help with building. Making the PowerPoint, business card, and company sign for the marketing piece of the challenge.

Once the teams are established, students are introduced to the client and the challenge. Students are given time to read and highlight the information needed to meet the client's needs. Students record the client information in their engineering notebook to summarize the important factors for consideration in addressing the challenge (see Figure 8.3). Engineering notebooks are a common tool in undergraduate engineering courses to document research, data, design drawings, and so on. We use engineering notebooks as a tool for students to record all steps of the design process and a tool for teachers to assess and monitor student learning. As a class, we discuss the key components of the challenge and generate a list of clarification questions for the client. This also provides students with an opportunity to review and clarify content vocabulary included in the challenge and provide the teacher with critical information on students' prior content knowledge.

**Figure 8.3.** Engineering Journal Entries Keep Students Focused on the Challenge and Constraints.

**STEP 1**-What am I being asked to do?

- ________________________
- ________________________
- ________________________
- ________________________

**Constraints**

Time: ____________________

Budget/Materials: __________

Finally, students are asked to complete a calendar for completing the design challenge. Each challenge takes approximately four weeks and a calendar helps students to break down the task into subtasks and address time-management issues (see Figure 8.4). The client information and calendar are part of the students' engineering notebooks. We focus on helping students to understand that their notebook is not only a record of their work but also part of the product being developed for the client.

**Figure 8.4.** Each Team Works Together to Create the Timeline for Building and Presenting Their Product.

## *STEP 2: Research*

Step 2 of the design cycle has several learning objectives:

- The students will learn to cite books and websites.
- Students will learn mathematics and science concepts and vocabulary needed to solve the problem.

During step 2, student teams determine areas of research needed to help create their final design. Students record their research in their engineering notebooks, including a minimum of three different research sources, before moving on to design phase. Each research section contains notes, a visual reminder, and a citation. The examples in Figure 8.5 (p. 130) are from challenges on designing a prosthetic limb and a bridge challenge. In addition to taking notes, students are encouraged to consider the materials they might use, how they might build the pieces, or additional questions they may have.

**Figure 8.5.** Examples of Student Research Notes, Visuals, and Citations From Engineering Notebooks

| Research Notes | Visual | Reference |
|---|---|---|
| prosthetic leg sockets need to be cushioned to protect<br>measure from knee to floor<br>ankle joint moves in flexion and extension | 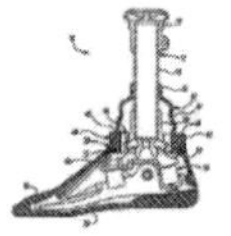 | *freepatentsonline. comfreepatentsonline.com*<br><br>Feb. 22, 2012 |
| Cables between the tower and anchorage.<br>Deck is suspended from cables by smaller cables<br>NOTE: what's an anchorage?<br>I could make my cables out of string or wire. | 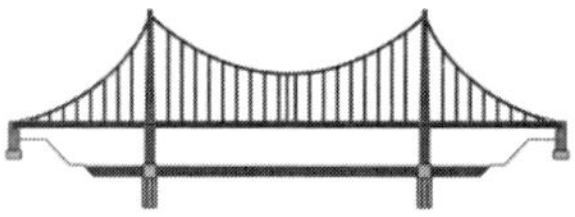 | *eduweb.com*<br><br>March 5, 2012 |

Although students do a great deal of independent research, this is also the perfect time to teach a "just in time" lesson. The challenge has engaged the students and now they are ready to learn the content needed to solve the problem. There will be plenty of opportunities for such lessons throughout the design challenge, so do not frontload all of the content. It is important to be prepared for content activities but the timing is critical—the content should be introduced when the students need it to solve a problem.

## *STEP 3: Design*

Step 3 of the design cycle has several learning objectives:

- Students will apply the information gathered during research to design solution ideas.
- Students will learn how to avoid plagiarizing existing products and processes.
- Students will learn how to draw to scale.

During the first stage of the design phase, the students use their research from step 2 to help them sketch product ideas or thumbnail sketches. The purpose of the thumbnail sketches is to flesh out ideas and perfect an idea. For example, our sixth-grade class was challenged to design and build furniture for dolls for a woman who provided free daycare for single working parents. The students had already researched commercially available furniture in step 2 and documented their work in the research section of their engineering notebook (see Figure 8.6). At this stage, copyright laws were discussed and it was stressed that to sell their product they must create an original idea.

**Figure 8.6.** Examples of Students' Step 2 Research for the Design of Doll Furniture

| RESEARCH NOTES | VISUAL | REFERENCE |
|---|---|---|
| Seats 3 dolls<br>Cushioned back, arms, and seat<br>Appears to have 9 legs<br>I could have a round, thin tube in the middle to glue the arms to. |  | *casasugar.com*<br>March 12, 2012 |
| Curved back<br>Pillows<br>Legs and couch look like wood<br>Covered in material<br>I could cut the shape of the back out of thick cardboard. I could paint the couch and make the pillows out of material | 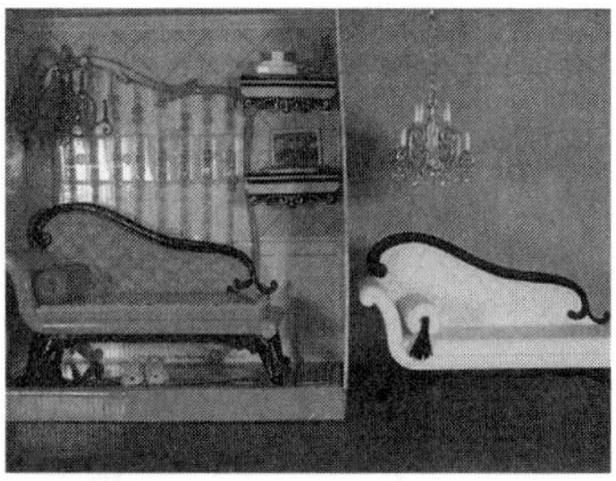 | *flickriver.com*<br>March 12,2012 |
| Made of wood and small chains<br>Need a lot of room for the legs to spread out<br>The seat looks thicker than the rest of the swing<br>I could use dowels for the frame, cardboard for the chair and string instead of metal chains |  | *etsy.cometsy.com*<br>March 12, 2012 |

In the design example shown in Figure 8.7, the student used elements from their first two researched products (shown in Figure 8.6) to design their own chair. On the right side of each design box, students are asked to reflect on their designs. The Materials box prompts student to think about both the physical properties and costs of the materials they will need. In the

**Figure 8.7.** Example of a Student's Individual Engineering Notebook Showing Possible Design Solutions to Present to Their Team

| | |
|---|---|
|  | **Materials** card board, box cutters, hot glue, material, brown paint for legs & arms, needle & thread |
| | **Design Strengths** Interesting design just like the client asked for |
| | **Design Weaknesses** curved legs & arms would be hard to make out of card[b] |

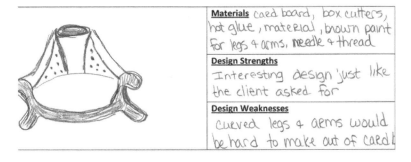

Design Strengths section, students are prompted to determine if their proposed design meets the client's needs. Finally, the Design Weaknesses section asks students to carefully consider their proposed design. This section is probably the most important section, yet it is the section most often skipped over by students. Many design flaws can be identified before construction of the product and problems pre-empted.

Students create three thumbnail sketches using different design elements from their research to create their own unique design solutions. It is important to check the students' work before they move on to the detail design phase. At this stage, the client also provides feedback before students move to the next step, again reinforcing that the chosen design must meet the client's needs.

Teamwork is an essential engineering skill, thus at this critical decision point the students are required to have a team meeting before moving on to the final stage in the design process. During the team meeting students present their thumbnail sketches to each other and discuss the pros and cons of each design, as well as the comments from the client. Often students will blend their ideas together to create an entirely new design for their client. Once the team has decided on a final design, the students begin stage two of this step: detail design.

During the detail design stage, the focus shifts from generating ideas to fine-tuning the details for the final design, including measurement and materials. Although we use a similar lesson on accuracy and measurement for all middle school grades, there are different needs at each age level. In sixth grade, the students struggle with three-dimensional objects. Having the doll in their hands to both measure and fit the furniture helps students to visualize and build in three dimensions. The focus is on the accuracy of the measurements rather than understanding how to design or build to scale. Each student is required to draw the detailed design in their engineering notebook; however, once the team starts building, it is the lead designer who makes a final copy of the detail drawing for the client (see Figure 8.8). Any changes made during the building process must be noted in the detail design. For each change, the engineer is required to circle the change, note the date of the change, the reason for the change, and initial to show who made the decision. Applying such small details goes a long way in providing a feel of authenticity for the students. It also holds the students accountable for their designs.

**Figure 8.8.** Engineering Teams Create a Final Scale Drawing for the Client

In seventh and eighth grade the students are more developmentally ready for the concept of scale. The products in the seventh grade tend to be larger pieces, such as adult-size chairs and models of entire rooms (see Figure 8.9). This allows the students to begin measuring items around the room such as doorways and tables, giving them a better understanding of actual room dimensions and size of items. For our seventh graders, it has been easier for them to do scale conversions if they measure in feet first. The scale in seventh grade is usually ½" = 1'. Once the students have their measurements in feet for their detail design, they are able to easily convert the feet to inches, especially if given ½" graphing paper. It helps to purchase examples of furniture in scales of 1/12 and 1/24 to give the students an understanding of the size they are building to. By eighth grade, however, we have found the students are ready to both draw and build to a scale. For this age group the students are given a scale of ¼" = 1' and use ¼" graph paper as they draw. After the detailed drawing is finished, the students list the necessary materials and tools needed to build the actual product.

**Figure 8.9.** Scale Model of an Earthquake-Safe Tower

Budgets are a critical feature of an engineering design lesson and also an opportunity to reinforce some mathematical concepts. Each project involves a budget for both materials and tools (see Figure 8.10). A large array of tools are available in the storeroom and students include the costs to purchase or rent equipment. Hardware materials range from cardboard (our primary building material), paint, paintbrushes, craft sticks, hot glue guns, construction paper, and any other material you might find in your backroom storage supply cabinet. Originally we rented items such as rulers and metersticks, but we found the students were more than willing to give up accuracy in measurements to save a few dollars. All measurement tools are now kept at a metrology station where a student is in charge of checking in and out items such as protractors, drawing compass, calipers, rulers, and measuring tapes. To reinforce the concept of percentages, students are required to calculate the taxes for each purchase and coupon discounts that are made available throughout the unit. As teachers, we feared the budgeting and purchase of materials added an unnecessary stress level, but the students found this to be a fun and worthwhile component.

**Figure 8.10.** Sample Budget Proposal Sheet, Included in the Student's Notebook

| Budget Proposal Sheet | | | |
|---|---|---|---|
| Categories | Items | Quantity | Price |
| Materials | | | |
| | | | |
| Glues Tapes | Items | Quantity | Price |
| Screws Nails | | | |
| | | | |
| Tools | Items | # of Days | Price |
| | | | |
| | | | |
| | | Projected Budget Needs<br>Tax Included: 0.0725 | $ |

### *STEP 4: Building and Testing*

Step 4 of the design cycle has several learning objectives:

- Students will learn how to test and revise their products.
- Students will maintain their engineering notebook.
- Students will apply vocabulary learned in discussions and notebook writing.

**Figure 8.11.** Students Learn to Use Basic Tools.

Before students begin building, it is important to teach tool safety for box cutters, hot glue guns, handsaws, and hammers (see Figure 8.11). It was also important in establishing our engineering program to have administrative support if a student needed to be removed from the class for safety reasons. We also use eighth-grade teaching assistants in the classroom to support students in the earlier grades.

Prior to building, students complete the second half of the challenge timeline, outlining a schedule from the client due date and working their way back to the current date. For any given building phase, students will need time for building a prototype, measuring product components, sanding, cutting, assembly, testing to ensure the product works, and painting. We also schedule time for marketing their work, described in more detail in the next section. If students fall behind schedule, they are required to communicate in a letter to the client why they are behind and the solution(s), which may include pushing back the due date or altering the product. A day or two later, we will tell the students the client's response, which might range from accepting the students solutions or providing a counteroffer. The students then use the information to rewrite their timeline.

During the building phase, students first make a prototype of their product. The purpose of the prototype is for the students to identify design problems before moving to large-scale construction. For example, Figure 8.12 shows prototype and final adult-size chair designs. Students were expected to apply knowledge of human anatomy for sizing the chair and understanding of force and tension on the structure when a person sits and gets up from the chair. Initially, students will complain about the prototype phase, as they want to jump immediately to constructing the final product, but as they start to see how many redesign issues are revealed during prototyping they will come to value this important step. Providing feedback during the prototyping step is critical to encourage better

**Figure 8.12.** Example of Prototype to Final Product

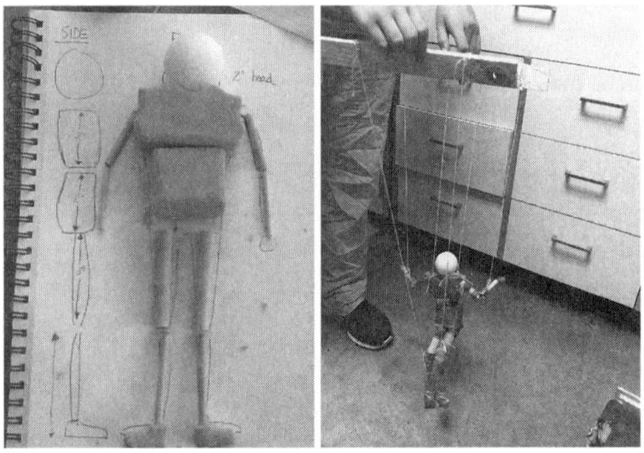

measurement and building skills. If possible, also ask the client to come in to provide feedback. Evaluating the products is very stressful for the students. They are either very proud of their work and will fear criticism, or they are embarrassed to show you their product. It is important to begin with what looks good before moving to areas that need improvement.

Throughout the building process, watch for opportunities to either reteach or bring in additional lessons. I try to begin each class with a 10–15 minute lesson related to mathematics or science content. This step is also a good time to introduce technology that can be used to test the various products. For example, we have successfully used Vernier sensors, earthquake simulators and maglev tracks.

## *STEP 5: Communication*

Step 5 of the design cycle has a couple learning objectives:

- Students will learn how to communicate their products to their client.
- Students will learn how to market their product to consumers.

**Figure 8.13.** Product Fair Display

Communication is a key skill in engineering and highlighted in both our state and national science standards (Minnesota Department of Education 2009; NRC 2012). We have developed a variety of mechanisms for verbal and written communication, as described below.

### Product Fair

A Product Fair involves students showing their work to other students in the school and sometimes upper elementary students from neighboring schools. The display includes the company name and logo, business cards, and a PowerPoint commercial to market their product (see Figure 8.13). As a marketing strategy to draw attention to their table, some students also bring candy or snacks. Teams have also brought in tablecloths and strings of holiday lights to wrap around their display tables. Student duties during the product fair include a greeter to shake the consumer's hand and offer them food or drink, engineers to show off the strengths of the product, a marketer to show the commercial, and an end-greeter to pass out business cards and thank the consumers for visiting the booth.

**Figure 8.14.** A Student Models Prosthetic Limb Design.

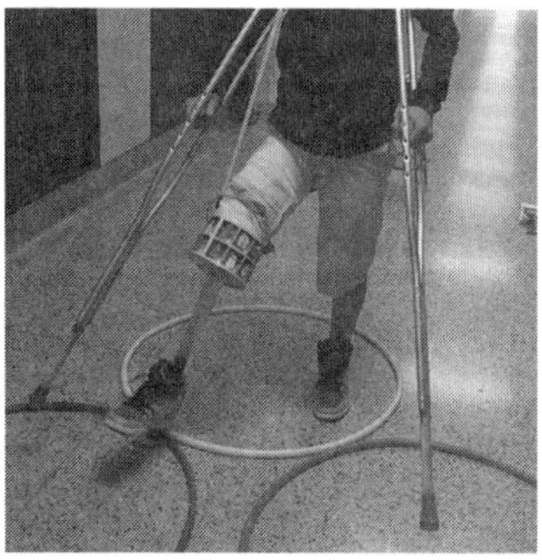

### Runway Show

For the runway show, the engineering students teamed with the theater students to put on a show. The first year the theater students set up the "runway" and invited an audience to watch as the engineers presented their prosthetic limb designs (see Figure 8.14). The following year the design challenge was to build a prosthetic leg for students in the theater class to use as they performed a dance. The challenge for the audience was to guess which dancer was wearing the prosthetic leg. Not easy to do when all of the dancers donned crutches and floor-length costumes!

### Portfolio Review

The students really enjoy working for clients outside the building; however, it is not always easy for engineers to visit the school in person. Therefore, the students have worked on portfolio reviews by submitting pictures of their work and copies of their engineering notebooks to a local engineer, Neil Maldeis. Throughout the process Mr. Maldeis encouraged students to e-mail him with questions or concerns and he ended the unit by evaluating each team portfolio. The students loved the real responses they received from Mr. Maldeis, especially comments relating to improving their work. Every team chose to keep their portfolio.

### Panel Review

Panel reviews are similar to a traditional classroom presentation where each team presents their work to the class. The critical difference is that they are also presenting to a panel of judges. This was the format we used when Channel 11 News visited the school to do video for a "What's Cool in Our Schools" feature on the eighth-grade engineering class (*www.kare11.com/cool-in-school/article/951362/148/Columbia-Academys-designing-students-are-whats-cool-in-school*). In this particular challenge, the student teams were hired by two architects to re-landscape the front and side of the school building. Each team presented their design using Google SketchUp, hand-drawn designs, and scale models of their work. The panel was made up of the architects, the building principal, and the district Grounds Director and Financier.

### Service Learning Projects

It is very important for middle school students to begin learning the value of giving back to their community. For example, a student who was passionate about the American Society for the Prevention of Animal Cruelty (ASPAC) asked me if there was anything we could do for the organization. It was the holiday season and she wanted to raise money to help animals in need. The students spent the month of December designing and building animals to scale decorated with a variety of holiday themes. For the product fair the students set out silent auction sheets. By the end of four different Product Fairs the engineers had raised $150 for the ASPAC. It was amazing to see such young children get into the spirit of giving.

### Engineering Fair

Our engineering fair is modeled on state science and engineering competitions. Each spring, our STEM students are given the opportunity to show their work to community scientists and engineers. Last year we invited over 20 professionals into the school to meet with students and judge their work.

### Community Presentations

As the program has grown we have sought presentation opportunities for students outside of the classroom. For example, in summer 2011, five students assisted in a presentation at the University of Minnesota K–12 STEM Education Colloquium. Seven additional students presented their winning Science and Engineering Fair products at the colloquium. Last February, three eighth graders had a chance to meet with school board members at the Minnesota School Board Association Conference to discuss their experiences in engineering.

## Assessment

Assessment should not be limited to formal assessments, rather assessment data should be continually collected to improve teaching practices, plan for curricular activities, and to develop self-directed learners. It is critical to assess the student's work as they progress through the challenge. Do not wait until the end of the unit to have the students turn in their notebook. The projects are too all-encompassing to put off assessing student understanding until the very end. Instead, meet with each student individually to assess their work at each stage of the design process. We make a point to speak to them from the viewpoint of the client; while students bemoan making changes for the teacher they are more than willing to make changes for their client. Additionally, this ongoing assessment allows us as teachers to develop "just in time" lessons to teach science or mathematics concepts needed to develop a successful product.

The budget grade is a pass/fail grade with a pass equaling all of the points allotted and a fail equaling 50% of the points allotted. If the students go over budget, even by a few cents, they receive a failing grade. The rationale for this is that they contracted with the client at a given budget amount. To earn back points teams can submit a budget proposal at the end of the challenge with a written section explaining actual costs, costs related to building problems, and a "sell estimation" of the product based on materials expense, labor costs, and research of the cost of similar products.

The building and testing phase evaluation relates to efficient use of time and keeping to the schedule. Every two weeks the students receive a "paycheck" that provides employer feedback on overall job performance (see Figure 8.15). Communication assessment occurs on three different levels. The students receive a teacher grade, a client grade, and a peer or consumer grade. Of the three grades the client grade is worth the greatest number of points. The use of paychecks reinforces the students' expectation of job performance in the work place.

**Figure 8.15.** Paycheck Grading System

| Pay Checks | |
|---|---|
| 10 – 9 points | You are an excellent employee. You are a value employee and will be promoted and receive a pay raise |
| 8 points | You are a good employee. You do your job well and are an asset to the company. You may not advance as quickly but I don't want to lose you. |
| 7 points | You do what you are asked, but you don't go above and beyond. If there comes a time when I need to lay off workers you will probably be the first let go. |
| 6 points | You do not do your job well. You are fired. |

## Practical Considerations

Implementing a program such as the Columbia Academy middle school engineering program requires a change in emphasis on teaching and assessment practices not only at the teacher level, but also at the school and district level. Administrators supported us through the development of partnerships and grants, professional development time for teachers in school-based teams and supporting co-teaching and curricular collaborations.

An effective engineering program also requires that teachers plan for investigations or engineering challenges that extend over multiple days or weeks; meaningful investigations cannot be fitted into a single class period. These investigations should not separate science knowledge and science process skills, as with meaningful inquiry activities, meaningful engineering design lessons integrate science content and process skills. Finally, engineering design challenges should promote teamwork and allow students to develop cooperative and collaborative learning skills. Teachers should plan to specifically teach and assess teamwork skills.

Using an engineering format is not only a new way of teaching but also a new learning experience for students. The first time the students are given an engineering challenge to read they will frequently feel overwhelmed. Remind them that the engineering process is designed to lead them to a solution; a solution that may be very different than their neighbors. As difficulties arise, encourage them to brainstorm and share ideas with their team. As the teacher, if you notice the students seem to be struggling and on the verge of wanting to give up, call for a gallery walk of each team's work. One person from each group stays behind and the students are allowed to ask questions and take notes as they observe each other's work. This simple activity will often give the students ideas and inspire them to continue working.

You will also find that students will be over anxious to start building. Remind them often that it is the design of the product that makes them an engineer! You will also notice throughout the unit that students may be unwilling to see a potential problem. Let them know that their ideas and creativity are valuable assets and are in fact, what the client is paying for.

Although using an engineering format of teaching may require more upfront planning time for the instructor, you will spend much more time watching your students grow and learn. You will find that companies and universities will be more than willing to work with your students in the classroom as clients and judges. We highly recommend that you contact local colleges and universities, as well as engineering businesses, for guidance and assistance.

## Success and Future Plans

Over the last four years student interest in engineering has grown so much that additional classes have been added to accommodate all of the students wanting to take the engineering elective. This year the program expanded to include a sixth-grade elective with plans to add a robotics elective in the fall of 2012. Revitalization is also underway at the high school. This past fall an engineering elective was offered to the students enrolled in the advanced ninth-grade physics course to provide students with experiences applying the physics principles learned in the science class. In the engineering elective the students designed and built cranes and roller coasters with the aid of a 3-D printer and laser cutter. Team teachers Matt Townsend and Luke Sands have plans to expand the elective to a full year in order to incorporate additional engineering challenges.

Family excitement over the engineering electives continues to grow each year. In fact, the program has attracted new students into the district. As students move through the engineering classes each year, parents feel strongly that the elective is preparing them for the future, as one parent commented:

*It is a wonderful opportunity for the students to be able to take this class in middle school. It will give them an advantage in college. It will also help them decide a career choice to pursue in college.*

Consequently, parents are choosing to keep their children in the district. Engineering has had a direct result on helping the district solve its number one problem, low student engagement. Because of this the strategic plan for the district has expanded to include STEM as one of its two main academic goals. As another parent commented:

*I think offering and introducing engineering to students at Columbia Academy is something that sets them apart from other middle schools. It shows me, as a parent, that the school district is making a commitment to our community. They are investing in our children BEYOND high school and see the value in getting our kids interested in college.*

While parental and student engagement and increased enrollment are positive indicators of success, it is also important to consider the impact of the engineering elective on student attitudes and learning. As one parent commented:

*He has really taken this class seriously. It has helped him with his math & science also. He's learning that he will need these classes past school, again – in the "real world."*

Another parent commented on his son's developing self-motivation and interest in STEM:

*My son is currently reading a book about Tesla, ON HIS OWN. I think his personal interest in the lives of engineers is growing.*

Finally, a third parent commented on her son's changing attitude toward group work:

*He is not particularly fond of group work, but I think he is learning how to cooperate and tolerate group work more because of the projects in engineering class. He may even be EXCITED about some of the group work projects.*

All of the students reported finding the engineering class interesting and engaging and only 4% of students said that they found the design challenges too unstructured and open-ended. Students appreciated the opportunity to "express my ideas" and most importantly students said that they liked the workplace orientation of the class. As one student wrote on the course evaluation, "Engineering is my favorite class, I love that it is one of the only classes that gives you a real life job perspective." Another student wrote, "I like how we get to work with a group of people and it feels like we are in an actual job. Many students reported that the STEM class had helped them to see engineering as a possible career choice as shown in the following quotes from two eighth-grade students:

- The class has definitely opened new options for me as engineering for a career and I go home some days so excited to tell my parents what I did in engineering.
- I have been in engineering for the last three years. It has helped me with teamwork and how to solve a problem. I tell my parents that the class is awesome and it has made me want to be an engineer.

The impact on student learning is documented here through standardized assessments of mathematics and science content. Columbia Academy administers the Measures of Academic Progressive (MAP) test in fall and spring of each grade for mathematics. Students' spring MAP scores in mathematics were compared for students taking the engineering elective and students selecting a non-engineering elective. Comparisons at each grade level showed a significant positive difference for students taking the engineering elective course, sixth grade ($p$ = 0.0013), seventh grade ($p$ = 0.00042), and eighth grade ($p$ = 0.00074). Similar comparisons of the engineering and non-engineering students were made using the state standardized test for mathematics, the Minnesota Comprehensive Assessment (MCA). Again, significant positive differences were noted at seventh grade ($p$ = 0.00629) and eighth grade ($p$ = 0.00321). MAP testing was only introduced for science this year for the eighth-grade class and a comparison of engineering and non-engineering students scores on the science MAP showed a significant positive difference ($p$ = 0.002). Finally, an in-class assessment of students' knowledge of the engineering design process was administered; 92% of students were able to identify and describe steps of the engineering design process.

In addition to the improvements in standardized test scores, students also spoke directly to the impact of the engineering design projects and their science and mathematics learning. A common area of mathematics learning identified by students was data analysis and measurement, with many students commenting that engineering helped them "in math using measurements." For example, a ninth-grade girl, stated:

*I thought engineering would be pretty fun, but it turned out to be amazing. I have used the five steps so many times in my life with problems, and it has taught me so much about taxing items, adding to the exact number and getting exact measurements.*

Another eighth-grade girl spoke to other areas in mathematics where the engineering projects aided her learning:

*I just started taking engineering this year because my friends enjoy the class and they recommended it to me. Being in the class has helped me with spatial thinking, geometry, and thinking beyond two dimensions.*

Similarly, students commented explicitly on how engineering had helped them to learn science concepts and made science "more interactive." For example, one eighth-grade student commented:

*Engineering has helped me solve problems by taking steps, and has helped me learn science by relating things to the real world's physics.*

Another eighth grader added:

*I took engineering because I like to build and design things. Taking the class has helped me to solve problems by giving me a process to go by. It has also helped me understand the science of things such as electricity and balance.*

Over the next year the district has plans to expand the engineering problem-solving approach into disciplines other than mathematics and science. Teams of teachers are currently meeting to look at the 5-step design process to determine how it can be used in a variety of classes K–12. Our goal for Columbia Heights students is to excel in skills related to creative problem-solving, teaming, and communication skills as they move toward college and beyond. For teachers this will require that we look at ways to support each other across subject areas and grade levels and look for ways to demonstrate to the students why education is important and how to apply what they have learned in creative and effective ways.

## Acknowledgments

Several individuals and organizations have assisted in the success of the Columbia Academy Engineering Program. Early encouragement, professional development, and curriculum resources were provided by the STEM Education Center at the University of Minnesota. Further professional development was provided through the Minnesota Region 11 Mathematics and Science Teacher Partnership (*www.region11mathandscience.org*). We have received more than $250,000 in grants from the Minnesota Department of Education to assist with the purchase of classroom materials, time for curriculum development, and technology integration such as laptops, sensors, and interactive whiteboards.

We thank Columbia Heights Teaching and Learning Director, Duane Berkas, and Columbia Academy principal, Mary Bussman, for their continued support of STEM including the development of the new engineering course and support in obtaining grants. Finally, we acknowledge mathematics teacher Emily Christianson for her excellent contributions to the integration of mathematics concepts into the engineering design challenges.

## References

Brophy, S., S. Klein, M. Portsmore, and C. Rogers. 2008. Advancing engineering education in P-12 classrooms. *Journal of Engineering Education* 97 (3): 369–387.

Bybee, R. W. 2010. What is STEM education? *Science* 329 (5995): 996.

Carlson, L., and J. Sullivan. 2004. Exploiting design to inspire interest in engineering across the K–16 engineering curriculum. *International Journal of Engineering Education* 20 (3): 372–380.

Fortus, D., C. Dershimer, J. Krajcik, R. Marx, and R. Mamlok-Naaman. 2004. Design-based science and student learning. *Journal of Research in Science Teaching* 41 (10): 1081–1110.

Frykholm, J., and G. Glasson. 2005. Connecting science and mathematics instruction: Pedagogical context knowledge for teachers. *School Science and Mathematics* 105 (3): 127–141.

Harris, J., and A. Felix. 2010. A project-based, STEM-integrated team challenge for elementary and middle school teachers in alternative energy. In *Proceedings of society for information technology & teacher education international conference*, ed. D. Gibson and B. Dodge, 3566–3573. Chesapeake, VA: AACE.

Hirsch, L. S., J. D. Carpinelli, H. Kimmel, R. Rockland, and J. Bloom. 2007. The differential effects of pre-engineering curricula on middle school students' attitudes to and knowledge of engineering careers.

In *Proceeding of the 2007 Frontiers in Education Conference*. Milwaukee, WI: Institute of Electrical and Electronics Engineers. *http://fie-conference.org/fie2007/index.html*

Koszalka, T., Y. Wu, and B. Davidson. 2007. Instructional design issues in a cross-institutional collaboration within a distributed engineering educational environment. In *Proceedings of world conference on e-learning in corporate, government, healthcare, and higher education*, ed. T. Bastiaens and S. Carliner, 1650–1657. Chesapeake, VA: AACE.

Mehalik, M. M., Y. Doppelt, and C. D. Schunn. 2008. Middle-school science through design-based learning versus scripted inquiry: Better overall science concept learning and equity gap reduction. *Journal of Engineering Education* 97 (1): 71–85.

Minnesota Department of Education. 2009. *Academic standards in science: Draft two complete. http://education.state.mn.us/MDE/Academic_Excellence/Academic_Standards/Science/index.html.*

National Academy of Engineering. 2009. *Engineering in K–12 education: Understanding the status and improving the prospects*. Washington, DC: National Academies Press.

National Research Council (NRC). 2007. *Rising above the gathering storm: Energizing and employing America for a brighter economic future*. Washington, DC: National Academies Press.

National Research Council (NRC). 2012. *A framework for K–12 science education: Practices, crosscutting concepts, and core ideas*. Washington, DC: The National Academies Press.

President's Council of Advisors on Science and Technology. 2010. *Prepare and inspire: K–12 education in science, technology, engineering, and math (STEM) education for America's future. www.whitehouse.gov/administration/eop/ostp/pcast/docsreports*.

# STEM Literacy Through Science Journalism

## Driving and Communicating Along the Information Highway

*Wendy Saul and Alan Newman*
*University of Missouri, St. Louis*
*St. Louis, MO*

*Angela Kohnen*
*Missouri State University*
*Springfield, MO*

### Setting

In 2008, the University of Missouri–St Louis (UMSL) received a four-year DRK-12 grant from the National Science Foundation (DRL-0822354) entitled "Science Literacy Through Science Journalism." Our goal was to better understand if and how student-written science journalism articles offer a route to increasing science literacy. Student work focused not only on the reporting of science, but also attended to technology, engineering, and elementary statistics; as such, it exemplifies an integrated approach to STEM learning. The principal investigators on this project, commonly referred to as SciJourn, come from various disciplines. Alan Newman, who holds a PhD in chemistry and worked for nearly 20 years as a science journalist and editor for the American Chemical Society, was at the heart of the endeavor. It was his experiences and his understanding of the complex relationship between science and the communication of science that served as a model for all those involved. Wendy Saul, who is by predilection and training a teacher educator concerned with writing and reading, sought to attend to the ways the authentic practices that Newman embodied could make sense in the classroom. Joseph Polman, who identifies with the learning science community and writes about how science learning takes place in both formal and informal environments, focused largely on issues related to student engagement. Cathy Farrar, who left her high school science classroom to work on the project, attended mainly to the measurement of science literacy. She, like Angela Kohnen and Jennifer Hope, earned doctoral degrees as they studied the effects of the SciJourn project. Ideas and methods born from SciJourn were piloted, tested, and refined first in classrooms in the greater Saint Louis region and now nationally with teachers from Kentucky, Indiana, and Colorado. Although this work preceded the publication of the *Next Generation Science Standards (NGSS)* and the *Common Core State Standards* (*CCSS*), the goals and methods born from SciJourn are satisfyingly aligned with both of these documents: The call for more writing in science classes identified as key to the *CCSS* and *NGSS* has been explored through SciJourn for more than five years.

## Background

How does SciJourn build STEM capacity for both teachers and students? As Roger Bybee suggests, STEM education has many definitions, but at the heart of all STEM learning, is recognition of the need to prepare teachers and those whom they serve for the challenges of the 21st Century (Bybee 2010, 2013) and the work skills needed to succeed in that environment. According to the National Research Council (NRC 2010) success will depend on "adaptability; complex communication; nonroutine problem-solving; self-management/self-development, and systems thinking." (Bybee 2013, p. 66)

Clearly, creating a school program that develops these skills and abilities is no easy task. Curriculum writers are scrambling at this very moment to write units that call for more depth than breadth and that provide teachers with activities and the background knowledge to make such instruction possible. Although connections between and among STEM disciplines are natural in the world outside of school, in the land of curriculum development and state testing (a land where teachers are trained as experts in one discipline), curricular reform is proving more difficult. Moreover, in the 21st century we recognize that one can be knowledgeably conversant in all STEM areas. Although a "scientifically literate" high energy physicist probably uses science, technology, engineering, and mathematics to do her work, she may not know, nor be called upon to know, the difference between congestive heart failure and a heart attack. If a competent evolutionary biologist is at a loss to explain how astronomers spot black holes, what can we realistically expect of teachers and their students as they seek to learn enough science, math, engineering, and technology to be called STEM literate? In other words, the key to curricular reform may lie not only (or mainly) in defining the information every scientifically literate individual should know, but also in building from student interests in ways that connect those interests to the work of scientists, engineers, and mathematicians, all of whom use technology. The 2011 NRC report articulates this position clearly, by noting that effective STEM instruction engages students' interests and experiences, identifies and builds on their knowledge, uses STEM practices and provides experiences that sustain their interest (NRC 2011; BOSE 2013).

Our overarching goal in the SciJourn project was (and is) to prepare students for life beyond high school by asking them to take seriously questions they themselves deem important. As teens work as science journalists, they identify and articulate an issue of personal interest with a STEM connection, research that issue under the guidance of an experienced editor and their classroom teacher using multiple credible sources, contextualize the ideas they encountered—which means gaining some understanding of what is well-established and what is nascent science information—and fact-check their own and others' assertions. Looking back at the STEM criteria the NRC and Bybee identify, SciJourn activity clearly requires adaptability, complex communication, nonroutine problem solving, self-management/self-development, and some understanding of systems thinking.

Scientists, science educators, and teachers have long recognized certain problems endemic to high school science instruction:

1. Students are rarely taught to make science-based personal and civic decisions.
2. Science is too often presented as a static (and dated) body of knowledge rather than a dynamic and relevant enterprise.
3. In a textbook-centric environment, young people are rarely asked to combine and assess information from multiple sources.
4. How to navigate the landscape of science information, especially on the internet, is not typically well explained or taught. Students need to learn to differentiate between well-established theories, embryonic research, and opinions.
5. Little has been done to help young people recognize expertise and credible sources, and how to query experts in their communities.
6. There has been little crossover between what students learn in science classes and their real-world understandings.

To help us address these issues, we looked for a model. Who in the non-school world authentically practices the kind of inquiry we sought to teach in the classroom? Although there may be others who are skilled at locating and evaluating multiple credible sources and others who know how to negotiate the landscape of science information, we decided to build our curricular efforts on the work of science journalists—persons whose very livelihood depends on their ability to identify problems or topics of contemporary interest; search for and collect information from relevant, reliable sources; fairly reflect scientific debates and controversies; write up the appropriate information in a way that the general public can understand; forefront what is most important and new; and attribute sources so that others can trace back information and query the purveyors of information. Journalists also have other skills that we have come to value: They know how to interview, listen carefully, and take good notes. And journalists know the value of multiple perspectives—in this case nonscientists whose health, finances, political action, or consumer choices are affected by new scientific advances and discoveries—as part of the information they provide.

We specifically chose *science* journalists as opposed to journalism generalists for our model because we wanted exemplars who—because of their background knowledge—are able to contextualize ideas, in addition to being able to share what they have learned. In sum, by engaging high school students in practices related to science news writing, we sought to prepare them for the kinds of research and decision making they will need in their adult lives.

This organic, authentic approach to problem solving takes into account the way knowledge informs real-life decision making. The kind of educational approach we seek is consonant with Bybee's notion that first "places life situations and global issues in a central position and uses the four disciplines of STEM to understand and address the problem." (2013, p. 32). This dramatically contrasts with past curricular efforts that separate STEM ideas into discrete units: In a technology class we repair a computer, in science class we study disease transmission, and in math we practice reading and constructing graphs. SciJourn students, acting as science journalists, find that each STEM discipline informs the other. For example, the student-written article *Cell*

*Phones May Increase Risk of Brain Cancer* discusses the risks of phone use and epidemiological data while *Text Your Neck Off* explores how frequent texting leads to neck strain. This piece even includes an interview with a chiropractor. (See *www.scijourner.org* for all student written stories.)

Science journalists know that all elements of STEM are likely to inform their research; though not all elements may appear in their story; still they need to be understood. Thus, the story *Oceans in Crisis* explained how higher $CO_2$ levels lower pH and harm shellfish and other organisms, but didn't recount all the evidence for human-induced climate change. Similarly, a story on sunscreens focused on the possible cancer risk of one ingredient, but didn't dwell on all aspects of skin cancer that these products help prevent.

Since 2008, 51 teachers and more than 10,000 teens have worked SciJourn into their science classes at 34 schools and in a science museum's youth development program. Three parochial and 34 public schools, urban (11), suburban (17), and rural (6) were included in our research sample. These schools included wealthy middle class and underserved communities; student populations both fairly homogenous and diverse, and—as measured by testing—high achieving and "failing" institutions.

Students in these schools have critically read science news articles in the press and on the internet, and researched, written, and in some cases published original science news stories in a youth citizen science publication called *SciJourner* (*www.SciJourner.org*). We have also engaged in efficacy research, and published a book (Saul, Kohnen, Newman, and Pearce 2012) and articles to share our insights with a broader public (e.g., Polman, Newman, Farrar, and Saul 2012). The project has been integrated into high school classes in biology, chemistry, physics, environmental science, and applied biology classes, and in two middle school classes in Earth and space science. It has also been used in two English classes.

SciJourn activities were first piloted in a single classroom in 2008, and in the summer of 2009, with a group of underserved minority teens at the Saint Louis Science Center (SLSC). These teens became the first authors published in our online and print publication, and youth from SLSC continued to publish regularly during the next four years.

While activities with students were underway, our research team developed science literacy and science journalism criteria or guidelines (sometimes called standards) by studying the moves science experts made as they read material both within and outside of their area(s) of expertise. In addition, we used focus groups to better understand how science journalists evaluate and develop their own expertise and that of their readers. SciJourn criteria were first used in the summer of 2009 and have been revised multiple times as students worked with them and professionals reacted to them. The science literacy standards and science journalism criteria we employ mirror one another (see Table 9.1).

According to administrators, teachers, and students, the SciJourn writing criteria—upon which every published *SciJourner* article is assessed—forge the connection between science journalism, STEM, and science literacy. By conforming to our criteria for science journalism, students and their teachers have found a motivated context for learning the skills of science literacy as well as a vocabulary and structure that helps scaffold and assess growth.

We agree with Krajcik and Sutherland (2010) who suggest that certain curricular features "promote students' ability to read, write, and communicate about science so that they can

**Table 9.1.** Criteria

| Scientifically literate individuals can: | A good SciJourner article: |
|---|---|
| Identify personal and civic concerns that benefit from scientific and technological understanding. | Tends to be local, narrow, focused, and up-to-date and presents a unique angle; findings are meaningfully applied to personal or civic issues and readers' likely questions are anticipated and addressed. |
| Effectively search for and recognize relevant, credible information. | Uses information from relevant, credible sources including the internet and interviews. Successful authors use internet search terms and search engines effectively, privilege data from credible government and nonprofit sites and can justify the use of "other" sites, and locate and query experts and relevant stakeholders. |
| Digest, present, and properly attribute information from multiple credible sources. | Is based on multiple, credible, attributed sources. Sources are relevant and reliable; stakeholders with varying expertise and experiences are consulted; sources are identified and basis of expertise is explained; and all assertions, numbers, details and opinions are attributed. |
| Contextualize technologies and discoveries, differentiating between the widely accepted and the emergent; attending to the nature, limits, and risks of a discovery; and integrating information into broader policy and lifestyle choices. | Contextualizes information by telling why the information presented is important both from a scientific and societal viewpoint, and indicating which ideas are widely accepted and which are preliminary. |
| Fact-check both big ideas and scientific details. | Is factually accurate and forefronts important information; the science connection is evident, difficult concepts are explained, precise language is employed, quantitative measures are given in correct and comparable units, information is current, and captions and graphics are checked for accuracy. |

engage in inquiry throughout their lives." SciJourn (1) links new ideas to prior knowledge and experiences, (2) anchors learning in questions that are meaningful in the lives of students, (3) connecting multiple representations, (4) provides opportunities for students to use STEM ideas, and (5) supports students' engagement with the discourses of science. Our work also aligns with research on engagement (e.g., Schiefele, Schaffner, Moller, and Wigfield 2012) as well as ideas proffered in the CCSS to improve STEM-related reading and writing (National Governors Association Center for Best Practices, 2010).

*SciJourner* articles focus on STEM topics that young authors generate and find relevant, such as the health hazards of tattoos; environmental contamination in their community; the science of penalty kicks in soccer; and rare and common diseases afflicting friends, family, or themselves. Journalists—professional or fledgling—position themselves as needing to learn on behalf of a public who depends on them; students value the legitimate identity of one who "doesn't know" but is willing to find out. As such, students become science information seekers, researching STEM concepts and learning skills they need now and in the future.

The SciJourn approach may also complement textbook science. An article on "why tennis ball cans pop" is really a lesson in pressure differentials, but couched in a real-life example that is both relevant and interesting to write about and to read. A story on why sharks attack swimmers more often during new moons hinged on a discussion of how the Moon and Sun influence tidal heights.

Other news stories move science outside of the textbook but still deal with STEM. A surprising number of articles have focused on concussions received in high school sports, an issue that only recently is being addressed by state legislation. Another group of teens, living in a rural area, regularly describe experiences and concerns that differ from those of city dwellers, such as the reintroduction of elk populations or the growing problem of feral hogs. Cultural practices have also become the basis of stories: A student who found relief from headaches with an alternative medical approach learned from her Korean grandmother drafted a glowing story about the value of non-Western medicines. The editing process pushed her to be more critical by directing her to National Institutes of Health (NIH)-supported research. Her final story offered a balanced analysis, pointing out that research supports some alternative approaches and warns that others may not work at all.

Our approach is clearly aligned with what is sometimes called "science literacy-in-action" (Aikenhead, Orpwood, and Fensham 2011). As researchers, we concur with Feinstein (2011), who believes that scientifically literate behavior involves acting as a "competent outsider" to science. Competent outsiders are prepared to deal with the explosion of scientific and technical knowledge (Aikenhead, Orpwood, and Fensham 2011) by finding, evaluating, and making sense of new scientific and technical information that cannot be predicted or completely and comprehensively "taught" in high school. We seek, for instance, to help students assess credibility not simply in a yes-no binary, but rather to consider how well the citation functions for a given purpose. Current efforts to teach students to assess internet sites using rules, (such as ".org or .edu are good, .coms are bad") are, we believe, misguided. Even the more sophisticated rubrics or rules for assessing website credibility, such as looking for recent updates and authors with degrees, fail in part because they do not consider function. If one is looking for information on hybrid cars, checking out and comparing data from various manufacturers or talking with a mechanic is sensible. Similarly, Wikipedia, in the sciences, tends to be credible and generally leads to other citations.

The real problem, of course, is that the internet provides a basically "flat" science information landscape—all STEM information looks equally credible, at least to a novice. Novices searching the web tend to frame their questions colloquially (e.g., "why does it hurt when he pees?") A term like "renal cell carcinoma" leads to very different results.

## Insights and Examples From the Classroom

As the new *A Framework for K–12 Science Education* (NRC 2011) notes, the practice of helping students to obtain, represent, communicate, and present STEM information is especially challenging for secondary science teachers: "For science teachers to embrace their role as teachers of science communication and of practices of acquiring, evaluating, and integrating information from multiple sources and multiple forms of presentation, their preparation as teachers will need to be strong in these areas" (p. 114). The activities and research we report on here were designed both to share results from a more than four-year study and to guide future efforts around science literacy learning.

Please note that SciJourn is not a curriculum in itself, and the ideas embedded in the criteria are enacted in dramatically different ways by various teachers, but, over time, certain key elements seem to have emerged as central to the effective communication process. The following data are taken from interviews, surveys, and observations of teachers who were introduced to SciJourn concepts through professional development (PD) workshops in which they were required to write a news story themselves—choose a topic, research it, write a first draft of an article, and revise it after being edited by our professional editor. Many share the experience of being edited with their students, detailing how difficult it was for them to receive feedback and how delighted they were when they finally got published.

In the classroom, nearly all teachers launch the project with what is termed the Read-Aloud/Think-Aloud (RATA; Langer 2011; Pressley and Afflerbach 1995; Saul and Dieckman 2005), in which they make explicit the connections between a news article and what they have (or will) study in class. The SciJourn criteria often guide their "stopping places." For example: "How does this article on creating a new color relate to what we have been studying about [chemical] bonding? In other words, what's the context?" begins one chemistry teacher; "This information is attributed to the USGS [US Geological Survey]; who are they and should we trust them? Let's check them out," asks an environmental studies teacher; "This article says that some hummingbirds can beat their wings 100 times per second. Do you believe that?" says a biology teacher. "How can we check that claim?"

At the same time, students may be invited to read articles on *SciJourner.org* or peruse the print edition during class "down time." Typically, teachers will ask students to comment online or through class discussions about articles they have read: "Why did you find this one interesting?" "What sources did the author use, and why were these sources considered 'experts?'" "If you wanted to learn more about this topic, where would you go?" Students are also invited to do RATAs: "I don't understand this sentence," says one young man. "Is this internal defibrillator the same kind they have in airports?"

SciJourn teachers typically begin introducing RATAs to their classes long before students begin writing or researching. The RATA has a really low-entry cost in terms of time or preparation. It takes only a few minutes and is usually a story that the teacher would normally enjoy. And some, a group of teachers we call the low implementers, never move beyond RATAs.

To buy into the SciJourn model and move beyond the RATA, teachers must be committed to teaching science literacy. Shelley [a pseudonym], who formerly worked at the Food and Drug Administration and now teaches chemistry in a "reconstitution eligible" school, began reading science news articles with her students and was surprised by the kids' enthusiastic response. She was also shocked by their ignorance regarding science issues directly affecting their lives, as well as their poor research skills. When a student in another SciJourn teacher's class from the same school was published in *SciJourner*, Shelley realized "my kids can do that" and began working individually with students during their free periods or when they had completed other work. "For every kid that got published there was another kid who was inspired to do a little bit more," she said. And interestingly, once the work of one student from a given school is published in *SciJourner*, other students and their teachers seem especially inspired to submit articles and to work toward publication. Shelley thus became what we call a high implementer.

Jason, another high implementer who teaches physical science in a suburban school, values SciJourn because it gives him a way to put the science concepts he is responsible for into context. "In science we compartmentalize things, and when we compartmentalize, they [students] see no real connection between something that's at the cellular level, to something that's at the atomic level, to something that they know structurally. We never really help them to connect the dots." We call our approach to searching "developmental" and it can be seen as a form of controlled release.

Research in SciJourn classes often begins with the teacher simply telling students that certain sources, for example *Science News*, *NASA.gov* or *cancer.org* [American Cancer Society], are more credible than others, and then explaining the reasoning. Conversing with a 16-year-old about why advice based on recent NIH data is more trustworthy than advice from another online teenager or *buzzle.com* (which is one of many "content farms" on the web) is well worth a teacher's time. In SciJourn, successful teachers and students explore why *WebMD*, a dot com, is recommended and *naturalcures.org* is not. Our goal is to help wobbly information seekers get their footing, to steady themselves somewhere before venturing out on the internet plains. In a sense, we are modeling the pre-internet-era role of the reference librarian. Thus, SciJourn teachers are often heard saying things like "Did you look for medical information on *cdc.gov* [Centers for Disease Control]" or "ESPN has an excellent sports science section; did you check that site?"

SciJourn teachers, unlike many of their peers (but like scientists and science journalists) have actually embraced Wikipedia as a "good place to begin, but a terrible place to end." Through Wikipedia, one can follow information back to its source using the references at the bottom of the page. Vocabulary is another reason to check Wikipedia; search engine users typing in "life on other planets" are not only getting too many sites to check, but are also encountering far-fetched and distracting stories about aliens, whereas a term like *astrobiology*, culled from Wikipedia, leads down much more promising roads.

When a student encounters a website that doesn't fit easily into the "credible" category, reading the "About Us" section and checking the frequency of updates can prove helpful. So, too, the journalistic proviso of describing the site in a few words for the reader: "According to WWF, an environmental group ..." or "... says *pottersyndrome.com*, a website run by a mother who lost a child to the disease." It is then up to the reader to rate the credibility of the source.

This developmental approach grew out of an analysis of comments made by professional science editors. Professionals tended to be highly specific, particularly when addressing content. SciJourn's professional editor typically directs students to high-quality sources by name (e.g., "see American Cancer Society and NIH"), asks for very specific additional information ("Who says?"), deletes irrelevant or biased sentences, and questions the accuracy of information (Kohnen 2012).

By comparison, teachers' comments to students before SciJourn training were very general ("Good job") or they framed comments as questions ("Does this make sense?"). Over time, however, project teachers have adopted a more targeted style of commenting themselves and were able to change the way they responded to student writing in part because of the SciJourn criteria.

The act of writing is difficult for many students and time-consuming for classes on a tight schedule. To address these concerns, a number of SciJourn teachers (our middle implementers) have asked students to become a class expert on a particular topic and create a PowerPoint or Prezi presentation to inspire questions from their classmates. Such an activity teaches question-posing and enables the "audience" to learn not only new information, but potentially something about search strategies. In addition, such an activity requires students to combine—at least on a conceptual level—information from multiple credible sources and attribute that information in an authentic public forum. The PowerPoint also invites "fact-checking." What the PowerPoint does not do is encourage students to interview experts or stakeholders, develop a nuanced argument, or go beyond a cut-and-paste of information from the web. What we can say, however, is that teachers and teens regularly report that our journalism-based criteria give them a vocabulary and structure for exploring, researching, and evaluating science information. And nothing stops a student from taking the next step and turning their slides into an article.

Teachers have been surprised at the extensive, specific, and content-oriented feedback our professional editor provides and have sought to emulate his practices. Looking at the "track changes" on a submission, one teen described the feedback as "a few black words in a sea of red." In the beginning teachers worried: "You can't say this to kids!" as they read through responses that mirror the journalistic standards (e.g., "Who says?" "What's your source?" "You need a number here—try CDC."). However, when the students looked at such feedback, especially when they did so under a teacher's guidance with the standards in hand, they seem much less upset than their teachers had predicted. Many young people actually appreciate the feedback, calling it "more straightforward than a teacher's" and saying it makes them feel like "a real journalist" and "now I know what to do."

The science writing criteria have also led to two peer review protocols, the Science Article Filtering Instrument (SAFI) and our version of a SciJourn Calibrated Peer Review, a modification of the process described by Chapman and Russell (2005). Both of these filters align with the SciJourn criteria (Table 9.1) and help teachers focus on science content rather than writing mechanics. The SAFI, developed by Laura Pearce, also builds on the SciJourn ethics, which we call SLAP (see Table 9.2). These instruments are designed to steer teachers and peer reviewers away from generalized comments like "good start." Moreover, if there is a lack of attribution or signs of plagiarism, overt advertising, and misrepresentations of data or stereotyping, the article is immediately returned to the student without any further editing, saving the teacher time and giving the student a clear message.

**Table 9.2.** SciJourn's Journalistic Ethics

| |
|---|
| SciJourn authors do not<br>• Stereotype<br>• Lie<br>• Advertise or<br>• Plagiarize<br>Articles that do any of the above will be **SLAP**ped down! |

Publication in the *SciJourner* is a difficult but achievable goal. Unlike the DuPont Challenge Science Essay Contest (*www.thechallenge.dupont.com*), in which the chances of winning are beyond the skill level of even our most determined authors, any student willing to work hard enough will, in fact, get published. Our editor says that the most important quality for potential authors is "tenacity;" we work with students as long as they are willing to work with us. This tenacity was evident, for instance, in the young author who endured more than five content-based revisions of a story about how the EPA planned to clean up a superfund site in his poverty-stricken neighborhood. In the process he learned about PCBs, trichloroethylene, and thermal desorption technology. He also went from a C in chemistry to an A, in part, because he moved from the back to the front of the class so he could communicate more directly with his teacher and in part because he began to see himself as a "science guy."

Prior to involvement in SciJourn, science teachers reported receiving little to no training on teaching or responding to student writing (Kohnen 2013). Those who had training cited professional development focused on generic (not science-specific) rubrics, such as the 6+1 system (Bellamy 2005; Spandel and Stiggins 1997). Although we initially considered adapting the 6+1 rubrics for our purposes, we found these generic rubrics inadequate for guiding science learning because they focus largely on technique rather than STEM information gathering and evaluation. By contrast, SciJourn criteria have proved immensely useful in helping teachers plan and students learn science literacy skills.

We note that even those teachers who no longer actively use SciJourn in their classes are found emulating our editor's questions: "Who says?" or "What's your source?"

## Assessing SciJourn

### *The Science Literacy Assessment (SLA)*

Several formal studies have been undertaken to assess the impact of the project on student achievement, the largest of which is the Science Literacy Assessment (SLA) study led by Cathy Farrar. Her work is premised on the assumption that all of us are asked to deal with information about which we know little. Scientifically literate individuals are undaunted by a topic of scientific or technological interest and have ways to think about information gathering. They know what questions to ask and where to go to find out more. Such individuals know enough about a topic to get a foothold to begin their investigations. This is the skill set we hoped to see developing among students.

The SLA consists of four separate tasks (a total of eight were created so that a form A and B could be used). In the first task students were asked to read an article that was developed by a trained science journalist that included some widely recognized, high-quality sources of science information, such as the CDC, World Health Organization, or the Environmental Protection Agency, but also included unknown sources with no credibility clues. This article also used numerical data that were clearly suspect. Students were asked which sources were and were not credible and why they thought so. The second task involved looking at a captioned photo (in one form of the test the Gulf oil spill and in the other a volcano spewing ash) and asked to talk about local and global connections that could be made. The third task involved reading a health-related brochure and answering questions about who might benefit from reading this information and why, and in a fourth task, students were presented with a graph (Figure 9.1) that was incorrectly labeled and asked to "peer review" it for a friend. Questions on each task were designed to measure growth in a person's ability to:

1. identify personal and civic concerns that benefit from up-to-date scientific and technological understanding (coded as *relevance*);
2. effectively *search* for and recognize relevant, credible information (coded as searching);
3. digest, present, and properly attribute information from *multiple credible sources*;
4. *contextualize* technologies and discoveries, differentiating between those that are widely accepted and emergent, attending to the nature, limits, and risks of a discovery, and integrating information into broader policy and lifestyle choices; and
5. *fact-check* both big ideas and scientific details.

Prior to administering the SLA, the test was given to scientists and "competent outsiders" in order to develop scoring criteria.

| Aspect of Scientific Literacy | Sample Question | Example Low Score | Example High Score |
|---|---|---|---|
| Multiple credible sources | Is this source credible? Why or why not? | "If it's an institute, it's credible" | "I need to know how was the institute started? Who runs it?" |

This assessment was administered before and after SciJourn participation to 673 implementation and 241 comparison students during the 2010–11 school year and blind-scored by a trained scorer. All five aspects of scientific literacy (relevance, context, multiple credible sources, information seeking, and factual accuracy) showed statistically significant increases in the implementation over the comparison group ($F$ (1,912) = 181.347, $p < 0.001$). Further analysis indicated that teachers whose students had incorporated the writing and revision of science news into their curriculum (the high implementers) had the greatest gains overall; these students' understanding of factual accuracy, multiple credible sources, contextualization, and searching was especially improved (Polman, Newman, Saul, and Farrar, forthcoming).

## *Impact on Participating Teachers*

We also sought to understand how involvement in the SciJourn project affected the participating teachers and to that end collected the following data:

- Observational field notes during all professional development meetings
- Observational field notes in teacher classrooms
- Focus group discussions about role of writing in the science classroom
- Teacher-generated artifacts, including science news articles and revisions, lesson plans, and teacher responses to student work
- Interviews with selected teacher participants and their students

- Structured pre- and posttests of teacher responses to samples of student writing
- Surveys with questions about teachers' reasons for joining the program, priorities for instruction, and attitudes about writing and writing response
- Video-recorded teacher lessons

Qualitative textual analysis and critical discourse analysis were used to interpret the data. We concluded that involvement in SciJourn had the following effects on participating teachers: (1) an increased understanding and awareness of the features of science journalism; (2) a changed understanding of the role of writing in the science classroom; (3) a redefining of the teacher's classroom role; and (4) a change in priorities when responding to student writing. Each of these findings is discussed in more detail below.

### An Understanding of Science Journalism

Most of the participating teachers considered themselves able, even avid, consumers of science news. One participant had even submitted a science news article to the local paper. Yet most of the participants described the initial summer professional development workshop, in which they learned to produce science news themselves, as an eye-opening experience. "I didn't know the things we learned," a chemistry teacher with several years of experience as a chemist told us. As teachers looked carefully at the genre of science news, noticing the details of how an article is put together from the initial sentences to the attribution of sources to the ending, they learned to read in a new way. One described himself as "seeing" science news differently; others agreed, saying they couldn't hear or read a science news story in the same way again. In particular, teachers remarked on how important it was for a science journalist to make science news relevant to readers and how essential the incorporation of multiple credible sources of information was to the credibility of a news story.

The teachers' new understanding of science journalism seemed to be the result of two factors. First, the professional development was led by an experienced science journalist and editor (and, in some years, several other science journalists participated via Skype). Listening to PhD chemist and science journalist Newman do a RATA seemed particularly important. One teacher said that he was "profoundly hit when Alan [Newman] read to us" and that hearing "someone who has a depth of knowledge that's obviously far beyond mine reading and giving their opinions" was a pivotal moment in his decision to implement science journalism activities with his students. Another teacher drew parallels between the role of science journalists and the role of science teachers, describing both as being in the business of translating "science for the masses."

The second important factor that changed teachers' understanding of science news resulted from having to write a science news article themselves. Many of the participating science teachers did not enjoy writing in any genre, but most considered themselves capable writers, particularly of academic papers. However, they found that writing a science news article and submitting it to Newman for professional editing was very different from the writing they typically engaged in. Teachers described science journalism as writing that should "catch and hold interest," something they had not thought about in their previous writing endeavors. Most teachers wrote several drafts of their own articles before being published in the online news magazine. Through that experience, they came to understand and internalize the qualities of science journalism in a way that they had not before, despite their frequent consumption of the genre. In a survey of participating teachers given near the end of the program, writing a science news article was cited as one of the most important things a teacher should do before implementing SciJourn (learning to do a RATA was the other most frequently cited activity).

## The Role of Writing in the Science Classroom

Participating in SciJourn also changed the science teachers' ideas about the role of writing in their classes. Because teachers self-selected to apply to the program, they presumably had interest in using writing in their classes (and a survey of reasons for joining the program confirmed this), yet participating teachers also had limited prior experience teaching and assigning writing. An analysis of survey questions and focus group discussions concluded that, before enrolling in SciJourn, science teachers typically assigned short writing assignments that did not ask for much interpretation or even composing on the part of the students (i.e., writing out vocabulary definitions); these assignments required a single specific correct answer; and were uninteresting or even painful for the teachers to read. Assessing this writing was also described as an unpleasant experience involving a strict rubric and the correction of errors. Many teachers had abandoned previous writing assignments because they were "horrible" or "tortuous"; others were discouraged by rampant plagiarism. Very few described using writing for anything but an assessment tool of what had been already taught in class.

After participating in SciJourn, science teachers approached writing very differently. According to most of the participating teachers, one of the most important features of a SciJourn writing assignment was that it allowed student topic choice. Many teachers reported that students were eager to write about topics of personal interest, including health concerns and issues around sports or hobbies. Through this writing assignment, students and teachers could engage in relevant discussions of science, something that many of the teachers had been seeking for years. As a result, several teachers reported feeling closer to their students. Furthermore, these writing assignments were not designed for students to demonstrate what they had learned in class (and what the teacher knew); instead, they were an opportunity for students to learn something new and communicate that understanding to a less-knowledgeable audience. Finally, many teachers described designing their SciJourn writing assignments with their own writing struggles fresh in their minds. Rather than simply assigning the writing, collecting it, and correcting it, the teachers found ways to scaffold the SciJourn writing experience, spending time on writing process activities like finding a good topic, searching for information, writing early drafts, revising, and even peer editing.

## Redefining the Role of the Teacher

Prior to joining the program, participating teachers conducted their classes in a variety of ways, including both the traditional lecture/lab format and inquiry-based lessons. However, despite variations in teaching style, most participating teachers positioned themselves as the content expert in the classroom and several even used the educational cliché "sage on the stage" to describe their role.

Yet once the science teachers began bringing in high-interest science news articles to read aloud with the class and, even more dramatically, once they allowed their students to write about any science topic of their choosing, most teachers found themselves altering their classroom persona, at least for the duration of SciJourn activities. Students often raised questions about RATA articles that stumped teachers. In the words of one teacher, classroom discourse changed from students speaking to "show off or show what they knew" to everyone discussing "what we didn't know and how we could learn it." The teacher still had an important expert role to play in the classroom, but the expertise teachers drew on was often no longer about factual content. Instead, teachers used their expert knowledge about sources of information, credibility, and searching the internet to help students look for high-quality science information. Many positioned themselves as co-learner, particularly when questions arose about RATA articles.

## Student Engagement

A third set of studies focused on student engagement. Looking at the typical high school curriculum, one notes the lack of opportunity for students to connect science research to issues teens view as salient. Textbooks typically include examples almost as an afterthought and often in a language students find alienating (Lemke 1990). And yet we know that history is rife with examples of scientists whose work grew out of a personal experience or interest, from Gregor Mendel to Columbia University geneticist Nancy Wexler, who helped discover the location of the Huntington's disease gene that afflicted her mother. Over and over again in student stories, we have documented instances of young authors writing what appears to be a bland, five-paragraph essay that ends, almost as an apology or afterthought, with a personal connection—for example, a grandfather who finally quit smoking or the online computer game they love to play that could also model the spread of an infectious disease. Other times, we have discovered this connection almost accidently by querying students on why they chose a particular topic. As a result, one of our most common edits on articles that read like an encyclopedia or textbook has become, "Why are you writing about this topic?"

In the SciJourn project, we view engagement with science and technology as both an outcome and a driver of science learning. In PD sessions, teachers have regularly remarked on students' new, positive attitudes toward science. In one chemistry class, for instance, the teacher asked each student to choose an element on the periodic table, research it using SciJourn standards and write about it with an interesting angle. One student wrote about the importance of potassium to the human body, another about fluoride and tooth health; these articles were never published but the teacher was thrilled that the elements became more than a wall decoration.

At a suburban school across town, students were challenged to write multiple science news articles about a personally relevant issue in environmental science. One young author in the class, who already planned a career in engineering, researched "how green" his school really was, and the standards and costs for constructing green buildings. That story was published and for his next topic he chose how hazardous waste can be safely incorporated into cement.

For many of these students, the most difficult part of the assignment was directly interviewing an adult expert. In one case, a student found an "expert"—a state conservation agent with knowledge of blue tongue disease in white-tailed deer—who went to his church and learned that experts could be found in many different places, a notion that fits nicely with our ideas about science as part of a distributed community. The student who wrote about green schools interviewed the principal, confronting him with the question of why they weren't doing more.

Whether student engagement feeds teacher engagement or vice versa is a moot point. "I feel 10 years younger because they're all excited about the topic, and they're just shoving energy back at me," says one SciJourn teacher. "My students turned from 'Do I have to?' to 'I can' or 'Can I?'"

Using a mixed methods approach, Jennifer Hope has sought to describe the increase in science engagement through a Youth Engagement with Science and Technology (YEST) Survey and case studies of three classrooms. Results of the YEST survey were considered across the sample of implementation ($n$ = 368) and comparison students ($n$ = 101), as well as by level of classroom implementation. As expected, the greatest differences between the implementation and comparison groups were seen in the high implementation classrooms. Most dramatic were increases in engagement scores for classes in which students were required (by graded assignment) to write as well as provide peer edits four times during the school year (once each quarter). Persistent engagement in the practice of science journalism yielded the greatest influence on student participation, interest and identification with science and technology (Polman and Hope 2014).

Through extensive observations and phenomenological interviews (Seidman 2005), Hope has identified factors contributing to or detracting from potential engagement in science communication (as revealed in SciJourn). Not surprisingly, choice of topic proved to be key in sustaining interest and many students expressed delight and surprise at this rare freedom. Random choice proved to be less appealing (several teachers had students draw topic ideas from a hat). While several of the more tentative students appreciated the guidance and direction forced choice provided, several who viewed themselves as writers and who may have otherwise excelled in their role as science journalists were alienated by the approach. High school science appears to offer an uncomfortable challenge for students who do not view themselves as future scientists. Happily, students described SciJourn as an opportunity to reconnect with science in unexpected ways.

While expressing initial surprise at the quantity of writing encountered in SciJourn-infused science courses taught by high implementers, all of the students interviewed claimed to have discovered something new and exciting through their writing. For some, this was an informed scientific viewpoint on a topic of personal interest, such as "tennis elbow" or the role of dietary choice in managing diabetes. For others, SciJourn offered an opportunity to talk to family members about topics that were clearly difficult but of profound interest, including depression, cancer, or unexplained death. The authentic nature of the writing task—submitting work to an external editor for potential publication in a real news magazine—also fostered engagement for academically oriented students who viewed the opportunity as a resume builder on their way to college acceptance. A few students seemed to take SciJourn as an opportunity to use new technologies in the service of science, such as podcasts and videos.

## Conclusions

The opportunity to teach good science writing practices is presented to us through *Next Generation Science Standards* (*NGSS*) and the *Common Core State Standards* (*CCSS*); what becomes of that opportunity will depend on the resources and guidance we offer to schools. We conclude this chapter by offering lessons learned from the SciJourn project that we believe can inform these decisions.

## *Science Writing Assignments*

Based on our experiences in the SciJourn project, we argue that effective science writing assignments:

- Are discipline-specific, rather than generic (i.e., five-paragraph essays)
- Exist in the "real world"
- Demand that students exercise science-literacy skills such as those in the SciJourn guidelines
- Enhance, rather than detract, from science curriculum by giving science teachers a way to incorporate relevant, up-to-date science topics and authentic science writing in their courses
- Are accompanied by science-specific criteria and not by generalized rubrics that may limit responses and turn writing into an exercise in following formulas rather than critical thinking

We further recommend that administrators resist the temptation to incorporate schoolwide writing assignments and rubrics; we and others (e.g., Applebee 2012) argue that they will not help students gain disciplinary literacy skills.

## *Science Teaching*

Good science writing demands an understanding of and a privileging of science; good science teachers must be at the center of this instruction. SciJourn science teachers have shown us that effective science writing instruction:

- Emphasizes the reading-writing connection during which teachers model thinking and revision through RATAs and share their own experiences as writers
- Adopts a whole-to-part-to-whole approach to teaching by looking first at the genre as a whole (whether it be the science news story or the editorial or the lab report) and then breaking it apart
- Includes explicit direction, such as targeted internet searches (e.g., "go to the CDC's website") and "actionable" writing feedback (e.g., "add a context paragraph here that includes the number of people diagnosed with this condition")
- Allows students and teachers to learn about new, relevant science topics, thereby emphasizing the dynamic nature of science
- Includes opportunities for content-centered revision under the teacher's direction

We have also noticed that a well-designed science-writing assignment, with appropriate criteria, has changed science teachers' attitudes about writing in science classes. Rather than feeling like they were being asked to take on the English teacher's job, SciJourn teachers realized that their discipline- and genre-specific knowledge made teaching and responding to science writing something they were uniquely qualified to do.

### *Science Students*

Finally, SciJourn has taught us something about the young people that sit in our science classrooms today. Many SciJourn teachers joined the program worried about students who were apathetic and disconnected in science classes. As they implemented RATAs and other activities involving science journalism, these teachers noticed changes in their students. Specifically, they discovered that science students:

- Are interested in science issues that are unusual and/or directly affect their lives
- Want to read science news written by and for teenagers
- Can be motivated by the authentic opportunity to publish their work
- Appreciate direct feedback on their writing
- Enjoy becoming the class "expert" about a topic, especially when giving the opportunity and encouragement to learn more than even the teacher on a particular issue
- Will write and research about science when given free topic choice

The last point deserves special emphasis. While all teachers are, understandably, driven by curriculum guides and content that must be "covered," we have found that allowing students to research and write about any science issue inspires students. If we consider teaching science literacy the responsibility of all science teachers, restricting student topics to, say, chemistry or physics becomes less important than helping students find a topic where they will be motivated to do the research, writing, and revision that good science journalism takes and that is necessary to practice science literacy skills.

Finally, we have all been surprised by which students have the tenacity to become published *SciJourner* authors. The "best" science students by strictly academic measures have not always been successful in SciJourn activities. In fact, some of the most interesting stories have come from schools considered "failing" or classes designated for "non-college bound" or "remedial" students. Throughout our lives, all of us need STEM information in order to function as both intelligent consumers and decision-makers. Good science writing assignments enable teachers to trace the ability of students—our future citizens—to search for information, assess its value in particular and specific contexts, synthesize ideas from multiple sources, and judge the status of evidence within the science community and beyond. SciJourn has helped us to see the potential for and importance of reaching all science students and helping them to acquire the STEM communication skills needed to participate as active and responsible citizens.

## Acknowledgments

This article owes much to the work of Joseph Polman, now Professor, the University of Colorado at Boulder: Cathy Farrar, teacher at Rockwood-Summit (MO) Schools and Jennifer Hope, Assistant Professor at McKendree University (IL).

This material is based upon work supported by the National Science Foundation under Grant No. DRL-0822354. All statements are the responsibility of the authors

## References

Aikenhead, G. S., G. Orpwood, and P. Fensham. 2011. Scientific literacy for a knowledge society. In *Exploring the landscape of scientific literacy*, ed. C. Linder, L. Östman, D. A. Roberts, P. O. Wickman, G. Erickson, and A. MacKinnon, 28–44. New York: Routledge/Taylor and Francis Group.

Applebee, A. 2012. Great writing comes out of great ideas. *The Atlantic.* September 27. *www.theatlantic.com/national/archive/2012/09/great-writing-comes-out-of-great-ideas/262653*

Bazerman, C. 2009. Genre and cognitive development: Beyond writing to learn. In *Genre in a changing world*, ed. C. Bazerman, A. Bonini, and D. Fifueiredo, 279–294. Fort Collins, CO: The WAC Clearinghouse.

Bellamy, P. C. 2005. *Seeing with new eyes: Using the 6 + 1 trait writing model.* Portland, OR: Northwest Regional Educational Laboratory.

Board on Science Education (BOSE). 2013. *Monitoring progress toward successful K—12 STEM education: A nation advancing?* Washington, DC: National Academies Press.

Bybee, R. W. 2010. Advancing STEM education: A 2020 vision. *Technology and Engineering Teacher* 70 (1): 30–35.

Bybee, R. W. 2013. *The case for STEM education: Challenges and opportunities.* Arlington, VA: NSTA Press.

Chapman, O., and A. Russell. 2005. Calibrated peer review. *http://cpr.molsci.ucla.edu*

Feinstein, N. 2011. Salvaging science literacy. *Science Education* 95 (1): 168–185.

Kohnen, A. M. 2012a. A new look at genre and authenticity: Making sense of reading and writing science news in high school classrooms. Doctoral dissertation. St. Louis, MO: University of Missouri–St. Louis.

Kohnen, A. M. 2012b. Teachers as editors, editors as teachers. In *International advances in writing research: Cultures, places, measures* ed. C. Bazerman, C. Dean, J. Early, K. Lunsford, S. Null, P. Rogers, and A. Stansell, 303–317. Anderson, SC: WAC Clearinghouse/Parlor Press.

Kohnen, A. M. 2013a. The authenticity spectrum: The case of a science journalism writing project. *English Journal* 102 (5): 28–34.

Kohnen, A. M. 2013b. Content-area teachers as teachers of writing. *Teaching/Writing: The Journal of Writing Teacher Education* 2 (1): 29–33.

Kohnen, A. M. 2013c. "I wouldn't have said it that way": Mediating professional editorial comments in a secondary school science classroom. *Linguistics and Education* 24 (2): 75–85.

Krajcik, J. S., and L. M. Sutherland. 2010. Supporting students in developing literacy in science. *Science* 328 (5977): 456–459.

Langer, J. A. 2011. *Envisioning knowledge: Building literacy in the academic disciplines.* New York: Teachers College Press.

Lemke, J. L. 1990. *Talking science: Language, learning, and values.* Westport, CT: Ablex.

National Governors Association Center for Best Practices, Council of Chief State School Officers. 2010. *Common core state standards.* Washington, DC: National Governors Association Center for Best Practices.

National Research Council (NRC). 2010. *Exploring the intersection of science education and 21st century skills.* Washington, DC: National Academies Press.

National Research Council (NRC). 2011. *Successful K–12 education: Identifying effective approaches in science, technology, engineering and mathematics.* Washington, DC: National Academies Press.

O'Neill, D. K., and J. L. Polman. 2004. Why educate "little scientists?" Examining the potential of practice-based scientific literacy. *Journal of Research in Science Teaching* 41 (3): 234–266.

Polman, J. L., and J. Hope. 2014. Science news stories as boundary objects affecting engagement with science. *Journal of Research in Science Teaching* 51 (3): 315–341.

Polman, J. L., A. Newman, E. W. Saul, and C. Farrar. Forthcoming. Adapting practices of science journalism to foster science literacy. *Science Education.*

Polman, J. L., A. Newman, C. Farrar, and E. W. Saul. 2012. Science journalism. *The Science Teacher* 79 (1): 44–47.

Pressley, M., and P. Afflerbach. 1995. *Verbal protocols of reading: The nature of constructively responsive reading.* New York: Routledge.

Rutherford, F. J., and A. Ahlgren. 1989. *Science for all Americans.* New York: Oxford University Press.

Saul, E. W., and D. Dieckman, D. 2005. Choosing and using informational trade books. *Reading Research Quarterly* 40 (4): 502–513.

Saul, W., A. Kohnen, A. Newman, and L. Pearce. 2012 *Front page science: Engaging teens in science literacy.* Arlington, VA: NSTA Press.

Schiefele, U., E. Schaffner, J. Moller, and A. Wigfield. 2012. Dimensions of reading motivation and their relation to reading behavior and competence. *Reading Research Quarterly* 47 (4): 427–463.

Seidman, I. 2005. *Interviewing as qualitative research: A guide for researchers in education and social sciences.* 3rd ed. New York: Teachers College Press.

Spandel, V., and R. J. Stiggins. 1997. *Creating writers: Linking assessment and instruction.* 2nd ed. White Plains, NY: Longman Publishers.

# Urban STEM

## Watch It Grow!

*Mary Hanson*
*Humboldt High School*
*Saint Paul, MN*

### Setting

Humboldt High School is located on the West Side of Saint Paul, Minnesota. Enrollment consists of approximately 836 students in grades 7–12. The population is 40% English Language Learners and 95% are eligible for free-and-reduced lunch. The Special Education and DHH students make up a combined 25% of the student body. Humboldt is ethnically diverse: 41.1% Asian, 29.4% African American, 25.4% Hispanic/Latino, 8.0% Caucasian, and 1.9% American Indian. These percentages show an overage of 5.8%, which is attributed to some students claiming a mixed race status, which leads to double counting them. Statistically, these numbers reflect our rapidly changing world.

The case study described here involves my 50 accelerated chemistry students, grades 10–12. These students range in age from 16 to 18. Of these 50 students, 3 are in the top 10 of their respective classes, 24 have parents that are divorced or were never married, 20 live with only mom, grandma, father, or sister, 38 are on free-and-reduced lunch, 20 have no phone or internet at their residence, 6 have parents that have died, 4 have been arrested, and 12 have admitted gang affiliations. This intimate knowledge of my students was derived from regular personal communication with them.

The nature of our demographic and economic population certainly presents challenges. The overwhelming disenfranchised nature of this population suggests that their ability to perform is hindered, but workable solutions and resources are bountiful, regardless of a host of social, cultural and economic factors contributing to academic performance. As educators we often spend considerably greater time with these young people than their families do. This provides us with a real and tangible opportunity to gift our students with much needed skills for life.

### Program Overview

Gone are the days of presenting students with notes, answering questions from a chapter in the assigned text, taking a test, and moving on regardless of student performance. Gone also are the days of expecting students to arrive to class mentally, emotionally, socially, intellectually, and academically prepared to absorb a teacher's content delivery. Additionally, gone are the days of the teacher-centered focus of student achievement. So where does that leave us and more importantly where does that lead our students?

Education today is an enormous challenge. Educators hoping to be successful must stand firm in the belief that the students are always and infinitely more important than the curriculum. Educators must be present and engaged in the planning and execution of lessons as we expect our students to be. Without the prior establishment of solid, trusting relationships, student learning will falter.

Genuine rapport between students and teachers is the foundation that, when fostered throughout our educational time together, removes barriers to the learning process. Good teachers understand this but great teachers practice this religiously in an effort to ensure that students will not only grow, but rather flourish academically, socially, emotionally, and mentally, thus preparing intrinsically motivated young people. In the past, student merit, validity, and worth were measured solely by academic marks. Their future successes in life were determined by whether they could perform academically and assumed that their level of "life" successes would be determined by how much higher their grades were than their peers. For the teacher, this meant delivering content in an organized and structured manner, but ignoring important community and world connections that strengthen relationships between academic life and the world and naturally engage our students in a more rigorous exploration of science around us every day.

Education today charges educators with far more than delivering content if we subscribe to the broader purpose of cultivating thoughtful, responsible young people. We know that to solidify content knowledge we must first meet seemingly very basic, fundamental needs before we can achieve more lofty goals of actually educating our young people. Many of us, especially in the science disciplines, lose sight of this and proceed with content delivery and expect to see measurable growth. An indicator of a high IQ is actually someone who learns from mistakes and changes their methodology to achieve different, more desirable results. Science teachers of all people should know and understand this because we practice it routinely. We have all read and applauded brain-based research as it pertains to juvenile development, because it provides a framework for the new direction of STEM (Science, Technology, Engineering, and Math) education today. Intentional and creative uses of innovative brain-based research strategies are the footholds educators need to make measurable strides in student cognition. It is imperative that we, as educators, help our students learn to organize content and establish more solid neural connections between content, effectively facilitating more complex cognitive processing (Sousa 2006, p. 247). This can be accomplished by the deliberate application of STEM methodology.

## Arguments Against STEM

In some circles, the perception is that hands-on learning and inquiry-based learning is the same. They are definitely not the same thing! "Materials such as Styrofoam balls and toothpicks for building molecular models, dominoes, base ten blocks, tangrams, spinners, rulers, fraction bars, algebra tiles, coins, and geometric solids" (Thompson 2009), represent manipulatives used in hands-on learning. Manipulatives, while valuable, primarily benefit the tactile/kinesthetic learners. Hands-on learning meets these learners' needs by allowing them to experience their world via touch, through a preferred learning modality. In contrast, the inquiry approach as inquiry suggests, encourages students to explore self-generated ideas. "In an inquiry learning environment, students can use a scientific approach to learning in order to discover rules of the domain or solve problems. Inquiry learning processes include orientation, generating hypotheses, testing these hypotheses, and drawing conclusions" (Kuhn, Black, Keselman, and Kaplan 2000). The inquiry-based approach offers a stronger student-directed freedom to learning. STEM should be inquiry-based, and can significantly deepen the investment a student is able and permitted to make in his or her own learning process.

The argument that STEM initiatives apply only to the disciplines of science and math is invalid. By that token, some could argue that Dr. Angela Peery's "write to learn" strategies only hold value in English Literature classes (Peery 2009). Reading and writing do not discriminate themselves to one discipline since at our very core we all are reading and writing teachers prior to our chosen area(s) of licensure. Those that can read and write well can communicate, contributing to a desirable outcome of STEM implementation.

A common response to suggested STEM inclusion in the classroom is that schools must provide more technology and engineering offerings. While a step in the right direction, at best that simply provides more opportunities for students to gain exposure to these disciplines. It neglects the need for students to develop true competencies by applying STEM outcomes when examining global issues or interpreting data or synthesizing information. Aleman (1992) and Darling-Hammond (1994) asserted that "Employers are looking for employees who possess the skills that are taught in STEM programs, including creative problem solving, product building, collaborative team work, design, and critical thinking" (Ejiwale 2012). If we as educators do not provide these opportunities for our students, who will? Colleges and universities expect students to use them routinely, and employers will assume solid competence when hiring.

## What STEM Means

For roughly 20 years since the National Science Foundation's inception of STEM, educators have been trying to come to a consensus regarding what STEM teaching is, what it looks like in the classroom, and the manageability of STEM integration in the high school setting. The brightest, well-intentioned educators have worked tirelessly to provide workable constructs for the STEM process. Some schools even tout themselves as STEM schools. The work done over the past two decades has been very good, but has not moved us any closer to a universal definition or application of the STEM process. Presented here is a redefining of what STEM looks like and feels like in an urban high school classroom setting. This model works! It is engaging, meaningful, and fully complements the state-mandated standards that form the foundation of the science content to which educators are held accountable. The data presented shows measureable academic growth during the progression of a unit.

## What STEM Looks Like

So what does STEM look like and feel like in the high school setting? How do we make this a reality without it feeling like a top-down unilateral directive absent of the elements of accountability? How do we make this as real for the students as it is for educators? Most importantly, how do we do this without feeling like it is an additional responsibility imposed on us? Truly employing STEM methodology is not the addition of anything new. Rather, it is merely the same content in a new way. Any time we are told to do more, we take the focus off what we should be doing. We need to prune our content, focus on essential, imperative content and not add any new content. We simply need to adjust the way we expose our students to the content.

The contribution offered here is an explanation of what STEM means and how it looks in the classroom. We know what the acronym STEM stands for but what does it mean?

## *S = Science*

This is the science content dictated by state standards, including priority and supporting standards, benchmarks and so on. This is what we typically think of when we describe science education.

## *T = Technology*

The rigorous and regular use of all forms of technology at our disposal constitutes technology. Common technologies such as computers and interactive whiteboards are necessary, valuable, and engaging for students. However they are just the tip of the iceberg. Equipment such as probes, digital microscopes, document cameras, and shoot-and-share cameras are technologies with which many students have not had exposure, but are extremely valuable in helping bridge the gap, the disconnect between high school and industry. Technology includes using interactive technologies for educational purposes. Technology that students have had exposure to such as tablets and cell phones also can be of value if used in ways new to the students and in conjunction with appropriate web applications. One application that has proven engaging is a site called Poll Everywhere, a free interactive site where teachers post questions and students text in their responses via cell phones or the internet. Computers can be hooked up to the LCD projector so all students can enjoy posted questions. Not all students have cell phones. The sensitive and responsive approach to combat this reality is to establish cooperative learning groups where one member of the group has a cell phone. Students engage in a scientific conference and make one response after group collaboration. An example question could be "What is the most important chemistry concept you learned this week?" The educator circulates and eavesdrops on students' discussions as they formulate their responses and refine their thinking as a group. Instructions to student groups are that if their group response appears on the screen, they need to make another contribution so that we can capture the greatest number of thoughts and ideas as possible. Some of the responses students generated in this particular instance were the following:

- "Prefixes for covalent bonding"
- "Anions and cations"
- "We learned about writing and naming formulas for ionic and covalent bonding"
- "I learned the difference between ionic and covalent bonding"
- "We learned to determine which differences can be observed to classify bonds as either ionic or covalent"
- "We learned about the difference between covalent and ionic bonding and how to classify them by the periodic table and their anions and cations"

It is evident that the responses become richer as this process continues.

## E – Engineering

Engineering encompasses designing, building, assessing design results, modifying design results, and repeating the inquiry-based tactile/kinesthetic approach while applying coherent science concepts infused with a strong logic base. The logic element and students' abilities to apply logic to given situations is difficult to quantify. Likewise, measuring individual growth in this area is challenging. Hence the logic pre/post-assessments frequently administered, while not perfect, certainly gauge student cognitive standing and logic development.

Mayes and Koballa (2012) offer a workable list of engineering practices aimed at just these objectives. Summarizing their well-articulated process looks something like this in my classroom: Establish the problem, use models to visualize the problem, carry out an investigation, analyze the data mathematically, critically and logically, communicate findings orally and in writing, and determine solutions.

The engineering component of STEM is laborious and time-consuming but extremely critical. It cannot be accomplished independent of science content, technology tools for the assessment of scientific problems and measurement of the validity of solutions, or the infusion of mathematical rules/laws that govern our universe. As challenging as this might sound, this is the area where students can grow immensely. It is the area of STEM that bridges the gap between high school and industry and it is best aligned to our basic human instincts and desires. Additionally, the communication elements of STEM are emphasized here, whether they are oral or written. Oral communication, however, allows students an opportunity to navigate relatively challenging social matrices in a safe and healthy environment, contributing to the development of socially skilled young people. Equally as important to identifying workable solutions achieved during the design process, are opportunities to play "devil's advocate," by challenging the ideas of others, and developing deeper responses and solutions. Students deserve to practice counter-argument development along with classical argument defense. Students lack the skills necessary to advocate for themselves, which is a lifelong skill imperative to their social development. Engineering is the crux of STEM!

## M = Math

In a science class, math is not taught. Instead, mathematical concepts that students learn in their various math classes are heavily reinforced. Learning adventures that require mathematical aptitude suitable for varying cognitive developmental levels are always designed. Every inquiry must involve math to a degree. We should expect to hear students say things like "Is there any math in this? How can we use math to help us?" The students utilize math in classes every day and are expected to demonstrate competence in this discipline. Math and the multitude of applications and value it holds is not only inherently valuable but also critical, but the *math concepts themselves are not taught*.

## Psychology

Let's go back for a moment to our basic human instincts and desires, which could themselves be chapters alone. In addition to all that STEM is, STEM thinking is an innate and fascinating component of the human psyche, fundamentally driving our explorations of every encounter we have. The Engineering component of STEM is an organizational framework, which is laden with developmental psychological constructs. Just for a moment, lets confuse (and enrich) the experience even more. Adam Cash, author of *Psychology for Dummies*, shares information regarding understanding psychosocial development. It is a wonderful starting point for understanding the psychology of our adolescent population. In it, he presents Albert Bandura's social-learning theory of personality development, which claims that part of our developing personalities is dependent on "observational learning experiences from those around us" (Cash 2002, p. 146). The term *self-efficacy* is defined by Bandura as being largely based on the self-appraisal process, the process of "analysis of one's actions and the evaluation of successes and failures" (Cash 2002, p. 146). Sounds like the evaluation, analysis, and communicative pieces of a successfully implemented engineering endeavor, does it not? Bandura claimed that our perception of our capabilities stem (pun intended) directly from this. Our perception of our capabilities and our actual capabilities may not truly be aligned, much like a student's preferred learning modality might not be aligned with their strongest learning modality. In addition to presenting science content, explaining the application of the STEM model, and making psychologists worldwide cringe at these oversimplifications, a basic understanding of cognitive, behavioral, and psychosocial development is necessary. Where are our students socially, emotionally, and cognitively? Do we know? Should it matter? Should we be intrigued by their behaviors and what motivates them? The answers to these questions are paramount and educators should make a practice of understanding as much about our student populations as possible. "Expert teachers don't just observe student behavior; they work to understand the affect that drives behavior so they can guide students in a positive direction" (Tomlinson 2010, p. 16). Understanding our students psychologically also positions us to practice tolerance and patience when working to support them during their academic growth.

## Instructional Program

The term *STEM* has been casually tossed around in certain circles, but to what end? Currently we have educators who claim an understanding of STEM, but have not put the initiatives into practice. It is often met with much resistance unnecessarily. If more educators understood the definition of STEM, the potential value these practices provide for our students, and the ease with which STEM components can be applied within their own classrooms, STEM practices would become more widely embraced. As educators claiming to prepare our students for the 21st century, we need to reflect on our pedagogical practices in hopes of employing this educational practice. Imagine for a moment that we completely understand our charge as educators of knowing the science content standards; understand a rigorous approach to applying STEM methodology within each unit of study; can embrace student learning styles and interests to target instruction to meet specific individual student needs; and could distinctly understand the cognitive, social, and developmental psyches of our students. Wouldn't that be a powerful experience? Educators need to take the initiative to develop solid science content understanding; apply STEM methodology routinely, deliberately, and with intentionality; help strengthen student logic skills; commit to understanding psychology; and attempt to bridge the gap between high school standards and industry practices. Benefits for our students are a sustainable program that fosters academic growth, critical thinking skills, social development, confidence and competence, and a resilient persona. Taylor and Thomas (2001) concur with the importance of developing resiliency in students, ultimately aiming "to gradually allow students to take some, then more and more, control of their own learning" (p. 12). We need to present ourselves as unified and consistent in reaching our students socially and emotionally, we must satisfy our students' intellectual hunger, and we must offer challenges beyond what our current construct suggests is sufficient.

It is imperative that STEM practices become the norm. Further, the non-inclusion of STEM practices in classrooms is a huge disservice to the developing minds of our young people. As such, an intrinsic need exists to synthesize thoughts and ideas surrounding STEM implementation in the classroom. What does it look like and feel like in action? It is my hope that some clarity surrounding STEM implementation will be gleaned from this chapter. This chapter is devoted to showing an understanding of and a working model of STEM practices. Employing STEM methodology is an achievable, measurable, and sustainable objective. An implementation of STEM practices in one unit of study in chemistry will be shared, emphasizing the exposure that students need to technology, but not necessarily just new, novel, cutting-edge technology that the majority of us do not have a sufficient budget for. Emphasis on the engineering design process, which involves creativity, logic, and social communication skills while encouraging personality development, will also be addressed. This entire endeavor begins and ends with the students. Building a solid rapport, understanding and embracing distinct learning styles and multiple intelligences, accounting for student interests, and making concerted efforts to engage students on every level is of utmost importance. STEM methodology encompasses all this and benefits our students in incalculable ways.

The primary message to be conveyed is that STEM education is an achievable objective. It is applicable to students during their high school careers. It is necessary and critical to their lifelong development cognitively and it is manageable in the high school setting. Student achievement is measurably greater and deeper using STEM methodology, provided we first commit ourselves to our ultimate goal as educators: to provide rich and rigorous experiences that will, in a very tangible and measurable manner, foster the development of citizens who are ready and able to be productive and responsible contributors to society. That goal must envelop the forefront of our minds and dominate our thinking as we design learning tasks that address each of the STEM competencies, while supporting and enriching our state-mandated standards and benchmarks. It must be a clearly synergistic relationship.

The connection between technology and other school-taught disciplines is that often times, technology is perceived as an island functioning independently. Understandably, the argument can be made that teachers of technology are generally not licensed to teach science or math and vice-versa. However, through interdisciplinary collaborations, the technology of STEM initiatives can be incorporated into virtually every classroom. The depth of content understanding a student, and quite frankly an educator, can experience through a creative collaboration using standards-based curriculum and content is valuable to achieving each of the STEM goals. The value cannot be ignored or dismissed as we have a duty, a responsibility, and a moral obligation to provide these connections to our students. Even the most intrinsically motivated students need our guidance to help them discover these connections.

## Student Engagement

The question of how to engage our students must be considered. It is a learned art stemming from years of trial and error. When articulating the magic that happens in the classroom when students are authentically engaged and applying their knowledge can be challenging. Most educators recognize when it is happening and truly dedicated educators quickly seize those precious moments and encourage the involvement and deepening of student understanding. The balancing act of presenting yourself as a competent educator with a commanding presence without becoming a dictator of sorts is not only a precarious balancing act, but also a struggle for most educators. For those who can do it well, there are seamless transitions, meaningful dialogue among students, evidence of higher-order thinking, a deeper content of understanding and an observed ability on the part of the students to make significant interdisciplinary connections outside the classroom environment. All of this transpires simultaneously during the class period,

Gauging the level of engagement of students is frequently determined by a host of observational facts, such as time on task, asking thoughtful questions, initiating meaningful dialogue with classmates without prompting, and bombarding the teacher with questions whenever they experience confusion or excitement about a topic. More importantly are the behavioral pieces exhibited by students that show a high level of engagement during class, as well as taking their learning outside the classroom. My students are engaged in both the science content and the overall experience when the following occurs:

- Students come to class early.
- Each day about 50% ask "What are we going to do today?" If I see other students in the hall prior to their scheduled hour, they ask the same thing.
- Students deny passes to guidance saying "I don't want to go."
- When class is over, many stay late asking "Can you write me a pass to my next class? I want to stay and finish this."
- "Can we take another formative assessment today because I get it now!"
- When students turn in formative assessments, they say "Check it now. Did I get it?" with the enthusiasm of a child ready to open a present.
- They stop by between classes asking "How did I do on the assessment? Was I proficient this time?"
- Frequent comments at the beginning of class, "How much do you have to talk today because I want to get started."
- They ask me if they can come in and work during their lunchtime or during their study hall—choosing learning time over time to socialize.
- When students come to class early, they immediately begin writing problems on the board for their peers to solve, prior to the start of class.
- Without prompting I hear students say to their peers, "Give me your work. I want to evaluate it." Very rich discussions transpire as I listen to them peer evaluate and defend their positions.
- Students typically turn in work before the scheduled due date, setting the pace for their own learning.

- Students often ask, "What are we going to learn next? Can I get it so I can do it over the weekend?"
- They check for interdisciplinary connections asking "Can I go next door and ask the teacher if this has anything to do with biology?"
- They take ideas home. "You said they put salt on the road to make the ice go away but when I put in on my steps it just made holes in it and my mom got mad that I used all the salt."
- Parent comments during conferences frequently involve them saying: "Tell me more about_______. My son came home and can't stop talking about it."
- I currently have nine TAs (teacher assistants,) five males and four females. Eight of them are current students who have chosen to come back for an additional class to TA. The majority of the time they are working on their class work rather than TA responsibilities.
- As inevitably happens from time to time, a student will be asked to leave another class for a variety of reasons and told to go to ISS (in school suspension) for the remainder of the hour. They somehow gravitate to my room saying "I don't want to go to ISS. Can I just stay in here and work?"

Oftentimes, the students are so engaged and intrinsically motivated that with a small amount of guidance, the classes can run themselves. Allowing students to work out their designed problems that they write on the board for peers, while not a component of the lesson plan for the day, needs to be honored and contributes to student engagement. Allowing students to sit in my class and work over lunch, for example, allows them to continue their learning, which is completely of their choosing, and demonstrates a high level of engagement. Establishing a climate, immediate feedback on formative assessments, building confidence in students by affirming what they have done successfully, showing students areas for improvement and allowing them to try again in that exact moment, encourages students to help them deepen their intrinsic motivation. This translates into a level of commitment to science that maintains a positive association between science and academic performance. All of these student-generated comments are a direct result of the inquiry-based STEM methodology implemented in my units of study. STEM reaches students on a much different level than the strict delivery of science content alone. It frees students to explore their interests in a constructive and healthy way.

## Evidence of Success in Classrooms

Good teaching in STEM practices begins with getting to know the students. In order to be an effective educator, we need to know our students' learning preferences, their learning capabilities, and their interests so that all three can be married into a harmonious and unified learning endeavor. Frequently, when examining students' multiple intelligences, we will see that their preferred learning modality often is not in keeping with their strongest learning modality. Some may argue that spending educational time making these determinations is a waste of time. See it rather as another learning opportunity, for students as well as the educator. It effectively allows educators to assign learning groups based on preferred modality, strongest capability, or an even richer grouping of students representing each of the intelligences and preferences. The heterogeneous grouping creates a deeply rounded learning cohort where each student has his or her own unique contributions to make to the group. Engagement and motivation increases naturally. Determining student interests and engaging those interests when appropriate shows you care and helps establish a solid rapport necessary for student investment in learning objectives. In short, we should aim to do everything in our power to show students that their academic success, personal growth, and cognitive development is valued above all else.

It is widely accepted that we each possess a dominant learning style preference. It is also widely accepted that each person is unique and harnesses their preferences differently. There is strong disagreement on exactly how to determine preferred learning style modalities. My approach gives the educator a snapshot of the cognitive processes that each of the students possess. The assessment that I have used for the past six years comes from David Sousa's (2006) book, *How the Brain Learns*. This assessment is designed to determine one's sensory preferences, has been proven reliable in terms of individual results and individual student perception regarding their strengths.

## Student Successes

The photo below displays the classroom learning styles. Each class is color coded by sticky notes and lined up horizontally so that data can quickly be analyzed visually. Students also can see this every day, which takes the onus off educators when explaining why we need to do what we need to do. Transparency is critical to give meaning to students. The results of the learning styles questionnaire administered and pictured below produced 16.7% auditory learners, 22.9% visual learners, and 60.4% tactile/kinesthetic learners in my classes. When structuring lessons, this display shows by hour what is the most prevalent learning modality, and quick determinations regarding an approach that would be most conducive to student learning can be made. Additionally, student names on sticky notes show who in each class is most comfortable with each learning style. This visual display also holds immense value for another reason. On the occasions when more oral direction is necessary, we can, with purpose, place one auditory dominant student in each group to be assured that someone in that group will have heard and processed the instructions.

Inherent in science and engineering is a need for well-developed, logical minds. In addition to the learning styles preferences, a logic pre-assessment is commonplace in my practice. This assessment is comprised of math, logic, and verbal questions that gauge student abilities in each of these areas. The data is analyzed individually and collectively by class to paint a picture of student cognitive capabilities. Throughout our educational journey, strategies to increase student thinking and reasoning abilities are routinely practiced. For example, my practice involves a logic question starter each day. Some educators call this the "Do Now" approach (Lemov 2010, pp. 152–153). The questions get the students thinking as they enter class. Ideally it takes about 3–5 minutes for individuals to answer, followed by a class discussion. "Do Now" questions can be anything that requires no assistance on the part of the teacher. My questions focus heavily on math, logic, and verbal reasoning skills, rather than specific science content or a previous day's review. Learning style preferences and capabilities coupled with daily logic class starters set the stage for the learning that happens as the year progresses. When our educational time together draws to a close, the same logic pre-assessment is used as a postassessment and again the results are analyzed to measure the effectiveness of this strategy.

The growth seen is statistically significant as the data table (Table 10.1) and graph (Figure 10.1, p. 173) indicate. The data calculated using my accelerated chemistry population is as follows: The average growth between pre- and postassessments was 6.11 points. The standard deviation was calculated to be + or –7.34 points. 68.2% of scores fell within one standard deviation below the mean, 18.2% of scores fell within one standard deviation above the mean and 9.1% of scores fell within two standard deviations above the mean; 4.5% of scores remained the same between pre and postassessments, but did not worsen. Even though the majority of student growth scores fell within one standard deviation below the mean, scores represent positive growth. While my students never achieve Mensa-level standing, they leave with a stronger skill set than when they came into the course.

**Table 10.1.** Pre- and Postassessment Scores From Accelerated Chemistry Students

| Subject | **Logic Pre-Assessment Data** | **Logic Post-Assessment Data** |
|---|---|---|
| **Math** | 62.8% | 78.4% |
| **Logic** | 54.1% | 71.8% |
| **Verbal** | 58.0% | 82.5% |

Analysis of the postassessment increases shows the following: 97.7% of students increased their score between pre- and postassessment. The remainder had postassessment scores remaining the same, but not worsening; 47.5% increased all three score areas of math, logic, and verbal reasoning; 40.0% increased in two areas, and 12.5% increased in one area. Additionally, growth experienced between genders shows that females had a slightly higher increase in overall growth compared to males, interesting and worthy of further exploration. Students were so engaged when they completed their postassessment that many said things such as "How did I do?" "Am I proficient now?" "Correct mine right now for me!" "How much did I improve?" Many students wanted to stay late to get their results, which were provided immediately. Providing immediate feedback continues the intrinsic motivation and self-directed learning behaviors that educators should make a priority to instill in their students. One of the primary skills of a self-regulated learner is that they practice "self-judgment skills" (Cullen, Harris, and Hill 2012). Requesting

**Figure 10.1.** Pre- and Postassessment Scores From Accelerated Chemistry Students

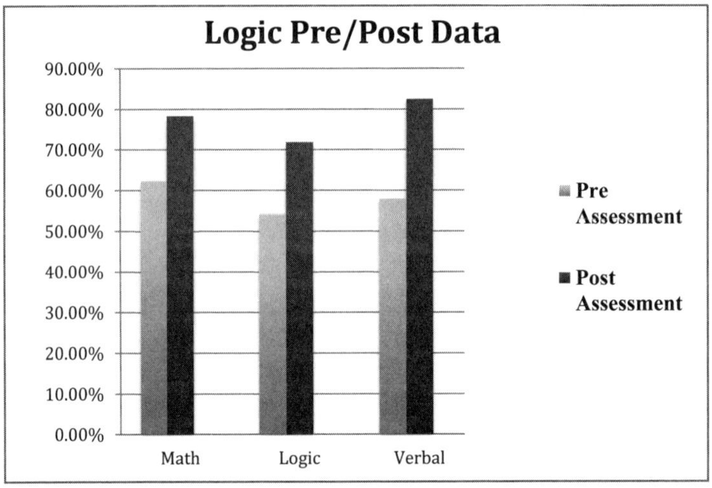


performance results clearly demonstrates growth toward becoming self-regulated learners, thus validating this class practice.

Why the heavy emphasis on logic? There are several reasons. The ability to think critically depends on a person's ability to think logically. Being able to think, according to independent educational consultant and PhD Susan Brookhart, "means students can apply the knowledge and skills they developed during their learning to new contexts" (Brookhart 2010, p. 5). The "E" in STEM requires an aptitude for logic if we expect our students to grow toward becoming innovative and creative. Additionally engineering provides an outlet for the critical-thinking opportunities we want for our students. Engineering naturally encourages this creativity and inquisitiveness, which we desire in our classrooms and require in industry. Cullen, Harris, and Hill (2012) spend a great deal of time emphasizing the role of creative thinking, stating that "Creative thinking thrives in environments that offer individual freedom, alternative thinking, safety in risk-taking, and collaboration and teamwork," practices inherent in STEM implementation. Additionally, preparing our students for the 21st century by offering opportunities for students to develop "social skills, non-routine problem solving, self-management/self-development and systems thinking" (NRC 2010) is completely in keeping with the STEM philosophy.

It would be irresponsible to not mention the reality that a sizeable disconnect between the science content standards we teach and what is practiced in industry exists. As educators we have a duty and responsibility to close that gap as much as possible. Emphasizing logic reaps its rewards in science classes, but also significantly benefits students' cognitive capacities in other disciplines and arenas of their lives.

The question posed might be, "What is your definition of a student's developmental level mathematically? Is it merely a matter of checking class schedules and transcripts to see where students are and have been?" The answer is no. Students' math RIT scores, known as the Rasch

Score, developed by Danish mathematician Georg Rasch, must be checked and fully examined. The Rasch scale is an equal interval achievement scale used to measure academic growth over time (*www.NWEA.org*). The Rasch scale, or RIT ranges, is accompanied by a list of competencies to be mastered as provided by Descartes. Descartes lists the RIT range for a student, as well as "next steps to master" section. Analyzing and charting student mathematical development is a tedious process. It takes time, but is necessary to match student competencies with science concept development necessities. An overall class average and list of abilities that students should be able to perform, along with a next steps abilities list, can be gleaned using these data. These ability lists guide the educator to the areas where efforts should be concentrated. Additionally, each student's mathematical strengths and weaknesses are clearly evident, which chart the course and provide the roadmap for when, and to what degree, individual student challenges should be presented. Using this knowledge during STEM unit implementation provided the growth data displayed below in Table 10.2 and Figure 10.2.

**Table 10.2.** Ionic/Covalent Bonding Assessment Data

| | Preassessment | Formative Assessment | Postassessment |
|---|---|---|---|
| **Proficient** | 0% | 59.2% | 87.1% |
| **Close** | 0% | 14.2% | 8.9% |
| **Far** | 0% | 15.7% | 3.0% |
| **Needs Time** | 100% | 11.0% | 1.0% |

**Figure 10.2.** Ionic/Covalent Bonding Assessment Data

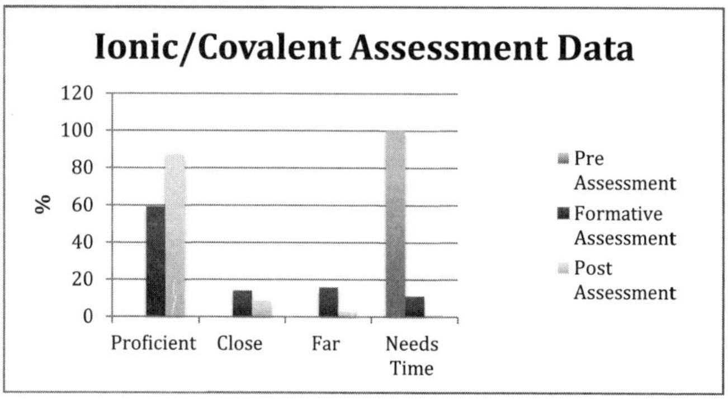


Statistical analysis of the above data is as follows: Average growth experienced with content was 15.83 points. The standard deviation was calculated to be + or – 16.60 points. 47.6% of growth scores fell within one standard deviation below the mean, and 52.4% of growth scores fell within one standard deviation above the mean. In content assessment, 100% of the students experienced growth. The minimum growth was 2 points and the maximum growth was 25 points. STEM implementation clearly produces quantitative measurable results in terms of student academic performance.

Appendix A illustrates how each of the STEM components are naturally embedded within the unit of Ionic and Covalent Bonding in chemistry. The chemistry standard provides the foundation for the incorporated STEM components. It is important to note that all of the National STEM standards, National Science Standards, Science as Inquiry standards A1–A6 generated but never fully adopted in 2006 naturally apply to explorations such as these. Additionally, the current *NGSS*, *Next Generation Science Standards*, adopted April 9, 2013, have been incorporated into this unit.

## Student Testimonials

The Ionic/Covalent Bonding assessment data collected during this STEM implementation speaks loudly regarding the successes and deepening of content understanding students were able to demonstrate. While impressive, it is important to take this a bit further by asking students for written responses to the unit. Self-reflection and performance analyses are critical components of STEM for the students, and information gleaned from student responses aids educators in making future instructional decisions that will have a greater and more targeted impact on student learning and academic achievement.

Students were given the following prompts:

- In my opinion, what is the value of STEM?
- How has STEM helped me succeed in this unit?
- How can I claim that my learning was my own?
- What was Ms. Hanson's role in helping me learn?
- How has this type of learning prepared me for college?
- What will I never forget about this learning experience?

These questions were typed and students were encouraged to respond freely, taking all the time they needed to process. Some of the responses generated are provided below.

- "The value of STEM is very important because it's a hands-on learning experience that makes me think more and makes me connect it to my own experience in learning."
- "I think STEM helped me connect everything and show how they are related with each other. Ms. Hanson teaches us by figuring things out ourselves and using our own strategies to figure things out. She doesn't give us the answer."
- "I think that the value of STEM is to help us prepare ourselves to solve situations and explain with more detail about how we achieved our results."
- "This type of learning prepared me to work individually and investigate using other resources outside the classroom and experience my own way."
- "STEM gives me the ability to learn freely and makes me remember what I am learning."
- "STEM allowed learning to occur with interactions and intriguing lessons. The benefit was invested learning and collaboration with my classmates."
- "It made me think more."
- "STEM helped me a lot because I had the chance to have my own opinion and come up with my own questions and experience experimenting and go deeper with my work."

- "STEM helped me succeed by controlling my own learning. It gave me skills to control my own learning. Ms. Hanson was there to provide explanations and keep me on track of working on my experiment."
- "Ms. Hanson's role in helping me learn was to explain the steps of doing research. It helped me understand my learning style and how to work in a group and individually."
- "It helps us to work in a deeper way so that we can successfully reach the goal of knowing the topic."
- "The lab reports helped me to organize all my thoughts and what I learned, what my results mean and how I came to that result."
- "STEM helped me learn how the structure of how technology, engineering and math work to make me succeed. It prepared me to be ready and be on my own and more outgoing and be brave."

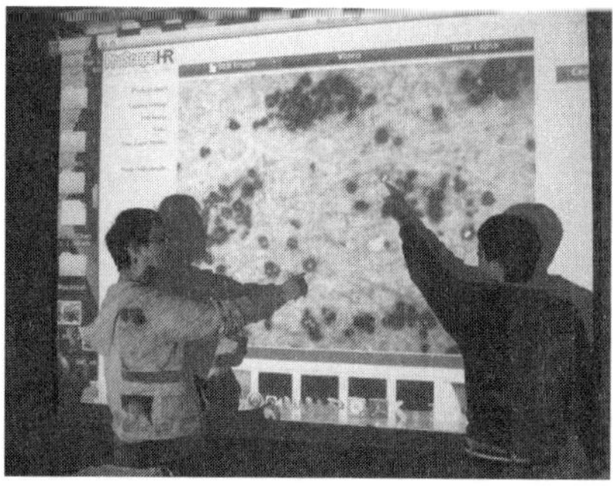

Examining crystal structure with the Proscope digital microscope

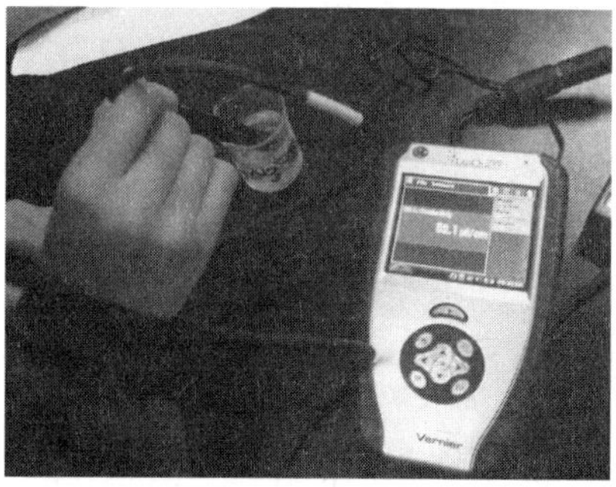

Research and measuring conductivity using the Vernier LabQuest technology

Presenting of the fuel source Xylene

## Next Steps

To achieve complete buy-in to any program or directive, the benefits must be explicitly stated to all parties involved. Frequently buy-in to STEM initiatives is met with great resistance by educators because it is viewed as applying exclusively to the disciplines of math and science. On the surface that may be the appearance, but the reality is it's simply untrue. Science and math are integral parts of every discipline taught K–12. We need to apply STEM initiatives in an effort to bridge the gap between the island of technology and all else we expect our students to master.

The area of engineering also presents some barriers to complete implementation of STEM. Unfortunately, not much progress has been made in this area since Rodger Bybee made us aware of the limited presence of engineering in our school systems, despite its direct link to both problem solving and innovation (Bybee 2010, p. 1). We must commit to a more intentional and active linking of engineering practices in the workforce with STEM implementation in educational institutions.

STEM initiatives are innovative and creative, but underused. Until they become a directive in the formal sense of the word, many educators may not be willing to examine the full value that this approach has to education. It could be argued that many educators, in many disciplines are using some STEM applications, but not labeling them as such; thus only using them at a superficial level.

STEM processes and applications hold infinite value in the classroom. They produce self-reliant and competitive young people prepared to contribute to a successful global economy, in a world where the job market is catastrophically dwindling daily. We cannot turn our backs to this reality and must be willing to employ STEM initiative, and practice them fully. We then will be able to look at the values STEM initiatives hold through a new and clearer lens, and be able to provide enriching and meaningful experiences for our students. Cohesive and applicable STEM strategies mean a commitment to better practice and when that happens, true, measurable growth will be experienced.

Why is it important to marry science with technology, engineering, and math? The answer is because it produces academically well-rounded students. Much like colleges look for well-rounded applicants that are academically successful, play on a sports team, have a job, and volunteer in their community, STEM clearly, harmoniously, synergistically, and in a cognitively and psychosocially healthy way, stretches students to become scientifically literate. STEM students understand the interrelatedness between technology to explore science concepts, engineering to use logic and creativity for solving problems, and math applications outside of math class alone. STEM can take the stagnated content standards and apply them in an inquiry-based tactile/kinesthetic manner that reaches all preferred learning modalities and multiple intelligences. STEM makes it clear to students how each component complements one another.

## Conclusion

It is time to have an authentic conversation about STEM initiatives and their inclusion in the classroom. Teaching chemistry, physics, and forensic science places me at a distinct advantage to apply STEM initiatives because the most popular school of thought is that STEM initiatives belong in science disciplines. This is certainly true, but is only a surface view of what STEM truly entails. The reality is that the prevailing and predominant concepts and skills presented as part of the STEM framework can be implemented into every discipline and at every grade level. This must become a measurable and sustainable practice initially, and secondly we need a consistent and commonly understood measure of accountability of STEM practice implementation by educators. The growth process in education begins with teams of committed educators dedicated to individual and interpersonal development and comes to fruition when we witness the successes of our students.

As educators, our job descriptions define our positions as ones that deliver state-mandated content in an effort to prepare our students for their next level(s) of learning. We must be able to attain measurable and consistently reproducible results in order to recognize our strengths and weaknesses as educators. It is reasonably presented and accepted that our main charge is to deliver content and measure student understanding. When students fall short, we do not punish or condemn, we try again, humbling ourselves to the reality that in most situations, we the educators are to blame. Historically this thought process and cycle has been the norm. True education though, is a much different undertaking. True education results in much more than good grades or increased scores on high-stakes tests. True education transforms student thinking, student engagement, and disciplined and methodical problem-solving abilities. "Educated people reason, reflect, and make sound decisions on their own without prompting from teachers or assignments" (Brookhart 2010, p. 6). This should be the end goal for all students. Helping guide students toward these abilities must be our primary focus and one that is in direct keeping with STEM ideology. Direct, targeted, embedded STEM practices provide the connective links necessary to achieve high-quality results during student educational experiences, without which, our educational system will likely become increasingly more fragmented and splintered. Through direct, disciplined implementation of units such as these, collectively we will come to trust and depend on the applicable STEM examples brought forth decades ago. This approach works and more importantly, it is an effort to satisfy the curious human minds of our students. STEM allows us to take full advantage of individual student uniqueness, harness those gifts, and put them to constructive uses.

## References


Aleman, M. P. 1992. Redefining "teacher." *Educational Leadership* 50 (3): 97.

Brookhart, S. M. 2010. *How to assess higher-order thinking skills in your classroom.* Alexandria, VA: ASCD.

Bybee, R. W. 2010. Advancing STEM Education: A 2020 Vision. *Technology and Engineering Teacher* 70 (1): 30–35.

Cash, A. 2002. *Psychology for dummies*. New York: Hungry Minds.

Cullen, R., M. Harris, and R. R. Hill. 2012. *The learner-centered curriculum: Design and implementation*. Hoboken, NJ: Jossey-Bass.

Darling-Hammond, L. 1994. Will 21st-century schools really be different? *Education Digest* 60: 4–8.

Ejiwale, J. A. 2012. Facilitating teaching and learning across STEM fields. *Journal of STEM Education: Innovations and Research* 13 (3): 87–94.

Fernandez, J. J. M., J. Poniachik, L. Poniachik, T. Sole, R. Marshall, and K. Richards. 2003. *The giant book of mensa mind challenges*. New York: Sterling.

Kuhn, D., J. Black, A. Keselman, and D. Kaplan. 2000. The development of cognitive skills to support inquiry learning. *Cognition and Instruction* 18: 495–523.

Lemov, D. 2010. *Teach like a champion: 49 techniques that put students on the path to college.* San Francisco: Jossey-Bass.

Mayes, R., and T. R. Koballa. 2012. Exploring the science framework: Making connections in math with the Common Core State Standards. *Science and Children* 50 (4): 8–15.

Morella, M. 2013. Many high schoolers giving up on STEM. *www.usnews.com/news/blogs/stem-education/2013/01*

National Research Council (NRC). 2010. *Exploring the intersection of science education and 21st century skills: A workshop summary*. Washington, DC: National Academies Press.

NGSS Lead States. 2013. *Next Generation Science Standards: For states by states*. Washington, DC: National Academies Press. *www.nextgenscience.org/next-generation-science-standards*

Peery, A. B. 2009. *Writing matters in every classroom*. Englewood, CO: Lead+Learn Press.

The RIT Scale. (n.d.). *www.nwea.org*

Saab, N., W. van Joolingen, and B. van Hout-Wolters. 2012. Support of the collaborative inquiry learning process: influence of support on task and team regulation. *Metacognition and Learning* 7 (1): 7–23.

Sousa, D. A. 2006. *How the brain learns*. 3rd ed. Thousand Oaks, CA: Corwin.

Taylor, E., and C. Thomas. 2001. Resiliency and its implications for schools. *Journal of Educational Thought* 36: 7–16.

"Text Message (SMS) Polls and Voting, Audience Response System | Poll Everywhere." *Text messages (SMS) Polls and Voting. Audience Response System | Poll Everywhere.* N.p.,n.d. Web. 21 Dec. 2012. *www.polleverywhere.com*.

Thompson, C. J. 2009. Preparation, practice, and performance: An empirical examination of the impact of Standards-based Instruction on secondary students' math and science achievement. *Research in Education* 81: 53–62.

Tomlinson, C. A., and M. B. Imbeau. 2010. *Leading and managing a differentiated classroom.* Alexandria, VA: ASCD.

**Table 10.3.** State, National, and Next Generation Science Standards Guided Inquiry STEM Lesson Connection Grid

| | | | | |
|---|---|---|---|---|
| *Current MN State Standard*<br>*9C2.1.2.3*<br>*Use IUPAC nomenclature to write chemical formulas and name molecular and ionic compounds, including those that contain polyatomic ions*<br><br>**Must extend this to understand properties and characteristics of ionic and covalent compounds.**<br><br>**Guiding Question:**<br>**How can we experimentally distinguish ionic compounds from covalent compounds?** | *National STEM Standards*<br>*B2. Structure and property of matter*<br><br>*NGSS*<br>*HS. Structure and Properties of Matter* | *National STEM Standards*<br>*3A. Technology transfer occurs when a new user applies an existing innovation developed for one purpose in a different function (Benchmark G)*<br><br>*5C. With the aid of technology, various aspects of the environment can be monitored to provide information for decision making (Benchmark I)*<br><br>*19F. Chemical technologies provide a means for humans to alter or modify materials and to produce chemical products (Benchmark Q)*<br><br>*NGSS*<br>*HS-PS1-3.*<br>*Plan and conduct an investigation to gather evidence to compare the structure of substances at the bulk scale to infer the strength of electrical forces between particles* | *National STEM Standards*<br>*B. An ability to design and conduct experiments, as well as to interpret data*<br><br>*G. An ability to communicate effectively*<br><br>*I. A recognition of the need for and an ability to engage in life-long learning*<br><br>*NGSS*<br>*HS-PS1-1.*<br>*Use the periodic table as a model to predict the relative properties of elements based on the patterns of electrons in the outer energy level of atoms*<br><br>*HS-PS2-6.*<br>*Communicate scientific and technical information about why the molecular-level structure is important in the functioning of designed materials* | *National STEM Standards*<br>*1C. Compute fluently and make reasonable estimates*<br><br>*2C. Use mathematical models to represent and understand quantitative relationships*<br><br>*3A. Analyze characteristics and properties of two and three dimensional geometric shapes and develop mathematical arguments about geometric relationships*<br><br>*5A. Formulate questions that can be addressed with data and collect, organize, and display relevant data to answer them*<br><br>*5C. Develop and evaluate inferences and predictions that are based on data*<br><br>*NGSS*<br>*HS-PS1-3.*<br>*Plan and conduct an investigation to gather evidence to compare the structure of substances at the bulk scale to infer the strength of electrical forces between particles* |

**Table 10.3** *(continued)*

| | Science | Technology | Engineering | Math |
|---|---|---|---|---|
| | The science inherent in this unwrapped chemistry standard:<br><br>• writing formulas<br>• naming formulas<br>• ionic compounds<br>• covalent compounds<br>• monatomic ions<br>• polyatomic ions<br>• properties | Use Vernier probes to measure conductivity of known ionic and covalent solutions.<br><br>Research accepted literature values of known substance melting points and boiling points.<br><br>Use knowledge gained to determine if unknown solutions are ionic or covalent. Justify thinking in a written response.<br><br>Use Proscope digital microscope to examine crystal structure of ionic and covalent substances. Compare and contrast physical appearance. What makes these substances behave as they do? M.p., b.p., solubility, conductivity etc. | Choose a chemical that would be suitable for each of the following applications:<br><br>• Food preservation<br>• Fuel source<br>• Pharmaceutical Industry<br><br>1. Are these chemicals ionic or covalent?<br>2. What information or research led you in this direction?<br>3. Why are these chemicals suitable for the above applications?<br><br>Justify reasoning in a written response<br><br>Are there any flaws in your logic?<br><br>How can your logic be more sound?<br><br>What further explorations are necessary? | Determine the molar masses of the substances chosen for the engineering application. What relationship, if any, exists between molar mass and ionic or covalent compounds?<br><br>Use model kits to build 3-D models of the substances above. Are there any structural similarities between ionic and covalent compounds?<br><br>Explain your suppositions in writing.<br><br>Presentation of findings—explain your discoveries to another. Your arguments must be clear and logical. |

Mary Hanson-2013

# Ecology Disrupted

## A Model Approach to STEM Education That Brings Ecology, Daily Life Impact, and Scientific Evidence Together in Secondary School Science Classrooms

*Yael Wyner*
*City College of New York*
*New York, NY*

## Setting

This chapter describes a STEM curriculum for secondary school students developed as a collaboration between the City College of New York and the American Museum of Natural History, a premier national natural science museum located in New York City. Now being disseminated online through the Museum's website (*www.amnh.org/explore/curriculum-collections/ecology-disrupted*), the curriculum was initially used in New York City public schools. The curriculum was designed, tested, and revised over the course of four school years (Fall 2007–Spring 2011) and was shaped through the museum-university partnership and through the extensive input of the approximately 75 New York City secondary school teachers who implemented the resources in their classrooms. The real-time classroom experience of the teachers enhanced curriculum development to better meet the needs of students and the learning goals of the project. The teachers represented a cross section of ages and ethnicities, mirroring the composition of the urban landscape. Teaching experiences ranged from 2 years to 30+ years. Teachers improved the program in small and large ways, including redesigning curricular components and suggesting numerous additional scaffolding opportunities and activities to access students' prior knowledge.

## Collaborators

Over a period of four years, I led a large team of people in developing and testing this STEM curriculum. The developers include staff at the American Museum of Natural History, university faculty, and consultants from science curriculum development companies. A panel of experts comprised of science curriculum developers, a learning disabilities specialist, a high school biology teacher, and education faculty advised the development and testing of the resources. During the first two years of the project, the program was tested in the classrooms of 29 secondary school teachers. These teachers taught in a wide range of New York City public schools, including schools identified as struggling as well as schools identified for gifted learners. The teachers also implemented the program in a variety of courses including general life science, general science, biology, and environmental science.

Focus group and interview feedback from these teachers was critical for revising the curriculum and determining what worked and what did not in the classroom. In fact, following the focus group feedback sessions from the first two years of testing and development, two of the testing teachers continued to contribute to the program design by participating in a full-day co-design meeting with curriculum developers. In the project's third year an additional six teachers piloted the revised curriculum in preparation for wide scale field- testing. These teachers, along with the 39 teachers who completed field-testing, used the program exclusively with ninth-grade biology students, unlike the teachers who tested the program in the project's first two years who taught every secondary school grade level. The final program product is therefore optimized for urban ninth-grade biology students, but contains adaptations for use with younger and older learners.

The teachers who used the program represented New York City's ethnic and cultural diversity, including African American, Asian, Latino, and white teachers. The students were also diverse. In the final project-testing year, Latino students were the most well-represented (31.7%), followed by Asian students (19.9%), African American students (17.3%), white students (16.9%), and students identifying themselves as other (14.1%).

## Responding to the Goals of the *National Science Education Standards*

We developed this STEM curriculum to help secondary school ecology and human impact instruction better meet the content and inquiry goals set forward in the *National Science Education Standards* (1996) and reiterated in the National Science Teacher Association (NSTA) position statement on teaching science and technology in the context of societal and personal issues (2010). This curricular program seeks to involve students in the STEM issues of today in order to develop a future citizenry that is able to engage in 21st-century challenges (Bybee 2010). This project contextualizes STEM as engaging students in the scientific process through analyzing evidence and drawing conclusions. These skills serve students' needs by increasing their understanding of what counts as evidence for a claim.

The major goal of this STEM program is to use ecology content in the context of personal and social perspectives to highlight the relationship of ecological function to daily life and environmental impact. The program uses data from published scientific studies to enhance student learning of the complex interplay amongst daily life, environmental impact, and ecological function. Two fundamental principles guide this program: (1) Ecology and environmental impact will be more meaningful to students if they are contextualized into their daily experiences. (2) Learning scientific concepts is more meaningful in the context of scientific research; scientific evidence provides the context for the scientific content.

To meet these goals, the program emphasizes an understanding of the concepts that underlie scientific knowledge rather than simply emphasizing facts and information. The program uses published scientific studies to integrate the processes and products of science into the same conversations, rather than keeping scientific information separate from the processes of science. Students using this program must make conclusions based upon multi-day investigations of evidence. They must not only communicate their claims in writing, but they must justify their claims to their peers based on authentic data. They then apply concepts that they learn through their own investigation of data to other data-based scientific studies about environmental issues.

## Meeting the Guidelines Set Forth in the *Next Generation Science Standards*

These curricular resources were also designed in the context of the *Next Generation Science Standards*. As such they address the three major dimensions of science learning: (1) Scientific and Engineering Practice, (2) Crosscutting Concepts, and (3) Life Science Disciplinary Core Ideas. The practice of science is fundamental to the curriculum. Grounded in the analysis of published scientific data, students must ask questions, analyze and interpret data, construct explanations, and engage in argument from actual evidence. Since these units closely track authentic scientific studies, the curriculum by its very nature immerses students in scientific practices (NGSS Lead States 2013).

The close examination of human impact on ecosystems means that students engage in two key crosscutting concepts: (1) *Cause and effect: Mechanism and explanation* and (2) *Systems and system models*. In this curriculum students examine the unintended consequences to ecosystems that result from daily human actions that disrupt normal ecological function. This method explores the relationship between cause (daily human action) and effect (disrupted ecosystems). It also helps students understand the complex interactions within ecosystems, since students learn how even one small daily human action can affect multiple ecosystem components (NGSS Lead States 2013).

Finally, this curriculum immerses students in many of the content standards associated with *Life Sciences Learning Strand 2: Ecosystems: Interactions, energy, and dynamics* and *Life Sciences Learning Strand 4: Biological evolution: Unity and diversity*. Within these learning strands the standards address *Interdependent relationships in ecosystems, Human impacts,* and *Biodiversity and humans*. By explicitly using human impact to understand ecological function, these units link learning of ecology, biodiversity, and human impact standards together (NGSS Lead States 2013).

## Target Learners

Although optimized for ninth-grade New York City public school biology students, this program is suitable for all secondary school students. Highly scaffolded and designed around data, it can be used successfully with a diverse range of learners. Its step-by-step scaffolding makes the curricular content accessible for young learners. At the same time, its grounding in data from published scientific research makes the program useful for all students, including more advanced learners. Middle school classrooms can expand the time that they devote to the program, while advanced high school students can focus on the in-depth data formatted in a manner appropriate for in-school learning rather than for scientific publication.

## Placing Ecological Science in the Context of Personal and Social Perspectives

Science touches our lives in multiple ways. It is for this reason that the National Science Education Standards emphasize the integration of personal and social perspectives into the learning of science. Understanding this interaction between science and our lives is important for developing the skills necessary for making informed decisions (BSCS 1993; NRC 2003) and for becoming scientifically literate. According to the National Science Education Standards, a major component of science literacy "is the knowledge and understanding of scientific concepts and processes required for personal decision making, participation in civic and cultural affairs, and economic productivity" (NRC 1996).

Others have emphasized the civic utility of scientific literacy by specifically defining civic scientific literacy (Miller 1998). A person who shows civic science literacy can understand "competing arguments on a given dispute or controversy," which means that they understand basic scientific ideas, the processes of science, and how science affects people on a personal and societal level. Central to this concept is the requirement that science be contextualized in society generally.

As our society becomes increasingly aware of our growing impact on the natural world, the contextualization of ecological science into our daily experiences also becomes increasingly important (NSF AC-ERE 2003; 2005; 2009). Linking ecology to daily life can civically engage a diverse range of learners to our growing impact on the natural world (SENCER 2012; Bjorkland and Pringle 2001; Hungerford et al. 2003). Furthermore linking ecology to daily life can increase student scientific literacy by improving learning of ecological science and environmental issues (Cordero, Todd, and Abellera 2008).

Connecting ecological function to the personal and social dimensions of science can help alleviate some of the documented disconnect that students show when connecting their own daily lives to the environment. STEM teaching and learning is about integration and making explicit connections between disciplines and between science and society.

Increasing the connection between the ecological life science content standards and the personal and social perspectives standards can help increase scientific literacy by helping students understand scientific concepts in the context of society. The *Next Generation Science Standards* do just that by embedding the personal and social perspectives of science into the life science content learning strands. Like the *Next Generation Science Standards*, our curriculum builds science literacy by enmeshing the personal and social perspectives of science in the life science content standards.

## Placing Ecological Science and Personal and Social Perspectives Into the Context of Scientific Inquiry

The final component necessary for STEM literacy is an understanding of scientific inquiry (NRC 1996; Miller 1998). It is for this reason that the curricular program described in this chapter is grounded in the interpretation and analysis of scientific evidence and data. A central component of the National Science Education Standards science literacy guidelines is "the capacity to pose and evaluate arguments based on evidence and to apply conclusions from such arguments appropriately" (p. 22)." Since the publication of the Standards, other science education documents also

emphasize the value of evidence in science (Schleicher 1999; Duschl, Schweingruber, and Shouse 2007; Abd-El-Khalick et al. 2004; Roberts 2007). The critical role of evidence and data were recently reiterated in the *Next Generation Science Standards* and *A Framework for K–12 Science Education*, on which the new Standards are based. Data and evidence are central part of the dimensions of STEM practices, particularly in the sections that emphasize analyzing and interpreting data and in engaging in argument from evidence (NRC 2011, NGSS Lead States 2013). This program focuses on the use of authentic published data for helping students to think as scientists. Already a valuable component of graduate science and medical education (Cave and Clandinin 2007; Iyengar et al. 2008; Kohlwes et al. 2006), published data are also used at the undergraduate level. Some researchers in what they call "adapted primary literature" have transformed published journal articles into a more accessible format (Falk, Brill, and Yarden 2008; Falk and Yarden 2009) and others have developed full-length semester courses that focus on in depth analysis of just a few journal articles (Hoskins, Stevens, and Nehm 2007; Hoskins 2008; Hoskins, Lopatto, and Stevens 2011) or research seminars by scientific faculty (Clark et al. 2009). All of these approaches have positively impacted student inquiry and critical thinking skills.
Researchers are now successfully bringing published data sets to secondary schools (Cary Institute 2012; Griffis, Thadani, and Wise 2008; Urban Advantage 2012). In one study at Oregon Health and Science University, researchers found that high school students enrolled in a class that incorporated data analysis from journal articles were seven times more likely to say that this class influenced their science career and life choices than more traditional science classes (Rosenbaum et al. 2007).

Data from published research or from large publicly available databases provide a scientific context for addressing personal and social perspectives of science through argumentation (Kerlin, McDonald, and Kelly 2010). When students use argumentation, they must justify their claims based on evidence, a process that they find difficult (Sadler 2004; Osborne and Patterson 2011). Incorporating evidence into scientific decision-making is still an area where many students need practice (McNeill and Krajcik 2007; Sadler, Chambers, and Zeidler 2004; Tytler 2001).
The difficulties students show in identifying and evaluating scientific data in current personal and social issues indicate why these resources are centered on the analysis of scientific data. In this program, students must use data to make arguments that support their claims

## Major Facets of the *Ecology Disrupted* Program

This curricular program has two overarching goals. They are:

1. Place the science content related to ecological functions into the same learning strand as personal and social perspectives topics (the role of daily life in human environmental impact) to help student learning of both topics.

2. Frame the learning of these topics in scientific inquiry, particularly the role of evidence and data in making scientific conclusions.

The *Ecology Disrupted* approach created a STEM program that used data to explicitly link daily life and environmental impact to ecology (Wyner and Desalle 2010). The goal of this method, developed as an NSF-funded initiative, is for students to learn about the importance and complexity of normal ecological function, by using data to study the environmental issues that result when daily life actions *disrupt* them. *Ecology Disrupted* uses the same intellectual approach that the field of genetics uses to understand gene function. Like geneticists who learn gene function by studying the changes in appearance that result from mutations that disrupt normal gene function, *Ecology Disrupted* students learn the complexity of functioning ecosystems by studying the environmental issues that result from daily life actions that disrupt normal ecological function.

This STEM curricular program implements this approach by using two case study modules with the students. These modules are based upon data from published scientific research (Kaushal et al. 2005; Epps et al. 2005). Each module is constructed around a question that asks students to link daily human actions to an environmental issue. One case study (7 lessons) asks students "How might snowy and icy roads affect Baltimore area's water supply?" and the other case study (6 lessons) asks "How might being able to drive from Los Angeles to Las Vegas in just four hours put the bighorn sheep at risk?" In these exercises students examine case study–specific data to learn how salting roads in Baltimore and how highways in the desert mountains between Los Angeles and Las Vegas disrupt ecological function. For example, the Baltimore case study is used to help students learn how salting roads for safe travel disrupts abiotic factors and water runoff in the Baltimore watershed, eventually leading to saltier drinking water supplies. The bighorn sheep example is used to help students learn how highways, built to connect Las Vegas to Los Angeles and help the Vegas economy, disrupt the bighorn sheep habitat, thus making it hard for sheep from different mountaintop populations to mate, and leading them to become inbred.

Students are then asked to consider sustainable solutions to both of these problems like using alternative solvents to melt snow and ice in Baltimore and elevating sections of Nevada's highways to allow sheep and other animals to cross under the highways in order to mate with animals on neighboring mountains. Finally, students are asked to apply the same *Ecology Disrupted* approach to other environmental issues that are caused by different human actions that disrupt the same ecological functions. For example, students learn how the environmental issue of light pollution caused by the basic desire for people to see at night, changes abiotic ecosystem components. This change in abiotic environmental factors harms living organisms like aquatic insects that lay their eggs at night. These insects perceive the artificially lit surfaces as water and consequently lay their eggs on dry land. After students connect the environmental issue to ecology and daily life, they once again develop sustainable solutions that recognize the human and ecological components of the environmental issues. In the latter example, students can research and describe new types of lights designed to reduce scattered light rays. These lights help people see better at night and also help to reduce the effect of artificial light at night.

A team of teacher and museum educators and biologists developed the modules that were field-tested in these urban classrooms. The modules represent our belief that STEM projects must use specific scientific reasoning in order to be successful. They are available from the American Museum of Natural History's website (*www.amnh.org/explore/curriculum-collections/ecology-disrupted*).

Each module is broken down into the components required to effectively implement the described case studies.

Module 1, " How might snowy and icy roads affect Baltimore area's water supply?" is comprised of seven lessons. They are:

| | |
|---|---|
| Lesson 1: | Setting the Stage |
| Lesson 2: | Exploring: Salt in our lives |
| Lesson 3: | Exploring: Water in our lives |
| Lesson 4: | How Do You Investigate and Represent the Data? |
| Lesson 5: | Representing and Making Meaning From Data |
| Lessons 6 & 7: | Ecology Disrupted: A change in any ecosystem factor, living or nonliving, can unexpectedly disrupt the ecosystem |

Module 2, "How might being able to drive from Los Angeles to Las Vegas in just four hours put the bighorn sheep at risk?" is comprised of six lessons. They are:

| | |
|---|---|
| Lesson 1: | Setting the Stage and the Scientific Process in Action |
| Lesson 2: | Exploring: The role of isolated populations in inbreeding |
| Lesson 3: | How Do You Investigate and Represent the Data? |
| Lesson 4: | Representing and Making Meaning From Data |
| Lessons 5 & 6: | Ecology Disrupted: Changes to habitat can unexpectedly disrupt populations |

In each module, the first lessons are designed to access students' prior knowledge about the topic. For example, in the "snowy and icy roads" activity students are asked to consider their own experience with snow, ice, and salt. Videos of the scientists are also used in the first lessons to introduce students to the scientists that conducted the research and to help students reflect upon the factors that motivate the scientists to ask and explore their research questions. The next set of "exploring" lessons is designed to familiarize students with content background that will help them understand the complexity of each case study. In the "snowy and icy roads" module students learn how salt varies in ecosystems in which they are familiar, observe the effect of salt on an eggplant, and explore an interactive about their water supply. In the "bighorn sheep" module students compare different examples about the impact of inbreeding in people and animals.

After students complete these "exploring" activities, they then consider the types of data they would need to collect to investigate the research questions, the type of evidence they would need to support their claims, and how they would represent their data. In the "salty and icy roads" case study students make graphs of the salinity data to compare annual and seasonal salt levels amongst forested, suburban, and urban Baltimore streams. In the "bighorn sheep" case study, students use maps to compare the level of breeding between populations that are and are not separated by highways (Figure 11.1, p. 190). In the next lessons, students use their represented data to justify their claims about the impact of snowy and icy roads on abiotic stream components or about the impact on bighorn sheep populations of highways that disrupt their habitat. This part of the STEM experience is explicitly recommended by the *Next Generation Science Standards*.

**Figure 11.1.** Gene Flow Between Bighorn Sheep Mountaintop Populations

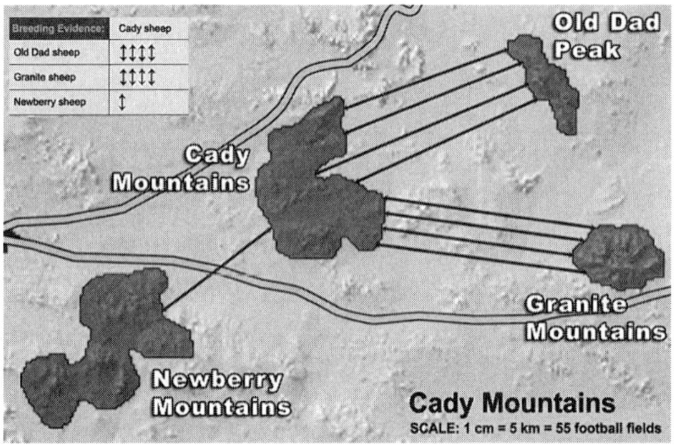


The arrows indicate levels of breeding and gene flow between sheep populations. Many arrows indicate high gene flow and breeding and few arrows indicate little gene flow and breeding between populations. Students draw the breeding level arrows onto the map and use that information along with data on geographic distance to predict the location of highways.

In the final set of *Ecology Disrupted* lessons, students explore other examples of ways daily life leads to environmental impact through unintentional disruption of ecological functions. They use Science Bulletins, which are brief media clips from the American Museum of Natural History to examine other ecological disruptions that are caused by daily life activity. In the "snowy and icy roads" case study, students pull out the aspects of daily life that disrupt ecosystem abiotic components and explain how those components impact the biotic components of that ecosystem. They also describe the data collected to support these findings. The additional examples that they consider are man-made surfaces and the urban heat island effect in Atlanta, artificial night light and aquatic insects, fossil fuel use from everyday life and climate change, and a video on the restoration of the Bronx River in New York City.

In the "bighorn sheep" case study, students pull out the aspects of daily life that can disrupt habitats and explain how the disrupted habitats impact local populations. They also describe the data collected to support these findings. The additional examples that they consider are the wood products that they use in their daily life and the declining monarch wintering habitat, fossil fuel use in everyday life and the Gulf oil spill, litter and its impact on seabird habitat and baby birds, and how roads they use for travel fragment the wood turtle habitat leading to wood turtle roadkill and decline.

Importantly, in each of these examples, students must consider solutions to the daily life-caused disruptions of abiotic factors and habitat function. To reduce climate change and oil spills, students suggest ways to reduce fossil fuel consumption. They also suggest innovations like green roofs to reduce the urban heat island effect in Atlanta and other human-built structures like malls and parking lots. Reduction of plastic trash and littering are choices that they suggest for reducing the impact of plastic on baby seabirds. Finally, they advocate scientific research on relevant habitats before road building to reduce their habitat fragmentation potential.

## Evidence for Success: Student Learning

The *Ecology Disrupted* program was field-tested in the final two years of the study. In the first year of the field test, teachers gave pre- and posttests before and after teaching their regular human impact curricular units. In the second year, the same teachers gave pre- and posttests before and after teaching the *Ecology Disrupted* replacement for their regular human impact curricular units. The pre- and posttests were designed to assess student knowledge of ecological function and human environmental impact in the context of daily life. The students in the classes of the participating teachers during the first year served as the control group, while the students in those same teachers' classes in the second year were in the treatment group.

### *Findings*

Assessment scores were calculated as the percentage of correct responses. On the assessment as a whole, the mean pretest score for the control group (50.9%) was significantly higher than those of the treatment group (47.1%) ($F$ = 19.65, $p < .01$, Eta-squared = 0.009) although the effect was small. However, on the posttests, the mean score of the treatment group (62.8%) was significantly higher than that of the control group (54.2%) ($F$ = 86.87, $p < .01$, Eta-squared = 0.037). On average, the treatment group got 15.7% more correct items on the posttest than the pretest, compared to just 3.3% for the control group (see Table 11.1).

Similar growth favoring the treatment group can be seen when breaking the assessments down into the subtests. Across both subtests, the treatment group made statistically significant greater gains than the control group ($p < 0.01$). Hence, being exposed to the *Ecology Disrupted* program has a positive and statistically significant impact on student assessment impact on student learning of ecological function and human environmental impact in the context of daily life.

**Table 11.1.** Pretest and Posttest Subtest Scores by Treatment-Control (% Correct)

| | Control Group ($n$ = 1,103) | | Treatment Group ($n$ = 1,131) | |
|---|---|---|---|---|
| **Overall** | | | | |
| | **Mean** | **Std. Dev.** | **Mean** | **Std. Dev.** |
| **Pretest** | 50.92 | 0.20 | 47.09 | 0.20 |
| **Posttest** | 54.24 | 0.22 | 62.82 | 0.22 |
| **Ecological Function** | | | | |
| | **Mean** | **Std. Dev.** | **Mean** | **Std. Dev.** |
| **Pretest** | 54.41 | 0.24 | 50.66 | 0.24 |
| **Posttest** | 58.24 | 0.24 | 67.61 | 0.24 |
| **Human Environmental Impact** | | | | |
| | **Mean** | **Std. Dev.** | **Mean** | **Std. Dev.** |
| **Pretest** | 46.87 | 0.24 | 42.97 | 0.22 |
| **Posttest** | 49.78 | 0.25 | 57.33 | 0.24 |

A student presenting her STEM reasoning for which plants were grown in a saline environment

## Teacher Feedback

Teacher feedback on their curricular implementation experience shows that teachers value the program for the ways it meets the curricular goals described above. Specifically, teachers described how the program was important for helping their students learn the connection between daily life, human environmental impact, and ecological functions, and for placing science learning in the context of data. Additionally, teachers explained how the videos of scientists helped students relate to the process of doing science.

## Ecological Function, Human Environmental Impact, and Daily Life

Many teachers, including these four teachers below, commented on the ways they thought the program facilitated explicitly linking human environmental impact and ecological function into one study topic. They described how ecology could be used as a lens or the language through which to understand human environmental impact.

A student completing a worksheet during the Ecology Disrupted lessons

The teachers' comments included:

- I really like the emphasis on the concept that humans change abiotic factors that then impact biotic factors. I think I will continue to use this as the "lens" through which ecology or human impact is taught.
- I liked how the unit was structured where you had the sort of in-depth look at one issue … and then all these other little issues … like the winter roads and then the heat island effect, so it kind of left you with the impression that we've learned about humans impacting abiotic factors in this context in depth and it leaves you with this impression there must be a whole bunch of other little stories like this out there … I don't think I've stressed … that language before where the environmental impact is about humans changing abiotic factors in the environment that impact biotic factors. I thought that was really smooth and well done and I think that came across really, really well.
- It became language we were speaking by the end of the units … I never really used that specific language before. I mean I would teach them abiotic vs biotic, never those connections; cause and effect.

One teacher even attributed high student performance on the ecology and human impact components of the final statewide biology exam to this ecology-human impact connection.

- I want to echo that because it [*Ecology Disrupted* section] was also my favorite part, but I think the real indicators to whether or not they appreciated it is that they actually wrote answers that were applicable on the [statewide biology exam] from the unit.

Some teachers highlighted the effectiveness of the program at helping students to personally connect to human environmental impact. These two teachers described how their students were able to identify and relate the human impact connections to their daily behavior.

- My students pointed out and gave their parents a stern talking to about the dangers of road salt and its effect on the environment. Was the unit effective? I have some parents who would say so!
- This was probably my favorite part out of everything. I loved this … liked the fact that not only were we looking at the impact, but why is it? Why are we creating these artificial services? Why are we littering in the Bronx River? So that students could see how their actions have consequences.

## Evidence, Data, and Personalizing Science

Teachers described the ways in which grounding these modules in case studies of published scientific research helped student STEM practice and understanding of authentic science.

This teacher summarized how all the parts of the units worked together to allow students to make conclusions from data.

- I liked the cohesiveness of the units, that everything worked off one basic concept and scientific study, that students got a chance to hear from a scientist working on a study that was accessible to students (fieldwork rather than creating models or mixing chemicals in a lab, or working on a tiny aspect of a chemical pathway, etc.), that students got to think through an actual experiment, and most impressive, work with actual data, and come to conclusions based on their data analysis.

Other teachers highlighted the data analysis from each unit that they found to be particularly effective for STEM learning. These teachers described the effectiveness of the bighorn sheep mapping activity for getting students to think about the impact of highways on sheep mating.

- So many great discussions arose from the introduction of these units. My students were excited every day to see real-world scientists, research, and data and especially enjoyed working on the maps; figuring out the place where the highways interfered with the breeding/mating behaviors of the sheep—they were engaged and had lots of questions.
- The graphs and maps of different mountaintops worked best. My students are English language learners so this was good for them.

- The maps, they loved the maps! It was amazing. I have the good fortune to have a smart board. At the end, I had that final map and had them draw. The class map ended up being the map on the smart board, and then the way they were able to figure out where the roads were. It really worked well. I mean the idea that we were trying to get across actually came to fruition. It was beautiful on the smart board. So it worked out really well

- Other teachers highlighted the utility of the graphing for connecting students to data and STEM process skills. My boss came in to observe me ... and it was the graph, the benchmarks, and she absolutely loved the fact that it was real data, real scientists ...
- Actually, I have enjoyed teaching this material to my students. Their graphing accomplishments were absolutely magnificent. Many of the graphs, which contrasted and compared Baltimore's forested, suburban and urban areas were done by hand and with colors. At least two of my students utilized technology in order to generate bar graphs for each studied area.

Finally, some teachers described how the videos of the working scientists helped students connect to STEM on a personal level.

- They were fascinated by the two interviews; they just really loved talking about feelings and reasons and personal histories and how people get over death and mourning; they just love talking about that stuff. And they're really connected with—to scientists on that level first, so our discussion time was actually extended for those parts because they just wanted to keep sharing.
- My students found the video profile of Dr. Kaushal [to be] very interesting. As we discussed Dr. Kaushal's motivation for his research, one of my students commented that he too, even as an immigrant, can someday be like him.
- I really liked the Bronx River movie and some of my students really identified with some of the people in the video, so it was great. And it was a positive thing.

## Conclusions and Next Steps

This STEM curricular program was successful at increasing student learning of human environmental impact and ecological function. The evidence of this success is in the statistical data on student achievement. The results demonstrate how this approach can be useful for connecting ecology science content with technology, mathematics and the personal and social perspectives content of the national science education standards. Furthermore, teacher feedback on the utility of using published data, points to how this method might be useful for helping student learning of the nature of science.

The *Ecology Disrupted* model can be expanded to contextualize other scientific studies of environmental issue research into other ecological concepts. For example, studies on the impact of combined sewer overflows on dead zones and algal blooms can be used to illustrate the connection of daily life to the nitrogen cycle or studies of overfishing of cod; a fish used in school lunch can be used to connect daily life to food web stability in ocean ecosystems. Additionally, the framing of these case studies in scientific data has the added benefit of immersing students in evidence required to make scientific claims.

The effectiveness of this STEM program for embedding the practice of science into all of its activities shows how it also meets the aims of the *Next Generation Science Standards* (NGSS Lead States 2013). These new standards emphasize the integration of the three major dimensions of science learning, crosscutting concepts, the practice of science and disciplinary core ideas into one study topic. The emphasis of the *Ecology Disrupted* program on the effect of daily life on ecological function shows how the program integrates the cause and effect crosscutting concept of science. The grounding of STEM concepts in data and evidence shows how the practice of science is central to the program. Finally, the emphasis on ecosystems shows how the program addresses the life science core idea of ecosystem interactions. This program is well-positioned to address both the *National Science Education Standards* and the *Next Generation Science Standards*.

## Acknowledgments

I would like to thank the people who contributed to developing and testing the Ecology Disrupted resources, including, S. Gano, D. Silvernail, A. Bickerstaff, M. Chin, S. Bothra, S. Ramnath, S. Fotiadis, R. DeSalle, J. Becker & B. Torff. I would particularly like to thank J. Koch for her contributions to this project and for her careful review of this manuscript. I also want to thank all the teachers and students who tested and used these materials in their classrooms. This work is supported by the National Science Foundation (NSF; 733269; 0918629). All opinions expressed are those of the author and not of the NSF.

## References

Abd-El-Khalick, F., S. BouJaoude, R. Duschl, N. G. Lederman, R. Mamlok-Naaman, et al. 2004. Inquiry in science education: International perspectives. *Science Education* 88 (3): 397–419.

Amherst Regional Public Schools. 2006. District scope and sequence. *www.arps.org/USERS/ms/pricen/scidepartment/District Scope and Sequence grade 7 one page.doc*

Arms, K. 2008. *Environmental science*. New York: Holt, Rinehart, Winston.

Biological Sciences Curriculum Study (BSCS). 1993. *Developing biological literacy.* Colorado Springs, CO: BSCS.

Bjorkland, R., and C. M. Pringle. 2001. Educating our communities and ourselves about conservation of aquatic resources through environmental outreach. *Bioscience* 51: 279–282.

Bybee, R. 2010. Advancing STEM education: A 2020 vision. *Technology and Engineering Teacher* 70: 30–35.

Cary Institute of Ecosystem Studies. 2012. The changing Hudson project unit plans. *www.caryinstitute.org/educators/teaching-materials/changing-hudson-project*

Cave, M. T., and D. J. Clandinin. 2007. Revisiting the journal club. *Medical Teacher* 29 365–370.

Clark, I. E., R. Romero-Calderón, J. M. Olson, L. Jaworski, D. Lopatto, and U. Banerjee. 2009. Deconstructing scientific research: A practical and scalable pedagogical tool to provide evidence-based science instruction. *PLoS Biology* 7 (12): e1000264.

Cordero, E. C., A. M. Todd, and D. Abellera. 2008. Climate change education and the ecological footprint. *Bulletin of the American Meteorological Society* 89: 865–872.

Duschl, R. A., H. A. Schweingruber, and A. W. Shouse. 2007. *Taking science to school: Learning and teaching science in grades K–8*. Washington, DC: National Academies Press.

Epps, C. W. P. J. Palsbøll, J. D. Wehausen, G. K. Roderick, R. R. Ramey, and D. R. McCullough. 2005. Highways block gene flow and cause a rapid decline in genetic diversity of desert bighorn sheep. *Ecology Letters* 8: 1029–1038.

Falk, H., G. Brill, and A. Yarden. 2008. Teaching a biotechnology curriculum based on adapted primary literature. *International Journal of Science Education* 30: 1841–1866.

Falk, H., and A. Yarden. 2009. "Here the scientists explain what I said." Coordination practices elicited during the enactment of the results and discussion sections of adapted primary literature. *Research in Science Education* 39: 349–383.

Filho, W. 1996. Eurosurvey: An analysis of current trends in environmental education in Europe. In *Environmental issues in education,* ed. G. Harris and C. Blackwell, 225–239. London: Arena.

Griffis, K., V. Thadani, and J. Wise. 2008. Making authentic data accessible: The sensing the environment inquiry module. *Journal of Biological Education* 42: 119–122.

Hoskins, S. G. 2008. Using a paradigm shift to teach neurobiology and the nature of science—A C.R.E.A.T.E.-based approach. *The Journal of Undergraduate Neuroscience Education* 6 (2): A40–A52.

Hoskins, S., D. Lopatto, and L. M. Stevens. 2011. The C.R.E.A.T.E. approach to primary literature shifts undergraduates' self-assessed ability to read and analyze journal articles, attitudes about science, and epistemological beliefs. *CBE Life Sci Educ.* 10 (4): 368–378.

Hoskins, S., L. Stevens, and R. Nehm. 2007. Selective use of primary literature transforms the classroom into a virtual laboratory. *Genetics* 176: 1381–1389.

Hungerford, H. R., R. A. Litherland, R. B. Peyton, J. M. Ramsey, and T. L. Volk. 2003. *Investigating and Evaluating Environmental Issues and Actions: Skill Development Modules*. Champaign, IL: Stipes Publishing Company.

Iyengar, R., M. A. Diverse-Pierluissi, S. L. Jenkins, A. M. Chan, L. A. Devi, E. A. Sobie, et al. 2008. Inquiry learning: Integrating content detail and critical reasoning by peer review. *Science* 319 (5867): 1189–1190.

Kaushal, S., P. M. Groffman, G. E. Likens, K. T. Belt, W. P. Stack, V. R. Kelly, et al. 2005. Increased salinization of fresh water in the northeastern United States. *Proceedings of the National Academy of Sciences* 102 (38): 13517–13520.

Kerlin, S. C., S. P. McDonald, and G. J. Kelly. 2010. Complexity of secondary scientific data sources and students' argumentative discourse, *International Journal of Science Education* 32: 1207–1225.

Kohlwes, R. J., R. L. Shunk, A. Avins, J. Garber, S. Bent, and M. G. Shlipak. 2006. The PRIME curriculum: Clinical research training during residency. *Journal of General Internal Medicine* 21 (5): 506–509.

Loughland, T., A. Reid, and P. Petocz. 2002. Young people's conceptions of environment: A phenomenographic analysis. *Environmental Education Research* 8: 187–197.

McNeill, K. L., and J. Krajcik. 2007. Middle school students' use of appropriate and inappropriate evidence in writing scientific explanations. In *Thinking with data*, ed. M. Lovett and P. Shah, 233–265. New York: Taylor & Francis Group.

Middletown City School District. 2010. Grade 9 biology: Marking period 4 living environment. *www.middletowncityschools.org/Portals/0/resources/SBL/9-12Curriculum/Science/Grade9BiologyMP4.pdf*

Miller, J. D. 1998. The measurement of civic scientific literacy. *Public Understanding of Science* 7: 203–223.

Miller, J. R. 2005. Biodiversity conservation and the extinction of experience. *Trends in Ecology and Evolution* 20 (8): 430–434.

National Research Council (NRC). 1996. *National science education standards*. Washington DC: National Academies Press.

National Research Council (NRC). 2003. *What is the influence of the national science education standards? Reviewing the evidence, a workshop summary*. Washington DC: National Academies Press.

National Research Council (NRC). 2011. *A Framework for K–12 science education: Practices, crosscutting concepts, and core ideas*. Washington DC: National Academies Press.

New York City Department of Education. n.d.. *NYC high school science regents scope and sequence*. New York: NYC Department of Eduction. *http://schools.nyc.gov/Documents/STEM/Science/HSScienceSSRegents.pdf*

NGSS Lead States. 2013. *Next Generation Science Standards: For states by states*. Washington, DC: National Academies Press. *www.nextgenscience.org/next-generation-science-standards*

NSF Advisory Committee for Environmental Research and Education. 2003. Complex environmental systems: Synthesis for Earth, life, and society in the 21st Century, a report summarizing a 10-year outlook in environmental research and education for the National Science Foundation. *www.nsf.gov/geo/ere/ereweb/acere_synthesis_rpt.cfm*

NSF Advisory Committee for Environmental Research and Education. 2005. Complex environmental systems: Pathways to the future. *www.nsf.gov/geo/ere/ereweb/acere_synthesis_rpt.cfm*

NSF Advisory Committee for Environmental Research and Education. 2009. Transitions and tipping points in complex environmental systems. *www.nsf.gov/geo/ere/ereweb/ac-ere/nsf6895_ere_report_090809.pdf*

National Science Teacher Association (NSTA). 2010. NSTA Position Statement, Teaching science and technology in the context of societal and personal issues. *www.nsta.org/about/positions/societalpersonalissues.aspx*

Osborne, J. F., and A. Patterson. 2011. Scientific argument and explanation: A necessary distinction? *Science Education,* 95 (4): 627–638.

Roberts, D. A. 2007. Scientific literacy/science literacy. In *Handbook of research on science education*, ed. In S. Abell and N. G. Lederman, 729–780. Mahwah, NJ: Lawrence Erlbaum Associates.

Rosenbaum, J. T., T. M. Martin, K. H. Farris, R. B. Rosenbaum, and E. A. Neuwelt. 2007. Can medical schools teach high school students to be scientists? *The FASEB Journal* 21 (9): 1954–1957.

Sadler, T. D. 2004. Informal reasoning regarding socioscientific issues: A critical review of research. *Journal of Research in Science Teaching* 41: 513–536.

Sadler, T .D., F. W. Chambers, and D. L. Zeidler. 2004. Student conceptualisations of the nature of science in response to a socioscientific issue. *International Journal of Science Education* 26: 387–409.

San Bernardino City Unified School District. 2008. Biology scope and sequence high school. *www.sbcusd.k12.ca.us/DocumentView.aspx?DID=5734*

Schachter, M. 2005. *Environmental science*. New York: AMSCO School Publications.

Schleicher, A., ed. 1999. *Measuring student knowledge and skills: A new framework for assessment*. Paris: Organization for Economic Cooperation and Development.

SENCER. 2012. Science Education for New Civic Engagement and Responsibility. *www.sencer.net*

Telluride Mountain School. (n.d.). Scope and sequence. *www.telluridemtnschool.org/documents/INTERMEDIATESCOPEANDSEQUENCE2007.pdf*

Tsurusaki, B. K., and C. W. Anderson. 2010. Students' understanding of connections between human engineered and natural environmental systems. *International Journal of Environmental & Science Education* 5 (4): 407–433.

Tytler, R. 2001. Dimensions of evidence, the public understanding of science and science education. *International Journal of Science Education* 23: 815–832.

Urban Advantage. 2012. Urban advantage middle school science initiative. *www.urbanadvantagenyc.org/home.aspx*

Wyner Y., and R. Desalle. 2010. Taking the conservation biology perspective to secondary school classrooms. *Conservation Biology* 24: 649–654.

# Mission Biotech

## Using Technology to Support Learner Engagement in STEM

*Troy D. Sadler*
*University of Missouri*
*Columbia, MO*

*Jennifer L. Eastwood*
*Oakland University Beaumont School of Medicine*
*Rochester, MI*

*William L. Romine*
*Wright State University*
*Dayton, OH*

*Len Annetta*
*George Mason University*
*Fairfax, VA*

### Setting

We, as the authors, are a group of university STEM education researchers who came together around common interests in students' STEM learning and the use of innovative technologies to provide contexts for learning. The project reported on in this chapter resulted from a National Science Foundation grant related to generating student interest in STEM, learning of biotechnology concepts, and the use of educational games to support these goals. The project was based at a major research university, but the work was focused on high school STEM teachers and students and took place in high school science classes throughout the state of Florida.

The project came together initially as a collaboration between two of the authors: Troy Sadler and Len Annetta. Troy was interested in engaging students in issues-based STEM learning experiences but dissatisfied with traditional ways of accomplishing this goal. Len was interested in exploring the potential of games to transform students learning. The two of us began envisioning a game-based virtual environment to engage students in issues-based STEM learning, and these efforts culminated in an NSF grant to design and study a biotechnology-focused virtual-learning environment. As this project unfolded, we were lucky to work with many talented high school teachers, scientists, and education researchers. Jennifer Eastwood joined the project team as a post-doc and coordinated research efforts associated with teachers' implementation of the game. William Romine, who has expertise in measurement and statistics, started working with us to help with the validation of content tests and quantitative data analysis. Together, along with the contributions of many others, we created *Mission Biotech*. The text that follows will introduce the *Mission Biotech* project, discuss how it responds to the new calls for STEM education, and provide evidence of its effectiveness for supporting STEM learning.

To provide a picture of a typical context in which our project unfolded, we present the vignette below. The vignette is a composite story that reflects common trends we observed across dozens of implementations of the *Mission Biotech* project.

*Upon entering the room, I noticed a few teams of researchers gathered around computers. I walked by one computer screen and noticed a series of graphs and tables, and one of the researchers referred to "the CT for H5N1." I moved onto another group and listened to a discussion in which one researcher discussed the need for conducting "reverse transcription prior to setting up the PCR for the HIV screen." One of her colleagues asked why this intermediate step would be required for the HIV screen when they had not needed it with the adenovirus screen. When there was a lull in the conversation, I pulled one of the researchers aside and asked for an interpretation of what I was witnessing.*

*Researcher:* A bunch of people seem to have contracted some sort of illness and are showing respiratory problems, but the doctors are stumped by what is causing it. We think we may be dealing with a new strain of a virus causing the problems.

*Author:* How will you know what the virus is?

*Researcher:* We're using qPCR to determine whether the patient blood samples contain viruses with HIV, H5N1, or adenovirus sequences.

*Author:* When you say sequences, are you referring to DNA sequences?

*Researcher:* Well, a virus like adenovirus has DNA, but viruses like HIV and H5N1 are RNA viruses. That's why we were talking about whether or not we needed to set up reverse transcription reactions for some of the tests prior to running the qPCR. But to answer your question, yes we're looking for DNA sequences, or in some cases DNA sequences that would complement a virus' RNA sequence.

The use of real-time polymerase chain reaction (qPCR; also known as quantitative PCR) to make viral diagnoses is fairly standard practice within medical biotechnology facilities. In such contexts, researchers use procedures like reverse transcription and interpret data like "CT values" (i.e., the PCR cycle at which a reaction amplifies enough DNA to exceed a predetermined critical threshold) as a part of their everyday work. What makes this vignette interesting is that the "researchers" cited were high school students. These students were *not* a select group of high achievers participating in a summer enrichment program that embedded them within working laboratories. This vignette unfolded in a biology classroom situated in a regular neighborhood high school. One may justifiably point out that regular high school biology classrooms do not have thermocyclers and other equipment needed for qPCR. Neither did this particular classroom: The thermocyclers, micropipettors, viral primers, and other equipment were made available to the student researchers through a virtual environment. In fact, the narrative which established the need for students to pursue these investigations; training on what PCR is, how to do it, and how to interpret results; as well as tools for keeping track of progress were embedded within a serious educational game, *Mission Biotech* (MBt), that took place in a computer-mediated virtual environment.

## Overview of *Mission Biotech:* How MBt Fits With National Goals

For decades, policy documents related to science education have highlighted the need for science learning experiences that help students understand and appreciate the significance of science and technology for modern life (AAAS 1990; NRC 1996). More recent documents, *A Framework for K–12 Science Education* (NRC 2011) and the *Next Generation Science Standards* (NGSS Lead States 2013), have reiterated this call with more explicit attention to engineering and technology. An underlying idea of these reform documents is that goals for science education (and STEM education more generally) should include helping students use evidence to address challenging issues and to consider ways in which science, technology, and engineering can be used to solve personal and social problems. Despite a long history of calls for making science more applicable (DeBoer 1991), many students progress through school science with very few opportunities to witness the meaningful interaction of science, technology, and society and even fewer opportunities to meaningfully engage in science and engineering practices in the context of significant societal issues (Dorph et al. 2011).

One of the challenges faced by today's teachers and curriculum designers is the rapid pace of progress and sophistication in modern science. It is difficult to keep up with new developments in science and even more difficult to feature cutting edge research in science classrooms. This presents a paradox in that a goal for science education is to help students understand and engage in the use of science for the resolution of important issues; and yet, for many of today's most relevant issues, the underlying science is difficult to feature within classrooms. Biotechnology serves as a good example of this phenomenon.

## Biotechnology

Advances in molecular biology and genetics have completely transformed the study of life sciences. The development of ideas, tools, and techniques brought about by this revolution has resulted in what is generally referred to as biotechnology (Thieman and Palladino 2003). Even the most casual observer of today's schools would quickly recognize that most life science classrooms have not kept pace with these changes. Many students never have opportunities to think about, much less actually engage in, practices with the genetic technologies that play such a significant role in modern science; the emerging sciences of genomics as well as bioinformatics, and the recent move toward computer-based biological and systems thinking, are left out of most classrooms entirely. Of course there are several very good reasons why biotechnology tends not to be featured prominently in science education. Chief among these reasons is the availability of the tools that make biotechnology possible. Biotech equipment and tools are costly and well beyond the meager budgets available to most teachers. Safety issues provide additional challenges for teachers hoping to incorporate biotechnology. Using viral vectors and recombinant DNA may be standard practice in professional laboratories, but they create potential safety problems in classrooms.

In order to address the barriers that work against the integration of biotechnology in science classrooms, we looked toward new educational technologies. More specifically, we explored the use of computer-based games as a medium for introducing students to the concepts and processes of biotechnology and to relate these ideas and processes to issues of social importance. Over the past several years, educational technologists have written a great deal about the potential of games and virtual environments to support learning (Gee 2007). Researchers have documented ways in which computer-based games support student motivation to learn, critical thinking, and in-depth understandings of target content (Devane, Durga, and Squire 2010; Watson, Mong, and Harris 2011). This work has been conducted in a variety of disciplines, but science education has been a particular focus. Science-related games have been shown to help students learn standards-based science concepts, scientific inquiry processes and scientific habits of mind (Annetta, Minogue, Holmes, and Cheng 2009; Ketelhut, Nelson, Clarke, and Dede 2010).

## MBt and Target Learners

Given the significance of biotechnology and the challenges associated with teaching biotechnology as well as the potential affordances of games, we saw biotechnology as an ideal focal point for development of a new educational game. With support from the National Science Foundation, we created *Mission Biotech* (MBt), a computer-based game that challenges high school students to use biotechnology to address a social problem (an emerging viral epidemic) and in the process explore biotechnology ideas, processes, and procedures. The game puts students in the place of a researcher working to uncover the cause of a viral outbreak. Players work within a virtual laboratory modeled after an actual molecular biology research facility as they race against the clock to diagnose viral pathogens. (See Figure 12.1 for a screen shot from the game.) They collect information about viruses and the biotech procedures they will eventually use (e.g., reverse transcription and PCR) and interact with tools found in the environment, like micropippetors, sterile gloves, a centrifuge, water bath, and so on. To accomplish their "mission," players must synthesize their prior knowledge along with information provided in the virtual environment (which may build upon or reinforce material explored in their classroom experiences) to develop protocols for tasks such as the isolation of DNA from patient samples, PCR, and reverse transcription. Along the way, players have opportunities to ask questions of non-player characters, explore the virtual environment, uncover hidden clues, find information about viruses, and learn about science careers.

Because students participate in a virtual world, many of the barriers we mentioned, such as safety and cost, are no longer relevant. Teachers do not need to worry about having necessary laboratory equipment and reagents. Students do need access to computers that will run the game, but as computer hardware and networking become more ubiquitous, the availability of

**Figure 12.1.** Screen Shot of the Laboratory Found Within *Mission Biotech*

basic computer technologies is becoming less of a problem. Other barriers may still need to be negotiated, including finding time in an already crowded curriculum, but taking advantage of virtual environments permits access to much broader ranges of scientific experiences in a fraction of the time. For example, within MBt, students complete experiments in which they extract DNA, perform reverse transcription, and conduct and analyze qPCR. All of this transpires within a few hours of game play (about three 50-minute class periods). Even if the equipment and reagents were available for teachers, conducting these actual experiments would take weeks if not months of class time.

## Major Features of *Mission Biotech:* Game-Based Curriculum

Whereas some investigations of gaming in science education have focused exclusively on whether or how students learn through game play (Steinkuehler and Duncan 2008), our design and implementation work consistently considered MBt as a game to be situated within the context of science classrooms. As such, we developed MBt as a part of a game-based curriculum. The game itself was a focal point of the instructional unit, but there were many non-game elements that contributed to the overall learning experience for the students. Figure 12.2 (p. 204) provides a graphic depiction how MBt served as a focal point for the broader curriculum that included laboratory activities, teacher-directed lessons, and assessments.

The MBt curriculum outlined a two-to-three week unit (10–15 hours of instruction) designed for implementation in high school science classes. The materials were designed with enough

**Figure 12.2.** Design Model for *Mission Biotech* and Supporting Curriculum

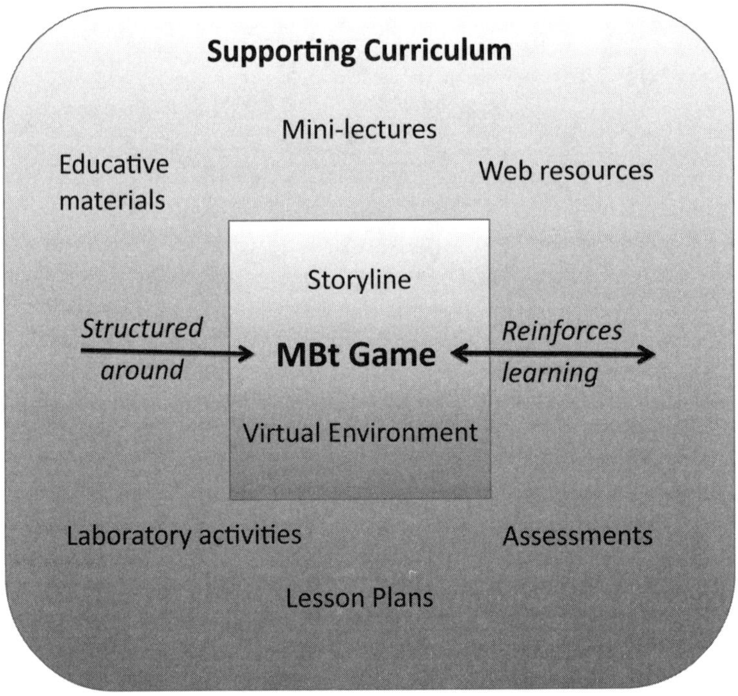


(A version of this figure is presented in Eastwood and Sadler 2013).

flexibility such that they could be used in a range of classes spanning life science–related disciplines (e.g., biology, anatomy and physiology, general science, and biotechnology) as well as academic levels (basic courses, honors courses, and Advanced Placement courses). The unit was aligned with eight science content standards from the state, Florida, in which our work took place (see Table 12.1). Typical classroom implementation involved student engagement with the game for significant amounts of time (approximately half of the unit) complemented by out-of-game experiences including small group activities, whole-class discussions, teacher-led mini-lectures, and laboratory exercises. For example, one of the first challenges students faced in the virtual environment was the need to extract DNA from a patient's blood sample. After the students negotiated this portion of the game, most teachers engaged them in a classroom-based lab activity in which they extracted DNA from fruit using commonly available household materials (Barko and Sadler 2013). The classes then discussed how the virtual and classroom protocols and products shared similarities and differences. Table 12.2 indicates a standardized instructional sequence for the implementation of MBt. This served as a model for instructional sequencing, and our partner teachers made implementation decisions based on the needs of their students, class schedules, and other local factors.

## Table 12.1. Florida State Science Standards Aligned With *Mission Biotech*

| Thematic Set | Standards |
|---|---|
| **DNA structure and replication** | SC.912.L.16.3: Describe the basic process of DNA replication and how it relates to the transmission and conservation of the genetic information.<br>SC.912.L.16.4: Explain how mutations in the DNA sequence may or may not result in phenotypic change. Explain how mutations in gametes may result in phenotypic changes in offspring. |
| **Transcription, translation, and protein structure** | SC.912.L.16.5: Explain the basic processes of transcription and translation, and how they result in the expression of genes.<br>SC.912.L.18.4: Describe the structures of proteins and amino acids. Explain the functions of proteins in living organisms. Identify some reactions that amino acids undergo. Relate the structure and function of enzymes. |
| **Genetic technologies** | SC.912.L.16.11: Discuss the technologies associated with forensic medicine and DNA identification, including restriction fragment length polymorphism (RFLP) analysis.<br>SC.912.L.16.12: Describe how basic DNA technology (restriction digestion by endonucleases, gel electrophoresis, polymerase chain reaction, ligation, and transformation) is used to construct recombinant DNA molecules (DNA cloning). |
| **Pathogens and immune responses** | SC.912.L.14.52: Explain the basic functions of the human immune system, including specific and nonspecific immune response, vaccines, and antibiotics.<br>SC.912.L.16.7: Describe how viruses and bacteria transfer genetic material between cells and the role of this process in biotechnology. |

Note: A version of this table is presented in Romine et al. 2013.

## Table 12.2. Standardized Instructional Sequence for the MBt Curriculum

| Day* | Primary Focus | Instructional Activity |
|---|---|---|
| 1 | Biotechnology tools, processes & safety | Student identification of equipment; mini-lecture (biotech tools, processes and safety) |
| 2 | Introduce MBt | Game play: begin level 1 (develop avatars; learn game controls and mechanics) |
| 3 | DNA extraction | Lab activity (extract DNA from strawberries); mini lecture (DNA location and function) |
| 4 | DNA extraction | Game play: complete level 1 (DNA extraction and introduce PCR) |
| 5 | DNA structure & PCR | Brief video and mini-lecture; Small group questions; whole class discussion |
| 6 | PCR | Game play: begin level 2 (extract DNA and conduct PCR) |
| 7 | PCR process | Small group activity: simulation of PCR process; Class discussion |
| 8 | PCR analysis | Game play: complete level 2 (conduct and analyze real-time PCR results) |
| 9 | PCR analysis | Mini-lecture; Students work independently to interpret real-time PCR data |
| 10 | Reverse transcription | Game play: level 3 (reverse transcription, conduct and analyze PCR results) |

*Each "day" represents approximately one hour of instructional time.
Note: version of this table is presented in Romine et al. 2013.

## MBt as STEM Education

Widespread agreement has emerged around the significance of STEM education, but little consensus exists around an operationalized definition for STEM education. In some cases, STEM is used as an umbrella term to describe anything related to science, technology, engineering, or mathematics. Other STEM initiatives represent a necessary merging of all four disciplines for the creation of completely interdisciplinary teaching and learning contexts. Our design and research team think that both of these ends of the spectrum for defining STEM are problematic. In designing MBt, we considered STEM from a more pragmatic perspective informed by how the disciplines may merge in meaningful ways within real-world environments. Biotechnology is a fundamentally multidisciplinary field, which draws on ideas and tools from multiple science disciplines (e.g., biology, biochemistry, chemistry, and physics). These tools are applied to problems and needs that stem from agriculture, medicine, and the environment. This application of scientific ideas to issues of human health, nutrition, food production, and so on is, by definition, a form of engineering; and the products that are developed through engineering processes are technology (NRC 2011). Therefore, biotechnology represents one of several emergent fields (including others such as chemical engineering, nanotechnology, and bioinformatics) that are natural contexts for STEM education.

In the case of MBt, the game and associated curriculum challenged learners to consider basic biological principles, molecular biology and biochemistry, in the context of a human health issue in which scientific ideas and technologies were applied to solve a problem (i.e., engineering). Mathematics was not an explicit focus of the MBt unit; although students certainly did need to engage mathematical ideas and processes as they interpreted results from qPCR, a central focus of the game. From our perspective, the multidisciplinary focus of MBt, even with the limited attention to mathematics, makes the unit a good example of STEM education.

## Teacher Roles

The roles played by teachers in the implementation of the unit represent another significant feature of MBt. The teachers with whom we worked and partnered to collect data featured in a later section of this chapter participated in a two-week summer institute. The summer institute provided professional development for biotechnology education in secondary schools and MBt was a featured element. During the institute, teachers played the MBt game, analyzed the supporting curriculum materials, and collaborated in small groups to create plans for implementing the curriculum in their own school contexts. Several teachers who did not attend the summer institute have used portions of the game and supporting curriculum, but the analyses offered later in the chapter are based on classes whose teachers participated in the professional development.

When teachers implemented the game with their classes, they took on several different roles as the unit unfolded. They helped to establish the context for game play and served as a resource for students as they navigated initial portions of the game. As students confronted new ideas and challenges within the game, teachers implemented a variety of strategies for supporting student meaning-making. Some of these activities included facilitating class discussions, coordinating laboratory investigations, conducting "just-in-time" lectures, and managing small group activities. How teachers enacted these roles varied based on their own teaching styles and the unique needs of their students. Some of these differences will be highlighted in the case summaries presented below.

## Evidence for Success

In order to assess the success of the MBt unit for supporting student science learning, our team collected multiple forms of data and conducted associated analyses. For this chapter, we provide evidence of (1) student learning of science content and (2) successful classroom implementation of the unit. The analysis of student learning is based on a quasi-experimental study in which pre- and post-intervention tests were used to document growth in students' understandings of the science content embedded within the MBt unit. Analysis of classroom implementation of the unit is based on three in-depth case studies conducted with teachers as they used MBt in their classrooms. In this section, we present highlights from these investigations.

### *Student Learning of Science Content: Data Collection and Analysis*

To investigate student learning associated with MBt, we administered two content assessments before and after implementation of the unit. The first instrument was closely aligned to the learning context; that is, assessment items mirrored tasks and challenges that students had experienced within the learning environment. We referred to this assessment that is "close" to the learning context as the *unit test*. We also administered a test that was constructed with publicly released items from state, national, and international tests. The items sampled aligned with the standards upon which the unit was based, but these assessment tasks were considerably farther removed from the curriculum than the unit test. We referred to this more "distanced" assessment as the *standards test*. The unit test contained 19 items with satisfactory reliability ($\alpha_{pre}$ = 0.715; $\alpha_{post}$ = 0.826). The standards test also demonstrated sufficient reliability ($\alpha_{pre}$ = 0.836; $\alpha_{post}$ = 0.853). Additional details regarding instrument construction, reliability and validity are presented elsewhere (Sadler, Romine, Stuart, and Merle-Johnson 2013). To analyze these data we used a repeated measures multivariate analysis of variance (MANOVA) with both unit and standards test scores serving as dependent variables.

#### Sample

We collected data from 642 high school students participating in 31 different biology classes. Ten teachers working in different schools around the state of Florida taught the 31 class sections. To differentiate among the different kinds of classes in which students were enrolled, we grouped the classes into three academic levels: general biology courses (Gen), honors biology courses (Hon), and Advanced Placement courses (Adv). These levels were used as grouping variables in the statistical analysis. The school settings included rural, suburban, and urban communities.

#### Results

Pre- and posttest means for the unit and standards tests are presented in Table 12.3 (p. 208). Multivariate tests suggest significant differences in student scores across time ($\Lambda$ = 0.671, F = 127.742, $p < 0.01$, $\eta^2$ = 0.329) and between academic levels ($\Lambda$ = 0.635, F = 66.63, $p < 0.01$, $\eta^2$

= 0.203). Univariate analyses of results from both the unit test and the standards test reveal statistically significant gains at $\alpha = 0.01$ (see Table 12.4). Figures 12.3 and 12.4 present graphic displays of score increases for both groups disaggregated by academic level. Careful analysis of these graphs suggests that whereas students in all three academic levels show similar patterns of improvement on the post-intervention unit test, there is an apparent interaction effect on the standards test. Students in the general classes made greater gains on the standards test over time than their peers in the two other groups. This trend is supported by Time x Level analysis presented in Table 12.4. These results suggest that students across all levels demonstrated statistically significant gains on both content tests, but that students in the lower academic level benefitted to a greater degree than their peers.

**Table 12.3.** Mean Unit Test and Standards Test Scores Across Ability Levels Before and After the Mbt Intervention

| Assessment | Level | Pre-test Mean (SD) | Post-test Mean (SD) | t | d |
|---|---|---|---|---|---|
| Unit test | Gen (*n* = 70) | 6.17 (3.15) | 10.13 (5.26) | 5.77* | 0.91 |
| | Hon (*n* = 248) | 7.95 (4.01) | 12.83 (5.34) | 12.77* | 1.03 |
| | Adv (*n* = 219) | 11.14 (5.03) | 15.28 (5.95) | 8.83* | 0.75 |
| Standards test | Gen (*n* = 71) | 5.96 (2.83) | 9.18 (4.82) | 5.37* | 0.82 |
| | Hon (*n*= 233) | 10.30 (4.66) | 11.61 (4.55) | 4.22* | 0.29 |
| | Adv (*n* = 220) | 14.22 (3.88) | 15.25 (3.08) | 4.50* | 0.29 |

*$p < 0.01$

Note: A version of this table is presented in Romine et al. 2013.

**Figure 12.3.** Comparative Gains on the Unit Test

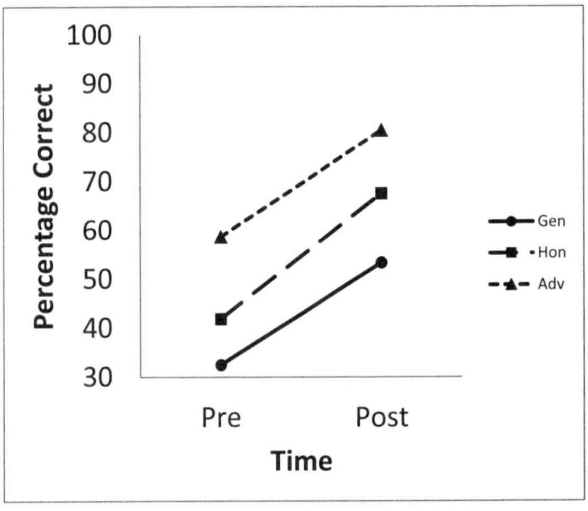


Note: A version of this figure is presented in Romine et al. 2013.

**Figure 12.4.** Comparative Gains on the Standards Test

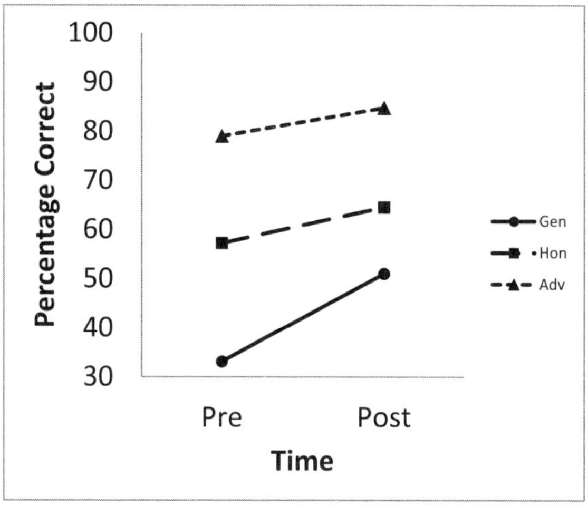


Note: A version of this figure is presented in Romine et al. 2013.

**Table 12.4.** Univariate ANOVA Tests

| Variable | Measure | SS | df | MS | *F* | *p* | h2 |
|---|---|---|---|---|---|---|---|
| **Time** | Unit test | 3753.35 | 1 | 3753.35 | 180.19* | 0.00 | 0.26 |
| | Standards test | 709.1 | 1 | 709.1 | 81.56* | 0.00 | 0.13 |
| **Level** | Unit test | 3344.56 | 2 | 1672.28 | 55.47* | 0.00 | 0.18 |
| | Standards test | 6695.7 | 2 | 3347.85 | 138.54* | 0.00 | 0.35 |
| **Time x Level** | Unit test | 58.87 | 2 | 29.43 | 1.41 | 0.24 | 0.01 |
| | Standards test | 152.45 | 2 | 76.22 | 8.77* | 0.00 | 0.03 |

*$\alpha$ = 0.01

Note: A version of this table is presented in Romine et al. 2013.

## *Classroom Implementation of MBt*

To complement the quantitative analyses just presented, our team wanted to build an empirically based picture of how the MBt unit was implemented in classrooms, teacher perceptions of the unit, and teacher assessment of student experiences with the game-based unit. To do this work, we conducted case studies with three teachers, and each teacher served as an individual case (Stake 2005). Primary data sources for the case studies included extensive classroom observations with associated field notes and teacher interviews. Other data sources used to supplement the analyses of these primary data included lesson-planning documents, teacher journals, and teacher surveys. A researcher who had not been involved with the design, professional development, or implementation of the MBt unit directed data collection and analysis.

## Case Study Teachers

The three case study teachers taught in different parts of the same state. Ms. Rand taught general biology to ninth and tenth graders; Ms. Cenna taught a mixed level (9h, 10h, and 11th grade) biotechnology course; and Ms. Denor taught honors biology for 9th and 10th grade. All three teachers participated in the summer professional development institute, were committed to integrating biotechnology in their classes, and worked in schools/districts that supported their use of innovative approaches for teaching science. While our participants were not representative of all life science teachers, they did reflect a group interested in biotechnology and committed to implementing innovation in their classrooms. Additional details regarding the three case study participants, their teaching contexts, and expanded descriptions of their interactions with MBt can be found in a recent article (Eastwood and Sadler 2013).

### *Classroom Implementation*

All three teachers successfully implemented the MBt unit with their classes. The case studies document ways in which the teachers made curricular enactment decisions based on the needs of their students and local contexts. Whereas all three teachers engaged their students in a combination of game play and out-of-game learning experiences, important differences emerged in terms of how the teachers facilitated these transitions. Ms. Rand and Ms. Denor consistently drew substantive connections between the students' in-game and out-of-game experiences. For example, Ms. Rand facilitated numerous whole-class discussions in which students were challenged to explain aspects of their game-based interactions in terms of ideas presented in mini-lectures or other class activities. In contrast, Ms. Cenna made very few explicit links between in-game and out-of-game experiences. She felt that students would make these connections on their own; however, observation of student conversations and work products suggest to us that they could have benefitted significantly from help in making these links. The MBt unit appeared to be most powerful in terms of supporting learning when the teachers embraced the notion of game as an anchor for student learning experiences that could be expanded upon with a range of pedagogical strategies employed beyond the game environment.

### *Teacher Perceptions of MBt*

All three teachers reported high satisfaction with the MBt unit. They uniformly endorsed the quality of the curricular materials. In particular, they appreciated the wide range of activities and materials provided, the educative supports built into the materials to help the teachers make sense of some of the more sophisticated aspects of biotechnology content, and the flexibility of materials that allowed them to customize the unit for the specific needs of their learners. They also noted the advantage of making use of a virtual environment that allowed their students to "see" STEM ideas and tools in more natural settings. Even Ms. Cenna, who taught biotechnology courses in a very well resourced classroom, commented on how the MBt was able to provide students access to tools and situations impossible to replicate in actual classrooms. Ms.

Rand expanded on this idea of the game being able to provide a specific context for her students to apply their ideas and build their understandings: "Instead of just referring them [students] to something abstract, we have something concrete to talk about ... and firsthand knowledge [for them to apply]." All three teachers appreciated the fact that student actions and decisions in the game elicited immediate feedback. That is, when students made mistakes in the game, such as not following a protocol or misinterpreting data, the game would let them know the nature of their mistake and allow for another attempt. Ms. Denor and Ms. Cenna talked about how this feature of the game was an important part of formative assessment that supported learning.

The three teachers also expressed some concerns related to the MBt unit, and these concerns tended to relate to the technology aspects of the unit. Although all three teachers were able to implement the gaming portion of the unit with their students, they cited concerns with the feasibility of incorporating games broadly in science classrooms. Ms. Cenna's room was equipped with a set of new laptops whose computing power exceeded the requirements for running the MBt games. However, her school's computer security protocols made it difficult to install the software in a way that would enable students to play the game (and save game progress) over an extended time period. The project team was able to work with the technology coordinator in Ms. Cenna's school to overcome this problem, but it raised a concern for her in thinking about broader implementation of the unit. Ms. Rand's school did not have sufficient computer hardware for a class of students to run the game, so she borrowed a classroom set of laptops from the project team. Here again, we were able to create a solution for this specific environment, but Ms. Rand was concerned about the challenges teachers may face without added supports.

### *Teacher Assessments of Student Experiences*

All three teachers asserted that the MBt unit supported learning for their students. They cited specific gains relative to content knowledge, science processes, and tools and purposes of biotechnology. The teachers reported that their students liked playing the MBt game as a part of class but that most students also experienced frustration at some times during game play. The teachers' interpretation of this frustration, however, varied. Ms. Denor cited some frustration as helpful because it reflected the nature of many scientific investigations. Ms. Rand suggested that her students who were "gamers" were well accustomed to challenging situations that might lead to frustration within games, so these students adapted more easily to game mechanics in the classroom. In contrast, Ms. Cenna expressed concern about the possibility of student frustration inhibiting learning.

Another point of divergence among teacher perspectives related to the extent to which the unit supported development of student interest in biotechnology and STEM-related careers. Ms. Rand and Ms. Denor felt that their students became more interested in STEM because of their MBt experiences. Ms. Denor reported that the game generated interest and excitement among her students and that it served to motivate student engagement in all aspects of the class. For example, she cited students being more motivated to excel in out-of-game activities associated with MBt because doing so would help them perform better within the game. Ms. Cenna did not think that the unit necessarily impacted her students' interest in biotechnology and STEM. However, it should be noted that Ms. Cenna taught biotechnology classes (whereas the other teachers had biology classes), and these students were more accustomed to exploring biotechnology themes and using associated tools.

## Interpretations

The research conducted in association with MBt provides quantitative and qualitative evidence supporting the idea that students can learn substantive science content in association with a game-based curriculum. The quantitative analyses of the unit test suggest that students made the kinds of learning gains expected of an extended (two-to-three weeks in length) unit that is functioning well (Ruiz-Primo, Shavelson, Hamilton, and Klein 2002). Results from the standards test provides evidence that the intervention has potential for supporting the kind of learning reflected through standardized achievement tests (Klosterman and Sadler 2010). Comparing trends in the standards test results among the three academic levels also suggests that the unit may have been more helpful for students in the lower level classes as compared to their peers participating in more advanced coursework. A possible contributor to the observed interaction effect (interaction of treatment effect with academic level) is that the standards test may not have been challenging enough to sufficiently demonstrate gains for students in the highest level classes (i.e, a "test ceiling effect"). However, given the results of students in the honors classes (the middle level), this ceiling effect seems unlikely to explain the complete differences. Based on our observations of classes engaged with the MBt unit, we think that a possible explanation for the difference in results relates to the fact that lower academic level students tended to be more willing to make mistakes and explore the game environment. Students in the higher-level courses tended to become frustrated more easily within the game and became uncomfortable when they did not get the right answer or know that they were making the correct choices. A level of uncertainty is a part of game mechanics, but this seemed to differ from some of the expectations held by students who had otherwise been very successful in their science classes.

Interestingly, some of the teachers who implemented MBt noted the same trend revealed in these data. During the summer institute, many teachers expressed concern that the unit would be too difficult for their general level students and did not think that these students would be successful with the game. In fact, many of the teachers with whom we worked (beyond the 10 whose data are included in this chapter) opted to implement the MBt unit only in their honors, Advanced Placement, or International Baccalaureate classes. However, among those teachers who tried MBt with their lower level classes, many reported that their students' performance in the unit exceeded their expectations. The quote below is taken from an email message from one of our partner teachers discussing the experience of one of her lowest performing students:

> *"[The student] is a freshman in regular biology … and really enjoyed the game. He was the first one to complete the first level, even before my AP biology students. He also has an F in the class. I can say that this [MBt] really has turned on a light for him … Thank you very much for this opportunity for my students. The younger kids got more out of it than the older kids for sure. Not what I would have expected."*

Not all students reacted as positively to MBt as the one referenced in this teacher's quote, but this episode provides a nice example of ways in which a game-based curriculum can support science teaching and learning. Games have the potential to address content that may otherwise be inaccessible, and in some cases, they may be able to reach students who may otherwise be unreachable.

Evidence from the qualitative analyses provides several insights with regard to design and use of game-based curricula. First, creating materials that enable the bridging of students' game experiences with classroom learning opportunities is important. Results from the implementation case studies suggest that some teachers need more support for effectively linking the in- and out-of-game experiences. This notion of making connections should be emphasized in game-based curriculum materials as well as professional development designed to support teacher implementation. Second, the MBt project provides evidence that creating a flexibly adaptive curriculum (Schwartz, Lin, Brophy, and Bransford 1999) that makes it possible for teachers to make curricular enactment decisions based on their local contexts can be successful. Models of educational innovation, which demand strict "fidelity of implementation" and therefore limit teachers' abilities to make informed curricular decisions do not adequately reflect the complexity of school environments and the professionalism of teachers. Finally, the teachers highlighted a number of aspects of game-based curricula that they deemed as particularly significant. In other words, these are the game-based curriculum affordances that they saw as being most important: interesting and motivating environment, numerous opportunities for formative feedback, and using the virtual environment of games to help students "see" parts of the STEM world that are otherwise inaccessible. Designers of games would be well advised to emphasize these aspects both in their development of game-based curriculum and their delivery of these curricula to teachers.

## Next Steps

Several teachers continue to use the MBt game and associated curriculum (available at *www.missionbiotech.com*), but one of the challenges with technology-based innovations is the constantly evolving nature of technology. We designed the MBt game about five years prior to the writing of this chapter, and not surprisingly, the technology behind game design and delivery has changed dramatically. For example, whereas most of today's new games can be accessed through tablet devices and are cloud-based, MBt is designed as software that resides on individual computers. Keeping pace with the progress of technology would require a level of resources not available to most education-based projects. Therefore, our team is exploring ways to create more flexible game platforms and yet still work within the constraints of modern schools. We are trying to use this approach to create new opportunities for students to experience game-based curricula and even to expand opportunities for students to be involved in the creation of STEM-focused games. With engineering design goals being called for in the *Next Generation Science Standards*, teaching students design of educational games is a natural progression of MBt. Initial results from two National Science Foundation projects suggest students learn more and show greater affective change through game creation as opposed to game playing (Annetta, et al., forthcoming).

Over the last few years, we have witnessed an unprecedented push for expansion and reform of STEM education. Concerns related to workforce development, economic growth, capacity to innovate, national security, and our nation's role within the international community are all cited as significant drivers for improving STEM education. From our perspective, it is essential that corresponding efforts help students to understand and experience dimensions of STEM as they exist in the world, which can diverge from the ways in which mathematics, science and engineering are commonly represented in schools. Fields like biotechnology, which naturally bring together science, engineering, and technology, are impacting modern society in profound ways, and we need to find better ways of exposing students to some of these developments. Technologies, such as educational games, can help accomplish these goals, and in this chapter, we have provided evidence that an educational game situated within a broader curriculum can effectively support learning within science classrooms. We encourage the STEM education community to continue considering ways in which technology, and games in particular, can be leveraged to support student learning and development.

## References

American Association for the Advancement of Science (AAAS). 1990. *Science for all Americans*. New York: Oxford University Press.

Annetta, L. A., J. A. Minogue, S. Holmes, and M. T. Cheng. 2009. Investigating the impact of video games on high school students' engagement and learning about genetics. *Computers & Education* 54 (1): 74–85.

Annetta, L. A., D. Vallett, B. Fusarelli, R. Lamb, M. T. Cheng, S. Y. Holmes, E. Folta, and B. Thurmond. forthcoming. Investigating science interest in a game-based learning project. *Journal of Computers in Mathematics and Science Teaching*.

Barko, T., and T. D. Sadler. 2013. Practicality in virtuality: Finding student meaning in video game education. *Journal of Science Education and Technology* 22: 124–132.

DeBoer, G. E. 1991. *A history of ideas in science education: Implications for practice*. New York: Teachers College Press.

DeVane, B., S. Durga, and K. Squire. 2010. "Economists who think like ecologists." Reframing systems thinking in games for learning. *E-learning and Digital Media* 7: 3–20.

Dorph, R., P. Shields, J. Tiffany-Morales, A. Hartry, and T. McCaffrey. 2011. *High hopes, few opportunities: The status of elementary science education in California. Strengthening Science Education in California*. Sacramento, CA: The Center for the Future of Teaching and Learning at WestEd.

Eastwood, J. L., and T. D. Sadler. 2013. Teachers' implementation of a game-based biotechnology curriculum. *Computers and Education* 66: 11–24.

Gee, J. P. 2007. *What video games have to teach us about learning and literacy.* New York: Palgrave Macmillan.

Ketelhut, D. J., B. C. Nelson, J. Clarke, and C. Dede. 2010. A multi-user virtual environment for building and assessing higher order inquiry skills in science. *British Journal of Educational Technology* 41: 56–68.

Klosterman, M. L., and T. D. Sadler. 2010. Multi-level assessment of scientific content knowledge gains associated with socioscientific issues-based instruction. *International Journal of Science Education* 32: 1017–1043.

National Research Council (NRC). 1996. *National science education standards*. Washington, DC: National Academies Press.

National Research Council (NRC). 2011. *A framework for K–12 science education: Practices, crosscutting concepts, and core ideas*. Washington, DC: National Academies Press.

NGSS Lead States. 2013. *Next Generation Science Standards: For states by states*. Washington, DC: National Academies Press. *www.nextgenscience.org/next-generation-science-standards*

Ruiz-Primo, M. A., R. J. Shavelson, L. Hamilton, and S. Klein. 2002. On the evaluation of systemic science education reform: Searching for instructional sensitivity. *Journal of Research in Science Teaching* 39 (5): 369–393.

Sadler, T. D., W. Romine, P. E. Stuart, and D. Merle Johnson. 2013. Game-based curricula in biology classes: Differential effects among varying academic levels. *Journal of Research in Science Teaching* 50: 479–499.

Schwartz, D., X. Lin, S. Brophy, and J. Bransford. 1999. Toward the development of flexibly adaptive instructional design. In *Instructional design theories and models: A new paradigm of instructional theory*, ed. C. Reigeluth, 183–214. Mahwah, New Jersey: Lawrence Erlbaum.

Scott, D. G., B. A. Washer, and M. D. Wright. 2006. A Delphi study to identify recommended biotechnology competencies for first-year/initially certified technology education teachers. *Journal of Technology Education* 17: 43–55.

Squire, K. 2011. *Video games and learning: Teaching and participatory culture in the digital age.* New York: Teachers College Press.

Stake, R. E. 2005. Qualitative case studies. In *The SAGE handbook of qualitative research,* ed. N. K. Denzin and Y. S. Lincoln, 443–466. Thousand Oaks, CA: SAGE Publications.

Steinkuehler, C., and S. Duncan. 2008. Scientific habits of mind in virtual worlds. *Journal of Science Education and Technology* 17 (6): 530–543.

Thieman, W. J., and M. A. Palladino. 2003. *Introduction to biotechnology*. San Francisco, CA: Benjamin Cummings.

Watson, W. R., C. J. Mong, and C. A. Harris. 2011. A case study of the in-class use of a video game for teaching high school history. *Computers & Education* 56: 466–474.

# Learning Genetics at the Nexus of Science, Technology, Engineering, and Mathematics (STEM)

*S. Selcen Guzey*
*University of Minnesota*
*St. Paul, MN*

*Tamara Moore*
*Purdue University*
*West Lafayette, IN*

*Gillian Roehrig*
*University of Minnesota*
*St. Paul, MN*

## Setting

The Region 11 Mathematics and Science Teacher Partnership (MSTP) has been funded through federal grants since 2008 to provide quality Professional Development (PD) for mathematics and science teachers in the Minneapolis/St. Paul metro region in Minnesota. During the past five years, the Region 11 MSTP has provided support in the form of workshops and Professional Learning Communities (PLCs) for more than 2,000 teachers in 258 schools from 48 school districts. The Region 11 MSTP involves a multiple institutional partnership that includes three higher education institutions, two regional education partners, SciMath MN (a nonprofit education and business coalition advocating for quality STEM education), and high-need school partners. Partners share the same visions and goals for improving STEM education in Minnesota, which forms a strong partnership. Each partner has identified roles and responsibilities; for example, the higher education partners are responsible for the design, implementation, and evaluation of the PD. Each year the PD deals with a different grade-level and subject matter focus. This chapter focuses on the PD module developed for life science teachers for grades 7–12 delivered in 2011–2012. In particular, the chapter provides information about the structure of the genetics section of the life science PD module that applied a context-based STEM integration approach and an accompanying research study that focuses on the impacts of the PD on teachers and their students.

## Overview of Program

Advancing the Science, Technology, Engineering, and Mathematics (STEM) education has become the center of educational reforms in the U.S. over the last decade (e.g., National Academy of Engineering [NAE] 2009; National Center on Education and Economy, 2007). It is argued that it is necessary to improve STEM teaching and learning in order to innovate and compete in the global economy (NRC 2011; The President's Council of Advisors on Science and Technology 2010). As shown by National Assessment of Educational Progress studies, STEM education in the United States is now behind compared to other nations. Thus there is a need to better prepare students in STEM fields and also enhance their interest in STEM majors to increase and strengthen the STEM workforce. Transforming STEM education in the United States will require many actions to be taken, such as establishing effective STEM-focused schools (NRC 2011) and preparing STEM teachers who have both deep content knowledge in STEM subjects and proficient pedagogical skills required to teach these subjects effectively (The President's Council of Advisors on Science and Technology 2010). To this end, the U.S. government has provided funding to better prepare STEM teachers so that they can prepare students with necessary knowledge of STEM disciplines and encourage interest in STEM majors and careers.

The overarching goal of the Region 11 MSTP is to increase student achievement in STEM disciplines through increasing teachers' content understanding and pedagogical skills through PD opportunities for mathematics and science teachers in grades 3–12. The success of the partnership is measured by: (1) the improvement in student achievement, (2) the number of teacher participants in professional development programs each year, and (3) improvement in participant teachers' content knowledge and pedagogical skills.

During the 2011–2012, the region 11 MSTP partnership provided three PD opportunities: Life Science for teachers in grades 7–12, Nature of Science and Engineering for teachers in grades 3–6, and Mathematical Reasoning with Rational Numbers for teachers in grades 6–8. The 6–8 mathematical reasoning with rational numbers module was developed using the Principles and Standards for School Mathematics (NCTM), the Curriculum Focal Points for Prekindergarten through Grade 8 Mathematics (NCTM), and the new Minnesota Academic Standards in Mathematics and provided experiences for mathematics teachers as they prepare their students for success in eighth-grade algebra. The nature of the science and engineering module was developed to help teachers explore and understand important science and engineering ideas related to the Minnesota Academic Standards in Science. Finally, the life science module, the focus of this chapter, was designed for teachers to advance their science content knowledge in the areas of evolution, ecology, and genetics and to improve their pedagogical skills to teach these content areas using integrated STEM education approaches.

The content for the life science PD module was aligned with the Minnesota Academic Standards and the *National Science Education Standards* (*NSES*; NRC 1996). Additionally, the PD was developed following the NSES standards for effective professional development programs (see Table 13.1).

**Table 13.1.** *NSES* Professional Development Standards and Actions Taken to Meet Them

| *NSES* Professional Development Standards | Actions |
|---|---|
| Professional development for teachers of science requires learning essential science content through the perspectives and methods of inquiry (Standard A) | 1. Teachers actively participated in inquiry-based activities in the workshops.<br>2. Workshop topics were engaging, interesting, and significant in science.<br>3. A variety of resources such as recent scientific information from the literature regarding the presented content shared with the teachers. |
| Professional development for teachers of science requires integrating knowledge of science, learning, pedagogy, and students; it also requires applying that knowledge to science teaching (Standard B) | 1. The workshop instructors modeled a range of instructional strategies (e.g., process-oriented guided inquiry) and teachers reflected on those strategies.<br>2. Research findings regarding students' misconceptions in content were presented in the workshops.<br>3. Teachers created change in their instruction through employing the teaching strategies modeled in the workshops. |
| Professional development for teachers of science requires building understanding and ability for lifelong learning (Standard C) | 1. Teachers worked collaboratively to design STEM integration lesson plans. They taught the new lessons and then reflected on them at the PLC meetings.<br>2. Recent research on teaching and student learning that is related to the workshop content was shared with teachers. |
| Professional development programs for teachers of science must be coherent and integrated (Standard D) | 1. A year-long program that included five face-to-face workshops and monthly PLC meetings was developed. |

## STEM Integration

As Bybee (2010a) argues, the term "STEM education" has been widely used; however, there has not yet been broad agreement among educators, policy makers, and researchers on what STEM education means. Most people refer to STEM education as teaching and learning in the individual STEM disciplines—science, technology, engineering, and mathematics. Here, we define STEM education as "an effort by educators to have students participate in engineering design as a means to develop technologies that require meaningful learning and an application of mathematics and/or science" (Moore et al., forthcoming). There is an emerging consensus that STEM disciplines should be taught and learned with an integrated approach. An integrated STEM approach, which involves an interdisciplinary approach, requires teachers to organize and teach content from multiple STEM disciplines around a real-world theme.

Incorporating engineering and technology into science standards is a response to the need for integrated STEM education. The authors of *A Framework for K–12 Science Education* (NRC 2012) promote the integration of technology and engineering into science. "Engineering, technology, and the applications of science" is one of the four core ideas, in addition to physical sciences, life sciences, and Earth and space sciences to be taught in K–12 science classrooms. In addition to developing and applying science knowledge, students in science classes should learn "how science is utilized, in particular through the engineering design process, and they should come to appreciate the distinctions and relationships between engineering, technology, and applications of science" (NRC 2012, p. 143). According to the *Framework*, engaging in science and engineering practices helps K–12 students understand how scientists and engineers work, the link between engineering and science, and crosscutting concepts and disciplinary ideas of science and engineering. The following eight key practices are identified in the *Framework* as important aspects of science and engineering that should be a central part of K–12 science and engineering curriculum: (1) asking questions (for science) and defining problems (for engineering), (2) developing and using models, (3) planning and carrying out investigations, (4) analyzing and interpreting data, (5) using mathematics and computational thinking, (6) constructing explanations (for science) and designing solutions (for engineering), (7) engaging in argument from evidence, (8) obtaining, evaluating, and communicating information (p. 49). The *Next Generation Science Standards* (*NGSS*) also emphasize the critical role for integrating engineering and technology in science education (NGSS Lead States 2013.)

STEM integration can be challenging for science teachers since there is not a single approach that defines how to integrate STEM disciplines into science instruction and there is no real STEM "curriculum" (e.g., textbooks) available for high school science teachers. There are two different STEM integration models: content integration and context-integration (Moore et al., forthcoming). The content integration focuses on merging multiple STEM disciplines. On the other hand, the context integration uses context of one or more STEM disciplines and focuses on the content of one discipline. As Bybee suggests, by following a context-based integrative STEM education approach, science teachers can use engineering or technology-related contexts in areas such as "personal health, energy efficiency, environmental quality, resource use, and national security" (2010b, p. 996) to define science content to students using appropriate student-centered instructional strategies. In our PD, we also used the context-based STEM integration approach by focusing on genetics as our content and using genetic engineering as a context to make the content more relevant.

## Major Features of the Instructional Program

For the improvement of STEM education, a coherent, sustained PD is necessary (The President's Council of Advisors on Science and Technology 2010). To address this need, we developed the life science module that applied a context-based STEM education approach. Professional development experiences, such as the one described here, can help teachers to attain the goal of the *Framework* and *NGSS*.

The life science module consisted of five full-day workshops and professional learning community (PLC) meetings. Table 13.2 shows the structure and timeline of the PD. A total of 113 teachers attended all three workshops in the summer of 2011 and then each teacher selected two of the three content areas for extended learning for workshop day 4 in the fall of 2011 and workshop day 5 in the spring of 2012.

**Table 13.2.** Life Science Module Timeline and Overview

| | Summer 2011 | Fall 2011 | Spring 2012 |
|---|---|---|---|
| Day 1: *Evolution I* | X | | |
| Day 2: *Ecology I* | X | | |
| Day 3: *Genetics I* | X | | |
| Day 4: Extended Learning, *Evolution II, Ecology II, or Genetics II* | | X | |
| Day 5: Extended Learning, *Evolution II, Ecology II, or Genetics II* | | | X |
| PLC Meetings | | X | X |

In this chapter, we only focus on one part of the life science module: genetics workshops (genetics I and II) and PLC meetings. In the face-to-face genetics workshops, STEM disciplines were integrated and several PD activities were developed using STEM-related issues. PLC meetings for genetics were designed for teachers specifically to explore ways to integrate STEM into their teaching. In the following sections, the structure of the face-to-face workshops and PLC meetings are discussed.

## Face-to-Face Genetics Workshops

In the face-to-face workshops, two specific contexts were used for STEM integration: (1) health and (2) the innovations in science, technology, engineering, and mathematics (Bybee 2010a). Personal, social, and also global issues regarding the selected contexts were addressed during the workshops. A variety of educational technology tools were also embedded in the workshop activities. Table 13.3 shows the workshop activities and also indicates how the activities aligned with Minnesota science education standards.

**Table 13.3.** Alignment of Activities With the Minnesota Standards

| Workshop Activities | MN Science Education Standards |
|---|---|
| Genetics I: Beta-Globin gene activity | Gene structure |
| Genetics I: Microarray case study | Gene expression, technologies, and advances |
| Genetics I: A case study of sickle cell anemia | Inheritance, mutations, pedigree analysis, genetic diseases, and genetic testing |
| Genetics I: Caenorhabditis *elegans* lab | Mutations, genetic engineering |
| Genetics I: Plasmids | DNA processes, genetic engineering |
| Genetics II: Paperclip PCR | DNA replication, biotechnology risks and benefits |
| Genetics II: Genetically modified organisms (GMOs) debate | Biotechnology risks and benefits, genetic engineering |
| Genetics II: GMO lab | Technologies and advances, genetic engineering |

Genetics I Workshop focused on the molecular basis of inheritance, flow of genetic information, mutations, gene expression, and genetic diseases. The Genetics II Workshop focused on genetic engineering and biotechnology. In Genetics I, teachers completed a sickle cell anemia case study activity where they performed a pedigree analysis of a family and a dry laboratory activity, which were used to simulate sickle cell anemia with DNA restriction enzymes and gel electrophoresis analysis. During the case study activity, teachers explored the inheritance of the disease, the structure and function of hemoglobin protein in healthy people and patients, DNA sequences of normal and abnormal hemoglobin genes, malaria resistance and sickle cell mutations, and comparison of frequencies of those in geographic areas of the world. At the end of the activity, teachers discussed the ethical concerns regarding genetic testing and the epidemiology of the disease providing an opportunity to reflect on the impacts of the STEM developments on the society. In Genetics II, teachers extended their knowledge of genetic engineering and the modern DNA technology used to insert foreign DNA into an organism of interest, the process of making recombinant DNA and producing genetically modified organisms. Teachers also engaged in a debate where they participated in a role-playing discussion on herbicide tolerant sugar-beet.

In both workshops, teachers explored innovations, recent developments, and research in the field of genetics. Research articles and web resources were shared with the teachers. The web resources were selected on the basis of two criteria: original information and accessibility (free of charge) and the presence of updated, accurate information. For example, teachers explored the National Institute of Health (NIH; *www.nih.gov*) where they can find accurate information about the recent development in the field of genetics. The Basic Alignment Search Tool (BLAST) on the NIH website were also introduced to the teachers so that they have access to a database of amino-acid sequences of different proteins or the nucleotides of DNA sequences.

## Professional Learning Communities Meetings

During the academic year, teachers and PLC facilitators met regularly for PLC meetings. PLC meetings were formed for teachers to discuss instructional strategies learned in the workshops, and most importantly to support each other in implementing new classroom activities practiced in the workshops. PLC meetings were included in the PD since it has been shown that through participating in PLCs, STEM teachers increase their understanding of their content area, feel more prepared to teach their content, and as a result they pay more attention to student understanding and reasoning, use more research-based methods for teaching content, and apply a variety of strategies for students to engage in problem solving (Fulton and Britton 2011). An overview of the PLC meetings is shown in Table 13.4.

Teachers who completed the Genetics II Workshop in September 2011 participated in the PLC meetings during the fall semester, while teachers who attended the Genetics II Workshop in January 2012 participated in the same PLC during the spring semester. The first and fourth PLC meetings were face-to-face meetings and the second and third meetings were online meetings. A Ning site was created to share resources and Adobe Connect was used for online meetings. Each meeting was approximately two hours and facilitated by the workshop instructor. The first meeting was dedicated to learning about the online tools that were used for collaboration and the online meetings. Learning about several new technology tools was challenging for

**Table 13.4.** General Outline for Four Sessions of the Professional Learning Communities

| Meeting | Activity |
|---|---|
| **Meeting 1, face-to-face** | Creating an online community: How-tos<br>Minnesota science and engineering standards<br>Forming groups and identifying the new STEM lesson plan ideas |
| **Meeting 2, Online** | **Instructional Strategies (Lesson I):** Discussing what/how to develop the first context-based STEM integration lesson. |
| **Meeting 3, Online** | **Instructional Strategies (Lesson I)- Results:** Reflecting on the classroom implementation of the first lesson.<br>**Instructional Strategies (Lesson II):** Discussing what/how to develop the second context-based STEM integration lesson. |
| **Meeting 4, face-to-face** | **Instructional Strategies (Lesson II):** Reflecting on the classroom implementation of the second lesson.<br>**Poster Presentation:** Creating and presenting a poster to reflect on the PD. |

some teachers at the beginning; however, after they practiced and explored the features, teachers started to feel comfortable with the tools and the online meetings.

In the second meeting, discussions focused on students' prior knowledge on genetics and common misconceptions that students often hold about genetics. As a part of their PLC requirement, teachers were asked to transfer new knowledge that they learned in the workshops into their classrooms. Thus, teachers developed and implemented context-based STEM integration lessons using genetics as a topic. In the third meeting, they shared and reflected on their second context-based STEM integration lesson. Many teachers expressed interest in working with other teachers to design the lessons; collaboration was encouraged as it could help teachers to have access to more resources and to design more effective lesson plans. The last meeting was dedicated for teachers to share their PD experiences with the principals and MSTP staff. Teachers presented posters about their context-based STEM integration lessons and student work from those lessons in the last meeting.

## Evidence for Success

We conducted a research study to explore the impact of the genetics workshops and PLC meetings on teachers' context-based STEM integration approaches and students' learning of genetics. From the 56 teachers who attended both Genetics I and II workshops, 15 teachers were willing to participate in the study. Of the 15 teachers, 12 were female and 3 were male. While 10 teachers had more than 10 years of teaching experience, 2 teachers had fewer than 5 years of teaching experience; the remaining 5 teachers indicated that they had 5–10 years of teaching experience. Five teachers taught middle school life science and 10 teachers taught high school biology. All 15 teachers had a masters' degree in education.

Data collections included three data resources: lesson plans, meeting notes from PLC meetings, and student assessments. Lesson plans and the meeting notes from PLC meetings were analyzed to investigate the first research question which was: How do science teachers apply context-based STEM integration approaches after they participate in PD that was developed using a context-based STEM integration approach? Student pre- and postcontent tests were analyzed to study the following research question: What are the effects of a PD developed using context-based STEM integration approach on students' understanding of genetics?

## Context-Based STEM Integration Lesson Plans

As a requirement of their PLC, teachers developed and implemented two new context-based STEM integration lessons focused on genetics and then uploaded those lessons on Ning site. The 15 teachers who participated in the study developed a total of 20 lesson plans—several teachers worked in groups of two to three to develop lesson plans. These lessons were categorized in three groups (see Table 13.5): (1) health and disease—genetic diseases, genetic testing, and ethical issues; (2) innovations in science, technology, and engineering—genetic engineering, biotechnology risks and benefits; and (3) DNA structure and function—replication and inheritance.

**Table 13.5.** Lessons Submitted to the Ning Site

| Group | Titles of the Lessons |
|---|---|
| **Health and disease** | BRCA1: Gene in breast cancer, Understanding genetic tests to detect BRCA1 mutations, Genetic testing, Rare genetic disease research: Who should pay?, Exploring genetic testing, Albinism, Cystic fibrosis, Different perspectives on genetic engineering |
| **STEM innovations and genetic engineering** | Restriction enzyme activity, GM foods, genetic engineering and cancer, Genetics in today's world: GMOs, To catch a thief: A restriction enzyme activity |
| **DNA structure and function** | DNA fingerprinting, DNA extraction, DNA forensics, Inheritance, Genetics with a smile, Dragon genetics, Extract your own DNA |

Information was provided about one lesson from each group to illustrate teachers' approaches to developing context-based STEM integration lessons and to show the differences among those lessons. *BRCA1: Gene in Breast Cancer* lesson in the *Health and Disease* group focused on breast cancer. Two teachers working at the same high school collaboratively designed this lesson. Teachers found the necessary information regarding the breast cancer genes (BRCA1 and BRCA2) and the DNA sequences of those genes from National Institute of Health (NIH) website. Then using that information, the teachers wrote a case study of a family with a history of breast cancer. Teachers provided students the pedigree charts and Punnett squares for the family to track down breast cancer gene mutations across the generations of the family. Then, they asked students to decide who in the family should be tested for the breast cancer mutation. The teachers also provided DNA sequences of breast cancer genes of the each family member so students could find the mutations. This provided students opportunities to explore STEM developments (bioinformatics) and interpret genetics information.

In the second group, *STEM Innovations and Genetic Engineering*, three high school biology teachers designed a *Restriction Enzyme Activity*. In this activity, students explored transforming genes from one organism to another using a plasmid. The classroom experiment required students to follow a procedure to transform E. coli with a gene that codes for green fluorescent protein, which in real life can be found in jellyfish. In this experiment, students specifically explored genetic engineering techniques such as using plasmid DNA to transform genes to bacteria.

In group three, *DNA Structure and Function*, four lessons focused on DNA extraction. The lesson entitled, *Extract Your Own DNA*, is offered here as an example. *Extract Your Own DNA* was designed by a middle school life science teacher. In this lesson, students first extracted their own DNA. Then through the completion of an online activity, students learned about doing DNA fingerprinting by creating a fingerprint in a virtual lab (NovaLab: The Killer's Trail). Afterward, students used this fingerprint to solve a virtual crime. Students compared the DNA fingerprint they created to seven suspects to nab the perpetrator. The students explored the DNA structures and technologies used in DNA forensic science.

Sample lessons from each group showed that the lessons in each group were different from each other and the differences clearly represented teachers' context-based STEM integration approaches. The biggest differences among the three groups of lessons were the topics. As emphasized earlier, lessons in the first group focused on genetic diseases, genetic testing, and ethical issues, while lessons in the second group were related to the innovations in science, technology, and math; genetic engineering; and biotechnology risks and their benefits, and lessons in the last group focused on DNA structure, function and replication, and inheritance. Another difference among the three groups was the lessons in the first and second groups that were newly developed lessons by the participant teachers. Teachers developed those lessons after participating in the first and second genetics workshops. Lessons in the third group, on the other hand, were modified existing lessons. For example, DNA extraction is a common lesson in which students usually extract DNA from food or cheek cells. However, teachers modified this lesson to include a crime investigation to teach about biotechnological applications that utilized DNA extraction. Finally, the challenges or problems that students engaged in lessons in each group were age appropriate; however, the level of content presented in lessons in group one was higher than the content that students were introduced in lessons in the second and third groups. The majority of the middle school science lessons (e.g., *Extract Your Own DNA*) were characterized as group two and group three lessons.

## PLC Meeting Notes

The workshop instructor facilitated the PLC meetings, and during the second and third meetings she held discussions around the context-based STEM integration lessons that teachers developed and implemented. The conversations were recorded and then transcribed. Two themes emerged from the analysis of the transcriptions: (1) sharing challenges, and (2) sharing teaching resources. These themes are discussed below.

During the meetings, while teachers shared how they implemented context-based STEM integration lessons about genetics, they also shared the challenges they faced when creating or implementing those lessons. For example, after a teacher shared her DNA extraction lesson another teacher opened a discussion on what to use to extract DNA and what to do when students completed DNA extraction.

*Ms. Tank:* What other sources do you use to extract DNA?

*Ms. Smith:* Wheat germ from an organic grocery store also works. It can't be roasted/toasted.

*Ms. Bakkum:* We use strawberries.

*Mr. Weise:* Banana.

*Mr. Jurney:* This year, students extracted DNA from their own cheek cells.

*Ms. Tank:* What do you guys do when DNA is extracted? I always struggle with this. I have them answer questions, but never know if I should do more with it.

*Facilitator:* That's a good question! Ideas?

*Mr. Jurney:* I do not go further.

*Facilitator:* When students ask questions about how we know it is DNA, what is a good response? I would talk about the structure of DNA and ask students the role of each material that they used in the experiment.

*Ms. Tank:* Yes, we talk a little about the materials that we used to extract the DNA and what role they played (the soap, alcohol, and so on).

*Ms. Smith:* Same as what [Miss Tank] says.

*Ms. Bakkum:* Does anyone do anything with it once it's been extracted and spooled? We've thought about running it through gels, but we run dyes instead.

*Ms. Smith:* Yes, you can use the sample for gel electrophoresis.

*Facilitator:* But it is too long to move through the pores. You need to cut the DNA with restriction enzymes.

At the beginning of the following excerpt, a middle school teacher shared a challenge that she had with her middle school students' knowledge of and misconceptions about genetics. Another teacher suggested that she uses case studies to help middle school students to understand genetics conceptually.

*Ms. Sahlin:* Seventh graders are almost a blank slate when it comes to genetics and sometimes I feel like they have misconceptions about everything. They often don't know what genetics is. They are very interested though, so it is a great unit! I think just giving them a basic understanding of genes, heredity helps with a lot. Nature vs. nurture

*Mr. Janssen:* I use case studies. I find that they are a good tool to identify misconceptions and allow students to relate the material to their lives. They often have stories to relate to the class about the different genetic conditions that we discuss in class.

*Ms. Watters:* Where do you find case studies to use for this?

| | |
|---|---|
| *Ms. Eaker:* | I would love to see the case studies. |
| *Mr. Janssen:* | Karen and I have files full. |
| *Ms. Eaker:* | Cool! |
| *Ms. Sahlin:* | This sounds like something that middle schoolers would really get into. |

Above quotes showed that PLC meetings provided valuable opportunities for the teachers to share their challenges about teaching about genetics. The following section presents the second theme, sharing teaching responses.

Teachers shared a variety of teaching resources in the PLC meetings. The following excerpt details discussions held around resources to teach mitosis and meiosis:

| | |
|---|---|
| *Mr. Frith:* | We did some AP Bio activities with protein synthesis by building paper DNA strand and then transcribed and translated it. The visual manipulative for meiosis, protein synthesis, mitosis is always the best way to have them show you if they know it. |
| *Ms. Blakemore:* | We used fun magnetic beads to model Mitosis and Meiosis and it worked well, especially showing crossing over. |
| *Ms. Kroc:* | That sounds awesome, is that a Lab Aid kit? I saw something like that but didn't know if it was any good. |
| *Ms. Blakemore:* | Yes, I would suggest that you provide them with yarn or paper to show spindle fibers and cells because that isn't provided. |
| *Ms. Kroc:* | Thanks! |

In another example, teachers discussed resources to assess student posters on genetic diseases.

| | |
|---|---|
| *Mr. McKenzie:* | In the past I have lectured on genetic diseases. I am looking forward to having the students discover them on their own this year. What do you do? |
| *Ms. Greenwald:* | I was thinking of having them doing a poster and then having what's called a gallery walk where they go around and look at each project and answer questions about the projects. I included that sheet on the uploaded assignment. |
| *Ms. Brown:* | If they know that everyone is going to see it, they tend to do better too. |
| *Ms. Greenwald:* | Yup. A peer review. |
| *Ms. Brown:* | I like to have them "grade" others too—for the peer feedback. |
| *Facilitator:* | I do have a rubric that you can use to grade student posters. |
| *Ms. Greenwald:* | Do you have a copy of that rubric by chance? |
| *Facilitator:* | Yes I do. I will post it on Ning. |

*Ms. Carlson:* Thank you that would be a good resource to have.

*Ms. Greenwald:* There is a fun way to present material as well—it is called Prezi. A free website that can be used for presentations and it is a good use of technology.

Joining PLC meetings was a great opportunity for the teachers to increase their repertoire of teaching resources. Teachers valued using new and different resources and tools to design and teach context-based STEM integration lessons. Group one lessons, which were discussed earlier (e.g., *BRCA1: Gene in breast cancer)*, for example, were developed by mainly using the new tools that were presented in the workshop and PLC meetings.

Overall, sharing how implementation of genetics lessons went in the classroom, challenges faced during the implementation, and resources used to develop the lessons were very helpful for the teachers to improve their pedagogical skills. Teachers felt that they were a part of a learning community and actively participated in discussions. Finally, they were comfortable discussing their lessons and sharing their ideas.

## Student Assessments

The 15 teachers who participated in the study also administered a student content test at the beginning and at the end of their context-based STEM integration lessons. The lessons took two to four class periods to implement during one-to-two-week-long genetics units. The student content test included 10 multiple-choice questions, which were chosen from AAAS Science Assessment Project (*http://assessment.aaas.org*), and the questions focused on the common misconceptions that students hold in genetics and the topics presented at the face-to-face workshops. The majority of the questions asked about DNA molecule and the flow of genetic information (e.g., "Humans, dogs, and trees are all living things. In which of these organisms would you find DNA molecules?" "How does genetic information in a fertilized egg cell (an egg cell that has combined with a sperm cell) compare with the genetic information in the skin cells of the adult organism that develops from the fertilized egg cell?").

A total of 957 students completed the pre and posttests. Students' scores were reasonably normally distributed (see Figure 13.1). A reliability study of the students' posttest scores indicated that the instrument had reliability as measured by coefficient alpha of 0.63.

**Figure 13.1.** Summary of the Test Scores

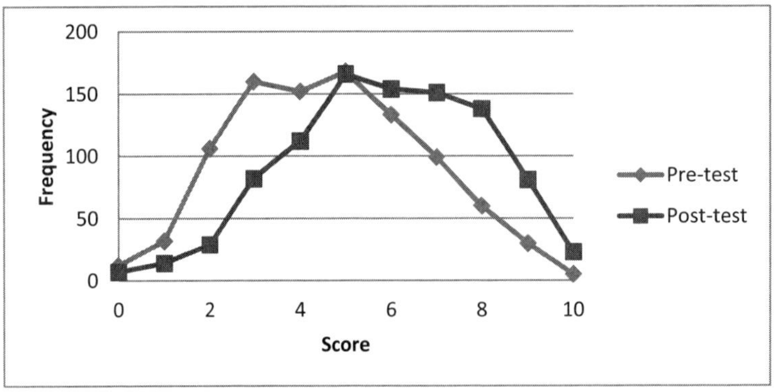


A paired *t*-test was used to compare the mean difference between pre- and postcontent tests. It was found that students significantly increased their understanding of genetics, $t(957) = 18.050, p < 0.05$ (see Table 13.6).

## Table 13.6. Paired Sample Statistics

| Test | Mean | *N* | Std. Deviation | Std. Error Mean |
|---|---|---|---|---|
| **Posttest** | 5.8934 | 957 | 2.09780 | 0.06781 |
| **Pretest** | 4.6635 | 957 | 2.07688 | 0.06714 |

While many teachers indicated that questions were appropriate for their high school students, some middle school teachers expressed concerns about several of the questions on the test. Middle school teachers found that a majority of their students had misconceptions about genetics. Thus some of the questions did not show any improvement. For example, Question 10 measured students' understanding about structure of DNA ("A student wants to build a physical model of DNA as a part of his science course assignment. Which of the following things is necessary for him to build a DNA model?"). It was found that many middle school students had a misconception about building DNA models and they, in both pre- and posttests, indicated that the parts of the model should exactly look like the parts of real DNA. The test results were shared with the teachers and teachers expressed that next time they will spend more time in the areas that students hold misconceptions.

The research study largely focused on the effects of the context-based STEM integration activities on student understanding of genetics; however, teachers also expressed that the lessons had positive results on students' teamwork skills and engagement in classroom activities. Since the majority of the lessons required students to work in teams, students became skilled at teamwork. Additionally, several teachers believed that STEM integration lessons enhanced student interest and engagement. For example, teachers who used genetic diseases and genetically modified organisms as a context to teach genetics found that their students' involvement to the classroom activities increased since these topics are very related to students' everyday life.

Teachers also stated that embedding genetic engineering into science classrooms helped many students increase their understanding about engineering and engineering practices. For example, ethics plays a vital role in engineering—particularly in genetic engineering. As a part of the *Genetic Testing* lesson students involved in an activity, in which they considered ethical issues regarding a scenario about genetic testing. Students read the following scenario and wrote a position statement regarding the issues in this scenario from the perspective of one of the six individuals: a religious person, a lawyer, a bioethicist, a molecular biologist, an expectant father with an inherited heart condition that may be passed on to children, and a eugenicist.

> *A couple has undergone genetic testing and discovered that both parents are carriers for cystic fibrosis, a disease characterized by a deficiency of certain enzymes needed for digestion and respiratory health. Some specialists have recommended against the couple's having children, stating that cystic fibrosis is a serious disease that usually shortens an individual's life span. On the other hand, great advances have been made in research, and people with cystic fibrosis are now living longer lives of a higher quality than ever before. In addition, it's possible that a cure for cystic*

*fibrosis could be found within the next 25 years. The couple is trying to decide what to do and is interested in the viewpoints of other people.*

Engaging in the above activity allowed students to evaluate ethical concerns regarding genetic testing. The issue is "personally meaningful and engaging to students, require the use of evidence-based reasoning, and provide a context for understanding scientific information" (Zeidler and Nichols 2009, p. 49). The activity and the prior classroom instruction on genetics helped students explore the issue of ethics in genetic engineering and develop a position on genetic testing.

## Summary of the Findings

The Region 11 MSTP genetics workshops and PLC meetings applied a context-based STEM integration approach so that teachers learned how to integrate technology and engineering concepts meaningfully in their science teaching. Furthermore, the PD provided science teachers a strong understanding of genetics and a repertoire of appropriate lessons to help their students to build accurate and conceptual understanding of genetics. Teachers expressed views that the PD had great influence on their knowledge and practices, and improved their students understanding of genetics and genetic engineering.

## Conclusions

Through working with hundreds of science teachers in this and other Region 11 MSTP PD series, we have found that science teachers can develop and implement science lessons that apply a context-based STEM integration approach following participation in a professional development program that focuses on STEM integration. This study also showed that genetics is an excellent topic for science teachers to develop context-based STEM integration lessons. Additionally, the study findings demonstrate that developing a coherent and sustained professional development progress and providing ongoing support for teachers as they integrate STEM in teaching are necessary to improve STEM education.

As a next step, we aim to reach more teachers through creating an online environment where teachers can find STEM activities, share their experiences as they integrate STEM, and collaborate with other teachers, content experts, and staff developers. Finally, having necessary materials and equipment is essential for teachers to integrate STEM, thus teachers who participated in the PD were provided with a variety of classroom materials, educational technology tools, and curricula. We aim to consult with more teachers and provide support and materials for them to advance STEM education.

## References

Bowling, B., E. Acra, L. Wang, M. Myers, G. Dean, G. Markle, C. Moskalik, and C. Huether. 2008. Development and evaluation of a Genetics literacy assessment instrument for undergraduates. *Genetics* 178 (1): 15–22.

Bybee, R. W. 2010a. Advancing STEM education: A 2020 vision. *Technology and Engineering Teacher* 70 (1): 30–35.

Bybee, R. W. 2010b. What is STEM Education? *Science* 329 (5995): 996.

Carr, R. L., L. D. Bennett IV, and J. Strobel. 2012. Engineering in the K–12 STEM standards of the 50 U.S. states: An analysis of presence and extent. *Journal of Engineering Education* 101 (3): 539–564.

Fulton, K., and T. Britton. 2011. *STEM teachers in professional learning communities: From good teachers to great teaching*. Washington, DC: National Commission on Teaching and America's Future.

Moore, T. J., M. S. Stohlmann, H. H. Wang, K. M. Tank, and G. H. Roehrig. forthcoming. Implementation and integration of engineering in K–12 STEM education. In *Engineering in precollege settings: Research into practice*, ed. J. Strobel, Ş. Purzer, and M. Cardella. West Lafayette, IN: Purdue University Press..

National Academy of Sciences (NAS). 2006. *Rising above the gathering storm: Energizing and employing America for a brighter economic future*. Washington, DC: National Academies Press.

National Research Council (NRC). 1996. *National science education standards*. Washington DC: National Academies Press.

National Research Council (NRC). 2011. *Successful K–12 STEM education: Identifying effective approaches in science, technology, engineering, and mathematics*. Washington, DC: National Academies Press.

National Research Council (NRC). 2012. *A framework for K–12 science education: Practices, crosscutting concepts, and core ideas*. Washington, DC: National Academies Press.

NGSS Lead States. 2013. *Next Generation Science Standards: For states by states*. Washington, DC: National Academies Press. *www.nextgenscience.org/next-generation-science-standards*

The President's Council of Advisors on Science and Technology. 2010. Prepare and Inspire: K–12 Science, technology, engineering, and math (STEM) education for America's future. *Tech Directions* 70 (6): 33–36.

Zeidler, D. L., and B. Nichols. 2009. Socioscientific issues: Theory and practice. *Journal of Elementary Science Education* 21 (2): 49–58.

# STEM Education in the Science Classroom:

## A Critical Examination of Mathematics Manifest in Science Teaching and Learning

*Todd Campbell*
*University of Connecticut*
*Storrs, CT*

*Stephen B. Witzig*
*University of Massachusetts Dartmouth*
*Dartmouth, MA*

*David J. Welty*
*Fairhaven High School*
*Fairhaven, MA*

*Margaret M. French*
*University of Massachusetts Dartmouth*
*Dartmouth, MA*

### Setting

Our work used a teacher professional knowledge lens as a framework. This framework is built on the tenets that teaching requires more than just knowing the subject matter; it is a process that involves the ability to apply knowledge from several different types of domains. In 1986, Lee Shulman proposed a model suggesting that teachers combine both subject matter content and pedagogical knowledge together into a specific knowledge for teaching that he called pedagogical content knowledge (PCK). PCK is knowledge "which goes beyond knowledge of the subject matter per se to the dimension of subject matter knowledge *for teaching*" (Shulman 1986, p. 9). PCK includes knowledge of the most useful forms of representations for a given subject area, knowledge of what makes topics easy or difficult for students to learn, and knowledge of a set of specific strategies to make the subject matter accessible to the students. This is contrasted with what Shulman (1986) called knowledge of general pedagogy, which was referred to as "pedagogical knowledge *of teaching*" (p. 14, italics added). Knowledge of general pedagogy (general PK) includes knowledge of classroom practices and management.

In science education, Magnusson, Krajcik, and Borko (1999) proposed a model of PCK for science teaching and provided the following definition:

> *Pedagogical content knowledge is a teacher's understanding of how to help students understand specific subject matter. It includes knowledge of how particular subject matter topics, problems and issues can be organized, represented, and adapted to the diverse interests and abilities of learners, and then presented for instruction. We argue that pedagogical content knowledge ... is integral to effective science teaching.* (p. 96)

Magnusson, Krajcik, and Borko (1999) note that teachers who draw upon multiple knowledge domains will be more effective than those teachers whose knowledge is limited. These domains, however, are all specific to teaching a specific science topic in a specific science discipline. Similarly, in mathematics education, the field has extended Shulman's initial ideas into a construct known as mathematical knowledge for teaching (MKT) (Ball, Thames, and Phelp 2008). This construct is helpful for understanding the knowledge that a mathematics teacher has for teaching mathematics concepts, though is limited if one begins to use this to explore teachers stepping outside of their content area. The thought of integration among STEM fields then, becomes problematic if we are to use PCK or MKT as a lens to study teacher knowledge.

In the case of STEM education, and especially in our work considering how mathematics and science coalesce in STEM learning, we contest that this is a new, unique domain of knowledge that requires an integrated approach across disciplines. And, this makes us question how prepared we as science educators or science teachers are (given our formal training as well as our experience as practicing educators/teachers) to implement STEM curriculum? More specifically, if science teachers possess PCK for teaching topics such as heredity, molecules, organisms, energy, or forces and interactions, as just a few examples, what is our expectation for with respect to PCK for teaching technology, engineering, or mathematics? These questions have served as a catalyst for our group's work and investigations and are, we feel, the most salient questions we seek to illuminate and begin to answer in this chapter.

## Science Educators Examining STEM Education: Who We Are and What We Have Done

This work was informed through a partnership between university science educators and a high school science teacher who has a dual role as a science department chair. Our group formed as the two university science educators started new positions at a university in the northeast. The high school science teacher/science department chair in our group was formerly called upon to teach science teaching methods courses at the university before the university science educators' arrival and also served on the hiring committee charged with bringing the two science educators to the university. Because the science teacher/department chair had previously taught the science teaching methods courses at his high school, as the two university science educators took over these courses, they decided to keep them in the local high school and as such the science teacher/department chair (Dr. Welty) agreed to serve as the host for these courses at the school. This led to many discussions about science teaching and learning, which up until this point for all of the authors involved in this study, had been their almost exclusive focus. But, because of the recognition of the importance and challenges of considering STEM education, this became a central focus of many of our early meetings. In fact, before conceptualizing a way to work at characterizing STEM education in practice, we had many conversations about curriculum, teaching, and learning. In sum, all of these conversations led us to identify an AP Biology course as an ideal site to establish our footing and collaborate to better understand STEM education and the successes and challenges of enacting STEM education curriculum. The AP Biology course was chosen as a fitting focus for our work since the curriculum was recently revised with a large emphasis on mathematics integration in science and there were two sections of the course being taught in the upcoming school year.

With the AP Biology course as a focus for our team's efforts in STEM education, our team met regularly to discuss lesson planning, scope and sequence of instruction, and student/faculty roles. Units were identified throughout the school year where mathematics manifested in science instruction. And, because we wanted to gather evidences of successes and challenges in STEM education teaching and learning, student lab notebooks and reflective quizzes were analyzed across units. Additionally, one unit, an inquiry into Beer's Law, was chosen for an in-depth, extended investigation. The unit included three separate laboratory investigations designed to scaffold student learning. Audio/video recordings of the unit from both sections were transcribed and analyzed along with research field notes, student work/assessments, and teacher reflections. All of this was done to begin to answer our own questions about teaching and learning in STEM education. The following questions guided our work:

1. In what ways and where (across units and within a unit) can mathematics manifest in the science curriculum?
2. What role can the science teacher play in cultivating students' facility in constructing and using mathematics and computational thinking?
3. What roles do students play in utilizing mathematics to make sense of the science content?

## *Question 1*

We begin with question 1: In what ways and where (across units and within a unit) can mathematics manifest in the science curriculum? The new AP Biology curriculum (College Board, 2012) places more emphasis on students doing science through the practices of science with student experimental design focusing on their questions. In addition, there is greater emphasis placed on mathematics related to biology. Table 14.1 (p. 236) shows where and how Dr. Welty identified what he saw as obvious entry points into the biology curriculum with mathematics.

As can be seen in the Table 14.1 (p. 235), there were several places where mathematics was used as a central practice in understanding important biological concepts across the curriculum (i.e., 10 units throughout the year). One example of using mathematics in the AP biology curriculum

**Table 14.1.** Mathematics Situated Within the AP Biology Curriculum

| Big Idea (Syllabus) | Lab | Biological Concept |
|---|---|---|
| Interactions | Properties of Water | Solutions |
| Interactions | Transpiration | Stomata and water |
| Interactions | Dandelion Population | Sampling and Population Size |
| Interactions | Osmosis | Osmosis and Water Potential |
| Information | DNA Gel Electrophoresis | DNA Fragment Size |
| Energy | Toothpickase Kinetics | Catalytic rate, Vmax, Km, and inhibition |
| Energy | Catalase Kinetics | Enzyme catalysis rate, Vmax and Km |
| Energy | Cellular Respiration | Respiration Rate |
| Energy | Photosynthesis | Photosynthesis Rate |
| Interactions | Beer's Law | Identify of unknown |

had students trying to find the surface area of a leaf for a transpiration lab. For this lab, students traced on paper the contours of 3 Xenia leaves, cut the leaves out, and found the mass of the paper. Students then created a standard curve by cutting out paper squares and finding the mass of 5 cm × 5 cm, 7 cm × 7 cm, 10 cm × 10 cm, and 15 cm × 15 cm squares. Students graphed the mass on the $y$-axis, the area of each square on the $x$-axis, and determined the slope. Then they could determine the area of the leaves by using the $y = mx$ formula of leaf area = slope × paper mass. Another example lab had students determining the area of an unusually shaped field for a dandelion population survey. An approach similar to the transpiration lab was used for an unusually shaped field on the school grounds. This time, a Google Earth satellite image was used. Students set up a standard curve using a scaled area per mass of paper cut from a printed image and determined the slope of the line. The mass of the cut image of the field was used to determine the area of the real field. Additionally, the slope intercept equation was also used to determine the rate of enzyme catalysis using catalase, determining the size of a Polymerase Chain Reaction (PCR) product following amplification of students' cheek cell mitochondrial DNA, and the rate cell respiration of an organism.

As can be seen, mathematics was found to be an important component of learning. Noticeably as the more broad function of mathematics were identified within the AP Biology curriculum, mathematics was conceived of as a tool for graphing linear functions to use the slope-intercept equation and Chi-square analysis. Dr. Welty explains his rationale for this focus in the following:

> *My mathematical objective for the new AP Biology curriculum was graphing linear functions to use the slope-intercept equation and Chi-square analysis. There were 10 topics where linear functions could be used to analyze data from biological experiments. Students enter high school with mixed messages concerning graphing. Prior to high school math and science, students are instructed to graph the measured quantity on the* y*-axis and the manipulated quantity on the* x*-axis. The slope can then be found by rise of the* y*-axis over the run of the* x*-axis, or "rise over run." In high school*

*the middle school terminology is related to dependent variable on the* y*-axis, independent variable on the* x*-axis, and slope is the change in y divided by change in* x. *The next priority is to get students to see the linear relationship between y and x is the slope-intercept equation,* y = mx + b. *For Chi-square analysis, the objective was to expose students frequently to data they could analyze.*

In addition to graphing linear functions using the slope-intercept equation and Chi-square analysis, we also found that the following three constructs of the science practice *using mathematics and computational thinking* were found throughout the AP Biology curriculum:

- Using mathematical, computational, and/or algorithmic representations of phenomena or design solutions to describe and/or support claims and/or explanations.
- Appling techniques of algebra and functions to represent and solve scientific and engineering problems.
- Applying ratios, rates, percentages, and unit conversions in the context of complicated measurement problems involving quantities with derived or compound units (such as mg/mL, kg/$m^3$, acre-feet, and so on) (NGSS Lead States 2013, p. 10).

This can be seen in Figure 14.1 (p. 238), as a student's work from a lab completed early in the curriculum is shared. Table 14.2 (p. 238) reveals how, in this example of student work specifically, these three constructs for grade 9–12 students from the *NGSS* identified as important facets of *using mathematics and computational thinking* were found in the laboratory experience. These three constructs were also found in other units throughout the AP Biology curriculum.

What was not found across the AP Biology curriculum were the following two constructs of mathematics and computational thinking:

- Creating and/or revising a computational model or simulation of a phenomenon, designed device, process, or system.
- Using simple limit cases to test mathematical expressions, computer programs, algorithms, or simulations of a process or system to see if a model "makes sense" by comparing the outcomes with what is known about the real world (NGSS Lead States p. 10).

Instances of using simulations or computer programs were found in the curriculum in ways that seemed effective and timely given the conceptual focus at the time, but these instances fell short in what we felt was complete alignment to these final two missing constructs. As an example, a PHET simulation (*http://phet.colorado.edu*) was used as in one lab during the Beer's law unit. In this lab, students used the Beer's law simulation to help them better understand such things as how percentage transmittance or absorbance changes as the concentration of a solution is increased or as the cuvette width holding solutions is changed. Additionally and aligned with other units where mathematics was found, they also used linear functions from graphing solution concentrations and absorbances derived from the simulation to identify the slope of the function or molar absorptivity of the solution being studied. Like in other units, this was seen as a valuable experience that served as an intersection for mathematics and science learning and engaged students in the first three *NGSS* mathematics and computational thinking constructs identified earlier, but it did not seem to aptly describe the final two constructs. This

**Figure 14.1.** Student Work From Respiration Lab Completed Early in the Curriculum

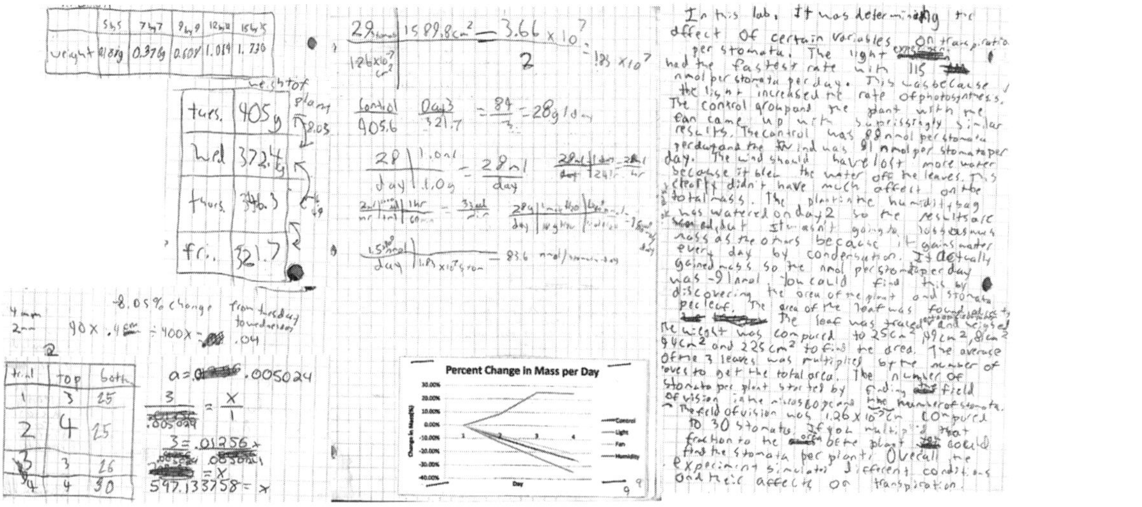


**Table 14.2.** Manifestations of Constructs of Using Mathematics and Computational Thinking

| Construct | Explanation of How It Is Found in Student Work From Figure 14.1 |
|---|---|
| Use mathematical, computational, and/or algorithmic representations of phenomena or design solutions to describe and/or support claims and/or explanations. | This can be seen as the student draws on the results from completed computational representations in making a conclusion. |
| Apply techniques of algebra and functions to represent and solve scientific and engineering problems. | This can be seen as the student is determining the concentration of respiration per stomata in a day. |
| Apply ratios, rates, percentages, and unit conversions in the context of complicated measurement problems involving quantities with derived or compound units (such as mg/mL, kg/m$^3$, acre-feet, etc.). | This can be seen as the student is determining the concentration of respiration per stomata in a day. |

was especially true since students could not be found creating or revising a computational model or simulation of a phenomenon, even though they were found using them. Additionally, while students were found reflecting on whether findings emerging from simulations made sense, they were not found comparing simulation findings with findings and data found in the real world.

These final two constructs are identified in the *NGSS* as important constructs of mathematics and computational thinking for 9–12 grade students (NGSS Lead States 2013) and examples of how these have been used in science instruction can be found across science education literature. One such example, Kohler, Swank, Haefner and Powell (2010), documents how they used mathematical modeling to engage students in creating and revising computational models as they learned about brine shrimp movement. Additionally, Mayes and Koballa (2012) share how they see investigating climate change with grade 12 students that has them examining a scatter plot of state temperature data. In these investigations, they ask students to provide a power function model or exponential model for the data. Through this, they describe rich discussions of which function is the best model for the data. In this activity, students are seen comparing their models with data collected from the natural environment (i.e., state temperature data) to test the validity of their expressions/models. While these are just two examples of these two constructs, what should be noted from the examples, is how mathematics educators or teachers were also involved in the planning and teaching of the examples shared. This was the case in Kohler, Swank, Haefner and Powell (2010) since the authors Kohler and Swank are mathematics educators who partnered with Haefner and Powell, biologists. And, Mayes and Koballa (2012) were also comprised of both a mathematics educator and science educator or scientists (Mayes is a mathematics educator, while Koballa is a science educator). Nothing definitive can be concluded here about how often and in what context these final two constructs are most likely found, but we did note that they were absent from the science learning that served as a context for our work, where we did not have a mathematics educator partnering with us. This leaves us wondering whether these two constructs may be found more frequently within science classrooms when collaborations are forged between math and science educators. In this, we wondered whether a math educator who has developed PCK around mathematics teaching might be more attuned to the need and importance of these facets of mathematics and computational thinking, while also being able to identify effective strategies for making these two constructs more prevalent within the science curriculum.

Finally, more generally in our context, it appears that mathematics and computational thinking emerged in the science classroom in a consistent pattern within the curriculum. In the AP Biology curriculum, the teacher reported how he recognized the power of helping students develop facility in graphing linear functions to use the slope-intercept equation and Chi-square analysis. In this, he noted how he identified specific concepts within the curriculum that he thought lent themselves to engaging students in this manner. This consistent pattern or strategy developed by the teacher was found in all 10 units identified where mathematics was found. In these instances, the teacher sought to engage students in science laboratory investigations where they were generating data to allow them to compare two variables graphically and through Chi-square analysis as an important mechanism for developing student understanding of a concept and as a mechanism for developing their facility in using graphical representations for making estimations and predictions.

## *Question 2*

In response to question 2 (What role can the science teacher play in cultivating students' facility in using mathematics and computational thinking?) we found that the teacher, at least in our context, played an intentional role rooted in instructional strategies, articulated learning objectives, and past experiences. This can be seen in the following reflection offered by Dr. Welty:

*I use both scaffolding and spiraling to assist students in constructing and using mathematical representations and models. Early in the AP Biology curriculum, I introduced both the slope intercept equation and Chi-square analysis. I knew both would be needed throughout the course. I wanted both approaches to resurface often, so students could apply the math in different and unexpected circumstances.*

As was shared in the reflection and found as different units were examined, Dr. Welty recognized and purposefully put into place structures for scaffolding and spiraling to cultivate students' facility in using mathematics and computational thinking. The scaffolding was seen as Dr. Welty generally followed a consistent instructional design/strategy:

*I used the following strategy of introducing an experimental protocol, which the students validate. Next, the students design their experiment to answer the question of their choosing. This is followed by a period of time where students optimize their experiment. Once the optimization phase is producing reliable results, students conduct their experiment and collect data for analysis.*

In this sequence, students had the opportunity, through a more teacher-centered approach early as Dr. Welty introduced an experimental protocol, to see an effective models of experimental design coupled with guided application of the slope intercept equation and Chi-square analysis. This was followed by students designing their own experiments as the instruction moved toward a more student-centered approach that could be supported by the teacher-centered guidance provided earlier. Others have noted the benefit of such scaffolded approaches in science education (Duffy et al. 2013; Slater, Slater, and Shaner 2008) and it appears that Dr. Welty has taken advantage of this in supporting science and mathematics learning.

When considering spiraling, which for the purposes of this chapter refers to giving students multiple meaningful opportunities to experience and build facility in practice, this can be seen as Dr. Welty shared how he "wanted both approaches to resurface often, so students could apply the math in different and unexpected circumstances." And also noteworthy of the spiraling units that Dr. Welty identified, was their dual focus on mathematics objectives (i.e., slope intercept equation and Chi-square analysis) and science objectives.

One other teacher role that emerged in this current context was the identification of dual purposed and complementary objectives in mathematics and science. This was seen in Dr. Welty's class as he identified a need for multiple experiences for students to develop their capacity for experimental design, while simultaneously recognizing and planning for developing students' abilities with graphing linear functions to use the slope-intercept equation and Chi-square analysis.

In addition to looking across the entire AP Biology curriculum, to answer this question we felt a need for close examination of a specific unit. Given this, we observed each day of the final unit in the AP Biology curriculum. This afforded us the opportunity to examine the very specific instructional moves of the science teacher, while also examining the impact of these moves on students' learning in the science classroom. Additionally, this allowed us to take video and transcribe audio from this unit for subsequent analysis. Based on our analysis of the Beer's Law unit, we were able to identify another important role played by Dr. Welty in cultivating students' facility in using mathematics and computational thinking in science: navigating what sometimes appeared to be tension between specificity required for learning and doing mathematics and learning the big ideas of the science lesson. More about this role is described next.

One difference noted between mathematics and science more generally is the need for specificity in mathematics. Wilson, Boatright, and Landon-Hays (forthcoming) explained this as they observed middle school mathematics and science classrooms to examine the similarities and differences in the social semiotics used to communicate these two disciplines. With respect to mathematics, they found that that much of the social semiotics or disciplinary ways of communicating were focused on specifics or specificity. This was seen as mathematics teachers were found using pointing as the most frequent gesture for accompanying other modes of communication to draw students' attention to important parts of numbers, charts, or shapes. In this they pointed out how "[m]uch of teachers' verbal discourse drew students' attention to observable representations as teachers pointed to specific aspects of them, such as specific decimal points, specific points on graphs, specific digits within written numerals, or specific vertices on three-dimensional objects" (Wilson, Boatright, and Landon-Hays, forthcoming). In this it was recognized that specificity was important in mathematics, since, for example, not being specific about the location of a decimal point in a number may change the meaning of the number by as many as 10 times for each decimal place.

This is contrasted with what Wilson, Boatright, and Landon-Hays (forthcoming) identified as a difference as teachers in science classrooms were observed. In science classrooms, instead of drawing students' attention to specifics of numbers, graphs, and shapes to ensure no meaning was lost because of lack of specificity, they found teachers using gestures such as pointing to connect two or more iconic representations (e.g., speech and a visualization or speech and spatial locations) that provided a depth of communication focused on understanding not found possible with only one representation, speech, for example (Wilson, Boatright, and Landon-Hays, forthcoming). What this seemed to imply is that the numerous mechanisms of communicating and understanding in one discipline, such as mathematics, is distinguishable from the numerous mechanisms and conventions of communication in another, such as science. These differences found by others are highlighted here because we believed they may have emerged in the Beer's Law unit as we observed Dr. Welty. In this, Dr. Welty possessed pedagogical content knowledge (PCK), training, and insight into the conventions of science, especially with a PhD in a science discipline, but only possessed limited PCK, training, and insight into mathematics. Because of this, we questioned, based on events observed in the classroom, whether this caused tensions when there was a need for simultaneously representing the disciplines and conventions of both disciplines concurrently. In the end, we believe Dr. Welty was able to navigate these tensions successfully to meet most of his learning objectives, but we questioned the extent to which his deeper grounding within the discipline of science might have played a role in how these two disciplines came together.

These differences across the disciplines of mathematics and science seemed to present noticeable tension at times within the Beer's law unit. This was seen as there was some initial confusion during the first lab in this unit when students were completing serial dilutions. As we observed the students throughout the class period, there were numerous times when students seemed confused regarding the specific mathematics behind the dilutions they were making. This is exemplified as Dr. Welty was found sharing the following at one point in the lesson: "Always keep in mind where you are with your concentrations okay because sometimes you kind of get lost there. You're doing your serial dilutions, but you're not thinking about the concentrations." Additionally, we noted that instead of referring to the dilutions by their concentrations, generally, students were almost exclusively referring to the different dilutions by the cup number in which they were held.

For the most part, we noted that Dr. Welty seemed to address any confusion experienced by the students on a one-by-one personal basis as he asked students to repeat themselves and rethink what they said. As we reflected on this, we realized that students seemed to have a qualitative grasp of the basis of serial dilution, even though many were struggling with specifics of the exact concentrations of each dilution. In the end, as we examined the science learning objectives of the lab activity (i.e., determining the limits of sight and taste senses), we realized that in this specific laboratory students could grasp the big idea of the lesson without being inhibited by the lack of specificity required in many mathematics classrooms. That is, since students realized that a dilution had been completed as they moved from the original concentration across the six cups and that each cup was more dilute than the one before it as a small amount was being taken from the cup before it and mixed with water. Through this, students were able to identify the limits of the different senses by understanding the concentration of each cup relative to the one before and after it.

Dr. Welty seemed to achieve his science learning instructional objective of helping students understand the limits of certain apparatus like senses, but did recognize the tension caused by the serial dilution lab and his lack of focus on the specificity necessary in mathematics instruction, as is seen in the following reflection:

> *Students were confused by my loose use of terminology of how to dilute, I used both "part-to-part and ½." For instance, I would say "do a 1:1 and a ½ dilution" for the same thing. I may have also misspoke from time to time by saying a 1:10, when I meant 1:9 for a 1/10th dilution. This was one of the key problems.*

On another occasion during a different lab activity in the Beer's law unit the specificity that is important in most mathematics learning could not be sacrificed. This can be seen as Dr. Welty directs a group of students who are unable to get a linear graph as needed to identify an unknown solution:

*Dr. Welty:* Okay, so you hadn't diluted these enough, okay. So what you may want to try to do, start off with a 1 to 10. Okay, or 1 to 9 or a tenfold dilution. Then do your 1 to 1s, so you're diluting by ½. Always keep in mind where you are with your concentrations okay because sometimes you kind of get lost there. You're doing your serial dilutions, but you're not thinking about the concentrations [noting the need for students to pay attention to the specifics required in mathematics].

As can be seen in this example interaction, which was characteristic of others found within the same lab, Dr. Welty was cognizant of the need for specificity and pushed students to attend to this in ways similar to what might be expected when dealing with mathematics in a math classroom.

Given the tensions that emerged during the Beer's law unit, while it is difficult to say with certainty, evidence could be found to support the notion that the science teachers' role in cultivating students' facility in using mathematics and computational thinking was understanding and attending to the conventions of mathematics, as well as how to navigate conventions that differ when comparing mathematics and science. We found that Dr. Welty was able to do this to meet his science objectives. In this, we noted that there were times when he did not push the students in pursuing the specifics of mathematics that may have been expected in a mathematics classroom. In these cases, specificity was not important and pushing students for specificity may have detracted from reaching the science learning objectives. This is juxtaposed with other times when specificity was required in mathematics (i.e., unknown solution lab) and at those times, Dr. Welty was found providing the needed attention to this area in ways that were more reflective of the specificity one would expect in a mathematics classroom.

## *Question 3*

As we began to address question 3 (What roles do students play in utilizing mathematics to make sense of the science content?) we found that students' roles included a problem-solving attitude, their problem solving relied heavily on prior experience, they used both qualitative and quantitative skills in the laboratory, and they were able to transition from data collection to data analysis. Dr. Welty established a norm in his classroom, as evidenced above, for setting students up for success with the proper learning tools necessary to work as independent scientists. This is evident in the following reflection from Dr. Welty:

> *The main role students play in utilizing mathematics to make sense of the science content is a problem solving attitude. Students often do not slow down and think about what prior knowledge they have relative to the experiment. They lack experience in designing experiments and are still forming their understanding of a biological concept. For this reason, instructional scaffolding provides students the support they need by prompting the most appropriate prior knowledge, recommendations of the experimental approach, and review of the biological concepts. Students often select a qualitative approach first for an all or none expectation. They do not select a quantitative due to lack of experience of quantitative analysis. Scaffolding provided through the pre-lab, purpose, and objectives guide students to select a quantitative experimental design.*

The scaffolding that Dr. Welty provided throughout the semester fostered the students' problem-solving attitude and helped them to connect their prior knowledge to current investigations.

The students' problem-solving attitude manifests throughout the semester in several distinct ways. The students were well-versed in how to incorporate the mathematical functions in computer software such as Excel into their everyday lab experiences. In one lab, Dr. Welty commented about this to the students, "I have to tell you, with you it is like automatic guys, you are in a lab, and you turn on Excel!" During most observations, the students were witnessed using calculators, Excel, as well as paper-and-pencil calculations to assist them in making sense of the science concepts. This required problem solving. During one lab, the students discovered that Excel was inadvertently rounding their calculations, thus influencing the results of their investigations. The students figured this out, and worked out a solution:

| | |
|---|---|
| *Dr. Welty:* | Did you guys use the slope, did you determine from your calculation? |
| *Student:* | I used my calculator, because the computer will say 0.009, but I got 0.0096. |
| *Dr. Welty:* | Okay, So it's not showing the 4th [digit]? |
| *Student:* | Yeah, but it's a considerable difference. |

Here the students learned that Excel was only showing three digits past the decimal point, which they learned was a "considerable difference" when they used those values to calculate their slopes. Thus, using a problem-solving attitude, they used the calculators to better represent the data they were collecting.

The students' problem-solving attitude was particularly evident during their final class presentations on the Beer's Law investigation. Here, a student works through some of the challenges they experienced with their own data, but also working with the data obtained from other groups in the class:

> *One of the big problems is, when we share the information of everyone's data, not everybody labeled the same things in the same ways, so when we got it we had to figure out which wavelength was which. For example one group did the actual numbers for the waves like 660 instead of the color, we had to go back and try to figure that out. And then also technical issues with graphing, things like Excel, we've put in the data and it didn't necessarily read it the way we wanted it to so we had to switch the columns the other way around and kind of organize the data on a different way to get the graph the way we want it to look like. [Switches slides in presentation] So this was all of the data and from these graphs we were able to look and see that the green wavelength had the best sort of data to look at the linear. [We] couldn't really use blue because that didn't really work out. Red obviously didn't work because the solution was right so we didn't really have enough data points to read. So, out of all of them, green had the most linear data.*

The students experienced the Beer's Law investigation much in the way that scientists experience science—the "right answer" was not known, there was no expectation that their investigation would definitively identify the unknown berry extract, and they had to rely on teamwork and others' results to make sense of what they were finding. Thus, the students had to rely on problem solving to navigate their way through the data analysis. This problem-solving attitude was fostered by the effective scaffolding that Dr. Welty set up throughout the school year and was critical for the students to make use of mathematics to help them understand and apply Beer's Law.

The students were encouraged, again, through scaffolding provided by Dr. Welty, to rely on prior knowledge to make sense of current ideas and investigations. This was seen in the Beer's Law Unit, when a student asked the question, "Is there a unit for absorbance?" Rather than just provide the student with the answer, Dr. Welty engaged the students in a discussion that has them recall their prior knowledge of mathematical logarithm functions:

| | |
|---|---|
| *Dr. Welty:* | Okay, he just asked a very good question. Does absorbance have a unit? If you take a log function of something, does it have a unit? |

*Students:* No.

*Dr. Welty:* No, yeah. So absorbance and pH don't have units. Good question [to student], I was waiting for someone to ask that.

Dr. Welty anticipated the students asking this question, and related this to their prior understanding of pH. They understood the reasons that pH did not have units, and therefore, can now apply that knowledge to the current lab on absorbance studying Beer's Law. Students' prior understanding of a slope of line also weighed heavily in their ability to understand and apply mathematical concepts to understand the science behind Beer's Law. Here the students used their prior knowledge of mathematical concepts of slope and the equation of a line in order to make sense of the absorbance data they were recording from the spectrophotometers. The connection between the slope and how to apply the slope of a line with Beer's Law related directly to the students' investigation of the unknown berry solution.

Students used both qualitative and quantitative strategies in the laboratory to help them make sense of the data. Dr. Welty noted that students tend to first rely on qualitative observations before resorting to quantitative calculations. During our observations of the Beer's Law laboratory, this was apparent. The students were tasked with determining the concentration of an unknown berry solution given three other colored solutions. We witnessed the students first comparing colors of the solutions, and even smelling the solutions, before setting up their serial dilutions to obtain numerical absorbance data. Student 1 commented, "That one that smelled like cherry medicine." Student 2 replied, "It smelled like Kool-Aid." And in labeling the serial dilutions, some students labeled their test tubes 1, 2, 3, 4 instead with the proper dilution value 1:10, 1:100, and so on.

The final significant role that students took on that we aim to describe here was their role as data analyzers. The students moved from not only collecting data, but toward analyzing the data in meaningful ways. Dr. Welty shares in the following reflection that this does not come naturally for students, and one that science teachers are tasked with navigating carefully so that students do not get turned away from science as a possible career path:

> *Students also lack experience in data analysis and statistics, so they do not or poorly design experiments that allow them to analyze results. They either need to be guided toward better design or provided the time to repeat an experiment for better analysis. It is the time constraints of the school year that prevents the teacher from providing the necessary time needed for the latter approach. The other dynamic related to time is, we are focused on students having success and enjoying science so students will select STEM fields. If students get frustrated, then we fear they will select another field of study. However, students need to be shown that science can be frustrating and time consuming. If the feeling of solving the problem, learning something and discovering something novel that no one else has ever seen before is a great feeling, then science might be a field for you.*

Dr. Welty works throughout the school year to provide structure for students in the lab so that they can build an appreciation for science. Because of time constraints, however, students often lack experience in data analysis. This is where scaffolding on the part of the teacher plays a large role. As depicted above in Table 14.1, Dr. Welty prioritized mathematics in his biology class throughout the school year. In early investigations Dr. Welty encouraged students to collect and analyze data, but not at the rigor that was expected by the end of the year. The early laboratories were more structured, where the Beer's Law investigation was more open-ended and much closer to how science is actually done outside of school contexts. Figure 14.2 shows examples of student data analysis from an early investigation to compare to the analysis from the final investigation on Beer's Law.

As can be seen in Figure 14.2, the students' attention to detail with respect to data analysis in the Beer's Law investigation was evident in the mathematical representations that the students prioritized in the laboratory notebooks. Early on in the plant leaf surface area lab, the student did not label the axes, hand-drew the representations, and did not provide an explanation as to how she interpreted the representations. Later on in the school year, as evidenced in the Beer's Law investigation, the student generated graphs in Excel, properly labeled axes, and used these graphs to interpret the science principle under investigation. The student provided an explanation as to how she interpreted the data using the graphs to identify the unknown berry extract.

We have identified four roles that students embodied as they used mathematics in their biology class to make sense of science content: (1) they had a problem-solving attitude, (2) their problem solving relied heavily on prior experience, (3) they used both qualitative and quantitative skills in the laboratory, and (4) they were able to transition from data collection to data analysis. These roles, however, were largely shaped by the culture that the teacher established in the classroom. Dr. Welty provided ample scaffolding for students to support students in these roles. Dr. Welty sees his role as teacher extending far beyond his classroom. Dr. Welty is not only trying to get his students to understand mathematics in a way to apply this to the science in his classroom, he is trying to get his students to apply these principles to life in hope that they find value in pursuing a career in STEM fields.

## Implications and Future Directions

Like many others nationally and internationally, we recognize this moment in STEM education as a unique historical moment. It is a time in history when STEM education has received extensive monetary and political support (OSTP 2011). This is occurring at the same time that the newest national standards documents are calling for a more integrated approach to teaching and learning across all subject areas (NGSS Lead States 2013; NRC 2012; National Governors Association Center for Best Practices and Council of Chief State School Officers 2010). This chapter was written to help establish a foundation for understanding the role and limits of professional knowledge in supporting teachers' enactment of interdisciplinary instruction, like many envision for STEM education.

In this chapter, we have identified where and how mathematics manifest in science and the roles teachers and students play in integrated STEM education teaching and learning. The evidence of the success found in Dr. Welty's classroom and through our collaborations can be seen as mathematics was integrated throughout the science curriculum (i.e., 10 units in the AP Biology Curriculum). Additionally, we found alignment between the ways in which mathematics manifest across the AP Biology curriculum and the *NGSS*-identified science and engineering

**Figure 14.2.** Mathematical Representations Compared Across the School Year

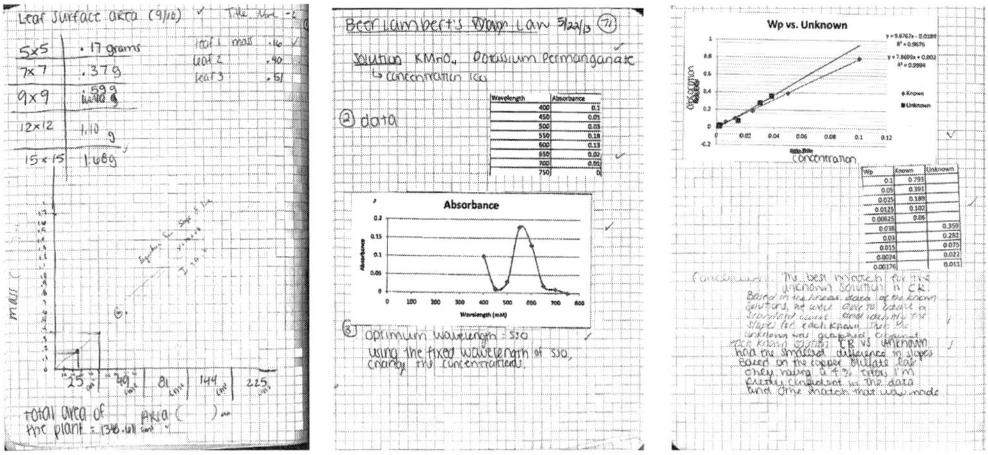


The left panel is an excerpt from an early investigation in the school year on plant leaf surface area. The middle and the right panels are excerpts from the same student's laboratory notebook from the Beer's Law investigation.

practice of using mathematics and computational thinking. These findings left us with a sense of both how mathematics was effectively being integrated in science teaching and learning, and where opportunities for additional integrated approaches remain.

As we considered the role science teachers play in cultivating students' facility in using mathematics and computational thinking, when we found mathematics playing an important role in the science curriculum, this was grounded in teacher intentional instructional strategies, articulated learning objectives, and past experiences. The intentional instructional strategies, which took the form of faded scaffolding, provided opportunities for needed teacher scaffolding of student learning and for spiraling important mathematics and science concepts across the entire curriculum. Within the spiraling units, identifying dual-focused units that targeted both mathematics and science objectives were a central role of the science teacher. So, one other teacher role that emerged in this current context was the identification of dual-purposed and complementary objectives in mathematics and science. Another important role for the teacher that we identified was understanding and balancing the congruence, as well as the tensions between the numerous mechanisms of communicating and understanding in one STEM discipline, such as mathematics, and distinguishable features of the numerous mechanisms and conventions of communication of another STEM discipline, such as science).

In investigating the role students play in using mathematics to make sense of the science content, we found students using problem-solving attitudes that relied heavily on prior experience. Additionally, the students' role involved using both qualitative and quantitative skills in the lab, and transitioning from data collection to reasoning about the data through data analysis. As alluded to earlier, we found that the teacher played a significant role in scaffolding these experiences for students, both through active engagement with the student through all phases of their work (e.g., introducing ideas, serving as a sounding board and consultant during investigations) and through providing multiple experiences throughout the AP Biology curriculum for students to experience and build facility in their roles as learners using mathematics to make sense of science.

We believe that further informing questions about professional knowledge are at the heart of the value of the work we have accomplished in this chapter. As an example, when considering the impact of teacher professional knowledge (i.e., PCK) on developing and/or enacting integrated STEM education curriculum, we believe our work offers some insight into the preparation of teachers (given their formal training as well as their experience as practicing teachers) to implement a STEM curriculum. Additionally, identifying areas where authentic and natural intersections across the disciplines also seemed to play a role in determining the readiness of the teacher for developing and enacting integrated STEM curriculum. Conversely, one area that we found that may take away from teachers' ability to enact STEM curriculum, was the limited training of the teacher across disciplines. So, while we believe there are many supports already in place that position a science teacher for effectively implementing STEM education curriculum, at the same time limited disciplinary content knowledge and training in the field of mathematics, or other STEM disciplines more broadly, likely places some limits on the preparedness of science teachers.

As revealed in this chapter, there are many reasons for the attention STEM education has garnered over the last two centuries. Among these are the integrated nature of how societal problems can be approached with the interdisciplinary knowledge and affordances offered by the STEM disciplines. In submitting our work for critique here to the STEM education community, we are hopeful that it is seen as informative, both as a naturalistic depiction of how mathematics manifests in science specifically, and as a more general depiction of how two STEM disciplines come together in the context of a classroom led by a science teacher. This work was informed through a partnership between university science educators and a high school science teacher. It came only after extensive conversations about curriculum, teaching, and learning. Like the interactions that led to the collaborations that resulted in this chapter, we hope that our work will support a more broad conversation beyond our group about STEM education into the future regarding the insights and issues we have raised and those that may have escaped our attention.

## References

College Board. 2012. AP biology: Course and exam description. *http://media.collegeboard.com/digitalServices/pdf/ap/IN120084785_BiologyCED_Effective_Fall_2012_Revised_lkd.pdf*

Duffy, A. M., P. G. Wolf, J. Barrow, M. Longhurst, and T. Campbell. 2013. Ecological investigations within an interactive plant community simulation. *Science Scope* 36 (8): 42–51.

Kohler, B. R., R. J. Swank, J. W. Haefner, and J. A. Powell. 2010. Leading students to investigate diffusion as a model of brine shrimp movement. *Bulletin of Mathematical Biology* 72: 230–257.

Magnusson, S., J. Krajcik, and H. Borko. 1999. Nature, sources, and development of pedagogical content knowledge for teaching science. In *Examining pedagogical content knowledge: The construct and its implications for science education*, ed. J. Gess-Newsome and N. G. Lederman, 95–132. Boston, MA: Kluwer.

Mayes, R., and T. Koballa Jr. 2012. Exploring the Science Framework. *Science & Children* 50 (4): 8–15.

National Governors Association Center for Best Practices and Council of Chief State School Officers (NGAC and CCSSO). 2010. *Common core state standards*. Washington, DC: NGAC and CCSSO.

NGSS Lead States. 2013. *Next Generation Science Standards: For states by states*. Washington, DC: National Academies Press. *www.nextgenscience.org/next-generation-science-standards*

Office of Science and Technology Policy (OSTP). 2011. Description of 5-year federal STEM education strategies plan: Report to Congress. Washington, DC: National Science and Technology Council.

Shulman, L. S. 1986. Those who understand: Knowledge growth in teaching. *Educational Researcher* 15 (2): 4–14.

Slater, S. J., T. F. Slater, and A. Shaner. 2008. Impact of backwards faded scaffolding in an astronomy course for pre-service elementary teachers based on inquiry. *Journal of Geoscience Education* 56 (5): 408–416.

Wilson, A. A., M. D. Boatright, and M. Landon-Hays. Forthcoming. Middle school teachers' discipline-specific use of gestures and implications for disciplinary literacy instruction. *Journal of Literacy Research*.

# Mixed-Reality Labs

## Combining Sensors and Simulations to Improve STEM Education

*Edward A. Pan, Jennifer L. Chiu, Jie Chao*
*University of Virginia*
*Charlottesville, VA*

### Setting

The Curry School of Education at the University of Virginia (UVA) has a long history of working to improve education, dating back to its inception in 1905. The Concord Consortium (CC), founded in 1994 and based in Concord, Massachusetts, is a non-profit laboratory that develops educational technology for science, mathematics, and engineering. The UVA and CC have been collaborating on the Mixed-Reality Labs (MRL) project since 2011. The CC team is responsible for development of the educational technology and accompanying curricula. The UVA team is responsible for conducting evaluation (classroom trials) and the needed educational research.

### Overview

In 2008 the National Science Foundation (NSF) published a report on cyberlearning, which is learning supported by computer and communications technologies (Borgman et al. 2008). The report identified seven opportunities for action. One of these was to "seize the opportunity for remote and virtual laboratories to enhance Science, Technology, Engineering, and Mathematics (STEM) education" (p. 36). Another was to "investigate virtual worlds and mixed-reality environments" (p. 38).

The Mixed-Reality Labs (MRL) project is a collaboration between the CC and the UVA that aims to enhance STEM education through the use of cutting-edge cyberlearning technologies. The MRL team investigated how combining sensors and simulations can help transform STEM education in high school and middle school settings. Thus far, we have conducted field trials in public middle and high school science classrooms across two states. The curricula, assessments, and technologies have been iteratively refined from the valuable insights gained from these trials and interventions. This chapter provides an overview of results from these implementations and trials.

Students spend much of their time these days engaging with computer-based (so-called *virtual*) content. Virtual environments allow students to interact and do things that they would not otherwise be able to do in the real world (also known as the *physical* environment). Examples of virtual environments that are useful in STEM education are virtual laboratories. In these virtual labs, students interact with computer simulations. They conduct science experiments much as they would with traditional physical science apparatuses. The virtual environment allows students the capability to do things such as alter time (pause, rewind, and fast-forward), easily modify variables and observe results of those modifications, and collect data in an automated way that facilitates multiple representations of those data (such as in real-time graphs).

Another aspect of virtual labs that is particularly useful to educational research is the ability to automatically collect data on student interactions with the simulation. This has potential of providing targeted feedback to teachers and students based on student performance and understanding. Virtual labs provide these kinds of inherent capabilities for learning that are very difficult or impossible to achieve for students working with purely physical lab equipment.

The Task Force on Cyberlearning also identified *mixed-reality* as another exciting area to investigate. Mixed-reality involves combining elements of virtual environments with elements of the physical environment. In the report, the task force described a hypothetical scenario where students engage in a physics experiment that is manifested as a computer simulation that projects a visual display onto the floor below the students. Motion sensors that track student actions drive the simulation and provide feedback to the students. Thus, students use different modes of interacting with the experiment than in traditional (physical) or purely virtual scenarios. The technology would allow teachers and researchers to be able to investigate how students learn using these different modes of interaction.

The Mixed-Reality Labs project directly addresses these challenges posed by the NSF report on cyberlearning. We are investigating how to combine virtual and physical labs to enhance STEM education, initially focusing on science education at the high school and middle school levels. We believe that mixed-reality labs can help students see the world as scientists do by providing an explicit link between the real world (sensors that input data from their surroundings) and accompanying expert models of the phenomena (computer simulations). For instance, our mixed-reality technologies enable students to touch a piece of metal and see how energy flows from their hands to the metal. Another one of our labs enables students to physically push on a virtual container of air and see how gas molecules create pressure through collisions with container walls.

**Figure 15.1.** The Gas Frame

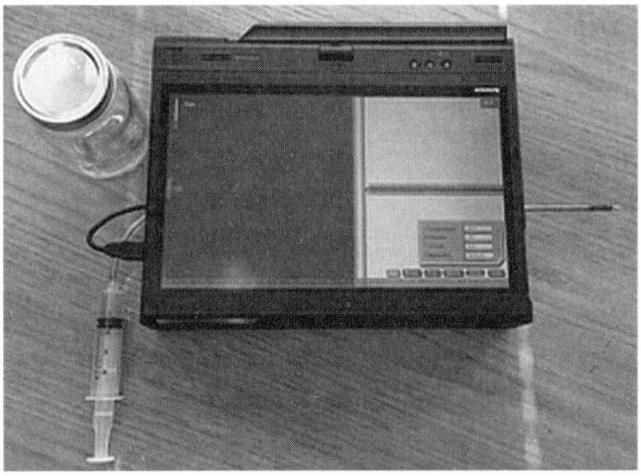

This device has a syringe molecule injector on the left, spring piston controller on the right, and a jar for holding hot/cold water placed next to the temperature sensor located in a housing upon which a tablet PC sits.

We have developed an assortment of labs that include topics in chemistry (gas laws and heat transfer) and physics (force and motion). One of the most promising mixed-reality technologies is the *Frame* (Xie 2012; see Figure 15.1), which is a customizable enclosure for a tablet computer that mates typical sensors used in science classrooms (e.g., pressure, temperature, and force sensors) with freely available accompanying computer simulations. Students manipulate physical controls (like pushing or pulling on a syringe to inject or remove virtual gas from the simulation) to drive a dynamic real-time simulation that depicts gas molecules bouncing around inside a piston chamber (see Figure 15.2). The virtual laboratory environment allows students to see the unseen: they can observe the behavior of gas at a molecular level and learn the underlying kinetic molecular theory explanations for their common everyday macroscopic observations of the behavior of gases.

**Figure 15.2.** Students Interact With the Frame Through Physical Controls.

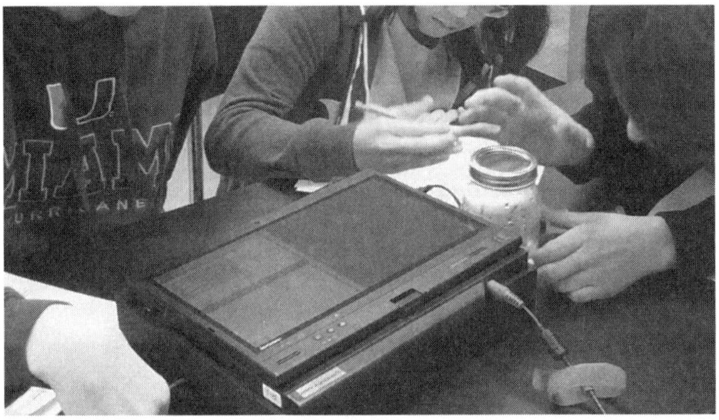

Here a student increases the temperature of the gas simulation by bringing a hot jar next to the Frame, while another increases the pressure of the system by applying a force through a spring that is directly connected to the virtual piston.

The MRL effort aligns with current STEM education reform efforts as described in the *Next Generation Science Standards* (*NGSS*; NGSS Lead States 2013). Although the *NGSS* focuses on atomic-level interactions at the high school level (whereas MRL depicts gases at the molecular level), there is a move toward particulate-nature explanations in physical science in both high school and middle school, with molecular-level explanations beginning in middle school. Classroom test results from the MRL project suggest that high school students still experience difficulties with these fundamental topics, as evidenced by the appearance of common alternative ideas in their explanations of the behavior of gases. MRL can play a role in helping to transition the current state of students' science learning toward the goals outlined by the *NGSS*. Additionally, MRL directly connects *NGSS* dimensions of scientific practices, crosscutting concepts, and core ideas.

## Major Features

### *Bridging Microscopic and Macroscopic Worlds*

Students have a tendency to isolate ideas in science. For example, students learn about atoms and molecules with abstract symbols and representations without connecting those letters and numbers to the everyday (macroscopic) experience of matter. Students need opportunities to connect these isolated islands of knowledge into an integrated network. Dynamic molecular visualizations can act as a bridge to connect these abstract ideas and representations in science to macroscopic phenomena. By providing dynamic molecular visualizations, MRL labs represent abstract symbols and equations of the gas laws in a more concrete form: moving balls on a computer screen. MRL acts as "science goggles," enabling the student to see the underlying molecular phenomena as a scientist does while facilitating inquiry learning at the molecular level.

What sets MRL apart from other dynamic visualizations is that MRL is a dynamic simulation that responds in real-time to physical inputs from students, allowing rapid micro-experiments. Many excellent dynamic simulations exist (for example, PhET; see *http://phet.colorado.edu*). In MRL, students interact with the simulations through physical interfaces. These physical interfaces are not merely physical controls such as buttons, but rather they are physical devices that directly map to natural behaviors. For instance, a syringe is a familiar device for injecting and removing substances. The Gas Frame uses a syringe to inject and remove gas molecules from the simulation. Rather than just pushing a button or clicking on some onscreen control with a mouse, the student uses the *same action* with the syringe to inject and suck out molecules that they would use to inject or suck out a real (physical) gas or liquid.

MRL uses these kinds of physical controls or interfaces, as opposed to onscreen virtual controls, to help students connect the simulations to the real world. Physical interfaces can help trigger students to think about phenomena in terms of their common everyday experiences rather than as some abstract idea playing out in a computer program. Preconceptions that students have about the scientific phenomena are more likely to be brought to bear as they interpret what they are seeing, including any alternative ideas they may have. In this way, MRL activities offer students ways to engage with the simulation that can directly address intuitive ideas they may have. MRL addresses the entrenched issue that students learn science as facts and rules isolated from their everyday knowledge and irrelevant to their daily lives.

## *Evolution of the Science Lab*

MRL is built upon previous attempts to combine beneficial characteristics of physical and virtual laboratories. Our Mixed-Reality Laboratories help to establish a new category of science laboratory practices.

In the past two decades, the debates around science laboratories have moved from the necessity of laboratories (Hofstein and Lunetta 1982) to choosing between physical and virtual laboratories (Triona and Klahr 2003) to combining the two (Zacharia 2007). Through these investigations, the science education community has come to the understanding that these two types of laboratories have complementary, beneficial characteristics for a variety of learning objectives (Olympiou and Zacharia 2010). Effective instructional design takes advantage of the special characteristics of each form of lab to meet targeted learning objectives (Olympiou and Zacharia 2012). Recent empirical studies suggest that sequentially conducting physical and virtual experiments can help students gain greater understanding of certain science topics compared to either virtual or physical labs alone (Jaakkola and Nurmi 2008; Zacharia 2007; Ünlü and Dökme 2011). However, research also suggests that students experience difficulties making connections between physical labs and virtual labs, particularly for subject matter that involves abstract concepts and unobservable processes (McBride, Murphy, and Zollman 2010). Innovative research in clinical settings has demonstrated that using mixed-reality approaches to model phenomena can be particularly beneficial to develop deep science understanding (Blikstein, Fuhrmann, Greene, and Salehi 2012; Blikstein and Wilensky 2007). This kind of side-by-side virtual and physical approach helps students engage in authentic scientific practices by guiding students to develop or interact with virtual models of phenomena based on real-world data collected through sensors.

Our mixed-reality approach builds upon this prior work by connecting physical and virtual laboratories through communication between sensors and simulations. Such a configuration synchronizes the observable physical phenomenon and its underlying visualized concepts and processes, creating a single coherent representation of both the perceptual world and the conceptual world for students. The physical interactions activate students' intuitive ideas of the targeted phenomenon, including bodily representations of targeted concepts (such as temperature as perceived by human skin) and tacit schemata developed through common experiences (for example, hot things expand, cold things shrink; diSessa 1993). These ideas importantly guide students to interpret the phenomenon and the visualization of science concepts in the simulation (Bransford, Brown, and Cocking 2000). Such a cognitive state enables students to connect, contrast, and reconcile their intuitive ideas and given science concepts—an important process to develop robust and integrated science knowledge (Linn 2006).

The simultaneity of the mixed-reality technology eliminates the need to remember experiences with one system when interacting with the other, and thus reduces extraneous cognitive load and spares working memory for deeper learning (Chandler and Sweller 1991). We have demonstrated these hypothesized benefits of the Mixed-Reality Labs in a quasi-experimental study (Chao et al. 2014). Specifically, students developed better mental connections between the macroscopic and microscopic properties of gases and constructed more sophisticated mental models of complex phenomena than they would under traditional instruction.

## Teachers as Partners in Learning

Teachers acted as design partners, collaborators, and facilitators for the MRL project. Participating teachers reviewed and helped revise prototypes of the technology and the curriculum. These teachers provided valuable feedback that was incorporated in designs and implementations of the MRL activities. Teachers acted as facilitators during classroom interventions, assisting with integration of the MRL activities into their lesson plans, as well as introducing the subject matter, administering assessments, and providing feedback during and at the conclusion of the classroom sessions.

We provided three teacher workshops prior to, during, and after implementation of MRL activities in classrooms. During the workshops, participating teachers learned about the theoretical foundation of our project, reviewed and critiqued newly developed mixed-reality technology and curriculum, and also analyzed their students' laboratory notes and assessment results. We facilitated ongoing discussions of how to best support integratiing the mixed-reality technology into teaching goals.

We also provided one-on-one training and consultations for all participating teachers before and after implementing the mixed-reality technology in their classrooms. Typically, a researcher visited the teachers one week prior to classroom implementation, reviewing any updates of the technology and curriculum as well as the implementation protocol. The researcher also sought observations and comments from the teachers immediately after the classroom implementations.

## Evidence for Success

We collected a variety of data, including:

- Student pretests, posttests, and delayed posttests
- Student lab worksheets
- Video recordings of students during the labs
- Automated screen captures of students' interactions with the MRL labs
- Automated log data of students' interactions with the virtual simulation
- Automated log data of student interaction with the physical probeware
- Interviews with a subset of students after they used the Gas Frame
- Survey responses from students after they used the Gas Frame
- Feedback (comments and suggestions) from the teachers, before, during, and after the classroom interventions

These data are used to evaluate the project's performance based on student learning and to inform future refinements to the technology as part of the curriculum. Preliminary analyses suggest that students are learning better using the Gas Frame than students with traditional instruction (sections taught by the same participating teachers that did not use the Gas Frame), and that the Gas Frame was generally well received by the students. We summarize some of the major evidence collected.

### *Pretest and Posttest Results*

We administered pretests, posttests, and delayed posttests with items adapted from validated sources such as AAAS Project 2061, TIMSS, and Concept Inventories. Items asked students to select a multiple-choice answer describing the behavior of a particular scenario (a different scenario for each question) and to explain their answers. In general, statistical analyses showed that students in the treatment group tended to perform the same as students in the control group on the multiple-choice items, but performed better than the control group regarding their explanations. Students in the MRL group included more normative ideas (especially ideas relating to a normative molecular model of gases) in their explanations. On the surface, students performed the same on recall items but differently on the explanation items that required a deeper understanding of the topic.

### *Student Approval*

MRL was generally well received by students. Students were enrolled in eighth-grade standard and honors-level physical science courses in a district with 26.9% of the student population on free-or-reduced-cost lunch. Student survey results ($n$ = 55) indicate the majority of respondents (69%) say they would recommend the use of the Gas Frame to others. Further, 24% were neutral, and only 7% did not recommend it.

Student comments elaborated on their approval of the Gas Frame. Some examples:

- "It was very interesting to be able to play around with the computer. The technology was very advanced and made the experiment more enjoyable."

- "I liked the interactive materials such as the spring and the touch pad to change the temperature. These provided us with tangible objects that changed what was happening on the screen, and it made it more engaging, and thus easier to learn from."
- "The gas frame made it much easier to understand the air around us isn't just empty space. It provided the background for why the books functioned the way they did in the experiment. In addition, because we could seemingly physically change the temperature, pressure, and volume, the experiment resonated more than if we simply clicked buttons on a computer."

In addition to approval of the visualization provided by the dynamic simulation, students also appreciated the physical controls, describing their experience as interactive and hands-on, as in the following student quotes:

- "[I liked] the fact that we could see molecules in action and control certain variables physically such as temperature and pressure."
- "I like that we were working with something that's not just a computer screen. We actually got to add and subtract molecules physically."
- "I liked how it had a thermometer that changed temperature."
- "I liked how there was a physical pump attached that made more molecules be used."

We interviewed a subset of the students to get more in-depth feedback from them regarding the Gas Frame. In the following transcribed session, we asked two students after they had used the Gas Frame to respond to a hypothetical case where we converted the physical controls into onscreen virtual controls:

*Researcher:* So what if we replace all this stuff [the physical controls] with some buttons on the screen?

*Student A:* I wouldn't.

*Researcher:* Say, pumping in molecules and sucking out molecules?

*Student A:* I wouldn't like that because, like the whole, well we have buttons to press already, sometimes I get bored with buttons, so if you actually have something else to do, it gets a little less boring than just pressing buttons. That's why. I think I like the ideas of this spring, you actually have to get up and get hot water and cold water and come back, make it more seems more of experimenting, yeah.

*Researcher:* How about you?

*Student B:* Yeah, I think of the same thing. I feel like if it were replaced buttons, you will still understand it as well, but it did add some, like, fun to it, knowing that you were actually changing it, instead of just clicking a button that the computer does it.

*Researcher:* Except for the fun part, do you think it impacts your understanding of the concept? If we were using buttons, you still be able to understand all the concepts.

*Student B:* I feel like I would be able to understand the concepts as well, but like here you can actually see that you are shrinking the volume, or adding more pressure, so that's helpful.

Here we see that the students think that the physical controls actually enhanced their learning. But it's not just the physicality of the controls, as buttons can be physical, but rather the fact that the physical actions are the same as those they would do in a physical experiment. Student A expresses this when comparing pressing buttons to having to actually manipulate physical objects (hot and cold water), and describes it as being more like conducting an actual experiment. Student B describes a benefit from seeing changes but is attributing this to the physical controls. Although one can see changes in a purely virtual simulation, this student appears to believe that somehow these changes become more salient in a system with physical controls. This suggests that perhaps the students are more attentive and engage more with the simulation when they can physically interact with simulations, as opposed to clicking with a computer mouse.

In separate interviews with different students, the students also suggested that the physical environment enhanced learning from the computer simulation. In the following excerpt, a student describes how the temperature sensor allowed the student to actually *feel* the temperature as well as seeing representations of temperature on the screen:

*Researcher:* Anything else that you think is particularly something you like or enjoyable?

*Student C:* I like that, I thought it was good that instead of having like a temperature bar or something, you actually had a sensor, that, I think that made it easier to, like, correlate what is happening when you could physically also tell what temperature was as opposed to just having a number you would change.

*Researcher:* So you mean if I replace the sensors and also this stuff with some buttons on the screen, say, increasing temperature, or pumping in air, a button for that, a button for pushing the bar. It would not be as, as interactive or?

*Student C:* Yeah, I am saying that it's easier when like, when we students like when we can physically do something, it's easier to, it makes a stronger connection on it, instead of just change, sliding a bar that says temperature.

In the following excerpt, the student elaborates on the difference between pressing buttons and using a natural interface:

*Researcher:* So what if you just replace all these objects with buttons on the screen, say pushing or pulling the piston, a button for adding molecules or removing molecules, do you, would you prefer that because buttons are pretty easy to touch, uh, or do you still prefer using the real objects?

*Student D:* I have to say I think I still prefer the real objects. Yeah, I guess it's because like I have been saying it's new, so instead like oh I am hitting a button, and now I am seeing it and do it. It actually feel like, instead of like clicking something in the computer because we do WISE activities, and instead of clicking like, oh I click this place on the computer and this happens, you actually feel that you are doing it, yeah.

*Researcher:* OK, but do you think it will change how you understand the concepts?

*Student D:* Uh.

*Researcher:* Like, if you use buttons, you would understand it less, or?

*Student D:* Maybe, I don't know, probably not as much, maybe a little bit.

*Researcher:* You think you would understand less than interacting with real objects?

*Student D:* Yeah, a little bit less.

*Researcher:* Why? Tell me why.

*Student D:* I am a hands-on learner. I like being able to, like, see it and piece things together, so instead of just hitting a button, I like actually feeling like you are doing it, or actually doing it, it is so much better than hitting a button, and then it happens, and then you are like, wait, why did this happen?

*Researcher:* So, you mean when you hit a button, you don't really feel like there is a connection with what the button means, right?

*Student D:* Right, it is kind of like hitting a button you are told to hit. And with this you had to think about it, OK, while I have to pull this spring, because it is moving this thing on the computer, so you know.

An interesting point that Student D makes is how having to manipulate the spring (which influences the position of a piston in the simulation and required constant attention) actually forced the student to have to "think about it." In other words, the physical interface caused the student to actually think about how the physical action affected the display on the screen, and therefore increased engagement with the simulation. This also appears in another interview with two other students:

*Student E:* I thought it was cool, like, you could use the pump and the spring to interact with it, or like, you feel like you are actually putting in the gas molecules, instead of like just press a button, oh, there are my gas molecules.

*Student F:* Yeah, you can just see exactly what you are doing, exactly any time, you could see it.

Student F highlights the benefit of having direct, simultaneous feedback from the simulation to the physical input. Students using the Frame can instantaneously observe the effects of their physical manipulation. This is a key advantage of mixed-reality over other configurations, such as conducting a physical lab at the same time as a virtual lab, where attention is split and must rapidly change between two side-by-side apparatuses; or conducting physical and virtual labs in serial (one before the other), where you cannot see microscopic changes as an instantaneous effect of a physical interaction. The *direct link* between *natural* physical manipulation and the microscopic-level effects made visible by the computer simulation is something that the students themselves recognize as important for their learning.

## Lessons Learned

We learned many things from conducting the MRL field trials. Some things that may be of interest to others who are working to enhance STEM education with technology include the following:

- *Natural physical interaction can stimulate student engagement and attention.* As evidenced by student feedback on surveys and in interviews, students felt that they were more engaged with the simulation when they were performing physical actions similar to conducting a physical experiment. Mixed-reality approaches that use physical controls with virtual dynamic molecular simulations may lead to an increase in understanding through increased attention and engagement.
- *Physical interfaces can facilitate student immersion.* Physical interfaces allow students to enter a different frame of mind when interacting with the computer simulations. In effect, the students begin to role-play a simulated physical experiment, which means that encouraging the role-playing extends the simulation beyond the confines of the computer and into the real-world environment of the student. This expansion of the simulation may enhance the learning experience for the student.
- *Instantaneous feedback allows simultaneous cause-and-effect analysis by students.* Allowing students to simultaneously assess the effects of their actions is key not only to holding their attention but also to helping them to map effects to their causes. In particular, the combination of the physical interface and the instantaneous feedback enables students to conduct traditional physical labs and connect underlying, molecular causes to the observed effect. Typical physical labs with gas laws involve re-creating pressure, volume, and temperature relationships without any connection to the underlying explanation of the Kinetic Molecular Theory. MRL labs enable students to see why these relationships occur through the molecular visualizations, providing an explicit connection from the molecular to the real world.
- *Iterative refinement is necessary to improve student learning.* Not everything in our project turned out perfectly. Teacher and student input about various aspects of the hardware and software has informed various design decisions with the Frame. The technology, as well as the accompanying curriculum, has been meticulously refined using data gathered from classroom observations, interviews, and assessments.

## Next Steps

In the longer term, we hope to expand our MRL curricula and technology to encompass more topics within science and STEM. Our Mixed-Reality Labs enable students to integrate core practices, core ideas, and crosscutting topics as outlined in the *Next Generation Science Standards*. We believe the Frame technology in particular can be used across a variety of settings to help students learn science concepts, engage in engineering design, and incorporate mathematical problem solving.

We also hope to articulate design principles based on our classroom trials that will help others who are interested in developing mixed-reality approaches to STEM learning. We have used different versions of the Frame that provide either a simulation of a phenomenon at an unobservable scale (such as gas molecules) or provide a simulation of an unobservable process at the everyday level (such as energy flow). We have also experimented with different types and configurations of physical interfaces. We hope to communicate these design decisions with accompanying student data to help inform future developments in mixed-reality STEM technologies.

The Frame has potential to help students engage in engineering design practices as outlined in the *NGSS*. Other curricular projects that we have implemented use the context of testing air mattresses, bungee cords, and beverage mug designs. Although the current curriculum focuses on the scientific principles behind these products, we envision in the near future we will add design into these units to complement the existing context. For example, in our mixed-reality energy and heat lab, students explore common features of everyday mugs to learn about energy flow. We envision adding the ability for students to design their own mugs as part of the project.

Similarly, we are exploring ways that the Frame can help students learn mathematical concepts. With our gas laws lab, students construct graphs and trend lines for data gathered during the lab. In our force and motion lab, the visualization includes a graphical display of position over time. We believe that our mixed-reality technologies provide opportunities for students to make connections among science, technology, engineering, and mathematical concepts and are currently exploring how to support such connections.

## Acknowledgments

We would like to thank the participants in this project and the MRL research group for their help with the design, implementation, and analysis of this work.

This material is based upon work supported by the National Science Foundation under grants IIS-1123868 and IIS-1124281. Any opinions, findings, and conclusions or recommendations expressed in this material are those of the authors and do not necessarily reflect the views of the National Science Foundation.

## References

Blikstein, P., T. Fuhrmann, D. Greene, and S. Salehi. 2012. Bifocal modeling: Mixing real and virtual labs for advanced science learning. In *Proceedings of the 11th International Conference on Interaction Design and Children*, 296–299. New York: ACM.

Blikstein, P., and U. Wilensky. 2007. Bifocal modeling: A framework for combining computer modeling, robotics and real-world sensing. In *Annual meeting of the American Educational Research Association*. Chicago, IL: AERA.

Borgman, C. L., H. Abelson, L. Dirks, R. Johnson, K. R. Koedinger, M. C. Linn, C. A. Lynch, D. G. Oblinger, R. D. Pea, K. Salen, M. S. Smith, and A. Szalay. 2008. *Fostering learning in the networked world: The cyberlearning opportunity and challenge: A 21st century agenda for the National Science Foundation.* Washington, DC: National Science Foundation.

Bransford, J. D., A. L. Brown, and R. R. Cocking. 2000. *How people learn: Brain, mind, experience, and school.* Washington, DC: National Academies Press.

Chandler, P., and J. Sweller. 1991. Cognitive load theory and the format of instruction. *Cognition and Instruction* 8 (4): 293–332.

Chao, J., J. L. Chiu, E. A. Pan, C. J. DeJaegher, E. Hazzard, and C. Xie. 2014. The effects of mixed-reality laboratories on high school students' conceptual understanding of gas laws. Proposal submitted to the 2014 annual meeting of the American Educational Research Association, Philadelphia, PA.

diSessa, A. A. 1993. Toward an epistemology of physics. *Cognition and Instruction* 10 (2/3): 105–225.

Hofstein, A., and V. N. Lunetta. 1982. The role of the laboratory in science teaching: Neglected aspects of research. *Review of Educational Research* 52 (2): 201–217.

Jaakkola, T., and S. Nurmi. 2008. Fostering elementary school students' understanding of simple electricity by combining simulation and laboratory activities. *Journal of Computer Assisted Learning* 24 (4): 271–283.

Linn, M. C. 2006. The knowledge integration perspective on learning and instruction. In *Cambridge handbook of the learning sciences*, ed. R. K. Sawyer, 265–286. Cambridge: Cambridge University Press.

McBride, D. L., S. Murphy, and D. A. Zollman. 2010. Student understanding of the correlation between hands-on activities and computer visualizations of NMR/MRI. In *AIP Conference Proceedings* 225: 225–228.

NGSS Lead States. 2013. *Next Generation Science Standards: For states by states.* Washington, DC: National Academies Press. *www.nextgenscience.org/next-generation-science-standards*

Olympiou, G., and Z. C. Zacharia. 2010. Comparing the use of virtual and physical manipulatives in physics education. In *International Conference on Education, Training and Informatics. www.iiis.org/CDs2010/CD2010IMC/ICETI_2010/PapersPdf/EB672AG.pdf*

Olympiou, G., and Z. C. Zacharia. 2012. Blending physical and virtual manipulatives: An effort to improve students' conceptual understanding through science laboratory experimentation. *Science Education* 96 (1): 21–47.

Triona, L. M., and D. Klahr. 2003. Point and click or grab and heft: Comparing the influence of physical and virtual instructional materials on elementary school students' ability to design experiments. *Cognition and Instruction* 21 (2): 149–173.

Ünlü, Z. K., and I. Dökme. 2011. The effect of combining analogy-based simulation and laboratory activities on Turkish elementary school students. *Turkish Online Journal of Educational Technology* 10 (4): 320–329.

Xie, C. 2012. Framing mixed-reality labs. *@Concord* 16 (1): 8–9.

Zacharia, Z. C. 2007. Comparing and combining real and virtual experimentation: an effort to enhance students' conceptual understanding of electric circuits. *Journal of Computer Assisted Learning* 23 (2): 120–132.

# Integrating Interdisciplinary STEM Approaches for Meaningful Student Learning

*Renee M. Clary*
*Mississippi State University*
*Mississippi State, MS*

*James H. Wandersee*
*Louisiana State University*
*Baton Rouge, LA*

## Setting

Our EarthScholars Research Group first proposed an integrated, interdisciplinary approach to science teaching more than a decade ago, long before the STEM acronym was in vogue. Since scientists do not conduct research in isolation, nor only solve one disciplinary aspect of a problem at a time, we became strong advocates of *teaching* science the way that scientists actually *do* science. Founded in 2002 by university science education researchers in the southern United States, the EarthScholars team included a botanist/biology educator (Wandersee) and a geologist/geoscience educator (Clary) who sought to identify effective portals through which integrated, interdisciplinary geobiological science could be implemented in science classrooms. Our target audience ranges from middle school through high school and college science classrooms, and even includes inservice science teacher professional development programs. Through the years, we identified and researched interdisciplinary portals, and affirmed that an integrated approach results in effective learning gains for students and inservice teachers. Although we were not calling our research program "interdisciplinary STEM," we were effectively demonstrating that an interdisciplinary STEM approach is successful for science student learners.

In this chapter, we discuss our geobiological interdisciplinary STEM approach, its theoretical underpinnings, and its alignment with the *Next Generation Science Standards* (*NGSS*). We then focus upon three geobiological case study projects—petrified wood, coprolites, and the Global

Education Project Earth Wall Chart—and discuss some of the evidence that documents the success of our interdisciplinary, integrated approach.

## Overview: An Integrated, Interdisciplinary Geobiological Approach to Science Education

The STEM acronym is now referenced often, but its familiarity does not imply *understanding*, or even knowledge of the disciplines it encompasses. Many individuals have difficulties moving beyond compartmentalization to convey the interrelatedness of science, technology, engineering, and mathematics. Conversely, we rally against compartmentalization, asserting that the historical curricular treatment of science—in which STEM could be subdivided and taught as separate units—does not facilitate an authentic view of the nature of science by our students; nor does it reflect the interconnectedness of science within our globally integrated, 21st-century lives. In our research agenda, student interest, engagement, and success are built upon an integrated approach that is based upon the constructivist learning theory and incorporates active-learning strategies and inquiry-based activities. We advocate integrating all aspects of science content, instead of the instructional separation of science content and process skills.

Our methods for interdisciplinary STEM incorporate a visual approach as well as integrated geological and biological science content. Where appropriate, we include mathematics, environmental science, physics, and chemistry. Technology has also been a focus in order to illustrate how scientists come to know what they know, and how computer-assisted tools can deepen understanding. Our integrated approach does not end with traditional STEM disciplines, either. We often extend our science lessons into history of science, visual arts, communications, and the social sciences—in order to illustrate relevance, and offer multiple social and cultural contexts for our students' science learning.

### *Interdisciplinary Geobiology and the National Science Education Goals*

Although our geobiological approach was initiated under the guidance of the older *National Science Educational Standards* (NRC 1996), the interdisciplinary approach easily and effectively addresses crosscutting concepts of the *Next Generation Science Standards* (NGSS Lead States 2013), to facilitate student application of scientific concepts across disciplinary boundaries. The seven crosscutting concepts (patterns; cause and effect: mechanism and explanation; scale, proportion and quantity; systems and system models; energy and matter: flows, cycles and conservation; structure and function; and stability and change) have application in *all* disciplinary cores, including our targeted geobiological ones: Earth and space science, and life science. In our integrated approach, we used both Earth processes (e.g., fossilization, coalification, volcanic eruptions, and geochronology) and biological processes (e.g., fermentation, seed dormancy, dinosaur locomotion, and pollination) in our development of student curiosity-stimulating STEM activities. In addition to the projects of petrified wood, coprolites, and the Global Education Project Earth Wall Chart that we discuss in this chapter, we also researched amber, backyard citizen science, pumice, seed banks, gravel, pollen, sand, unusual fossils, and coal as bio-geo content portals for interdisciplinary STEM effectiveness.

## Researchers With an Integrated Approach

We founded EarthScholars Research Group to promote interdisciplinary approaches in science classrooms because we recognize that important biological and geological constructs do not exist in isolation, and that applications in natural Earth environments convey relevance for the content. We first identified the interconnectedness of our sciences' curricula, and proceeded to develop activities that would effectively integrate biological and geological learning in multiple educational settings from middle school, through high school and college science classrooms to professional development programs for inservice science teachers. We also sought to implement an integrated approach outside formal school environments, since informal science learning is the default learning environment for adult learners (Falk and Dierking 2002).

## The Geobiological Approach Within Multiple Science Audiences

Because our focus is on the *integration* of biology and geology as a way of improving science education, our research studies targeted multiple K–16 audiences, as well as the general public. Therefore, in addition to traditional classroom environments, we include online classrooms and informal science education sites (e.g., fossil parks, arboreta, zoos, state and national parks) in our research settings.

## Traditional Middle School, High School, and University Classrooms

Middle school (grades 6–8), high school (grades 9–12), and university college classrooms all became a focus in our mission to improve science education. We chose middle school science as our earliest grade levels since it is here that science typically becomes a regular, scheduled class period. (We also advocate greater inclusion of inquiry science activities in elementary classrooms; our professional development workshops with inservice teachers revealed that science is often sacrificed in elementary classrooms since it is not assessed on annual state tests at all grade levels, and in some cases, even when it is assessed, it does not count toward a school's score.)

Some middle school science classrooms adopt an "integrated" science approach; but, in our experience, "integrated" often translates into four separate units on Earth science, physical science, life science, and chemistry without actual interdisciplinary integration. By the time students schedule high school science classes, they are enrolled in separate biology, chemistry, Earth science, or physics courses. However, it is through the integrated, interdisciplinary science investigations—not compartmentalization—that students can gain a better understanding of the complexity of our planet (Orr 1994).

It is interesting to note that this compartmentalization of the sciences did not always exist in our schools. In the 19th century, students studied "natural philosophy," which included interdisciplinary scientific investigations that incorporated mathematics as needed to solve problems. Within our modern singular-discipline science classrooms, we fail to adequately convey the interdisciplinary scientific scope, as well as an accurate depiction of the *nature of science*, including the nonlinear methods and social and cultural contexts in which science advances. Over 50 years ago, Schwab (1962) lamented the "rhetoric of conclusions" in our science classrooms. An illustration of scientific divergent thinking and nonlinear progression is needed to convey an authentic view of science to our students, and avoid the "final form science" that exists in many classrooms (Duschl 1990, 1994)—from middle school throughout university courses.

### *Online Environments*

Our integrated geobiological science approach also necessitated research investigations in online classrooms. When online course options first emerged in the 1990s, many teachers were skeptical of their potential to provide effective learning environments. However, online classes are now readily available within university and secondary settings. In particular, inservice teachers reported that online learning offers multiple advantages for them with professional development, including convenience and greater variety of course options. Our integrated bio-geo research often targeted inservice teachers, enrolled in an online master's program.

Online courses differ from traditional classrooms in several important and obvious factors. Research demonstrated their effectiveness; they are dependent upon instructional strategies and best practices employed (Means et al. 2009; Tallent-Runnels, Thomas, Lan, and Copper 2006). Online science classrooms provide equitable, but different, paths to learning science content, even with an absence of the laboratory and field components found in traditional classrooms (Clary and Wandersee 2010; Johnson 2002; King and Hildreth 2001). We sought to determine whether an integrated geobiological approach was also effective in online environments.

### *Informal Science Learners*

Informal or free-choice learning undoubtedly provides the majority of the science learning opportunities for adults, but school-age students also engage in informal science learning *more often* than science in traditional school settings (Falk and Dierking 2002). Theoretical bases of informal learning have been investigated (Anderson, Lucas, and Ginns 2003; Falk 2001; Falk and Dierking 2000). Several opportunities have been found to exist for assimilation of informal learning environments within school science courses (McComas 2006). Fieldwork can serve as informal learning laboratory experiences for both university (McLaughlin 2005; Wandersee and Clary 2006) and high school students (Singer, Hilton, and Schweingruber 2005).

Informal science environments are typically interdisciplinary by their very nature. Informal sites can provide an authentic outdoor field or laboratory setting for integrated geobiological instruction that effectively extends classroom teaching. However, although integrated science opportunities exist, our students will probably not be aware of them without proper guidance. We advocate for the inclusion of informal sites in formal education, since field work and other informal learning opportunities can make lasting impressions on our students (Bernstein 2004), in addition to providing authentic research environments.

## Theoretical Underpinnings: Interdisciplinary Geobiological Approach and STEM Reform

Is there a theoretical rationale behind an integrated, interdisciplinary approach in science education? In addition to the historical context—in which natural philosophy was originally taught as in interdisciplinary, exploratory endeavor—there exist theoretical bases that support our integrated science approach. The learning theory of human constructivism, informed by research in cognitive psychology, serves as the theoretical framework upon which we design our investigations. We seek to *build* instruction upon what the learner already knows, focusing upon important constructs as opposed to isolated facts. As a botanist/biology educator and a geologist/geology educator, we identified important constructs and processes in our individual disciplines that have relevance across science, and accordingly designed our classroom activities for an effective approach that illustrates the integration of knowledge within modern society.

## *Human Constructivism and an Interdisciplinary Geobiological Approach*

Novak (1963) pioneered research into the learning theory of human constructivism, based on the cognitive psychology research of Ausubel (1963, 1968; Ausubel, Novak, and Hanesian 1978). This learning theory posits that humans seek to make meaning, and learning results when the meaning of experience changes. Knowledge is conceptual, with concepts based upon patterns that learners identify and label. *Meaningful* learning occurs when new concepts are connected to prior knowledge and experiences, which leads to cognitive restructuring (Novak 1998). Human constructivism has been advanced through research into the nature of science, cognitive psychology, and epistemology (Mintzes, Wandersee, and Novak 1998, 2000).

The goal for science education researchers and teachers is to promote meaningful learning within students, where students monitor and take control of their learning (Novak and Gowin 1984). When students have an accurate understanding of the nature of science and realize that knowledge evolves, they can also engage in mindful learning (Langer 1997) Meaningful, mindful learning is in staunch opposition to rote memorization, or the study of discipline-specific factoids and formulas without context or greater application beyond the science classroom.

Human constructivism promotes a "less is more" approach for science learning, in which a few selected scientific constructs are focused upon for deeper, meaningful investigation. The focus becomes substantive learning, through experience and restructuring of learners' cognitive framework. Instead of attempting to cross a "mile wide, inch deep" inclusive science curriculum, the learning environment promotes quality over quantity, and understanding over awareness. Learning is *integrated* and connected for meaningful learning.

Meaningful learning is not a passive process but involves thinking, feeling, and *acting* within the learner (Gowin 1981). DeBoer (1991, 2005) noted that from the 1950s, "inquiry" summed up the goal of science educators. Active learning in secondary and college classrooms helps the learner to learn (Michael and Modell 2003); active-learning strategies and inclusion of student research activities yield positive learning benefits (Felzien and Cooper 2005; Lawrenz Huffman, and Appeldoorn 2005). Field-based research also can yield significant student learning gains (Elkins and Elkins 2007).

## *Interdisciplinary Geobiological Science Approach and the* NGSS

An integrated science curriculum promotes an authentic view of science aligned to our modern world. Specifically, a geobiological approach effectively demonstrates the interdisciplinary nature of the crosscutting concepts of our *Next Generation Science Standards* (*NGSS*; NGSS Lead States 2013). This approach also aligns with constructivist science education reforms that promote student understanding over awareness, studying fewer concepts in depth (Mintzes, Wandersee, and Novak 1998, 2000); implementing inquiry in extended investigations (DeBoer 2005); facilitating communication of scientific results, and, in general, providing a more authentic science experience for our students for improved understanding of the nature of science. The integrated bio-geo approach, extended into interdisciplinary units, also reveals how STEM is relevant to global issues. Incorporating social and cultural contexts with real-world issues builds competencies that our future citizens will need to solve these issues (Bybee 2010).

It is important to note that we were addressing crosscutting concepts *before* they were identified in *A Framework for K–12 Science Education: Practices, Crosscutting Concepts, and Core Ideas*. (NRC 2012). In the older *National Science Education Standards* (NRC 1996), unifying principles were concepts that transcended individual disciplines. However, the developers of the new *Framework* recognized that students were not receiving adequate instructional support for connecting these unifying principles across disciplines; therefore, they sought to emphasize them to "elevate their role in the development of standards, curricula, instruction, and assessment" (NRC 2012, p. 83).

## *Teacher-Student Relationships in the Interdisciplinary, Active-Learning Classroom*

During integrated instruction, teachers become partners in science learning. Ausubel's dictum (1963, 1968) reminded us that the most important factor in classroom instruction is what the learner *already knows*; therefore, teachers must assess incoming knowledge and skills in order to determine how best to facilitate meaningful science learning experiences for their students. Each classroom contains students from a variety of backgrounds, and with differing skill levels. Instruction must be adapted to fit individualized needs and interests.

We focus on assessing the prior experiences and interactions of our students with respect to plants, the planet, and the weather (Clary and Wandersee 2006; Clary, Wandersee, and Sumrall 2013; Wandersee, Clary, and Guzman 2006), and we encourage our colleagues to do the same. Through Sense of Place writing templates, students reflect on childhood memories of colors, smells, and favorite activities associated with the scientific content to be investigated in the upcoming unit. The templates also provide students with opportunities to recall childhood experiences through mini-essay prompts, and with student permission, teachers can weave some of these experiences into the class, connecting the scientific content with students' personal lives. Therefore, teachers probe the background experiences and memories of their students, and by including these personal stories, ensure that the scientific content is relevant to learners.

In an integrated, reformed STEM classroom, teachers and students share the responsibility for science learning. Extended inquiry investigations address science content and make available platforms for review of background concepts, analysis, interpretation, and communication of results. We provide teachers and students with opportunities to research and communicate in the same manner as scientists. Although teachers facilitate classroom investigation, there is no predetermined outcome; students and teachers journey through the investigations as partners in researching and learning.

## Evidence for Success: Three Research Projects

In our geobiological research, we have investigated numerous interdisciplinary portals for effective science instruction, including biological topics (amber; backyard entomology; corn and biodiversity; fermentation; human-plant relationships; natural selection; seed banks; stromatolites; trackways), geological topics (coal/carbon cycle; earthquake fault zones; unusual fossils; geologic time; gravel investigations; petrified wood; sand; volcanic processes), communication strategies (argumentation, sustained climate change discussions; fishbone diagrams; online discussion; Sense of Place writing templates), and history of science investigations (controversies; historic visualization tools; polar exploration; historic scientists). Our data and analyses to date have indicated that our interdisciplinary STEM approach is effective in promoting and attaining cross-linked, meaningful science learning.

Many different types of data were collected in our integrated approach. We employed mixed methods designs (Tashakkori and Teddlie 1998; Creswell 1994) that used both quantitative and qualitative data collection tools. Pre- and posttest assessments, open-ended surveys, portfolios and projects, targeted examination questions, and archival data collection are all methods by which we gather data to analyze for our integrated geobiological impact on learner knowledge and affective outcomes.

We will overview designs, dimensions, and results of three selective research studies. Specifically, we highlight our research investigating petrified wood and coprolites as portals for geobiological integrated learning. We also include new visualization research on teachers' and students' perceptions of graphics contained in a compiled "supergraphic" (Tufte 1990), the Global Education Project (GEP) Earth Wall Chart.

## *Petrified Wood as a Geobiological Portal*

### Overview of Research

Petrified wood is found in every U.S. state and on every continent. Its unique properties (petrified wood looks like regular wood, but cannot burn) led us to investigate its usefulness as a geobiological portal to address important interdisciplinary scientific concepts, including geologic time and evolution (Clary and Wandersee 2007, 2012). We developed and validated the Petrified Wood Survey (PWS) to assess student understanding of fossilization processes, petrified wood's mineralogical components/chemistry, geologic time, and geographic occurrences. Our first research involved general college geology courses for nonscience majors ($n$ = 187, 190, 138) from 2003 through 2005, in which we probed petrified wood's effectiveness as a portal to increase student understanding of important interdisciplinary constructs. We next used the PWS in a junior level college design class in 2008 ($n$ = 25) as the basis for assessing science content gains during the creative design of an informal educational space that depicted geologic time (Clary, Brzuszek, and Wandersee 2009). Our last petrified wood research study involved teachers enrolled in professional development programs: Mathematics and Science Partnership (MSP) programs ($n$ = 41, 18, 28) and a state-funded summer enrichment professional development program ($n$ = 10). In the teacher professional development programs, we incorporated petrified wood activities similar to those used in college science classrooms, and also investigated the effectiveness of a fossilization replication activity.

## Data Sources of Petrified Wood Research

Our initial investigations involved informal education sites where petrified wood was publicly displayed. Although the Petrified Forest National Park in Arizona has the largest display of petrified wood in the United States, other smaller sites also display petrified wood in informal educational settings. At the Petrified Forest (Calistoga, California) and the Mississippi Petrified Forest (Flora, Mississippi) we observed visitors, coded and analyzed visitors' comments, and analyzed and interpreted signage and brochures for effectiveness for conveying important scientific constructs, such as the Big Ideas of Earth Science Literacy (Earth Science Literacy Initiative 2010).

We designed and validated a survey instrument that probed these scientific constructs, using petrified wood as a unifying concept (Clary and Wandersee 2007). In 2003, we administered the Petrified Wood Survey (PWS) as a pre- and posttest assessment to determine whether college students had made gains in these interdisciplinary constructs during the semester of an Earth history course for nonscience majors ($n$ = 187). The following year, we directly incorporated petrified wood in the Earth history course, using petrified wood specimens in the classroom, incorporating mini-laboratories that invited comparison and contrast of wood and petrified wood specimens, and offered an optional research project for students to investigate petrified wood on public display in a unique setting. (No two students were allowed to research the same site.) Using the PWS as a pre- and postsurvey, we also measured gains in interdisciplinary constructs ($n$ = 190). The third year of our investigation, we only used the PWS as a postsurvey to determine whether there was any testing effect in year 1 or year 2 with the survey instrument ($n$ = 138; Clary and Wandersee 2007).

We used the PWS in a junior-level design class at a research university in 2008. We probed pre- and posttest responses of students who were assigned to develop an informal education space that depicted the geologic time of Mississippi. Students ($n$ = 25) also produced a research report on geologic time, in addition to their final design solution (Clary, Brzuszek, and Wandersee 2009).

In 2011, we administered the PWS to groups of inservice teachers enrolled in Mathematics and Science Partnership (MSP) professional development programs and a state-funded science professional development program (Clary and Wandersee 2012). For the MSP programs, we incorporated petrified wood activities ($n$ = 18, 28), including wood and petrified wood comparisons, and a laboratory activity that mimicked the fossilization process that teachers could implement in their own classrooms (see Figure 16.1). Teacher reflection forms were used for data collection, and the PWS was administered to all groups ($n$ = 41, 18, 28, 10).

## Results From Users of Geobiological Petrified Wood Approach

Our research investigation with petrified wood integration revealed no testing effect with the use of the Petrified Wood Survey (PWS). The PWS was validated as an instrument which revealed students' misconceptions about geologic time, fossilization, and the properties of petrified wood. Our research revealed that significant gains ($\alpha$ = 0.05) were made by students in a non-science major Earth history course over the semester, regardless of whether petrified wood was directly incorporated as a geobiological portal in the classroom or not. However, when gains were compared across semesters, those students with an integrated petrified wood curriculum made even higher gains over the semester, which were significant ($\alpha = 0.05$). The greatest gains were made in student understanding of petrified wood's abundance, properties, and nature of its formation. At the end of the courses, students still struggled with geochemical concepts, such as the role of dissolved minerals and the lack of oxygen during petrified wood formation (Clary and Wandersee 2007).

**Figure 16.1.** Fossilization Replication

In MSP professional development programs, science teachers tested fossilization replication activities with magnesium sulfate and paper towels before implementing the activities in their own classrooms. Here, a middle school student (grade 7) is preparing "fossilized paper."

The junior-level design course students ($n = 25$) exhibited even greater gains in the PWS over the course of the semester. However, the sample size was small, and more research is needed to determine if the active research and design incorporation of petrified wood in an informal education space was responsible for those gains. Analysis of project designs against the PWS-integrated science content gains revealed that a minimum content knowledge of 75%, as measured by the PWS, was required for successful project designs. However, not all students with 75% content knowledge produced successful designs depicting geologic time (Clary, Brzuszek, and Wandersee 2009).

With inservice teacher groups investigated, we determined that all teacher groups ($n = 97$) exhibited higher content knowledge scores, as measured by the PWS, than students in the earlier research investigations. Teachers who were actively participating in a *geosciences* professional development program ($n = 18$) scored significantly higher than the other teacher groups. However, the fossil wood and wood sample comparison activities revealed that some teachers continued to exhibit misconceptions about composition and time differences between the formation of the specimens (Clary and Wandersee 2012).

The wood/petrified wood comparison activity was confirmed as an effective classroom investigation for probing fossilization, composition, and geologic time. With the fossilization replication laboratory, inservice teachers affirmed the activity's classroom use, but detected potential problems (the use of matches, time requirement, possible student misconceptions). However, the teachers proposed that the problems could be addressed through classroom planning and attention to the possible inaccuracies that could be conveyed in the investigation.

## Voices in Support of the Geobiological Petrified Wood Approach

Our anonymous end-of-course surveys and activity organizer/data collection forms recorded student and teacher comments and constructive feedback for improving our research and our classroom activities. Both student and teacher voices have also affirmed the importance of the integrated, interdisciplinary approach that has relevance beyond the classroom and includes local, familiar environments. The petrified wood inquiry activities in classrooms garnered positive comments. When asked what students enjoyed most about the class, several students affirmed, "The projects were the best … Don't get rid of the projects." One inservice teacher noted that for petrified wood activities, particularly the fossilization replication activity (shown in Figure 16.1), "Most students will enjoy literally getting their hands wet at the beginning; but, to then watch the process over several days and not really 'see' the change occurring from 'wood' to 'rock' until they attempt to light the log. I think there will be some amazed students and some disbelieving students" (Clary and Wandersee 2012).

## *Coprolites as a Geobiological Portal*

### Overview of Research

From the data analysis of our first petrified wood research investigation, we documented that students respond well to unusual fossils. We turned our attention to an even more unusual fossil, the coprolite. In rock and fossil presentations, in both traditional classroom settings and informal ones, we noticed that students had noticeable reactions to coprolites, or fossilized fecal material. Students asserted they were either "awesome" and "cool" or "absolutely disgusting" and "gross." We sought to determine whether a formal introduction of coprolites in science classrooms would be productive for students, and would pique students' interest for investigation of important integrated scientific constructs (Clary and Wandersee 2011). Our research included archival examination of William Buckland's papers and coprolite collection at Oxford University. We then designed an exploratory investigation with coprolites with inservice teachers enrolled in an online master's program ($n$ = 28).

### Data Sources of Coprolite Research

We conducted onsite research at the Oxford University Museum of Natural History to record an accurate history for coprolite research; we extended our investigation of geologist William Buckland's correspondence to other repositories of archived documents (e.g., National Museum of Wales, Geological Society of London, British Geological Survey). From our research, we constructed an interesting, stranger-than-life version of Buckland's coprolite research, which illustrated some of the earliest paleobiological investigations in the history of science. In 2007 and 2008, we implemented an ichnofossil investigation within online paleontology courses composed primarily of practicing teachers ($n$ = 28). We assessed incoming knowledge through an online formative survey. Teachers were next provided an interesting historical record of Buckland's coprolite research, and assigned an in-depth, three-week investigation into the scientific value of coprolites, including paleoenvironmental information provided by coprolite fossils. Teachers also developed coprolite investigations for their individual classrooms, aligned with educational objectives, and state and national benchmarks. In addition to project and portfolio assessment, we also probed teacher opinions of the coprolite activity and coprolite incorporation in their K–12 classrooms (see Figure 16.2) in an end-of-semester anonymous survey.

### Results From Users of Geobiological Coprolite Approach

The majority of inservice teachers could not identify coprolites as ichnofossils; nor did they exhibit any background knowledge of the history of science in ichnofossil research. However, following their coprolite project investigations, teachers performed well. In both years of the investigation, the mean Paleo-Waste project scores were high, at 89.9% ($n$ = 16) and 86.0% ($n$ = 12). Teachers produced classroom coprolite investigations that extended investigations to ecology, paleoenvironments, geologic time, food chains and food webs, fossilization processes, and evolution (Clary and Wandersee 2011). Analysis of open-ended anonymous responses revealed that only one teacher did not anticipate an interest in coprolites from science students.

**Figure 16.2.** Coprolite Investigation

Science teachers affirmed the interest of their students in coprolite activities and investigations, and designed activities for their students involving coprolites. A middle school student (grade 8) investigates various coprolites in an attempt to match the fossil with the organism that produced it.

### Voices in Support of the Geobiological Coprolite Approach

The topic of coprolites was well-received as an integrated geobiological portal for the classroom. A teacher reflected that "Students LOVE this topic. It's not only interesting in a geologic sense, but it has that 'Ewwwww' factor that sometimes can be a HUGE draw to students." Another teacher advised that we should "Keep the coprolite activity, it is one of those activities that some students will do that will hook them forever into science, especially geology" (Clary and Wandersee 2011).

## *Global Education Project (GEP) Earth Wall Chart*

### Overview of Research

The Global Education Project (GEP), a non-government Canadian organization, designs high-density graphic representations organized under a theme. Founded in 1998, GEP strives for impartial and comprehensive representation, using commentaries, graphics, and maps with cited sources (The Global Education Project, n.d.). We used the Earth Wall Chart (*Earth: A Graphic Look at the State of the World*), which incorporates 99 charts, maps and graphs in a 27"× 36" (68.6 cm × 91.4 cm) format. Additionally, the individual graphics are available on the internet for free educational access.

We identified the Earth Wall Chart as a potentially effective visualization tool since it aligned with the classic goals of Earth System Science education, to help students understand interactions between the atmosphere, hydrosphere, biosphere, and lithosphere (Earth System Science

Partnership 2012). Our Earth System is determined by the interaction of biological, chemical, and physical components and processes, further influenced by human interactions. Since the Industrial Revolution, human influences on the planet have escalated, leading some scientists to propose that we are currently in a new epoch, the Anthropocene (Crutzen and Stoermer 2000).

The GEP Earth Wall Chart can be described as "Anthropocene-Panoramic." We proposed that the wall chart can serve as the foundation to Earth System Science education since its graphs and data maps provide the necessary information to promote systematized and integrative thinking about our planet. Johnson (2006) noted that the three inseparable elements of Earth System Science education are state, process, understanding, and their interrelationship. All these elements are illustrated in the GEP Earth Wall Chart. Our research with the GEP Earth Wall Chart involved a graduate biogeoscience education class in 2008 ($n$ = 20), as well as middle school teachers of science in 2012 ($n$ = 17).

## Data Sources of GEP Earth Wall Chart Research

In Spring 2008, we required each student ($n$ = 20) to purchase a copy of the GEP foldable Earth Wall Chart. Throughout the semester, we systematically assigned weekly topical subsets of the chart's contents for focused study and discussion, through both electronic and in-class forums. The sequence of the chart section assignment paralleled the second class text, an online course developed primarily by Harvard University scientists, *The Habitable Planet: A Systems Approach to Environmental Science* (Annenberg Learner 2008). Michael and Modell's (2003) text was also used to implement active learning strategies. Quizzes, targeted examination items, and anonymous end-of-course survey responses directly linked to the Earth Wall Chart were cross-analyzed to assess overall learning impact of the course-integrated wall chart.

In 2012, we provided middle school teachers ($n$ = 17) of science (grades 6–8) with a laminated Earth Wall Chart. Teachers investigated the charts and determined their top three graphics (see Figure 16.3). Teachers then identified three graphic components that could be used in their science curricula, noting the competencies and benchmarks the graphics addressed. Teachers then developed an activity using at least one graphic component of the Earth Wall Chart. When these charts were displayed on classroom walls, teachers reported on the graphics that gained the most student attention, and the most comments.

## Results From Users of Geobiological GEP Earth Wall Chart Approach

With our graduate biogeoscience students ($n$ = 20), we found that the majority increased their personal knowledge of the distribution and limits of the Earth's resources (88%), and shifted their knowledge and interest horizons from regional or national to global (72%). A majority of students (65%) also recalibrated their professed personal consumption patterns to accommodate their new understandings of global resource realities, and expressed their intent to reduce their ecological footprints (68%). Students (81%) also confided that the Earth Wall Chart impacted their personal discourse positions on Earth-related issues, and heightened the feeling that they were familiar and necessary stewards of planet Earth.

In 2012, middle school teachers ($n$ = 17) confirmed the usefulness of the variety of graphics contained in the GEP Earth Wall Chart. When first asked to peruse the chart, teachers independently identified 18 graphic components that were of high interest, with the highest-rated graphics being Food Distribution, People Without Safe Drinking Water, Oceans, and the World's Top Military Spenders. However, four months later, teachers focused upon *different* graphics for use in their science classrooms, when they had to target those components that could be integrated in their classrooms and aligned with their state science framework. The Climate Change graphic received the most attention, followed by Historical Temperatures and Major Greenhouse Gases, and Surface Currents. There were several topics that tied for third place in classroom implementation, including Terrestrial Biomes, People Without Safe Drinking Water, Deforestation, and Solar Energy. The most popular topics addressed by teacher-developed activities were Solar Energy, Terrestrial Biomes, and Deforestation. We found it interesting that teachers identified different graphics (than the ones they found most interesting), when seeking graphics for immediate classroom use.

**Figure 16.3.** GEP Earth Wall Chart

Science teachers identified several graphics of interest on the GEP Earth Wall Chart, and were able to design lessons around the component graphics. Teacher-identified graphics of interest were not always included in lesson plans they designed, however. Teachers noted that their state-required competencies and benchmarks were often limiting variables.

Teachers reported that their students were very interested in the displayed GEP Earth Wall Chart, because "it shows so many different types of information in one space." Only one teacher reported that the graphic was not useful, and "students believe it is too busy with information that they do not get." Students' interest in the chart also departed from their teachers' interests, as well as teacher-identified graphics for classroom incorporation. The exception to this trend was student identification (in three teachers' classrooms) of Oceans, Surface Currents, Thermoclines, and Terrestrial Biomes as a favorite topic. However, the graphic is located in the center position of the GEP Earth Wall Chart, so placement may have some influence on both teachers' and students' choice. Students in three teachers' classrooms also liked the Space Debris graphic, while students in two classrooms voted People Without Safe Drinking Water and Life Expectancies as two interesting graphics. Our research affirmed that multivariate, interdisciplinary visualization tools like the GEP Earth Wall Chart can address integrated science concepts in classrooms. The chart also illustrates human impact on the planet, demonstrating relevance for viewers.

### Voices in Support of the Geobiological GEP Earth Wall Chart

Teachers who displayed and used the GEP Earth Wall Chart in their classrooms reported positive student responses. One teacher noted that the chart grabbed student attention because of "the size and the colorfulness." Students liked the interdisciplinary charts because "It has a lot of stuff on it." One teacher who developed a lesson with a chart graphic reported that "Students stated they liked the chart (Ocean Surface Currents, Thermoclines, and Terrestrial Biomes) because they understood what was being presented in the [classroom] activity."

## Dissemination and Feedback From an Integrated Geobiological Approach

The petrified wood, coprolite, and GEP Earth Wall Chart research studies we overviewed in this chapter are but a brief representative sample of our integrated geobiological research program. Our data and analyses confirm the effectiveness of an inquiry-based, integrated approach. We disseminate our research and findings through professional conferences at the local, regional, national, and international levels, with nearly 160 presentations to date. We have also disseminated our research through 50 peer-reviewed journal articles, 15 book chapters, and 45 electronic articles. The feedback we receive from our readers and participants confirm that our integrated approach to science education is relevant beyond our own classrooms. Teachers reported that they have incorporated our methods and activities for student success in their individual classrooms.

## The Future of the Integrated Geobiological Approach

Our research demonstrated that an integrated geobiological approach can be effective for addressing crosscutting concepts, connecting the curriculum to students' lives, and promoting a more authentic view of the nature of science. We have researched classroom and informal incorporation of various technologies, communication forums, and topics to serve as portals for interdisciplinary learning. These approaches, which incorporate active-learning and inquiry-based methods, have met with student success in increased content knowledge, application, and affective learning outcomes.

## The Next Research Investigations

To continue the progress made with our research into effective and meaningful interdisciplinary science learning, we must extend the current knowledge base and geobiological integrated applications with the youngest learners, or those in elementary grades. We specifically focused on middle school, high school, and university college classrooms (as well as informal environments) because our experience with elementary classrooms revealed that many schools lacked a regularly scheduled science component. However, science process skills should not begin in a late elementary grade, when science is first assessed on a state test. We must achieve greater inclusion of science in *all* elementary classrooms, using an active-learning, inquiry-based integrated approach that piques students' interest and demonstrates relevance to their lives.

We also propose that the science educational atmosphere is conducive to changes in assessment. Much of our high-stakes tests have evolved into objective items at an end-of-year state examination. In order to facilitate STEM reform, we must move beyond isolated assessment in high-stakes environments. Teachers should regularly implement formative assessments to illustrate the current skills and knowledge levels of their students, and then build upon these for meaningful science learning. Furthermore, process skills are not easily assessed in an objective, line-item fashion. Therefore, we anticipate greater use of hands-on performance tasks, portfolios, projects, independent calculations, engineered solutions, creative designs, and effective student communication of results as assessment items. Our students deserve feedback on their level of science process skills in order to conduct, apply, and communicate science.

We currently see one reform effort moving knowledge level content *out* of the classroom, and using classroom time and teacher expertise for application and problem solving. The "flipped classroom" model is under scrutiny, and we must determine whether it will be a viable, effective method for achieving meaningful learning opportunities for our students.

In addition to the new STEM research directions, we also must continue to research and reinforce our current geobiological integrated science, for better implementation within middle school, high school, and college classrooms. Capps and Crawford (2013) empirically confirmed that some of the best teachers cannot effectively implement reform-based teaching, and concluded that continuous professional development is needed. Therefore, we must strive for greater incorporation of field-based, hands-on scientific interdisciplinary science investigations, in order to provide authentic learning environments for better understanding of the nature of science and its application in real-world problems.

What should the current educational focus on STEM ultimately achieve for our students? It must facilitate an authentic view of the nature of science, within interdisciplinary and integrated STEM curricula. We cannot approach STEM as compartmentalized content. The new focus on STEM has the potential for meaningful learning, greater K–12 science engagement, and perceived relevance by our students to their own lives. STEM reform will result in students *engaged* in science, not simply reciting scientific factoids.

## Proposed Inclusion of Integrated Geobiological Science Programs

The data we collected and analyzed indicate successful outcomes with integrated, interdisciplinary approaches in science learning. Our research has been conducted with informal educational sites, middle school, high school, and college populations, and online learning environments as well as traditional classrooms. However we readily acknowledge that more research needs to be implemented to determine if the integrated geobiological approach will facilitate success in classrooms across the United States, and in a wide range of geographic and socioeconomic conditions. Not all teachers may be able to incorporate all classroom research investigations, nor be able to utilize informal educational opportunities outside the classroom; they may need to rely on virtual field investigations for scientific research and comparisons. Regardless of the classroom demographics, however, we strongly encourage teachers to use an integrated approach, and generate data for the students—with activities as simple as growing a plant, collecting local rocks, or procuring sand samples. Science does not have to be expensive, nor does it have to exist in faraway places, beyond local regions. This is what STEM reforms can accomplish!

## Hurdles and Concerns for an Integrated Geobiological Approach

According to the teachers with whom we have interacted, the biggest hurdle to implementing inquiry-based, integrated science in the classroom is *time*. Administrative support has also been an issue for some, but both of these hurdles typically involve very specific state frameworks that leave little time for creativity. With high-stakes tests, teachers must first address the content that will be assessed, before they can extend the content into an interdisciplinary investigation. Therefore, we advocate gradual implementation of the integrated bio-geo approach. Teachers must become comfortable with implementing interdisciplinary, open-ended investigations in the classroom, and managing the time commitments for extended classroom research and activities. Teachers must also gain familiarity in assessing students' science content and skills via methods other than the objective end-of-unit test.

We will continue our integrated geobiological research program for reformed STEM education, lest our students of the future enroll in separate science, technology, engineering, and mathematics courses. Our goal for science instruction is for students—our future citizens—to gain an authentic understanding of the integrated, interdisciplinary nature of science, with an awareness of the connectedness of STEM content with real-world applications.

## References

Anderson, D., K. Lucas, and I. Ginns. 2003. Theoretical perspectives on learning in an informal setting. *Journal of Research in Science Teaching* 40: 177–199.

Annenberg Learner. 2008. *The habitable planet: A systems approach to environmental science. www.learner.org/courses/envsci/index.html*

Ausubel, D. P. 1963. *The psychology of meaningful verbal learning*. New York: Grune and Stratton.

Ausubel, D. P. 1968. Educational psychology: A cognitive view. New York: Holt, Rinehart and Winston.

Ausubel, D. P., J. D. Novak, and H. Hanesian. 1978. *Educational psychology: A cognitive view*. 2nd ed. New York: Holt, Rinehart, and Winston.

Bernstein, S. N. 2004. A limestone way of learning. *The Chronicle Review* 50 (7): B5.

Bybee, R.W. 2010. Advancing STEM education: A 2020 vision. *Technology and Engineering Teacher* 70 (1): 30–35.

Capps, D. K., and B. A. Crawford. 2013. Inquiry-based instruction and teaching about nature of science: Are they happening? *Journal of Science Teacher Education* 24 (3): 497–526.

Clary, R. M., R. F. Brzuszek, and J. H. Wandersee. 2009. Students' geocognition of deep time, conceptualized in an informal educational setting. *Journal of Geoscience Education* 57 (4): 275–285.

Clary, R. M., and J. H. Wandersee. 2006. A writing template for probing students' geological sense of place. *Science Education Review* 5 (2): 51–59.

Clary, R. M., and J. H. Wandersee. 2007. A mixed methods analysis of the effects of an integrative geobiological study of petrified wood in introductory college geology classrooms. *Journal of Research in Science Teaching* 44 (8): 1011–1035.

Clary, R. M., and J. H. Wandersee. 2010. Science curriculum development in online environment: A SCALE to enhance teachers' science learning. In *Handbook of Curriculum Development*, ed. L. Kattington, 367–385. New York: Nova Science.

Clary, R. M., and J. H. Wandersee. 2011. A "coprolitic vision" for earth science education. *School Science and Mathematics* 111: (6): 262–273.

Clary, R. M., and J. H. Wandersee. 2012. The effectiveness of petrified wood as a geobiological portal to increase public understanding of geologic time, fossilization, and evolution. In *Earth Science,* ed. I. A. Dar, 46–70. Rijeka, Croatia: InTech.

Clary, R. M., J. H. Wandersee, and J. L. Sumrall. 2013. Sense of place writing templates: Connecting student experiences to scientific content before, during, and after instruction. *Science Scope* 37 (7): 63–67.

Creswell, J. 1994. *Research design: Qualitative and quantitative approaches*. Thousand Oaks, CA: Sage Publications.

Crutzen, P. J., and E. F. Stoermer. 2000. The "anthropocene." *Global Change Newsletter* 41: 17–18.

DeBoer, G. 1991. *A history of ideas in science education*. New York: Teachers College Press.

DeBoer, G. 2005. Historical perspectives on inquiry teaching in schools. In *Scientific inquiry and nature of science: Implications for teaching, learning, and teacher education*, ed. L. B. Flick and N. G. Lederman, 17–36. Dordrecht, Netherlands: Springer.

Duschl, R. A. 1990. *Restructuring science education: The importance of theories and their development.* New York: Teachers College Press.

Duschl, R. 1994. Research on the history and philosophy of science. In *Handbook of research on science teaching and learning*, ed. D. L. Gabel, 443–465. New York: Macmillan.

Earth Science Literacy Initiative. 2010. *Earth science literacy principles: The big ideas and supporting concepts of earth science*. *www.earthscienceliteracy.org/es_literacy_6may10_.pdf*

Earth System Science Partnership. 2012. Earth System Science Partnership (ESSP). Alfred-Wegener-Institut. *www.essp.org*

Elkins, J. T., and N. M. L. Elkins. 2007. Teaching geology in the field: Significant geoscience concept gains in entirely field-based introductory geology courses. *Journal of Geoscience Education* 55 (2): 126–132.

Falk, J. 2001. *Free choice science education: How we learn science outside of school*. New York: Teachers College Press.

Falk, J., and L. Dierking. 2000. *Learning from museums: Visitor experiences and the making of meaning.* Walnut Creek, CA: Alta Mira Press.

Falk, J., and L. Dierking. 2002. *Lessons without limit: How free-choice learning is transforming education.* Walnut Creek, CA: Alta Mira Press.

Felzien, L., and J. Cooper. 2005. Modeling the research process: Alternative approaches to teaching undergraduates. *Journal of College Science Teaching* 34: 42–46.

Global Education Project. (n.d.). Planet Earth: The executive summary. *www.theglobaleducationproject.org/earth/index.php*

Gowin, D. B. 1981. *Educating.* Ithaca, NY: Cornell University Press.

Johnson, D. R. 2006. Earth system science: A model for teaching science as a state, process, and understanding. *Journal of Geoscience Education* 54 (3): 202–207.

Johnson, M. 2002. Introductory biology on-line. *Journal of College Science Teaching* 31 (5): 312–317.

King, P., and D. Hildreth. 2001. Internet courses: Are they worth the effort? *Journal of College Science Teaching* 3 (2): 112–115.

Langer, E. J. 1997. *The power of mindful learning.* Reading, PA: Addition-Wesley.

Lawrenz, F., D. Huffman, and K. Appeldoorn. 2005. Enhancing the instructional environment. *Journal of College Science Teaching* 35 (7): 40–44.

McComas, W. 2006. Science teaching beyond the classroom: The role and nature of informal learning environments. *The Science Teacher* 73 (1): 26–30.

Means, B., Y. Toyama, R. Murphy, M. Bakia, and K. Jones. 2009. *Evaluation of evidence-based practices in online learning: A meta-analysis and review of online learning studies*. Washington, DC: U.S. Department of Education.

Michael, J., and H. I. Modell. 2003. *Active learning in secondary and college science classrooms: A working model for helping the learner to learn.* Mahwah, NJ: LEA.

Mintzes, J. J., J. H. Wandersee, and J. D. Novak, eds. 1998. *Teaching science for understanding: A human constructivist view*. San Diego, CA: Academic Press.

Mintzes, J. J. J. H. Wandersee, and J. D. Novak, eds. 2000. *Assessing science for understanding: A human constructivist view.* San Diego, CA: Academic Press.

National Research Council (NRC). 1996. *National science education standards.* Washington, DC: National Academy Press.

National Research Council (NRC). 2012. *A framework for K–12 science education: Practices, crosscutting concepts, and core ideas.* Washington, DC: National Academies Press.

NGSS Lead States. 2013. *Next Generation Science Standards: For states by states.* Washington, DC: National Academies Press. *www.nextgenscience.org/next-generation-science-standards*

Novak, J. D. 1963. What should we teach in biology? *NABT News and Views* 7 (2). Reprinted in *Journal of Research in Science Teaching* 1: 241–243.

Novak, J. D. 1998. The pursuit of a dream: Education can be improved. In *Teaching science for understanding: A human constructivist view*, ed. J. J. Mintzes, J. H. Wandersee, and J. D. Novak, 3–29. San Diego, CA: Academic Press.

Novak, J. D., and D. Gowin. 1984. *Learning how to learn.* Cambridge, UK: Cambridge University Press.

Orr, D. 1994. *Earth in mind: On education, environment, and the human prospect.* Washington, DC: National Academies Press.

Schwab, J. 1962. The teaching of science as inquiry. In *The teaching of science*, ed. J. Schwab and P. Brandwein, 1–104. Cambridge, MA: Harvard University Press.

Singer, S. R., M. L. Hilton, and H. A. Schweingruber, eds. 2005. *America's lab report: Investigations in high school science.* Washington, D.C.: National Academies Press.

Tallent-Runnels, M. K., J. A. Thomas, W. Y. Lan, and S. Copper. 2006. Teaching courses online: A review of the research. *Review of Educational Research* 76 (1): 93–135.

Tashakkori, T., and C. Teddlie. 1998. *Mixed methodology: Combining qualitative and quantitative approaches.* Thousand Oaks, CA: Sage Publications.

Tufte, E. R. 1990. *Envisioning information.* Cheshire, CT: Graphics Press.

Wandersee, J. H., and R. M. Clary. 2006. Fieldwork: New directions and examples in informal science education research. In *NSTA handbook of college science teaching: Theory, research, and practice*, ed. J. Mintzes and W. Leonard, 167–176. Arlington, VA: NSTA Press.

Wandersee, J. H., R. M. Clary, and S. M. Guzman. 2006. How-to-do-it: A writing template for probing students' botanical sense of place. *The American Biology Teacher* 68 (7): 419–422.

# Building TECHspertise

## Enhancing STEM Teaching and Learning With Technology

*Elizabeth Lehman and Deborah Leslie*
*University of Chicago*
*Chicago, IL*

### Setting

This chapter describes a partnership between the Center for Elementary Mathematics and Science Education (CEMSE), which is a research and development center at the University of Chicago, and a local suburban public school district. CEMSE offers strategic planning and professional development services for schools and districts that want to strengthen their PreK–grade 8 mathematics and science programs; creates instructional and planning materials for classroom use, professional development materials for teachers, and planning tools for district leaders; and researches questions of key importance and concern to practitioners and leaders of education improvement efforts. The partner school district enrolls over 5,000 students in seven elementary schools and two middle schools in suburban Chicago. The student body is approximately 20% minority and 9% low income.

The district contacted CEMSE in 2008 as part of their science curriculum adoption process for Grades K–5. The district was seeking meaningful, student-centered technology usage in inquiry-based science materials, but was concerned about a lack of such usage in the materials they were reviewing. They had found that the technology in the available materials was either nonexistent or primarily for teachers' use (e.g., interactive whiteboard presentations), whereas they were interested in a more integrated STEM program that engaged students as technology users. CEMSE supports and develops *Science Companion*, a PreK–Grade 6 inquiry-based science program and one of the options that was under consideration by the district. At the time, *Science Companion* lacked this type of student-centered technology use, but we recognized this deficiency and saw it as an important "next frontier" for the development of the curriculum. As a result, CEMSE was eager to partner with willing educators to further work in this arena.

The *Science Companion* curriculum was developed after the *National Science Education Standards* (*NSES*) were published (NRC 1996). As such, *Science Companion* meets many of the goals of the NSES for science education, including engaging in scientific processes and exploring the natural world, while simultaneously teaching the disciplinary content elementary students need to know. *Science Companion*'s focus on experiential, inquiry-based learning also engages students in many science and engineering practices, as articulated in *A Framework for K–12 Science Education* (NRC 2011) and *Next Generation Science Standards* (*NGSS*; NGSS Lead States 2013), such as asking questions, using models, planning and conducting investigations, and collecting and analyzing data. However, *Science Companion* was not written with current technology in mind and thus did not offer many opportunities for students to use technology to support their inquiry and scientific learning. As a result, it also met few of the *National Educational Technology Standards for Students* (NETS Project 2007). One goal of this partnership was to expand the focus of *Science Companion* to provide a truer, integrated STEM experience for students that would simultaneously provide content knowledge, engage students in science and engineering practices, and have students use technology in ways that enhance their work with the practices. A second, equally important, goal was to improve the capacity of teachers to thoughtfully integrate student technology use into their core instruction.

In partnership with the district, CEMSE engaged in an iterative process that began in the summer of 2009 and concluded in June 2013. Through the iterative process, each "technology-enhanced" unit was developed by CEMSE curriculum writers, piloted in a small number of classrooms by district teachers and students, revised based on learnings from the pilot, and then implemented in all classrooms in the district. In total, we iteratively developed student-centered technology activities for 17 Science Companion units (2 for kindergarten and 3 each for Grades 1–5). Ten technology-enhanced units were developed, piloted, and revised in 2010–11, with the revised units used by all teachers in 2011–12. An additional 7 technology-enhanced units were developed, piloted, and revised in 2011–12. All 17 technology-enhanced units were used by all teachers in the district in 2012–13.

## Project Description

This collaboration had three main sets of "doers": CEMSE staff, district leadership, and classroom teachers in the district. CEMSE staff developed technology activities that enriched the inquiry and the content within *Science Companion* units, leading to enhanced STEM experiences for students. The district's instructional leadership team recruited a cohort of 13 classroom teachers (called "TECHsperts") to pilot the technology-enhanced units, provided supports for the overall *Science Companion* implementation plan, and ensured that teachers and students had access to the necessary resources. The TECHsperts piloted the technology activities in their classrooms prior to districtwide implementation and provided feedback to the CEMSE development team, including thoughts on the materials' impacts on student learning, student use of scientific and engineering practices, and student engagement, among other things. Over the course of this project, TECHsperts also became leaders within the district around this initiative—sharing their experiences and knowledge with other teachers, conducting professional development sessions and ongoing support for other teachers in the district, and providing ongoing consultation and advice to the CEMSE development team and to district administrators. Following the districtwide rollout of the technology activities, CEMSE and district leaders also sought feedback from K–5 classroom teachers throughout the district.

Throughout this project, learning was expected at all levels. The CEMSE team expected to learn more deeply about the potentials and limitations of technology integration in elementary science classrooms, including: (1) which types of activities have the largest impact on engagement and learning, (2) logistical realities and constraints of student technology use, and (3) how to best communicate about technology activities with teachers and support both teacher and student learning. The district leadership expected to learn many details about the needs of teachers and students, including how to manage the sharing of a limited number of devices, the ideal number of devices in each classroom, the "tech support" needed within a building and across the district, the varied uses of different tools, how to use technology to promote inquiry-focused teaching and learning, and how to use technology to support differentiation. TECHsperts expected to increase STEM teaching and learning in their classrooms, learn about various technologies their students could use, gain confidence in their abilities, serve as teacher-researchers, and develop as teacher-leaders. The other teachers in the district expected to enhance STEM teaching and learning in their classrooms through the use of technology. Finally, we expected that students would learn STEM content and begin to master science and engineering practices in age-appropriate ways, as well as improve their understanding of how to use various types of technology. Further, we expected that student facility with technology would improve over time, as they gained experience with a diverse array of tools.

## Major Features of the Instructional Program

### *Framework Guiding Activity Development*

From the beginning, our driving question was, "How can technology enhance students' inquiry experiences?" As such, the CEMSE developers sought meaningful technology use by students, not simply "technology for technology's sake," to move toward a more integrated STEM program. This approach has three main implications. First, *how* and *why* students use a particular technology tool is more important than *what* tool. As various devices and software were considered, we recognized that, to a large extent, many specifics of the tool or program are unimportant—different brands or versions can replace each other with no impact on learning. Second, given the value of hands-on science experiences, technology needs to *complement* those experiences rather than be a substitute for them. Finally, significant portions of the technology must be integral, rather than supplemental, to students engaging in science and engineering practices in the classroom.

Guided by the above principles and the content and practices that are central to the lessons in *Science Companion*, we developed a framework of four categories of technology tools, each defined by how and why students use the tool. The four categories include tools used to: (a) collect and analyze data, (b) reflect on experiences and create representations of knowledge, (c) communicate and collaborate with others, and (d) discover and learn more about a topic (see Figure 17.1 on page 284). Each of these categories is described in more detail below, with examples of how they have enriched students' experiences with STEM content and practices.

#### Technology Tools to Collect and Analyze Data

Data are central to science and engineering, no matter the age of the investigator. Technology tools in this category make data collection and analysis easier, faster, and more precise than they would be without the tools. They allow students to see things they could not otherwise see and to collect, store, organize, and analyze various types of data (e.g., visual, auditory, and numerical).

**Figure 17.1.** Four Categories of Student Technology Use

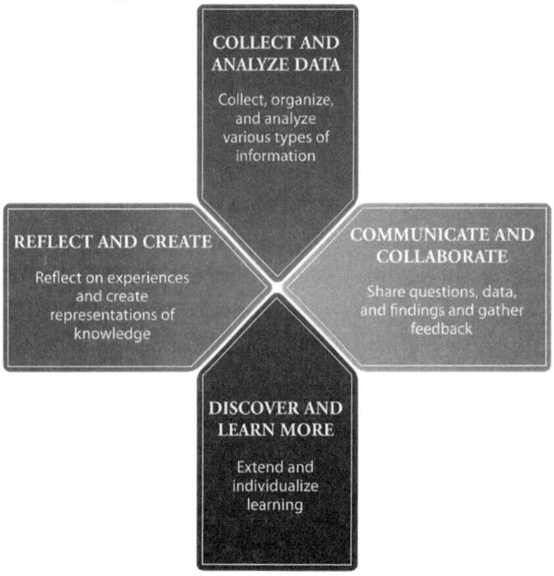


We use these four categories to guide the development of technology activities, with a strong focus on how and why students are using the tools rather than what tool is being used.

Putting such tools in the hands of elementary students helps them develop mastery of science and engineering practices, as well as an understanding of the important role of technology in science and engineering. Using data tools allows students to focus on the data and what they mean, rather than on laborious data collection processes that add little to students' understanding of the data or related concepts. Some examples of data tools in our technology-enhanced units are digital microscopes, digital cameras, time-lapse video creation software, temperature probes, light sensors, and graphing software.

### Technology Tools to Reflect and Create

Reflecting on scientific explorations and engineering design tasks is a key feature of *Science Companion* because of the importance of processing experiences to solidify learning. Reflection typically takes the form of a class "reflective discussion" after a hands-on exploration. Technology tools in this category have helped us broaden the options for and promote depth in student reflections. Our technology-rich lessons involve students in reflecting on their experiences in a variety of formats, not all of which involve verbal skills. Further, we have included creation tools in this category because they allow students to create representations of what they understand, which requires reflection. One key advantage of using technology to accomplish these types of learning goals is that students are able to incorporate real data, often collected using a data tool (see above), into their reflections. Some examples of reflection and creation tools are software for creating animated models (e.g., of how light travels in a periscope), digital graphic organizers (e.g., a digital Venn diagram comparing and contrasting humans, goldfish, and crickets), media software to compile photographs with explanatory labels and captions, and student-created videos with narration to explain the concepts presented. Student work produced with these tools also provides rich formative assessment data for teachers.

### Technology Tools to Communicate and Collaborate

Scientists and engineers are expected to communicate their findings and designs to others and are increasingly involved in collaborative research. These skills are promoted with collaboration tools. Collaboration tools allow students to interact with experts and other students who may not otherwise be accessible; to share questions, ideas, data, and findings; and to gather feedback.

Further, students and teachers can create ongoing records of scientific work and learning that can be shared electronically with families or others. Because collaboration tools allow students to share their data, questions, and ideas with others, there can be substantial overlap with data and creation tools, especially when digital products and data are shared online. Collaboration tools can include software for creating videos or podcasts, online discussion forums, and video conferencing.

### Technology Tools to Discover and Learn More

One challenge of teaching can be customizing teaching to the different interests of students. The tools in this category provide students with "discovery resources," which allow students to extend and individualize their learning while exploring and reinforcing STEM content. Examples of discovery resources include websites with interactive games, simulations, or animations; content-based sites with age-appropriate information; and video clips about scientific content.

## *Considerations Guiding Classroom Implementation*

Based on input from the TECHspert teachers, the technology activities CEMSE developed were considered (and offered as) "starting points" for teachers and students. Depending on each teacher's level of experience and comfort with technology, and the skills and interests of their students, they could use the suggestion as written or use it as a springboard for their own ideas. Both teachers and students went beyond expectations when they began using the technology, an achievement we attribute to significant planning and support for classroom implementation.

Training and support for all teachers was embedded in the implementation and development plan from the beginning. TECHsperts, district administrators, and CEMSE staff collaborated to craft a comprehensive professional development (PD) plan that covered basic implementation of *Science Companion* and, later, integration of technology activities. TECHsperts attended an introductory PD session at which we explained the philosophy of the technology activities, particularly how they aligned with the inquiry approaches for *Science Companion* and would enhance student work with science and engineering practices. The TECHsperts also received one-on-one time with CEMSE staff to focus on technology activities and tools specific to the units they were teaching. Each TECHspert also was assigned a CEMSE "liaison" who was his or her go-to person for questions and in-class support. When the technology-enhanced units were implemented districtwide, all K–5 teachers participated in technology-focused PD prior to using the technology-enhanced units. For these teachers, this corresponded to their second time teaching a particular *Science Companion* unit.

During districtwide grade-level PD sessions, teachers were introduced to the overall philosophy of *Science Companion* and the goals of the technology enhancements, and to the technology activities for the specific units they would be teaching. TECHsperts shared their experiences with the technology activities, as well as examples of student work. Teachers had time to discuss the benefits and challenges of incorporating technology, as well as time to plan collaboratively about which technology activities they wanted to incorporate and how to implement them in the classroom. This shared planning time was particularly valuable for teachers who wanted to make use of limited resources, such as a shared laptop cart with a classroom set of laptops for student use. The district administrators also set realistic expectations for how teachers should begin to change their practice to more fully incorporate technology: All teachers were expected to have a class website and complete at least two technology activities for each unit. The district devoted resources toward providing adequate devices and support for teachers to meet these expectations. For example, they purchased digital microscopes, tablets, and laptops as funds allowed. TECHsperts provided building and grade-level peer support, and CEMSE staff was available to teachers for additional planning or in-class support if requested.

## Evidence for Success

Over the course of this project, we collected several types of data to evaluate whether we were meeting our shared goals and to guide ongoing revisions to the suggested technology activities. CEMSE staff observed classrooms during science instruction, which included conducting informal conversations with students about their work and with teachers about their instruction. TECHsperts filled out logs to provide feedback on specific lessons and technology activities soon after teaching them, and participated in informal "focus group" discussions twice per year. CEMSE staff also had discussions with district administrators and TECHsperts throughout the project and reviewed student work and other items posted on the classroom websites of TECHsperts. The data collected were exclusively qualitative because that was appropriate and sufficient for the goals of this project. As such, no quantitative evidence is presented below. While specific, representative examples are given in the discussion below, it should be noted that the conclusions reached hold true across the district and not based on an individual teacher or classroom.

### *Students and Teachers Are Capable Technology Users.*

Unsurprisingly, we observed that students quickly learned to use new software programs and tools. Although teachers were sometimes hesitant to try new technologies (software, in particular), once they "took the plunge" they were almost invariably glad they did. After a couple years of experience with using classroom technology, one teacher noted, "There's less focus now on how to use the technology. It's more second nature." This statement reflected a generalized shift from focusing on procedures to considering how technology was impacting teaching and learning. Further, many teachers began using technology throughout the school day, not just during science time, after witnessing students' abilities and engagement, and experiencing successes along with the challenges. As for the students, in the third year of this project one teacher reported that she saw students growing in their skills throughout the year, "but also the students are coming in with more skills." This statement indicated that the technology experiences students had in prior years carried over into future years. Even the youngest students were able to independently handle tools like digital microscopes once teachers had taught and established norms, procedures, routines, and expectations for their use.

The technology activities gave students many opportunities to share their knowledge with each other and with their teachers. We observed and teachers reported that this new context also revealed student capabilities that may not have shown up in more traditional assignments. We often observed students working cooperatively to troubleshoot issues that emerged with the technology, indicating that students had a great deal of knowledge of both routine tasks (e.g., how to save a file) and more complex issues (e.g., uploading different file types to their personal blogs). In some schools, "student TECHsperts" emerged. These students served as in-house experts for both students and teachers. Some even led lunchtime study groups to teach other students how to use technology tools.

### *Students Think More Deeply About Scientific Concepts and Practices.*

Our data indicate that these technology activities enhanced students' experiences with STEM, leading to deeper understanding of both content and practices. One teacher said that more "juicy learning" was happening when her students used technology with their *Science Companion* lessons. Classroom observations and teacher reports indicated that students using creation and communication tools often thought about their work more deeply than when engaged in typical classroom discussions. Frequently used tools that provided evidence to support this conclusion were student-created videos and other media. For example, first-grade students used digital cameras as a data tool to document their findings and then wrote captions or recorded narration to explain their findings in slide shows. We observed that students who knew they would be recorded thought more carefully about the arguments and evidence they were presenting. In another example of this phenomenon, we observed fifth-grade students explaining energy transfers that occur in common toys adding additional details and examples to their "script" before recording, and revising and rerecording to make sure they said things correctly, until they conveyed exactly what they wanted. Similarly, one teacher reported that having to share photos of an investigation, "held students accountable for their observations and made them think critically in order to decide how they would demonstrate their learning for others."

One TECHspert who regularly had his fifth-grade students create videos and then post them online experimented with having an outside viewer (in this case, a CEMSE staff member) post comments on the student videos to see how having an "outside audience" might impact student thinking after the initial video creation process. In the comments, students were asked questions that, among other things, required them to provide additional evidence or clarify their explanations. Students then posted a comment in reply to the question. Students were excited to receive and reply to the comments, and replying provided an opportunity for them to revisit the concepts and their understanding of them after some time had passed. The TECHspert found that having students compose short replies to the question pushed their thinking a little further and did not require much additional time—a worthwhile and recommended trade-off.

Students in many grades created animated models to communicate their understanding of various concepts. Modeling, a difficult practice for many, was no longer such a mystery to these students and their teachers. For example, third-grade students created models to explain why we have night and day, and fourth-grade students created animations showing their understanding of how rivers form. Teachers and CEMSE staff observed students as they created these and other animated models, noting in particular how this experience differed from students drawing static models in their notebooks. First, the students thoroughly enjoyed making animations, even though they were focused on "school" topics and not everything in their imaginations. In contrast, drawing models in a notebook was less exciting and engaging for many students. Creating the animated models required more time than drawing them by hand, but that additional time was not exclusively due to the complexity of creating an animation and using the software. As they created the animations, students carefully considered if the animation depicted what they wanted to convey in the model. A second-grade teacher noticed that "the kids really paid attention to their observations as they created their animation" of how magnets attract and repel. This often resulted in students going back and modifying their animations in some way. In contrast, students rarely revisited hand-drawn models in their notebooks. With animated models, students appeared to think through the content and their understanding of it more deeply to make sure the animation reflected their mental model or previously constructed models. For example, a third-grade teacher noted that her students "were able to assess themselves in a way and compare their animation to the pictures we have been drawing in class. It helped them make sense of the closed and open circuit."

Additionally, in some cases data tools enabled students to experience content that was previously inaccessible to them. One example comes from a second-grade unit on motion. When exploring the pushes and pulls that result in motion and the effects of collision, students used a motion sensor that automatically graphs how an object's position changes over time. Even though these young students did not yet understand how to calculate speed as a rate or how to represent it on a graph, we observed that using the motion sensor allowed them to notice the correspondence between slower and faster movements and flatter or steeper lines on the graph. While students sometimes had difficulty seeing slight differences in speed when observing objects in motion, they easily saw the differences in the graphs produced by the motion sensor. Students also used the graphs as evidence to support their claims about how the strength of a force affected the resulting motion. Without the motion sensors, their evidence for such a claim was limited to descriptions of their visual observations.

### *Digital Microscopes Are Powerful Data Tools.*

Teachers and students both came to love digital microscopes and document cameras (a good stand-in when digital microscopes were not available or not practical) as data tools to support their STEM work. The TECHsperts chose the digital microscope as one of their top three technology tools that teachers should use. One kindergarten teacher called it "phenomenal" for her students. Classroom observations and teacher reports indicated that students of all ages were engaged by the clear, close-up views a digital microscope can provide and were eager to talk about what they saw. One teacher stated, "The dialog is so wonderful as they use this tool." For first-grade students examining recently hatched brine shrimp, details of the body were much easier to see with a digital microscope than with a hand lens or the naked eye. We observed that this increased level of detail prompted more questions from students, such as, "What are the hairs for?" Additionally, students were able to produce better scientific drawings because they could better see what they were attempting to draw. When the same students went on to observe the bodies of crickets, they made some exciting discoveries. One was that it was much harder to observe crickets with the digital microscope because it was hard to focus the image on the larger and more active crickets. This limitation of technology had not previously been acknowledged by the students. A number of classes also discovered that, "brine shrimp like light and crickets don't," a behavioral difference between the two that they probably would not have noticed using hand lenses alone. And when another animal, a goldfish, was added to the classroom, the students were able to design a simple investigation to determine if goldfish like light or not.

Some special needs students especially benefited from using the digital microscope. One special education teacher said, "The students couldn't manage regular magnifiers and close one eye. Their frustration levels were so high with regular magnifiers. The [digital microscope] was a wonder. The students being so interested, connected, and 'with me' was rare. It was a wonderful learning experience. It was very student-led."

Images taken with the digital microscope (i.e., data) became important sources of evidence for students' scientific arguments as well. Whereas individual students examining objects with hand lenses sometimes observed very different things, digital microscopes allowed a whole class to look at the same image when discussing what one student observed. When some painted lady butterflies died in their cage in one second-grade classroom, the class took the opportunity to study them more closely with a digital microscope. They got to see the various body parts up close and in more detail, and were able to compare and contrast different images from different individual butterflies. They then used these images as evidence that butterflies can have slightly different appearances, just as people do.

### *Document Investigations With Digital Photos and Video.*

Digital photos and video quickly became another favorite tool for both students and teachers, once teachers got over the fear of having students handle the devices. While students loved to take photos and video, their power lies in students using them to collect and discuss data and to show what they have learned. As one TECHspert advised, "Just be careful that we are not just taking pictures to take pictures and that there is a greater purpose." Thoughtful discussion about how the photos could be used as data and evidence for arguments helped students to focus their work. After her students used digital cameras to take photos of their investigation of a local habitat, one third-grade teacher noted, "Having the pictures to look at after we got back into the classroom helped enrich the discussion. Kids didn't have to remember everything they saw, and they could take a closer look at what we did find!" When a first-grade class studying wind and weather created a video of their wind tools in action, "it got them thinking about the direction of the wind and had them questioning why the arrow kept moving directions." Students in other classes used time-lapse video to collect observational data they could otherwise not access. For example, a third-grade class had a terrarium with hermit crabs that did little during the day, even though the students observed that the crabs' food was disappearing. When the students wondered if maybe the crabs were active at night, they were able to investigate their question by setting up a time-lapse video to see what the crabs were doing after school hours. The class was able to see, "lots of interesting behaviors not seen during the school day," and used the video evidence to support their conclusion that hermit crabs are nocturnal.

Using photos and video to collect data and document investigations also helped teachers with formative assessment. For instance, one third-grade teacher realized how student photos could help her see what she missed, "Many students made observations that we were able to share with the class, which I may not have even been aware of them making if it had not been documented and recorded." One first-grade class wrapped up a study of living things by taking a walk around the neighborhood to identify signs of spring. Small groups documented their findings with photographs; the class later chose to make photo collages to show what they had found. Each student chose one photo and composed text to explain the observation, then added it to the group collage. The teacher was able to use this quick task to assess students' level of understanding, with student captions ranging from "I saw many sticks on the recess field" to "I saw white flowers. These are signs of spring because I don't see some in the fall." Similarly, a fifth-grade teacher had students photograph different examples of energy around the school and sort the photos into groups to show different forms of energy. The resulting graphic organizers allowed him to assess his students' understanding of energy forms quickly and easily; he also noted, "My students enjoyed this lesson more than my teammate's kids because they were able to take pictures of the sources of energy rather than draw them."

### *Technology Can Support Reflection and Communication in the STEM Classroom.*

In addition to promoting reflection and sharing using various digital media, technology helped students develop their verbal and written communication skills. CEMSE staff observed many class sessions in which students created podcasts or brief videos to synthesize and record the findings of their scientific investigations prior to a culminating class discussion. When students had used a communication tool to document their thoughts prior to the culminating discussion, more students participated in the discussion and student contributions were less superficial. Further, students who were typically quiet in whole group discussions could contribute their voice in other ways. Teachers noted that when students used varied media to communicate their observations, findings, conclusions, and questions, their language was more clear, precise, and detailed and their ideas were more organized. When students in a fourth-grade class created videos about electrical hazards and safety, the teacher commented, "I think it was a great way for students to verbalize how to stay safe from electricity—the public service announcement was a real-life situation and brought what we were learning to real life applications." According to another teacher, using various technology tools was, "worthwhile for students to share their data and understand that communication of results is part of being a scientist and making discoveries. It also helped students compile their data and reflect on what they collected and why some groups [had different results]."

Additionally, we often observed that students' observations and explanations were more detailed and sophisticated when recorded orally than when students were asked to write them down. This was especially true for young students just learning to write and older students who struggled with writing. Although low literacy skills are often perceived as a barrier to STEM success, the use of technology freed these students from the burden of written communication and allowed them to demonstrate previously "hidden" abilities and achievements.

### *Nothing Works 100% of the Time, But Teachers Work Through the Difficulties.*

TECHsperts and other teachers noted that problems inevitably arose with technology, requiring teachers to be flexible with their plans—for example, going back to a low-tech activity if necessary.

The availability of technical support in the building was important to their willingness to try certain technology suggestions; if they did not have good support, teachers were less likely to implement activities they felt were complicated or required advanced skills. Further, teachers had different preferences and comfort levels with technology, so CEMSE staff found that the wording of activity suggestions needed to reflect that teachers have a choice in what they do in the classroom and that they can offer choices to their students. For example, in a second-grade study of rocks, one teacher had students make digital books about their rocks while a partner teacher had students create the book pages on paper. Similarly, a fifth-grade TECHspert found that some of her students loved creating digital drawings while some preferred painting with watercolors; she allowed students to use the medium of their choice to illustrate poems they had written about water. Teachers were supported in these choices by wording such as, "you may wish to have students create their books digitally" and "students may also create these drawings digitally."

We also observed that the more "reluctant" teachers responded well to discussions with their colleagues about the benefits and challenges of using technology in the classroom. Several opportunities for teachers to share with each other were built into the implementation plan and TECHsperts developed their leadership skills by facilitating these sessions, leading the way in sharing their practice with others. Additionally, we noted that teachers became more comfortable trying out technology activities after district administrators made it clear that they had reasonable expectations regarding changes in teacher practice. The success of this collaboration hinged on district administrators communicating their expectations around classroom technology use to teachers, technical support staff, and building leadership, and then supporting those expectations with concrete supports.

### *Students Don't Need 1:1 Devices for All Tasks.*

At the start of this project, many involved voiced concerns about the number of devices (e.g., laptops, tablets, and other mobile devices) that would be required in classrooms. Teachers were concerned about how they would manage students as they were using the devices. Technology support staff was concerned about keeping large numbers of devices updated and functional, and about having enough bandwidth to support many students working online at the same time. District leaders were concerned about being able to afford enough devices to meet the needs of teachers and students. We quickly realized that the district would not need to find a way for each student to have a dedicated device, either through a 1:1 deployment or a bring-your-own-device plan. While some TECHsperts had access to class sets of devices, others did not. But those with access only to a limited number of devices generally had no troubles implementing technology activities with students sharing devices. Teachers reported that sharing devices worked well because their classrooms were already structured for partner or group work, and the sharing further promoted collaboration skills in students. However, some tasks did work better when each student had access to a device, particularly when a teacher wanted to use individual student work for assessment. For example, to assess students' understanding of an investigation, one fifth-grade teacher had each student write a blog post interpreting a graph their group had created from their data. In this case, collaborative work would not have given the teacher an accurate assessment of each student's progress on graph interpretation.

Teachers reported that they quickly learned to identify which technology activities would be easier to accomplish with 1:1 devices and which would work well with shared devices. For many activities, 1:1 was only possible if teachers could schedule time in the school computer lab or reserve a laptop cart, which housed a class set of laptops that moved from room to room. Inability to access these resources sometimes prevented teachers from implementing a technology activity, but most found ways around the problem. Often, teachers adapted an activity to make it work for student groups working on shared devices. For example, teachers would ask students to collaborate on describing their investigation plan rather than having each student write a description.

It is also worth noting that, in many classroom observations, students were somewhat hindered in completing technology tasks in a timely manner because they were not proficient at typing. While they quickly mastered the features of software programs, the task of typing their thoughts into a document, blog post, or presentation was sometimes challenging and time-consuming. As a result, teachers in the primary grades were sometimes observed asking students to record themselves describing an image or graph rather than typing a caption. In upper grades, the student who was the fastest typist was often tapped to write for each group. Teachers reported that students improved at typing as they gained experience, but few became adept typists over the course of this project. The district now offers an online keyboarding tutorial that students can access outside of school time, which may help alleviate this problem.

### *Technology Isn't Always Better.*

An important finding from this project was that technology did not *always* improve student learning. This was manifested in several ways. First, it was often important and useful for students to do something "manually" before or in parallel with using the technology. For example, it worked well for students to observe objects with their naked eye and hand lens magnifiers *before* using a digital microscope. This supported their understanding of the digital microscope as a tool that works better than their eyes alone, rather than something magical, and helped them connect what they were seeing on the screen with the object in hand. Figuring out the best way and time to introduce certain technologies is an ongoing interest and area for learning.

Similarly, teachers and students both loved using a kid-friendly, web-based graphing program to graph their data. This program made it relatively easy to create and compare multiple graphs, allowing students to compare results from different variables (e.g., how different exercises affected heart rate) or designs (e.g., effects of different types of insulation on a thermos) because the teacher could display them all at the same time. However, teachers found that students could get distracted by changing the appearance of the graph (e.g., while making a 3-D bar graph, using fonts and colors that make it unreadable) unless they had clear guidelines. Teachers rarely encountered this problem when students created graphs by hand.

Additionally, several TECHsperts experimented with digitally creating word walls, KWL charts, and other graphic organizers and posting them to their class websites, rather than hanging them on their classroom walls. In general, teachers found these digital forms to be less useful than creating them on chart paper and hanging them on the wall because the digital forms were not as readily accessible to students. Tasks like this should likely be kept low-tech.

## Conclusions and Next Steps

This project demonstrates that successful partnerships between school districts and curriculum developers can develop into exemplary STEM programs. Our evidence indicates that thoughtfully engaging students as technology users in STEM settings can positively impact student engagement with and mastery of both content and practices. Educators can build from our experiences and new knowledge to incorporate technology into their pedagogy, even without a university or curriculum partner. The decision to incorporate technology into classroom instruction often requires balancing many factors, including instructional time and resources, before positive results can be achieved. Taking our findings into consideration should help educators at all levels find the appropriate balance needed to let students take advantage of what technology has to offer. Using technology in purposeful ways to help students understand STEM content and develop science and engineering practices is likely to become even more important as teachers, schools, and districts move towards implementing the *Next Generation Science Standards.*

Even though this collaborative project has ended, we expect that teachers in the district will continue to grow in their use of technology for STEM teaching and learning. Teachers and district leaders are particularly interested in using more technology for summative assessments in science, as well as other disciplines. For example, fifth-grade students who have been studying energy forms and energy transfers are assessed through a design challenge in which they design, but do not build, a Rube Goldberg device that utilizes a minimum number of energy forms and transfers. A group of teachers is currently studying the role of technology in such assessments. How could technology improve or change this assessment? Would it accomplish the same goals? What new challenges would it raise?

At CEMSE, we are motivated to build on the promising findings from this project. We are currently in the early stages of developing a new elementary STEM education program in which students will explore real-world, unsolved problems using a cross-disciplinary approach. They will use many of the technologies and other approaches discussed in this chapter to collect, share, compile, and analyze real-world data, collaboratively discuss and test ideas and solutions within and across classrooms, and learn and apply core and crosscutting STEM concepts and practices that are relevant to the problem. Students will work in a blended environment, sometimes completing hands-on, offline work and sometimes working directly on a collaborative, personalized online platform. We are eager to move from the promising technology "enhancements" described previously to an approach in which technology is integral by design. We are grateful for this productive collaboration, which has both informed and inspired us to take this next step.

## References

NETS Project. 2007. *National educational technology standards for students*. Eugene, OR: ISTE.

National Research Council (NRC). 2012. *A framework for K–12 science education: Practices, crosscutting concepts, and core ideas*. Washington, DC: National Academies Press.

National Research Council (NRC). 1996. *National science education standards*. Washington, DC: National Academies Press.

# Rolly Pollies, Bubbles, and Wheelies

## Inquiry STEM in the Early Childhood Classroom

*Fred Estes*
*Nueva School*
*Hillsborough, CA*

### Setting

Located in the San Francisco Bay area, our small K–12 independent school with approximately 530 students has a tradition of involvement with innovative programs, including social-emotional learning (SEL), the problem solving-thinking approach to math instruction, and design thinking, as well as inquiry science. Integrated themes and project-based instruction have also been hallmarks of our school's curriculum since our founding in 1967 with a mission to serve gifted and talented students. Thus, the philosophy and ideal of science, technology, engineering, and mathematics (STEM) curriculum neatly match our existing programs into which we have long worked to integrate the STEM disciplines. Our lower school has long been inspired by progressive, constructivist principles and, especially in early childhood education, by Reggio Emilia's child-centered approach and investigation-oriented methods (Edwards, Gandini, and Forman 2011). Accordingly, we create most of our own curriculum and do not use textbooks.

### Target Learners

Although our whole school is involved in STEM curriculum, this chapter focuses on the early childhood and the early elementary program in grades pre-Kindergarten through second; these grades often receive less attention in STEM discussions.

"Can students in the primary and early elementary grades really learn science and engineering?" This is a question I hear frequently as a science specialist teacher who works with young children. It is often closely followed by the query, "What do you do with them? They cannot read textbooks and they cannot sit still for lectures, so what do you do?" I am happy to explain to parents, community members, and other teachers that they are right that we do not read textbooks and I do not give lectures. We learn science by *doing* science. The soul of science is scientific inquiry and the investigation of authentic questions. The heart of engineering is making things to make things better. Our classes focus on integrated, cross-disciplinary inquiry-based investigations and projects in collaboration with their classroom teachers and other STEM specialist teachers (NRC 2000).

## Elementary STEM Team

A critical element of our success is a solid collaboration among the classroom teachers and teaching associates, the specialists in science, math, technology, and design engineering; as well as the students. Our core group at present is comprised of the classroom teachers, the associate teachers, the science specialists, the science associate/garden teacher, and the math specialists, with the others involved opportunistically. One of our future directions is to better align schedules to increase the opportunity for involving the other specialist teachers. In addition, the strong support of our elementary school division head and our executive director are vital. Many of our other faculty members are involved to one degree or another in our elementary STEM program, including our art and music teachers. Also, our whole staff is invariably helpful when needed during a project. Our indefatigable operations staff, in particular, goes out of the way for such notable events as the Behind the Scenes Infrastructure tours of our building and grounds, including the Pipe tour and the Electricity tour, as well as helping to clear nature trails, providing freshly mown grass for compost, and help in managing tools in our building projects. All of the students at the elementary level are involved in these STEM investigations, with some grade levels more deeply involved and with more experience.

## National Goals

Our STEM program in the elementary school is based and focused on investigations, projects, and Reggio-style studies, where students have the opportunity to observe, explore, hypothesize, test, question, and discuss the world around them to clarify their understanding and construct their own theories. These hands-on investigations proceed from student interests and curiosity, using inquiry methods and student-generated questions. Thus, our curriculum emerges from student interest, though it is molded by the pedagogical and content knowledge of our teachers, and informed by the national and state standards. In practice this means that while we give wide scope to pursuing questions and interests arising from students spontaneously, including allowing for some time to explore less traveled by or less fruitful pathways, we do use our hard-won knowledge of our students and our disciplines to guide, to suggest, and to facilitate investigation of avenues rich in possibilities and opportunities. Our division head calls this a "negotiated curriculum," since it honors the curiosities of our students while leveraging our skills and experience as teachers.

In addition, the *Next Generation Science Standards* (*NGSS*; NGSS Lead States 2013) serves as a beacon, helping us keep perspective on the longer journey into disciplinary mastery, indicating the overall directions and a map of the intellectual terrain without specifying step-by-step, turn-by-turn directions along the road. Our investigations and projects are a natural fit to the spirit and the letter of the *NGSS* and *A Framework for K–12 Science Education* (the *Framework*; NRC 2012). For instance, the *NGSS* stresses the horizontal integration of the science curriculum particularly with math, engineering, and technology courses, as well as with language and literacy skills, in addition to specifying vertical integration of STEM content and skill development.

For science and math, we have defined both content (disciplinary core ideas, in *NGSS*) and process skills (science and engineering practices, in *NGSS*) expected by the end of the elementary program and broken that down by grade level. These align with the *Framework* and the *NGSS*. The curriculum broadly outlines the learning activities (investigations, projects, and studies) students engage in at each grade level. The crosscutting concepts highlight the touch points among the STEM disciplines and the areas of natural convergence for partnership on our projects and investigations. The faculty collaborates to rethink the curriculum each year. Many learning activities will carry over but need adapting for this year's students. Some activities need major redesign, while some new activities are created to fill gaps or to respond to student interest. The best way to demonstrate the fit to the *Framework* is to describe an example of an inquiry STEM project and call out the linkages to the *NGSS* and the *Framework*. Let's examine a typical pre-Kindergarten (PK) project that results in a series of investigations (Estes and Fucigna 2012).

"Look! Look! The compost is getting hot again and that means the microbes are working!" exclaims Maya as she runs to post the latest temperature reading on the wall calendar. Our year in this classroom regularly begins by starting a compost pile that serves as a focus for classroom research and integrated STEM investigations throughout the year. For the last seven or eight years, we have built a compost pile to recycle the leftovers from the PK snack time. Originally, the idea of composting with students grew out of the interests of the teachers in engaging our students in a responsible habit with many learning possibilities. Our compost pile has become an ongoing community project, a link with families, an opportunity for practicing social-emotional learning, and a part of the STEM curriculum we call "global stewardship." This compost project is an example of the integrated inquiry projects at the heart of our STEM program and curriculum.

We often begin a study with an essential question intended to engage student interest and indicate the direction of the study: "How do these food scraps I'm holding in my right hand become this compost in my left hand?" As students share their ideas, we document their thoughts and preliminary theories on a Know-Want to Know-Learned (KWL) chart, pointing out that in scientific investigations we look for evidence to test our original idea and often revise what we originally thought we knew (NGSS Nature of Science). This initial discussion generally leads to other key questions: What is compost? What is the process by which food scraps become compost? Is *compost* the same stuff as *soil* or are they different? Have you ever composted before? Can everything be composted? What can and what cannot be put in a compost pile?

During this discussion we elicit their ideas and questions, which will serve to guide the investigations accompanying this ongoing project. This is exactly parallel to the process of discovery and knowledge building, where new knowledge is built on old, and it is presumed that there may be newer and deeper levels yet to be discovered that adults and older students use. It is also parallel to modern conceptions of science and epistemology as building better models, more robust explanations, and discovering most useful principles, rather than arriving at one final truth.

Following the discussion, the next part of the project is actually building the new compost pile. At this point, we are actually restarting our compost pile from last year. We divide up into rotating job teams to prepare the food scraps for composting, to empty out the finished compost from last year's pile, and to gather dead leaves and grass to balance out the carbon-nitrogen ratio of the pile. As we do this, new questions emerge. Our students ask, "Why are we cutting the food scraps into smaller pieces?" "What is this fuzzy gray stuff on these scraps?" "Why didn't this corncob and this big old potato rot?" "What [kind of food or what part of food] is this thing?" "Why do we need dead grass and leaves?" These questions are added to the KWL chart and we invite students to think about how we might go about finding the answers by trying things out, rather than just doing searches online for the answer or asking someone. Certainly, we can do firsthand investigations of some of these questions, we tell our students, and we get some great conversations about what is testable and what is not.

At this point, another part of the *Framework* comes into play, namely Obtaining, Evaluating, and Communicating Information. Real scientists, engineers, and technologists, we remind them, use not only the results of their own firsthand experiments, but they also seek, examine, and build on knowledge created by others. While we work very hard to foster our elementary students' views of themselves as primary investigators and to build their own trust in their own ideas and discoveries, this is a practical time to examine secondary sources for some of their questions that are not easily testable or to make use of multiple lines of evidence. For example, we may compare our findings and ideas with information gathered from a parent who has experience composting or from a website or a book. Depending on where the class is developmentally, it may be good to gather information from multiple sources and compare them. What is the same and what is different? What can we try out and see what we can discover? Young children are capable of critical thought and analysis, as long as the questions and the investigation are framed appropriately and the ethic of basing our conclusions on the best evidence we have available (Engaging in Argument from Evidence) is honored.

Also, as we build the new pile, a fourth team works with magnifiers and simple, sturdy microscopes, examining compost from the old pile to see how decayed matter looks as it decomposes. Students usually find some still-recognizable bits, such as eggshell fragments. The students also often see macroinvertebrate decomposers such as isopods, earthworms, and various insects. The invertebrates, the molds, and the decaying matter fascinate elementary students. As they observe, we discuss what we see, document their comments, questions, and ideas, and list the critters they encounter. They also document for themselves by drawing what they see. At some point, we may also have them take digital pictures, but we first want to sharpen their observation skills by drawing and developing their self-confidence as observers (Science and Engineering Practices). In addition, their drawings comprise part of the formative and embedded assessment discussed later. Their own drawings provide a basis for deeper discussions of what they saw and didn't see, than do their photos.

Monitoring the progress of the compost over the course of the year clearly develops a firsthand understanding of decay and decomposition, which advances the *Framework*'s disciplinary core ideas life science section concerning ecosystems: Interactions, Energy, and Dynamics. How does matter change as it decomposes? Does it look or smell differently? How does its structure change? How long does it take for matter to decompose? Do different types of matter decompose at different rates? If so, what differences seem to be important? Are there any factors that affect the rate of decomposition and, if so, what are the patterns? Does the amount of water in the pile? How about the position of the matter in the pile? Does it make a difference how often we turn the pile? Does the temperature of the pile make a difference? We also observe and learn about invertebrates such as isopods, slugs, and earthworms. What do they eat? What are their habitats and their body structures? How do these structures seem to help the organisms? What can we observe about life cycles? Are large and small versions of the same type of critter or adults and young or individual variation? Are there dead critters?

These understandings naturally link with life and Earth science standards. This early, hands-on inquiry approach to life science is the type of project encouraged by the *Framework*. For example, the composting process clearly involves the crosscutting concepts of Patterns, Cause and Effect, Systems, Energy and Matter, and Stability and Change. It touches on all four of the disciplinary areas: physical science; life sciences; Earth science; and engineering, technology, and the applications of science. Our investigations help us guide our students through the scientific and engineering practices of Asking Questions and Defining Problems; Planning and Carrying Out Investigations; Analyzing and Interpreting Data; Using Mathematics; Constructing Explanations and Designing Solutions; Engaging in Argument From Evidence; and Obtaining, Evaluating, and Communicating Information.

## Major Features

Our investigations are the heart of our early childhood and elementary STEM program, as it is in the active use of head and hands that people learn best; skills and knowledge are most effectively learned and extended in context and in application. We make these inquiries as authentic as possible and root them in student questions and interests. The curriculum is planned from top down with alignment to national standards, as well our own judgment, and it is then built from bottom up based on the investigations. The curriculum is also informed by reference to the Atlas of Science Literacy's conceptual strand maps (AAAS 2001, 2007); we create a framework from our goals and standards and we flesh out this framework with rich, inquiry-based investigations.

We use a mix of some of our own curriculum created over the years, as well as some of the well-designed and tested guided-inquiry curricula created by national development centers, though we always adapt and extend these curricula for a better fit for our students and their interests. We're very comfortable borrowing parts and pieces from these curricula to help us fabricate or extend an investigation; for example we may begin with a FOSS unit, such as Magnetism and Electricity as a backbone, but modify and adapt it extensively, adding or dropping lessons and activities, modifying the pacing or the delivery for increasing levels of differentiation or independent work.

Several other major features are integral to the success of our program. Although we do not use textbooks, we do some reading from well-written trade books that tie to our investigations, as well as short, timely, and appropriate articles from a variety of sources such as science magazines or websites. We have also long used class discussion for assessing general knowledge prior to a lab, as well as for summarizing and reviewing the key understandings from a lab lesson. In recent years, we have been spending more time on science discussions focused on making meaning, consolidating theories, and sharing ideas. Writing for analysis and reflection has always been integral to our program, and we are extending this by partnering more and more with the language arts strand of the homeroom curriculum in our shared concerns with analytical reading, writing, speaking, and listening. The *Common Core State Standards* support this approach, and literacy is one area we have grown and will continue to grow in our STEM program.

Integration of the curriculum is an important element of our program. Our projects call on skills and knowledge from all the STEM disciplines and most often include technology challenges. For example, the second-grade electricity unit culminates in students using what they have learned about electricity along with their own creativity to design and then build a functioning electrical switch. The big idea is to foster student self-efficacy in conducting investigations, as well as investigation skills. At the elementary level, particularly, this is more important than specific chunks of content, as the new standards make clear. Over the years, we have been increasing the integration of the individual STEM disciplines, using investigations, studies, and projects as the focal point. It is important to us that integration be organic and authentic, rather than forced or *pro forma*. Therefore, we proceed deliberately and seek opportunities to work together. We are prepared for the curricula to run in parallel, joined only at the top level of outcomes, rather than contrive artificial projects or disregard student interests. Authentic integration opportunities may come to fruition slowly, but the results are generally successful. The net result is more like punctuated equilibrium, evolution by jumps and plateaus, rather than linear progress.

These curricular projects, generally completed by pairs or small groups, promote teamwork and shared understanding. We consider taking the time to work through the inevitable tensions and misunderstandings inherent in teamwork to be a vital part of the process. Applying social-emotional learning (SEL) skills, which are part of our school's cultural DNA, as well as our curriculum, is a key 21st century competency. Social-emotional learning develops capabilities such as self-awareness, self-management, social awareness, relationship skills, and responsible decision making. Where possible we like the work of individual teams to come together and integrate into a larger work. For example, building our compost pile combines the work of several teams, each performing an essential component of the project. Self-management skills and self-regulation are essential to achieve high levels of independence and differentiation.

Design thinking and design engineering are also key features of our program. Our school has been a pioneer in teaching both design thinking and design engineering to elementary and middle school students. Briefly, design thinking is a method for using creative thinking grounded by research into the human context of the situation for solving practical problems to improve the lives of people. Design engineering is the process of building out those designs. Design thinking and design engineering are a natural fit to the STEM curriculum and we try to incorporate a design engineering component in all of our STEM projects. Beyond that, design thinking is a habit of mind and a great toolset for examining real-world problems and devising solutions. For example, last year our kindergarten classes became appalled at the situation of the Great Pacific garbage patch and began their own set of experiments to learn more about the effects of salt water on plastic decomposition. In addition, they used tablets to learn more about people and organizations working on the problem and undertook their own trash collection efforts. This cohort of young students already feels empowered to tackle large global problems with both analysis and small-but-meaningful local action.

Finding the time for emergent investigations and projects can often feel like a huge barrier. Traditional units have timelines attached to them that may often be fairly rigid. Investigations have their own natural lifecycle. They begin with students wondering about something that generates questions. Discussions bring the ideas and questions into focus. Building on what students already know, we plan tests and projects to discover what we don't know. This generally entails making or building things to help us with the investigation. As some questions are answered and shared in discussion, other questions naturally arise and some of these will be investigated as well. At some point, enough of the questions have been answered and the curiosity of the students begins to turn in different directions. This is a logical and organic point to wind up this particular investigation, summarize our learning, and begin to sniff out the next hot topic.

Since we are more concerned with depth and developing the ability of our students to solve complex, unstructured, and novel problems, we don't worry excessively about how long an investigation will take. We are comfortable adding class periods or lessons to a unit or an investigation, as long as it is accomplishing our big objectives of fostering student inquiry and learning. Integrating disciplines and skill areas also help us save time by "multitasking"; for example, second graders write a poem about their electricity project and communicate their ideas about the project, as well as demonstrate their literary skills and communicate their feelings. The same piece of work can be assessed by the science teacher for science content and by the language arts teacher for language skills. Students can develop their science and engineering practices and their process skills with nearly any meaningful investigation. We have also learned that it is generally possible to find ways to accomplish our major curricular goals for the core ideas within the context of almost any sufficiently rich investigation. We also make room for surprises, serendipity, and special opportunities.

To a large extent, we do build on the expressed interests of students in developing projects, as is the ideal of an emergent curriculum. We gather these interests not only from class discussion and the resulting documentation, but also from observing the students at play, from seeing what our students notice when we take an orientation and observation walk early in the year, from listening to what they talk about during transition times, and from asking them to tell us what interests them. We look at this information and try to find patterns and clusters. Sometimes there are lots of different ideas in a class with no particular topic standing out; sometimes there is a clear focal point, such as learning more chemistry. To the extent we can, we attempt to work with these points of interest and see how we can turn this into a study or investigation.

One example of this was the "Wheelie" investigation. A couple of years ago a growing group of students were making cars out of Lego blocks they called "wheelies" and racing them. They sometimes argued about whose wheelie was the best. This was easy to transform into an investigation in the physical sciences (Motion and Stability: Forces and Interactions and Energy). Did the height of the ramp affect how far a wheelie would go? Did height affect how far a wheelie would push a standard block at the end of the ramp (measure of force)? Did a longer but lower ramp give a further ride than a shorter higher ramp? How long did it take for a wheelie to travel a ramp? This is also clearly an engineering, technology, and applications project as well. What design traveled farthest? Which design traveled a standard distance in the shortest time?

This group also got interested in the question of whether three wheels were better than four wheels, which naturally lead to discussions of multiple variables and the need for controls and fair comparisons. These sessions often turned into sharing and collaborating on building better wheelies, which also advanced our goals for social-emotional learning (SEL) and this type of collaboration is also frequently mentioned as a goal for STEM education. While the wheelie project was a good example of a purely emergent student project, our compost project is an example of where teachers initiate an overarching activity but follow student interest within the large project.

This being said, not every student interest lends itself to a study, as it may be too abstract or require too much telling students information. Alternatively, it may be too difficult, too dangerous, or too short-lived to begin, to gather materials, to have meaningful investigations, and to discuss them. At the same time, we really don't force-fit an existing "traditional" unit or a personal favorite, such as magnets and magnetism, when there is not enough student interest to engender true curiosity. This is obviously easier in the early grades. On the other hand, the *Framework* and the *NGSS* are flexible about the specific curriculum and curricular materials used as long as the core ideas, crosscutting concepts, and the scientific and engineering practices are taught.

## Evidence for Success

There are several levels to evaluating a STEM program, including student learning, teaching methods, content and inquiry standards, program standards, and assessment philosophy and methods. Historically, assessing these areas is difficult as typically the most complex knowledge levels, the highest levels of skill, and long-term outcomes are the most difficult to evaluate. It is also more difficult to measure processes, such as teaching methods, than it is to measure outcomes, such as student performance on standardized tests. It is also important to keep in mind the differences between *formative* assessment that informs instruction during the process of learning and *summative* assessment that attempts to measure the final result of a set of instruction.

At our school, we use many ways to assess student learning in STEM investigations and projects. We use these multiple measures of learning formatively during the course of an investigation to assess the level of student and class understanding, to modify the instructional strategy or even the entire direction of the study, and to diagnose student progress on key knowledge and skill areas. These ongoing formative assessments, along with assessments of the final project or outcomes of the investigation, contribute to the qualitative narrative assessments we complete for each student as part of the school's overall evaluation process. Parents and students receive these narrative assessments at the end of each evaluation period, typically twice a year at the early childhood and elementary level.

These summative, qualitative narrative evaluations are quite comprehensive and tailored for each student. Typically, for a given assessment period, we may track a standard set of science and engineering practices, such as *recording all the data from an experiment* or *coming up with a test plan* to find the answer to a question. In addition, we would assess the level of understanding of key concepts for our investigations during the period, such as *knowing how to classify a new organism as an insect or not from defining characteristics* and *knowing how to separate soil into its component parts using water*.

The goal in the evaluation is to give a quick, realistic picture of where an individual student is on the continuum of learning for that grade as well as where the student should focus his or her efforts. For example, the report of one student may emphasize improvement on elaborating and detailing her observation of a specimen, while another might focus more on the opportunity of another student to improve his learning by engaging more fully with his lab partners and cooperating better to carry out lab procedures.

The big idea is to help students and their parents view learning as a long-term process of continual improvement, where progress may be uneven and opportunistic, rather than as a series of short-term "hoops" to be jumped though, but never used again. We want to focus on fewer and bigger ideas and on processes and practices that will carry a student from pre-Kindergarten through graduate school and into life as either a STEM professional or as a citizen who is able to evaluate and apply STEM knowledge. In addition to narratives from the classroom teacher, the student also receives narrative reports from all the other specialistS for a total of four or five typed pages, excluding course and curriculum description. From the STEM team in our school, the math teacher, the science teacher, and the technology teacher contribute narratives for the overall evaluation packet.

Currently, we provide these comprehensive narrative evaluations twice a year. These are supplemented with e-mails about specific projects or other student work. We address any significant skill deficiencies, learning issues, or behavioral concerns immediately with phone calls, e-mails, conferences, or some combination of all of these. This blend of immediate attention to serious or urgent issues, coupled with comprehensive summative evaluations of the semester's work seems to work well for our community.

One of our primary formative assessment tools is class discussion and individual dialogue about our activities. A main reason for assessment is to provide the feedback that is essential for learning, and class discussions enable very quick and immediate feedback about learning, as do individual conversations. As is traditional with the Reggio Emilia approach to making learning visible, we document extensively the class discussion, small group interactions, and individual dialogues at the PK level. This documentation, in the form of transcripts, journals, digital photos or movies, records as much and as accurately as possible the exact words and language of the students and of their teachers during the inquiry (Constructing Explanations, Arguments from Evidence). This provides the opportunity to review the conversation, examine the flow of the discussion, see what needs to be reviewed, corrected, revisited, or where next to direct the focus of learning. The Reggio Emilia philosophy is that students, even very young children, are active participants in their learning and an integral part of the collaborative partnership along with their parents and teachers in their learning. This documentation, along with serving as a type of embedded, formative assessment, and a way to review learning with students, also helps communicate with and involve parents in the life of the classroom through newsletters, blogs, and wikis. Teaching aides or parent volunteers can be a great help in collecting this type of data, leading to further and possibly deeper involvement. In addition, these data, these actual artifacts of the learning process, communicate very directly and deeply about their children's schooling and what they are learning.

We also use observation extensively throughout the elementary STEM program. During investigations, I carry around a blank tablet or a form with a few key evaluation items to observe my students using their skills and making conceptual connections in the middle of their investigations, which is one of the absolutely best ways to assess development. This allows me to focus and comment on what seems most important at the moment. I also have an internalized set of processes and practices I routinely examine for each student. I may put one or two skills and one or two concepts to focus on during a particular class on an index card for reference. After the class, I transfer my notes on a particular student to my computer log. As the semester progresses, I may notice that I need more information about what a particular student understands about magnetism, for example, and then focus more on observing and asking that student about magnetism in the next class.

Some teachers use an extensive checklist of observable objectives for observation. This allows them to select which items to evaluate opportunistically. Naturally, it is not possible to observe every student on every skill in every class, nor would there be the time to assimilate all observations even if it were possible to capture it all. Among the instructional decisions we need to make are when to observe and when to coach, who to observe, and what and how much to document the language of the students, as well as when and how much observe and note for which students. What makes this easier is having multiple STEM teachers observing on a target set of processes and concepts over time; this is yet another argument for collaboration among teachers.

Another way to make observations is to take photos and videos of investigations in process. They create visible traces of student learning and you can go back to them whenever you need to. Sometimes is it just easier to take one quick photo to help you recall what was happening in the classroom, rather than write a 1,000-word observational note to yourself. We have also found it helpful to students to view a set of photos taken from an investigation as an aid to their recall; this is especially true for younger students, who would not be expected to take notes. A short video or a slide show can serve as a springboard for discussion. For example, we can view a set of slides together and ask "Maria, tell us about what you were doing right then," or "Jackson, what were you thinking when you saw the liquid turn pink?"

One great way to involve students is to designate a student to take photos of the class investigation. You won't get exactly the photos you would have taken, but you will get a different perspective, and what someone chooses to photograph gives you an assessable insight into his or her thinking. This could be coupled with having students make entries into a class blog about their investigations as a way to incorporate both application of technology with literacy skills and to give students another avenue to "voice" their ideas.

In addition to giving good information about teacher learning, a video may yield helpful data to teachers about their teaching, which is why videos are included in the National Board Certification process. Along with seeing what students were doing while reviewing a video, I also sometimes note with chagrin when I talked too much or jumped in too early when a student was explaining something.

While process assessment is very important, so is assessing learning outcomes. Student work of all types is a rich vein for assessment, including journals; drawings; labeled diagrams; short-response questions; physical models; work shown in math problem solving; physical artifacts constructed by the students, such as a thermometer; graphs of experimental data; audio recordings of explanations; exit cards; or explanatory computer graphics. Written short-response questions and labeled diagrams give a quick check of specific content knowledge, whereas journals, drawings, physical models, and other constructed-response methods yield insight into larger conceptual structures. One STEM colleague has had great success in using audio recordings by giving students who cannot or will not write in depth a recording device and asking them to describe in their own words what they did, what they saw, and what they found out.

Using student drawings, physical models, and graphs as the basis for documented dialogues greatly increases the power of these assessments and their value for formative feedback as well as summative evaluations. One of the single quickest ways to understand someone else's experiment is to look at a graph of the data; routinely analyzing graphs helps students construct their own graphs and it also helps them understand the work of a classmate. One great feature of having students build actual devices and programs is that they give authentic, real-world feedback to the student; if a lamp won't light or a computer program won't run, then there's more work to be done.

Student self-assessments have long been part of our overall approach to evaluating students. Properly viewed, evaluation is an integral part of the learning process and the student is arguably the most important client of any evaluation process, much like a medical patient is the most important client of medical testing. Teachers are vital to helping students understand, interpret, and formulate learning plans based on educational evaluations, just as physicians serve those roles for patients in designing treatment procedures based on the results of medical testing.

One method we use after we are well into an investigation is to ask students to review a set of work and pick out one or two pieces of their best work, one or two pieces that show growth in what they know, and one or two pieces they think could be improved. The most valuable part of this is the discussion with the student about how he or she thinks about personal work. It quickly becomes clear whether we have communicated to the student what the learning goals are and what constitutes good work. Well-designed rubrics help with this, as do providing examples of work at varying degrees of proficiency.

Making an effective rubric forces me as a teacher to think deeply about what the learning goals are and what is evidence that a goal has been met. In providing examples of good work, I need to show a range of different types of acceptable work, lest students overgeneralize and fixate on "one right answer, one right way." Also, I find myself paying more attention to teaching metacognitive strategies helping students think about what and how they are learning; at the same time, my students are teaching me more about the efficacy of different teaching strategies and methods.

Standardized tests are another way to assess what students know and can do, though properly interpreting the results of these tests is fraught with difficulty. Our early childhood and early elementary students do not take standardized tests. Our school begins standardized testing in the third grade and we use the Comprehensive Testing Program (CTP4), which assesses student achievement in language arts and mathematics. When normed to other independent schools, our students' median scores are consistently in the top-tenth percentile, and by eighth grade, the median scores are in the top-two percentile.

Naturally there are many caveats to interpreting these results in terms of STEM achievement. First and foremost, is that there are many reasons for the success of our students on these tests in addition to the quality of their STEM instruction in previous years, including the fact that we admit high potential students who have enriched home environments and life experiences, as well as the work of the many non-STEM teachers they learn from. Also, the CTP4 was not designed as a test of STEM skills or content knowledge, but draws broadly from content indicated by national standards in language arts and mathematics. No part of the written CTP4 addresses technology learning or engineering, although the online version of the CTP4, which we do not use, does have a science section. Even with these important qualifications, we do believe that our strong early STEM program supports the achievement of our students on these standardized tests.

As an independent school, we undergo a rigorous accreditation process, which includes an extensive self-study, a visit by an evaluation team, and a detailed evaluation report to the board. These occur about every five years and I have participated in the last two. While there is no specific STEM section of the report, the overall results for the school and the STEM disciplines were gratifying, as well as enlightening. Part of the value of these reports is that the evaluation team makes many high-level suggestions for program improvement to a school. Their suggestions, reflected in the Next Steps section of this chapter among other ideas, urged devoting more time and effort to collaboration among teachers and greater levels of integration.

## Next Steps

Our elementary STEM program is continuing to evolve and develop. One main goal is to extend and to expand our integrated STEM investigations approach more broadly throughout our elementary division and the whole school. In our elementary division there are many innovative and integrated curricula that are not fully coordinated or aligned with STEM efforts. In a school like ours, where much of the innovation happens opportunistically as one or two teachers get an idea and begin to work together, the key to taking successful ideas to the next level is sharing and generating interest so other teachers want to become involved with the program. Our faculty has great latitude for curriculum and there are few top-down curricular mandates. To advance our elementary STEM program, our next steps are expanding the number of faculty involved with cross-disciplinary integrated curriculum, along with continuing to seek additional opportunities to collaborate among the STEM staff.

Given the great autonomy for instructional decisions granted to each faculty member and to grade-level teams, the connection of the STEM program to the classrooms varies by grade level. Our STEM program works best when classroom teachers and STEM teachers work collaboratively on common learning goals and, when possible, on shared projects. Building in joint planning time greatly facilitates this. When teachers share the learning goals for a given grade level, there is nearly always very significant overlap and that provides synergies for collaboration.

The school day is too short to have time for each area separately teaching slightly different versions of the fundamental skills of reading, writing, computing, problem solving, critical thinking, creative thinking, and effective speaking and listening. By aligning objectives and creating a set of shared goals and shared terminology, ineffectual duplication is reduced and more energy is directed at achieving these goals. There are also gains from working on a set of shared integrated projects, rather than separate projects in the classroom and in each of the disciplinary areas, where these projects truly work for all the areas and faculty. We strongly believe in integration, but also strongly believe in the integrity of each discipline, respect differing ideas and approaches, and value a multiplicity of methods, as no one learning method works best for everyone, teacher or student.

Literacy skills are a great opportunity for integrating and collaborating with the classrooms. There is great interest in linking literacy with STEM subjects and clear advantages to doing so (Worth et al. 2009). We began this initiative with second grade. The classroom teachers and the STEM teachers met and agreed on the overall goals and the broad outlines of the plan.

The new Seeds of Science/Roots of Reading curriculum is a natural fit with this work and provides a great platform for us. We picked Our Soil Habitats as a joint area of investigation as a starting point. The classroom theme connects to this location-based study and will extend to the early Americans living in this area who depended on the local ecosystem. In our modern world we still depend on the local ecosystem. Caring for our interconnected global network of ecosystems is a clear priority and this curricular strand links to our concern for Global Stewardship. For science, this focus works well with the *Framework* goals for second grade life and Earth sciences and with many excellent inquiry investigations. Our math specialists are excited about these investigations and see no end to the possibilities for applying math thinking and problem solving.

Our technology teacher found a simple web application we use that allows the second graders to observe, identify, and classify local plants and animals; to learn more about them through print and online media; to write up their observations and findings; and to share their work with others though a blog. We continue to find many ways to build and extend on this. Likewise, our design engineering team has long cooperated with science and the classroom on many projects. They are excited about exploring ways to teach and to apply design thinking and design engineering skills with the second graders on this investigation.

We are also busy finding additional trade books to accompany the Seed/Roots materials to meet the needs of a wide range of reading skill. One of the best means to increase interest and involvement in STEM has been to share the successes of the student projects with the larger community, including other faculty and parents. Using this evidence of success paves the way for greater and broader involvement, as well as deeper school commitment.

## Conclusions

For our school, integrating the STEM disciplines and integrating STEM work with the classrooms and the other specialty areas is a rewarding effort and a never-ending process. We see the benefits of common goals, integrated projects, and shared assessments. In a dynamic organization, changes in people and structures mean that we must never take what we have accomplished for granted, but must remain open to reinventing elements of our STEM program each year.

What generalities can we make from our experience? Our experience with STEM could be encouraging to other schools; this is not because we have made no mistakes, but because we keep trying and good outcomes result from that effort. We learned that while a fully integrated and fully deployed STEM program is desirable, great results could grow from two or more teachers with a powerful idea for helping students working together. Small changes and subtle shifts in curriculum and methods can make significant results and build the basis for further innovation. As Winston Churchill said, "While waiting to do everything, do *something*!"

Also, with STEM programs, as with all innovations, there will never be a final ending point where it is all done. The ongoing discussions with colleagues may at times seem never-ending and repetitive, but those conversations and even disagreements may also result in new insights;

we gain both a higher ground of integration, as well as a deeper grounding in our teaching practice. Finally, the efforts invested in in students are never wasted and the efforts to build and expand our STEM program benefit our students enormously.

## References

American Association for the Advancement of Science (AAAS). 2001. *Atlas of science literacy, volume 1.* Washington, DC: AAAS.

American Association for the Advancement of Science (AAAS). 2007. *Atlas of science literacy, volume 2.* Washington, DC: AAAS.

Edwards, C., L. Gandini, and G. Forman, eds. 2011. *The hundred languages of children: The Reggio Emilia experience in transformation.* 3rd ed. Westport, CT: Praeger.

Estes, F., and C. Fucigna. 2012. Compost: The rot thing for our Earth. *Science & Children* 50 (6): 47–51.

National Research Council (NRC). 2000. *Inquiry and the national science education standards: A guide for teaching and learning*. Washington, DC: National Academies Press.

National Research Council (NRC). 2012. *A framework for K–12 science education: Practices, crosscutting concepts, and core ideas.* Washington, DC: National Academies Press.

NGSS Lead States. 2013. *Next Generation Science Standards: For states by states*. Washington, DC: National Academies Press. *www.nextgenscience.org/next-generation-science-standards*

Worth, K., J. Winokur, S. Crissman, M. Heller-Winokur, and M. Davis. 2009. *The essentials of science and literacy: A guide for teachers*. Portsmouth, NH: Heinemann.

# A State STEM Initiative Takes Root, Blossoms

*Jeffrey D. Weld, Disa Lubker Cornish, and Erin Heiden*
*University of Northern Iowa*
*Cedar Falls, IA*

*Mari Kemis*
*Iowa State University*
*Ames, IA*

## Setting

Iowa's STEM story begins in 2008 when the three public universities, Iowa State University, the University of Iowa, and the University of Northern Iowa, established a collaborative partnership dubbed the Iowa Mathematics and Science Education Partnership. The partnership was directed by this chapter's lead author Jeff Weld. Despite the organization's success in reaching the goals of bolstered math/science teacher production, heightened faculty engagement with K–12, and invigorated networking across institutions, it became clear to leaders that a broader effort was needed for Iowa—one that would unite all institutions of higher learning with K–12, as well as with nonprofits, businesses, and government. Moreover, the effort needed a more encompassing vision than that of two academic disciplines. It needed a STEM (science, technology, engineering, mathematics) focus. And it needed to be measurable through well-defined benchmarks and metrics that would steer programming. Fortunately for Iowa STEM, ground-floor evaluative work on grant-funded programming of the math/science partnership was already taking place through coauthors Disa Lubker Cornish and Mari Kemis, who established benchmarks and metrics for Iowa's statewide STEM initiative launched in 2011. The evaluation is a collaboration across research centers at Iowa's three Regent's institutions, including the University of Northern Iowa's Center for Social and Behavioral Research (CSBR), Iowa State University's Research Institute for Studies in Education (RISE), and the University of Iowa's Center for Evaluation and Assessment (CEA) and Iowa Testing Programs (ITP). Dr. Cornish was succeeded in her STEM evaluation position in 2013 by coauthor Erin Heiden, who has overseen full implementation of the suite of monitoring and assessment activities steering Iowa STEM today. Together, Weld, Cornish, Heiden, and Kemis bring an historical and chronological, a practical and theoretical, and a detailed, as well as overarching perspective to the telling of Iowa's STEM story.

## What Is The Big Deal About STEM?

"STEM prepares innovators!" "STEM anchors the economy!" The cacophony around STEM today has succeeded in engaging all the right stakeholders for change. Our climate of *STEM-mania* (Sanders 2009) is a welcome clamor—a leverageable banner under which long-simmering science and mathematics education reform can finally penetrate all corners. But harnessing all the energy and good intention for productive work is the challenge of STEM leaders at the local, regional, state, and national levels of late. It begins with a common definition of STEM (notice that the need for tagging on "education" has become a redundancy, given that it is only in the context of education that the acronym has come to be used). The *Next Generation Science Standards* unite the disciplines of science and engineering like never before, and in Iowa, a consensus definition for STEM has arisen:

> *an interdisciplinary approach to learning where rigorous academic concepts are coupled with real-world lessons as students apply science, technology, engineering and mathematics in contexts that make connections between school, community, work and the global enterprise enabling the development of STEM literacy and with it the ability to compete in the new economy* (Tsupros, Kohler, and Hallinen 2009).

STEM literacy for all has in some circles become secondary to STEM excellence for some as the main driver of reform efforts. The dichotomy is framed as "all STEM for some" versus "some STEM for all" by advocates, the former inclined to pour our finite collective resources into developing the skills and knowledge of those who show a propensity for STEM (Atkinson 2012). Alternatively, science education authority Rodger Bybee, who implored us all to "move beyond the slogan and make STEM literacy for all students an educational priority," argues that STEM literacy should be a fundamental purpose of school programs in advancing STEM education (Bybee 2010). The prospect is compelling. It is a delightful exercise on the imagination to envision an emerging American generation possessing the habits of mind that spring from a sound STEM education. Picture these science and engineering practices, called for in *A Framework K–12 Science Education* and the *Next Generation Science Standards*, embodied in our children:

- Asking questions and defining problems
- Developing and using models
- Planning and carrying out investigations
- Analyzing and interpreting data
- Using mathematics and computational thinking
- Constructing explanations
- Designing solutions
- Engaging in arguments from evidence
- Obtaining, evaluating, and communicating information (Mayes and Koballa 2012)

To realize that promise, we need to progress beyond a common definition to strategies, tactics, and actions informed by research in a perpetual forward feedback cycle. What hangs in the balance for this generation of learners is personal and collective livelihood. What stands in the way are a lack of universally adopted and rigorous mathematics, science and technology standards (excitingly queuing the *Common Core State Standards* and *NGSS*), of qualified instructors, of postsecondary STEM preparation, of motivation and interest among our youth, and a lack of alignment of what's taught in postsecondary STEM and what is needed to thrive in STEM entry-level occupations (Thomasian 2011). That is why economic and social realities have become the rallying point for reform to STEM nationally and in Iowa (mirroring the condition of many other states):

- The top 10 bachelor-degree majors with the highest median earnings are all in STEM fields (Thomasian 2011).
- At all levels of educational attainment, STEM job holders earn 11% higher wages compared with their same-degree counterparts in other jobs (Thomasian 2011).
- STEM workers drive our state's innovation and competitiveness by generating new ideas, new companies, and new industries (Carnavale, Smith, and Melton 2011).
- Over the past 10 years, growth in STEM jobs in Iowa was three times as fast as growth in non-STEM related jobs (U.S. Department of Commerce 2011).
- Iowa's economy is anchored to STEM-based industry sectors, including advanced manufacturing, information technology, and biosciences, which depend on a steady supply of talent (Battelle 2011).
- Growth in Iowa's STEM workforce is projected to be 16% (nearly 58,000 in 2008 to just over 67,000) by 2018 (Carnavale, Smith, and Melton 2011).
- Coupled to the bright career horizons in STEM is an ominous trickle of talent into the field:
- Just 12% of Iowa postsecondary enrollees major in STEM fields (national average is 15%) (ASTRA 2011).
- Just 11% of Iowa's 2011 ACT test-takers scored college-ready in both math and science and expressed an interest in STEM majors (Triplett 2012).
- U.S. STEM degrees represent up to one-third of bachelor's degrees in some regions, but they represent more than half of the first degrees awarded in Japan, China, and Singapore (Thomasian 2011).
- Iowa's ethnic and racial minorities account for 93% of the state's population growth since 2000, but minorities are statistically about half as interested in STEM post-secondary majors (Grey 2006).
- 34% of Iowa eighth graders scored proficient on the math test of the 2009 National Assessment of Educational Progress (Hanushek, Peterson, and Woessmann 2012).
- 35% of Iowa eighth graders scored proficient on the science test of the 2009 National Assessment of Educational Progress (Hanushek, Peterson, and Woessmann 2012).
- 46% of Iowa eighth graders report that they "never or hardly ever" write reports on science projects and 37% report they "never or hardly ever" design a science experiment. (Change the Equation 2011).

So yes, STEM is a big deal. Our economic future as well as the quality of our children's lives relies on sound STEM education, while plentiful indicators point out that we are falling short. The President's Council of Advisors on Science and Technology (PCAST) recently advised that, "As the world becomes increasingly technological ... STEM education will determine whether the United States will remain a leader among nations and whether we will be able to solve immense challenges in such areas as energy, health, environmental protection, and national security" (PCAST 2010). Some see this challenge as our generation's moon shot or Manhattan Project. To couch this watershed moment in terms our kids can appreciate, perhaps the rapper Eminem said it best: "You only get one shot … you better never let it go; this opportunity comes once in a lifetime … yo."

## A State STEM Initiative Takes Root

The fourth graders in New Hampton, Iowa, are tying their math, science, and literacy lessons to a common theme: motion. A day at school may catch them calculating velocity, reducing friction, modeling automobile safety, and presenting technical reports that bring it all together. That is STEM in the classroom as experts define it—integrated, active, real-world problem solving. Learners in this style of education are more creative, collaborative, intelligent, and interested in STEM-based careers (Duschl, Schweingruber, and Shouse 2007). The scene in New Hampton is being replicated across Iowa in 900 locales as part of a cornerstone initiative of the Iowa Governor's STEM Advisory Council: Scale-Up. Its roots extend back to a wintry morning in early 2010 when founding thinkers assembled to re-imagine our state's STEM trajectory.

An eclectic group of professionals spanning K–12 education, informal and nonprofit youth agencies, business and industry, community college and university, and government and advocacy groups developed over the course of 2010 a STEM "roadmap" for Iowa. *The Iowa STEM Education Roadmap: A Strategic Plan for Science, Technology, Engineering, Mathematics Education* detailed where we stand as a state, why STEM should matter to Iowans, where we fall short, who should be at the table, and what we can do to get STEM right. The Roadmap was widely disseminated in early 2011 to leaders in business, government, and education, and then it was presented to the Governor. (*www.iowastem.gov/imsep/sites/default/files/2011stemeducationroadmap_finalrc51.pdf*)

In July of 2011, Iowa's Governor created through Executive Order #74, the Governor's STEM Advisory Council, charged with enacting the Roadmap's seven Targets:

1. Increased interest and performance of Iowa learners in STEM fields;
2. Increased emphasis on STEM fields in Iowa from PreK through 20;
3. More high quality STEM teachers prepared by Iowa's institutions of higher education;
4. An Iowa citizenry that recognizes the importance of STEM in leading productive lives and creating/sustaining a vibrant economy;
5. A national leader in STEM workforce preparation and retention in STEM careers;
6. Wide-scale partnership of Iowa's education systems and private enterprise; and
7. Coordinated, complementary and uniform STEM education opportunities across Iowa.

Similar grassroots STEM movements have unfurled in other states for all of the same reasons—an interstate and international competition for talent. Iowa's STEM Council began by cataloging best practices with the assistance of leadership organizations including the National Governor's Association, the Triangle Coalition for Science and Technology Education, STEMconnector, and the National Science Teachers Association's STEM Coalition. A number of U.S. states have mature

STEM initiatives under way, including Texas' T-STEM (*www.txstem.org*), North Carolina's STEM Learning Network (*www.ncstem.org*), Ohio's STEM Learning Network (*www.osln.org*) and the Massachusetts STEM Network (*www.mass.gov/governor/administration/ltgov/lgcommittee/stem*). In fact, Iowa leaders have relied heavily on guidance and gracious sharing on the part of officers of the Massachusetts STEM initiative, whose influence can be seen throughout the Iowa program.

## Creating an Organization

Forty-six Governor-appointed applicants serve rotating three-year terms on the Iowa Governor's STEM Advisory Council, co-chaired by the Lieutenant Governor and the president and chief executive of Vermeer Corporation, Mary Vermeer Andringa. Council appointees represent industry, education, government, nonprofit, and other sectors of the professional spectrum. An executive committee of 14 makes voting decisions and sets agendas. The Executive Order dictates that directors of state agencies (Workforce Development, Economic Development, Education), presidents of public universities as well as a private university and a community college, executives from industry, and leaders from K–12 comprise the executive committee of the STEM Council. The Council is served by an executive director and administrative staff.

High quality STEM education statewide requires a structure. Intra-Council governance as well as an external skeleton upon which to build was conceived of by thought leaders on the Council and informed by best practices observed elsewhere, notably Massachusetts. Figure 19.1 depicts the current organizational structure of the Council.

**Figure 19.1.** Iowa Governor's STEM Advisory Council and Committee

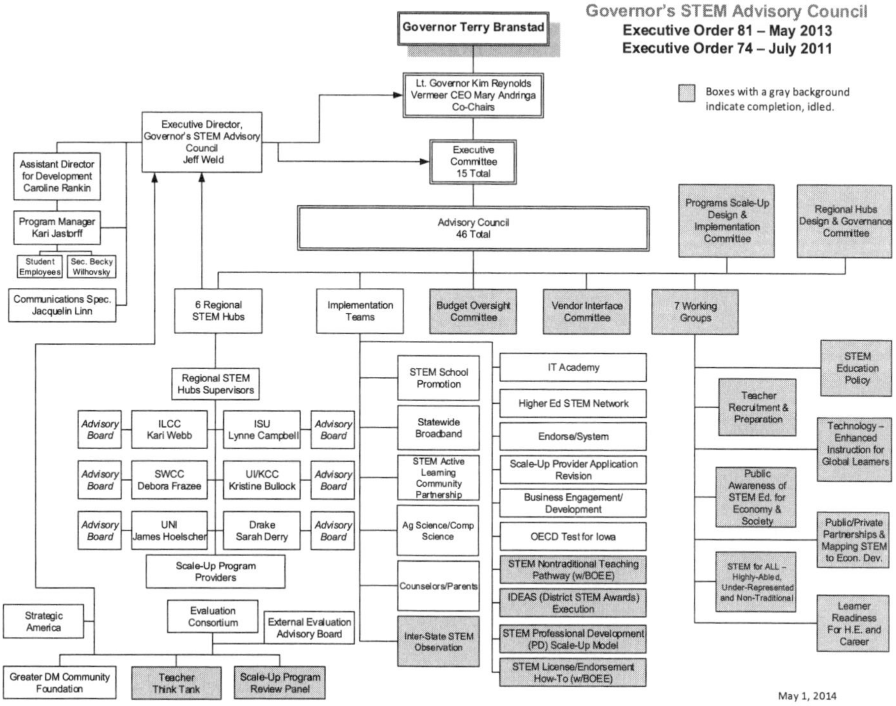

Iowa's Regional STEM Network (Figure 19.2), was created to serve as a skeleton for building capacity for statewide implementation of innovative solutions to come. Geographic rather than demographic (e.g., population centers) regions follow school district borders, dividing the state into roughly equal sixths. Each region is supported by a hub institution that won the right to serve in that capacity through a competitive bid process that required cost-sharing. Each hub employs a regional manager charged with carrying out the mission of the council locally. Managers are advised by six separate advisory boards consisting of representatives of K–12 schools, area education agencies, county extension offices, community colleges, public and private universities, informal science centers, nonprofit organizations, youth agencies, regional businesses, workforce and economic development departments, and local governments. All activities of regional managers and their boards are guided by the Governor's STEM Advisory Council.

**Figure 19.2.** Iowa Statewide Regional STEM Network

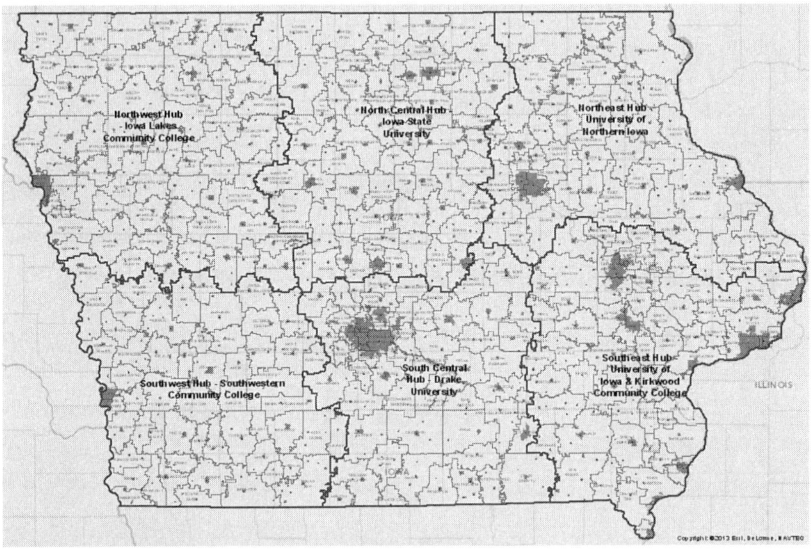


## Implementing Statewide STEM Reform

With roles defined and a structure in place, the Council set about the design and implementation phase. The Council further refined Iowa STEM Roadmap Targets to eight priorities at their first meeting in the fall of 2011:

1. Top Priority—Student Interest and Achievement
2. Technology Enhanced Instruction for Global Learning
3. STEM Teacher Recruitment and Preparation
4. STEM Learner Readiness for Post-Secondary Education and Career
5. STEM Education Policy

6. Public Awareness of the Importance of STEM Education for the Economy and Society
7. Public/Private Partnerships and Mapping STEM Education to Economic Development
8. STEM for All—the Highly-Abled, Under-Represented, and Nontraditional

Subgroups consisting of Council members plus expert outsiders pursued the development of strategic plans around these eight priorities over the course of the 2011–12 academic year and delivered to the executive committee their recommendations in May of 2012. Each group's report is viewable at *www.iowastem.gov/targeted_priorities*.

## Priority 1 on the Fast Track

Iowa's STEM Council members practice what Admiral Hyman Rickover termed *strategic impatience*, in that we know of the problem and of solutions, so why wait? The well-defined challenge around priority 1 is student disinterest and underachievement. Well-known solutions exist in the form of exemplary research-supported programming. Thus a committee was formed with the mandate to deliver to the Council a model that could fast-track the delivery of outstanding STEM programming to youth throughout the state. That model became known as Scale-Up, informed by a small-scale initiative piloted in Massachusetts. The Iowa model was delivered to the Council in April of 2012, recommending the issuance of a call-for-proposals from those who could demonstrate gains in student interest and achievement and increases in diversity of participants, among additional criteria. Of 38 proposals, 12 were selected by an expert panel of reviewers to be recommended for scaling across Iowa. Funds appropriated to STEM by the Iowa legislature drove the scaling process. The exemplary programs were presented to educators across the state who were then invited to apply to bring one to their classrooms or clubs (formal and informal STEM educators are eligible for Scale-Up, including scout troops, after-school clubs, and traditional classroom teachers). Over 900 educators delivered an exemplary STEM program to some 38,000 Iowa children during the 2012–13 academic year. There are computer programming contests, agricultural experiments, wind turbine modeling, family STEM festivals, robotics clubs, and more. And they're happening more often than not in the most STEM-deprived areas of Iowa thanks to the mission of the Governor's STEM Advisory Council to level the field of opportunity for all Iowa youth. The 12 programs of 2012–13 were:

- A World in Motion (AWIM): *www.awim.org*
- Fabulous Resources in Energy Education (FREE): *www.uni.edu/free-ceee*
- FIRST LEGO League: *www.isek.iastate.edu/fll*
- FIRST Tech Challenge: *https://sites.google.com/site/ftciowa*
- HyperStream, Technology Hub for Iowa's Students: *http://hyperstream.org*
- iExploreSTEM: *www.iexplorestem.org*
- KidWind: *http://learn.kidwind.org*
- Project HOPE (Healthcare, Occupations, Preparation and Exploration)
- The CASE for Agriculture Education in Iowa: *www.case4learning.org*
- State Science + Technology Fair of Iowa: *www.sciencefairofiowa.org*

- Partnership for Engineering and Educational Resources for Schools (PEERS): *www.peersprogram.com*
- Corridor STEM Initiative (CSI): *www.corridorstem.org*

A variety of measures were taken to establish the effect of this Scale-Up initiative. They are discussed in the State STEM Yield: Measures of Success section. It is important to note that the effort involved in identifying, marketing, packaging, implementing, and evaluating a program that touches almost 10% of educators and youth across the state in a timeframe of less than a year is a credit to the dozens of professionals practicing *strategic impatience* to improve STEM education for young Iowans.

## Long-Term Strategic Planning

The recommendations that came from first-year working groups on the Council's other seven priorities were condensed by administration office staff into a total of 19 non-overlapping and complementary tactical solutions. The Council's co-chairs, working with the executive committee, further filtered and combined recommendations into those most likely to endure. Sustainable success involves supportive policies, school cultures of innovation, modernized teacher preparatory pathways, stronger public-private partnerships, a more inclusive talent pipeline, and strong public buy-in. Long-term, sustainable fronts for the STEM Council are:

- Promote STEM-focused schools.
- Carry out a STEM public awareness campaign.
- Create the STEM teaching certificate at the secondary and elementary levels.
- Create a blended professional development model for education practitioners.
- Unite the informal network of Iowa to deliver on the promise of out-of-school STEM.
- Build a comprehensive online resource to serve STEM educators.
- Build access to high-speed internet for all schools in the state.
- Forge public-private partnerships that share both the financial and human resource responsibilities of a strong state STEM initiative.

Each of these fronts is being realized over the course of year 2 and beyond and each is being managed by an "implementation team" (purposeful language to differentiate from a study committee or working group). The STEM schools team, for example, after touring STEM academies of other states recently announced winners of a statewide call-for-proposals from schools wishing to rebrand and enact a STEM focus to their practices. Winning schools model practices in STEM pedagogy and curriculum integrating leading edge learning technologies in partnership with local business. The team overseeing a public awareness campaign has launched a youth STEM film festival, and recently awarded a cost-matched contract to a private advertising firm to build out a new brand featuring the tag line "Greatness STEMs From Iowans." A STEM license team is working with the state's teacher certification board to craft a STEM license model for teacher preparatory institutions to adopt.

Coinciding with the rollout of the *Next Generation Science Standards*, an implementation team is developing a universal, blended professional development model to help Iowa's teachers of mathematics, science, and technology to move seamlessly into teaching STEM. Council members representing the state's informal STEM network (zoos, science centers, museums, and so on) are bringing forth a plan for integrating the out-of-school learning assets of STEM into Council goals and activities.

An unprecedented coalition of education leaders and telecommunications executives are creating a plan for delivering high-speed broadband internet access to all of Iowa's schools.

And finally, business leaders and educators make up an implementation team to bring to the Council a blueprint for engaging business-school partnerships across Iowa, forged by regional STEM managers. Now just over two years since the launch of Iowa's STEM initiative, there is much to show, much to grow, and much yet to know.

## A State STEM Initiative Blossoms

Year 1, 2011–12, was the developmental year for Iowa's state STEM initiative. The following accomplishments were put in place over the course of the (unfunded) year:

- Council installed, executive committee named;
- Protocols codified including decision-making, succession, roles;
- Council goals and objectives (*priorities*) identified;
- Short- and long-term strategic plans drafted around each priority by Council working groups;
- Inter-university evaluation consortium named and strategies developed;
- Statewide regional STEM network conceived;
- Fast-track Scale-Up initiative conceived;
- Regional Hubs designated through a competitive bid process;
- The Council convened four times (one a state STEM Summit) and the executive committee nine times; and
- A legislative appropriation request of $4.7 million was authored by the Council to support year 2 plans.

The bulk of groundwork for Iowa's STEM initiative was laid in the winter and spring of 2012 (see Figure 19.3, p. 318).

Year 2, 2012–13, was the "blossom" year for Iowa's STEM initiative. With governance defined, a statewide network built, and tactical solutions to Iowa's STEM challenges delineated, the stage was set for what has become an energetic lift-off for Iowa's moonshot. The following accomplishments are a credit to the more than 300 STEM advocates serving on the council, regional boards, working groups, committees, and implementation teams that together make up Iowa's STEM initiative, as well as to the elected officials who shepherded through a $4.7 million appropriation:

- Regional managers installed at all six Hubs;
- Regional boards installed;
- Exemplary Scale-Up programs identified;

**Figure 19.3.** 2012 Timeline for Events of the Governor's STEM Advisory Council

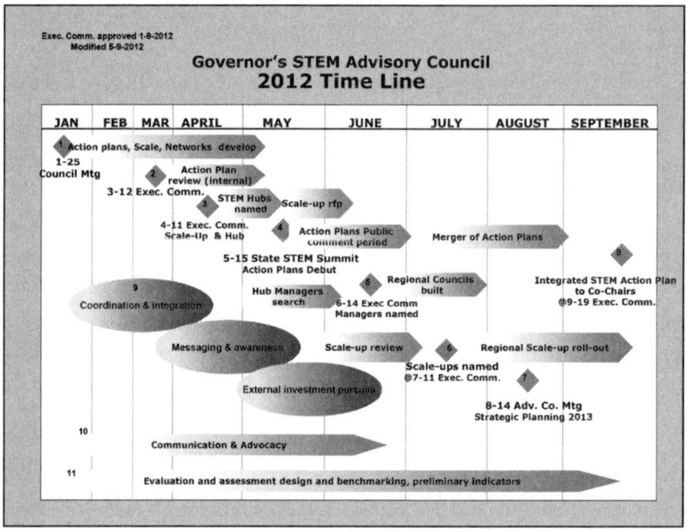


- Interagency contracts, agreements, and memoranda of understanding developed for Council-to-hub-to-educator regulation; and
- Policy for interweaving Council efforts with existent structures including area education agencies, county extension, 4-H, after-school networks.

## State STEM Yield: Measures of Success

Key to the success of any initiative focused on change is a robust and thoughtful plan for evaluation and monitoring. A statewide initiative such as the work of the Iowa Governor's STEM Advisory Council has certain characteristics that were taken into consideration as monitoring systems were developed. First, many separate, yet linked and overlapping, activities were taking place. Second, activities were reaching multiple audiences and levels (local, regional, state, and national). Third, many organizations and agencies were involved, many of which had a wealth of data regarding STEM in Iowa.

The Iowa STEM Monitoring Project (ISMP) was established in 2012 to identify and monitor changes in Iowa STEM on three levels. At its most broad, the project monitors Iowa STEM in the national context by comparing to other state initiatives and data collection efforts. At the state level, the project assembles and tracks indicators of progress toward Advisory Council goals and objectives. Within the statewide STEM initiative, the ISMP tracks the processes and impacts of Scale-Up programs and other regional efforts.

As the project name and purpose implies, monitoring of the Advisory Council activities in Iowa includes tracking national, state, and program data, analyzing data for trends, and observing the STEM landscape in the state in a systematic way. To that end, the ISMP is comprised of four components: (1) Iowa STEM Indicators System (ISIS); (2) Statewide Survey of Public Attitudes

Toward STEM; (3) Statewide Student Interest Inventory; and (4) Regional Scale-Up Program Monitoring. Figure 19.4 shows the Iowa STEM Monitoring Project Infographic. Each ISMP partner has specific areas of responsibility with areas of overlap. Collaboration among ISMP partners was key to the success of the ISMP in the first year.

**Figure 19.4.** Four Components of the ISMP

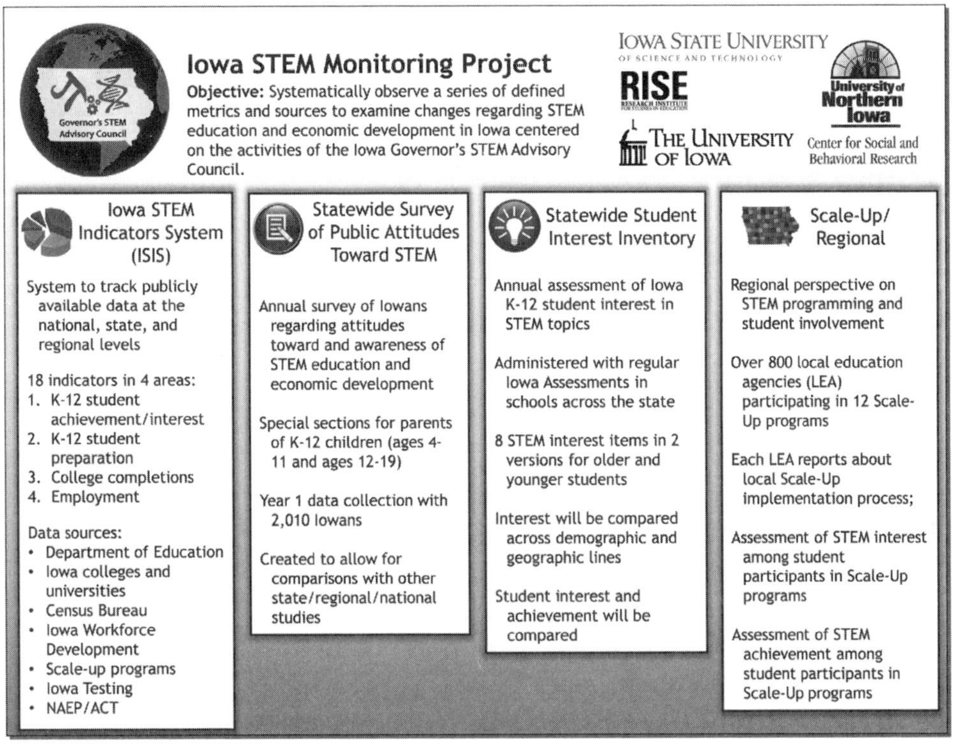


The ISMP was developed within an evaluation framework for the STEM initiative in Iowa (see Figure 19.5, p. 320). This framework included multiple levels of evaluation, additional resources leveraged in support of evaluation, and alignment of evaluation activities with initiative goals and priorities. This evaluation framework for the STEM initiative informed the ISMP that was implemented and is reported here. Selected year 1 results of the ISMP are also reported.

**Figure 19.5.** Iowa STEM Evaluation Framework

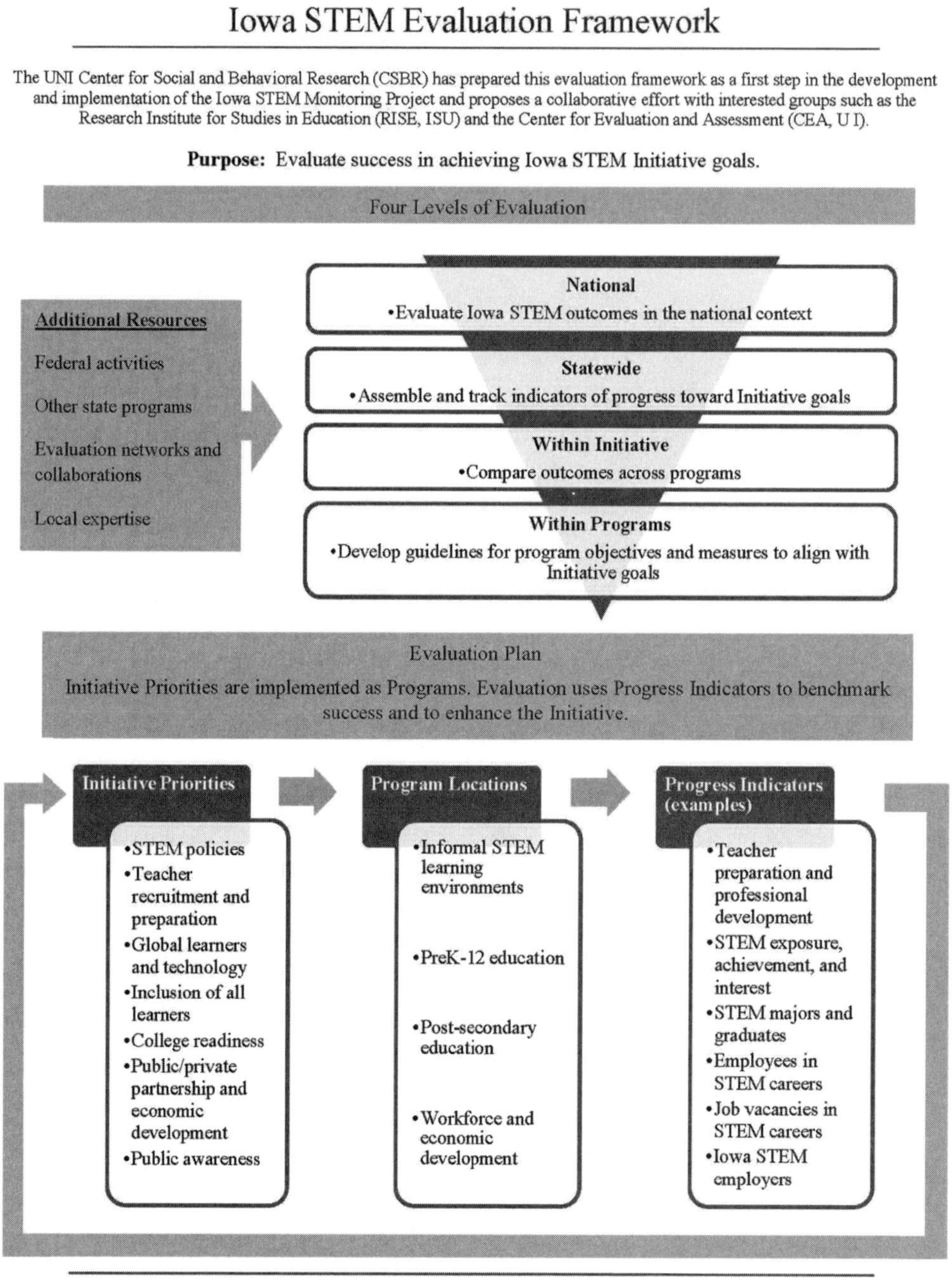
Iowa STEM Evaluation Framework
The UNI Center for Social and Behavioral Research (CSBR) has prepared this evaluation framework as a first step in the development and implementation of the Iowa STEM Monitoring Project and proposes a collaborative effort with interested groups such as the Research Institute for Studies in Education (RISE, ISU) and the Center for Evaluation and Assessment (CEA, U I).
Purpose: Evaluate success in achieving Iowa STEM Initiative goals.
Four Levels of Evaluation
Additional Resources
Federal activities
Other state programs
Evaluation networks and collaborations
Local expertise
National
•Evaluate Iowa STEM outcomes in the national context
Statewide
•Assemble and track indicators of progress toward Initiative goals
Within Initiative
•Compare outcomes across programs
Within Programs
•Develop guidelines for program objectives and measures to align with Initiative goals
Evaluation Plan
Initiative Priorities are implemented as Programs. Evaluation uses Progress Indicators to benchmark success and to enhance the Initiative.
Initiative Priorities
•STEM policies
•Teacher recruitment and preparation
•Global learners and technology
•Inclusion of all learners
•College readiness
•Public/private partnership and economic development
•Public awareness
Program Locations
•Informal STEM learning environments
•PreK-12 education
•Post-secondary education
•Workforce and economic development
Progress Indicators (examples)
•Teacher preparation and professional development
•STEM exposure, achievement, and interest
•STEM majors and graduates
•Employees in STEM careers
•Job vacancies in STEM careers
•Iowa STEM employers

## Statewide Student Interest Inventory

The Statewide Student Interest Inventory is administered using the Iowa Assessments. In Iowa, all students in second grade and above take the test annually. Two versions of the inventory were developed to accommodate grade differences in grades 3–5 and 6–12. In the first year, eight questions were pilot tested, asking about interest in the following STEM subjects (engineering, math, science, computers and technology, and a job that uses skills in STEM) as well as in other subjects (art, English and language arts, and social studies), using slightly different scales in the two versions. In the grades 3–5 version, the scale included "I like it a lot," "it's okay," and "I don't like it very much." In the grades 6–12 version, the scale was "very interested," "somewhat interested," and "not very interested." The data were collected and analyzed by Iowa Testing Programs. The interest inventory was developed in part to serve as a data source for both the Iowa STEM Indicators System and a way to compare students who participate in Scale-Up programs with all students statewide.

Among all students statewide who took the Iowa Assessments, interest in all subjects and STEM careers was highest among elementary students followed by middle school and high school students. Results are consistent with evidence that suggests overall academic motivation and interest in all subjects, including STEM, decreases over time from elementary to high school (Barber and Olsen 2004, Dotterer, McHale, and Crouter 2009, Eccles, Midgley, and Adler 1984).

## Iowa STEM Indicators System (ISIS)

The Iowa STEM Indicators System (ISIS) tracks publicly available data from a variety of sources at the national, state, and regional levels. The purpose of the system is to provide annual benchmarks on a variety of STEM topics in education and economic development by systematically assessing the progress and condition of the state's STEM landscape. ISIS was created to identify and fulfill the need for benchmarks related to a variety of subtopics in the area of STEM education and workforce development. Eighteen indicators were identified, with four primary areas of focus: (1) STEM achievement and interest among K–12 students, (2) STEM preparation of K–12 students, (3) STEM college completions, and (4) STEM employment (see Figure 19.6, p. 322). When possible, these indicators are analyzed to include comparisons across demographic, geographic, and other characteristics. A limitation of ISIS is the inability to report on all indicators at the same time annually. The variability in when data from these sources are collected, analyzed, and released publicly requires continuous tracking and updating. In addition, previously identified indicators may not lend themselves to ongoing surveillance throughout the ISMP after assessing the integrity and applicability of the data in providing useful benchmarks; new indicators may be identified as other data and data sources are identified or become available.

One of the first tasks in designing the system for monitoring STEM data was to define just what constitutes STEM. A review of literature and statewide STEM initiative websites did not result in a commonly used definition for STEM subjects that applies across educational levels, industries, or government agencies. Next, we consulted with Iowa Department of Education staff, selected mathematics and science teachers in Iowa, STEM hub managers, selected higher education faculty, and STEM Initiative project directors in other states and received suggestions for developing a definition. For the purposes of this project, the National Assessment of

**Figure 19.6.** Iowa STEM Indicators System

## ISIS Iowa STEM Indicators System

**Purpose:** Benchmark a variety of STEM topics in education and economic development by systematically measuring the progress and condition of the state's STEM landscape. The Iowa Indicators are focused on four primary areas: 1) STEM achievement and interest among K-12 students, 2) STEM preparation of K-12 students, 3) STEM college completions, and 4) STEM employment.

**STEM Achievement and Interest among K-12 Students**

A. STEM Achievement: Iowa Tests
   *Indicator 1:* Iowa student achievement in mathematics and science (scores and AYP).
B. STEM Achievement: National Tests
   *Indicator 2:* Iowa student achievement on NAEP mathematics and science tests
   *Indicator 3:* Number of students taking the ACT and average scores in mathematics/science.
   *Indicator 4:* Number of students taking advanced elective STEM courses in high school (AP/dual).
   *Indicator 5:* Predicted ACT scores among 10th grade ACT-Plan test-takers
C. STEM Interest
   *Indicator 6:* Percentage of ACT test-takers interested in majoring in a STEM area in college.
   *Indicator 7:* Percentage of Iowa 8th graders interested in STEM careers and educational paths (IHAPI).
   *Indicator 8:* Number/Percentage of K-12 students interested in STEM topic areas (as identified in ITBS interest inventory).

**STEM Preparation of K-12 Students**

A. STEM Teachers
   *Indicator 9:* Number of current Iowa teachers with licensure in STEM-related subjects.
   *Indicator 10:* Number of current Iowa teachers with endorsement to teach STEM-related subjects.
   *Indicator 11:* Number of beginning teachers recommended for licensure/endorsement in STEM-related subjects
   *Indicator 12:* Teacher retention in STEM-related subjects
B. STEM Educational Opportunities
   *Indicator 13:* Enrollment in STEM-related courses in high school.

**STEM College Completions**

*Indicator 14:* Number of college students who complete degrees in individual STEM majors (AA, BA, other).
*Indicator 15:* Number of college students who complete graduate degrees in individual STEM majors.

**STEM Employment**

*Indicator 16:* Percent of Iowans in workforce employed in STEM occupations.
*Indicator 17:* Job vacancy rates in STEM occupational areas.
*Indicator 18:* STEM workforce readiness (NCRC test-taking/scores)

Educational Progress (NAEP) definitions of STEM subjects seemed to be most applicable and appropriate and were adapted for monitoring Iowa's STEM Initiative.

ISIS data compilation and synthesis provided a wealth of information about STEM in Iowa. The first year of data provided an important baseline and examined many indicators retrospectively to allow for assessment of trends over time. Highlights from the data gathered for the indicators include:

- In the five years between 2008 and 2012, the number of Iowa students taking the ACT increased slightly. Mathematics and science scores remained relatively constant during that time period, as did the percentage of students meeting the math and science benchmarks for college readiness.
- Among all ACT test-takers who indicated interest in a STEM major in college, the largest proportion expressed an interest in pursuing degrees related to health sciences or technologies. A larger percentage of students planning on two years or less of college than students planning on four years or more of college expressed an interest in pursuing agriculture or natural resources conservation programs. Conversely, a larger percentage of students planning on four years or more of college expressed an interest in engineering.

- Among all students statewide who took the Iowa Assessments, interest in the four STEM subjects and STEM careers was highest among elementary students, followed by middle school and high school students.
- Regarding the number of STEM-related teachers in Iowa, overall, the numbers of STEM-related teachers in each category have been relatively stable over the past three years. Over 200 teachers held initial licenses, over 2,000 held standard professional licenses, and over 1,000 held master educator licenses.
- Iowa does not have a STEM endorsement for teachers at this time. Overall, the number of Iowa teachers with an endorsement to teach a science subject has decreased 8% in the past five years. The number of Iowa teachers with an endorsement to teach math has remained steady in the past five years (2008/09–2012/13). The number of science secondary endorsements appears to be declining, as well as subject-specific endorsements in biology, chemistry, and physics. However, the number of science middle school endorsements has been increasing.
- Nearly 10,000 post-secondary degrees ($n$ = 9,680) were awarded in STEM-related fields in 2011–12 from Iowa's four-year public and private colleges and universities and 15 community colleges.
- Among all occupational areas in the state, approximately 16% are anticipated to be within STEM sectors in the 10-year period from 2010 to 2020.
- From 2011 to 2012, there were an estimated 10,000 vacancies in STEM jobs statewide.

## Statewide Survey of Public Attitudes Toward STEM

One of the goals of the Iowa STEM Advisory Council was to raise public awareness of STEM education and workforce/economic development. To measure public awareness, the UNI CSBR initiated a statewide public survey of Iowans. The development of the survey was accomplished in several steps. First, a thorough search of the extant research on the topic was conducted to identify previous studies on the topic. Second, likely concepts to be included were compiled and presented to members of the Advisory Council through an online survey format. Members were e-mailed an invitation with a link to an online survey that contained a series of open-ended feedback questions. Third, a draft of the questionnaire was created, cross-walked with targeted priorities to ensure inclusion of relevant items, and reviewed by the ISMP partners. Once revisions were complete, the fourth step, programming and testing, was conducted.

The field period for the survey was July through September, 2012. Three sampling strata were used: general population, parents of 4- to 11-year-old children, and parents of 12- to 19-year-old children. The dual-frame sampling design included both landline and cell phone numbers.

The survey yielded 2,010 completed interviews. Data were weighted by demographic variables to better represent the adult population of Iowa. As part of the weighting process, case weights were calculated for each respondent to enhance the extent to which the sample is representative of the population on several key demographic characteristics. This weighting procedure includes adjustments for nonresponse bias and increases the match between the sample and the larger population. The weighted percentages are approximately equal to the percentage of people in the population for those demographic characteristics included in the weighting process; however, the weighted percentages for characteristics not included in the weighting process are not necessarily equal to the distribution in the population.

Moreover, one of the main reasons for conducting the survey was to estimate the attitudes, opinions, and behaviors of the population for which population values are unknown. These weighted data produce population estimates of the number of adult Iowans who likely hold a particular attitude or opinion or have engaged in particular behaviors. Descriptive statistics, including frequencies and distributions, were calculated for the total sample and for population subgroups based on gender, education, parent status, place of residence, and race for select questions in the survey. Unless otherwise noted, the term "percent" refers to the "weighted percent" and not the percent of survey respondents. Likewise, descriptions of findings are based on an analysis of the weighted data. All analyses were conducted in either SPSS or Sudaan. (See Figures 19.7–19.10, pp. 325–236.)

The public survey will be conducted annually to provide periodic cross-sectional measurements of public attitudes toward and awareness of STEM education and workforce/economic development in the state. Highlights of the findings from the statewide survey include the following:

- Only 26% of Iowans had heard of the abbreviation STEM. Recall was highest among Iowans with a four-year degree or higher and among Iowans with children in school.
- Although "brand awareness" of STEM may be low, 65% of Iowans said they had heard something in the past month about "improving math, science, technology, and engineering education" in the state.
- Most Iowans agreed that advancements in STEM will give more opportunities to the next generations (98%), increased focus on STEM education will improve the Iowa economy (86%), more jobs are available for people with good science and math skills (85%), and more companies would move to Iowa if the state had a reputation for workers with good STEM skills (76%). Two-thirds of Iowans (67%) say there are not enough skilled workers in the state to fill the available STEM jobs.
- The perceptions of Iowans closely reflect the actual national rank of Iowa students on standardized test scores, as Iowa's national rank on math and science do fall in the middle third nationally.
- Among parents of children ages 12–19, just 44% said their child has *a lot of interest* in STEM topics and 62% said their child is doing *very well* in STEM subjects in school.
- Nearly one-half (48%) of all Iowans said their child is being *very well prepared* in STEM subjects by the school he or she attends. However, only 37% of parents living on a farm or in a small town responded that way, compared to 62% of parents in cities.
- After high school graduation, 83% of parents said their child is likely to attend a two-year college or four-year college/university and 59% said their child is likely to pursue a STEM career.

**Figure 19.7.** Have You Heard? Percentage of Iowans With STEM Awareness (%)

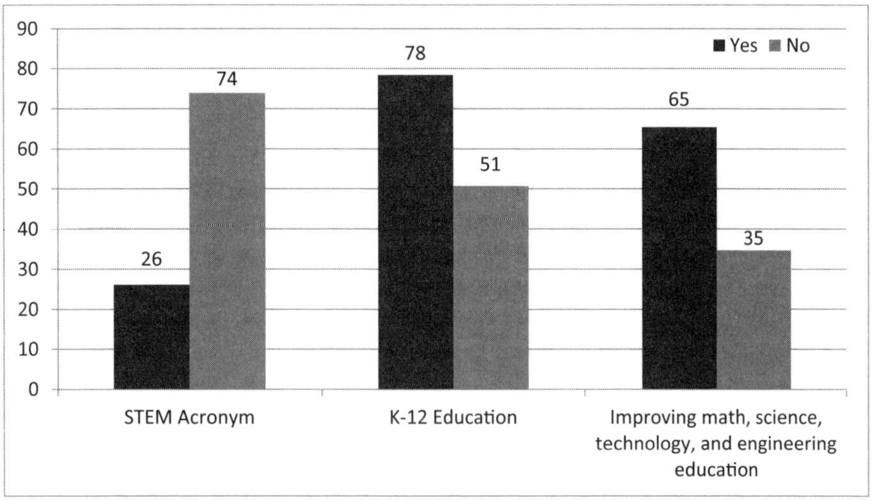


**Figure 19.8.** Percentage of Iowans Who Feel There Are Enough Skilled Workers to Fill Available STEM Jobs

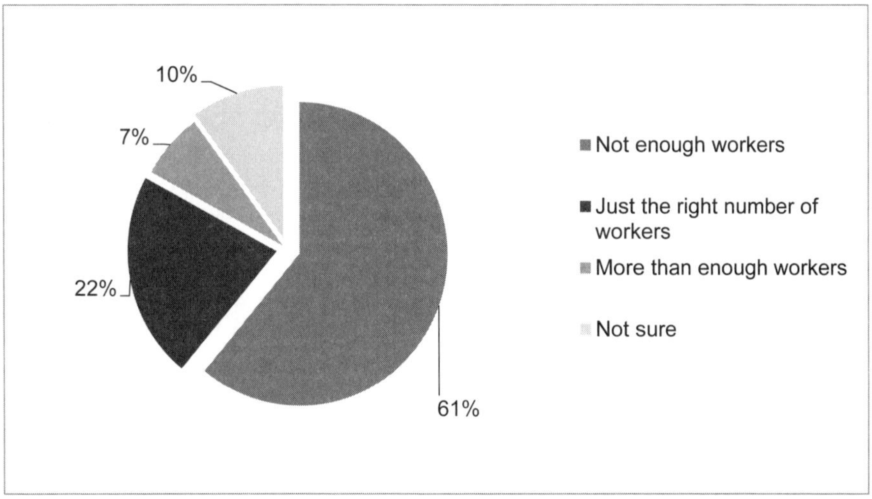

**Figure 19.9.** Attitudes About STEM Education (% Agree)

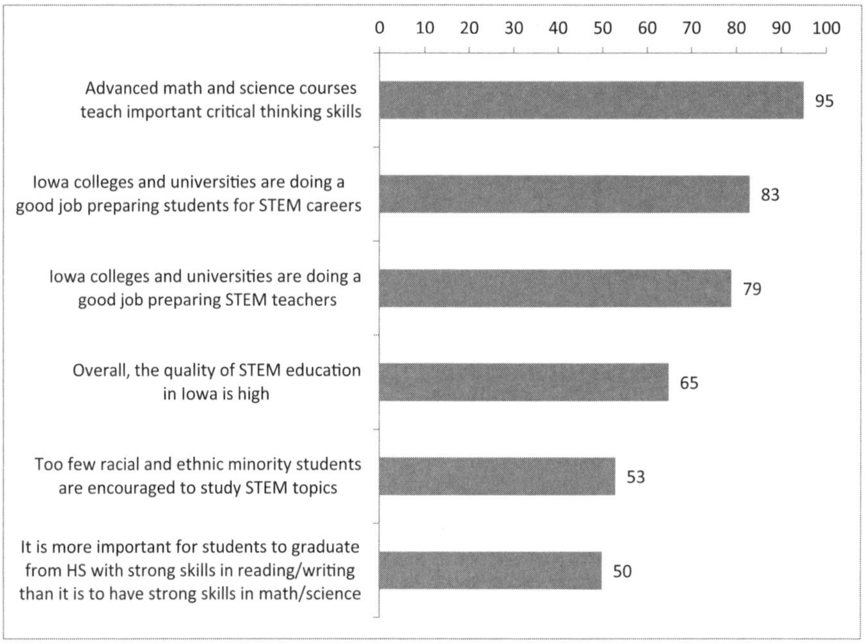


**Figure 19.10.** View of Quality of Education in Schools (Percentage of Iowans)

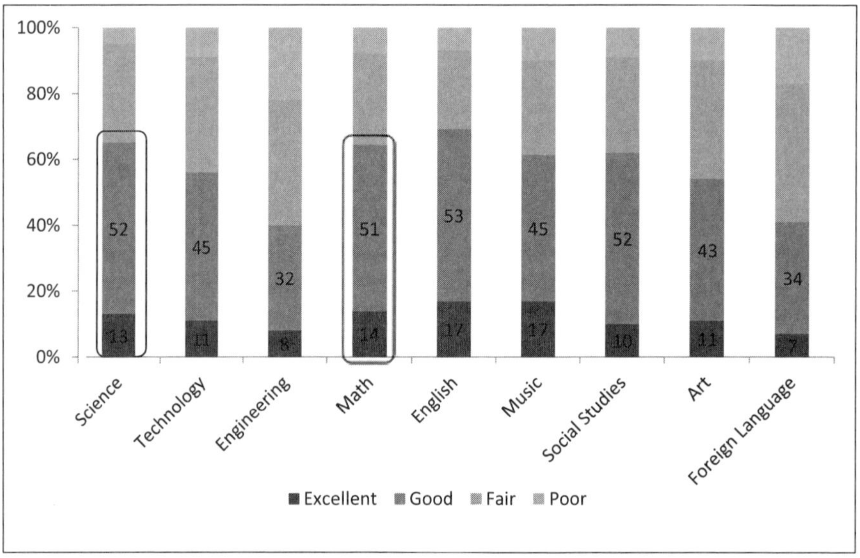

## Regional Scale-Up Program Monitoring

As part of the Iowa STEM Monitoring Project, the following three submissions were required from all LEAs implementing a Scale-Up program: (1) a Report of Process and Outcomes (RPO), (2) a Report of Participant Information (RPI), and (3) completed student questionnaires.

The RPO is an online report that is submitted by teachers and leaders implementing a Scale-Up program. The general purpose of the RPO is to inform the ISMP by providing the project partners with consistent information across all Scale-Up programs implemented in the regions. The RPO includes brief questions about Scale-Up program implementation and outcomes.

In addition, any LEA implementing a Scale-Up program working directly with students in grades K–12 or working with teachers who have a class of K–12 students was required to submit one RPI. The purpose of the RPI was to provide information about each Scale-Up participant (or Scale-Up student) and to match Scale-Up participants to their records within the statewide data set of students who have taken the Iowa Assessments standardized tests. To protect the confidentiality of Scale-Up participants, the information used to match Scale-Up participants was submitted directly from the LEA to Iowa Testing Programs using a secure web-based interface (with security similar to an online banking website). Identifying data from the RPI was not shared with any other entity. De-identified and aggregated interest and achievement scores of participants across programs enabled comparisons between Scale-Up participants and other students in the state.

Last, a short student questionnaire was created for completion by all students who were served or impacted by Scale-Up programs with the purpose of assessing self-reported changes in STEM interest as a result of participating in the Scale-Up program. Following each Scale-Up program, teachers and leaders were asked to have students complete a brief, seven-item questionnaire to assess student interest in STEM topics and careers. The questionnaire was administered via paper and pencil by the teacher or group leader. The questionnaire asked the student to indicate his or her change in interest across STEM topics and in STEM careers after participating in the Scale-Up program compared to the beginning of the fall/semester. Change in interest was measured on a 3-point scale using the response choices of "less interested," "just as interested," and "more interested." In addition, the survey asked for demographic information about gender and age. Three versions of the instrument were created to accommodate different grade levels, and the instrument was pilot tested with the target audience during development. The student survey was to be administered on the last day/session of the program/semester (or as close to that day as possible). Descriptive statistics were used to analyze the data from the student survey. The percent of students who indicated they were "more interested" in STEM topics was compared across three grade groups (elementary vs. middle school vs. high school). Significant differences were assessed using chi-square tests. All analyses were conducted in SPSS.

### *Report of Process and Outcomes (RPO)*

Two-hundred eighty-three (283) Scale-Up programs reported, documenting 10,046 participants in four different categories: (1) K–12 students; (2) parents; (3) teachers; and (4) "others," which included community members/partners, engineers, business mentors, and preservice teachers. All Scale-Up programs involved K–12 students, with the exception of one program that included college students. Additionally, over 75% of the programs included teachers, and approximately one-third of the programs included parents and others. The teachers who participated in the Scale-Up projects primarily taught courses related to STEM. However, some LEAs reported teacher participants that taught courses such as language arts, reading, and social studies. Many teachers taught multiple subjects.

Highlights from the findings related to the RPO include the following:

- *Customization.* Many participants reported that they stayed on schedule. Reasons given for deviations to timelines and plans included setbacks due to bad weather, late arrival of materials, other lessons that interfered with STEM programming, and lack of clarity about expectations and student schedules. Additionally, approximately one-third of the LEAs customized their Scale-Up program in order to serve unique local needs. Some of the customizations included adjusting lessons to fit grade level, adjusting or eliminating lessons due to time constraints, adding field trips, and utilizing different materials than those provided in the kits.
- *Experiences with service providers.* Over 50% of the LEAs reported having positive experiences with their service providers all of the time. They had adequate contact with the service provider, they received materials and resources in a timely manner, the service provider was responsive to questions and needs, and the partnership met overall expectations.
- *Collaboration.* Over 40% reported collaborations with in-school groups, and approximately one-quarter of Scale-Up programs collaborated with out-of-school, community, or volunteer groups. Participants described in written comments collaborating specifically with other teachers from a variety of different grade levels and subjects, school administrators and staff, experts from local colleges and universities, Iowa State extension offices, and parent volunteers. Participants also collaborated with 4-H programs, local businesses, college and university staff, and other local and regional teams in the area.
- *Local involvement.* Over 40% of LEAs reported receiving media coverage and community support, and about 60% of LEAs reported a local interest in STEM Programming. Other sources of local involvement included support from business and industry and receiving additional funding or resources.
- *Challenges, barriers, and successes.* Some of the challenges and barriers reported were being first-time coaches or teachers, the financial rules of the grant (i.e., reimbursement instead of being paid upfront), implementation taking away from classroom time, learning new technology and being familiar with new materials. Regarding recommendations, many respondents mentioned building a network of fellow teachers, engineers, industry volunteers, other regional and state teams, and local colleges and universities that helped smooth the implementation process.
- *Observed Outcomes.* LEAs positively reported on the observation of outcomes as a result of the Scale-Up programs, with 96% of them responding that the outcomes they observed met their expectations. Areas that fell short of expectations included: some stu-

dents were not motivated, time constraints, lack of support and training for participants, and insufficient organizational and leadership skills.

- Over 80% of the LEAs reported observing an increase in both awareness and interest in STEM topics, while over 50% of the LEAs reported observing an increase in awareness and interest in STEM careers. Approximately 40% of LEAs observed increased student achievement in STEM topics and more than a third reported increased interest in post-secondary STEM opportunities. About one-fourth reported that they had established partnerships between schools and local businesses. A few respondents also noted other observable outcomes, included students who experienced increases in confidence, critical-thinking skills, and interest in technology and science.

## *Report of Participant Information (RPI)*

Overall, student information was submitted to successfully match 4,482 Scale-Up participants to their Iowa Assessments data. The proportion of Scale-Up participants expressing interest in STEM subjects and careers was compared to the proportion of statewide test-takers that expressed interest. Select findings include the following:

- In each of the grade groups, the percent of Scale-Up students who said "I like it a lot" (Grades 3–5) or were "very interested" (Grades 6–12) was higher than students statewide.
- Comparing Scale-Up students and students statewide, the relative difference between Scale-Up students was smaller in elementary and middle school, with larger differences between the two groups in high school.
- Notably, interest in STEM subjects decreases for both students in Scale-Up programs and statewide from elementary into high school.
- However, interest in having a STEM job increased for Scale-Up students from elementary into high school (from 48% in Grades 3–5 to 53% in Grades 9–12), but decreased for students statewide (from 44% in Grades 3–5 to 38% in Grades 9–12), respectively.

## *Scale-Up Program Student Survey*

Student questionnaires were completed by students following participation in a Scale-Up program. A baseline survey of student participants was not completed, limiting the ability to show differences in student interest before and after Scale-Up program participation. LEAs implementing Scale-Up programs returned 7,729 student questionnaires. Of these, 4,181 were male (54.4%) and 3,505 were female (45.6%). The average age of participants was 11.3 years. Elementary students had the largest group of participants at 38.3% of the total sample ($n$ = 2,955), followed by middle school students (33.6%, $n$ = 2,588) and high school students (26.8%, $n$ = 2,063), respectively.

- Following Scale-Up Program participation, a significantly larger proportion of elementary students said they were more interested in STEM topics and in STEM careers compared to middle school and high school students.

- Significant differences were found in the percent male versus female students who said they were "more interested" following Scale-Up Program participation in all grade groups, but was most pronounced among students in grades 9–12 (see Figure 19.11).

**Figure 19.11.** Percentage of Males and Females in Grades 9–12 Who Were "More Interested" in STEM Subjects or Careers After Participating in the Scale-Up Program

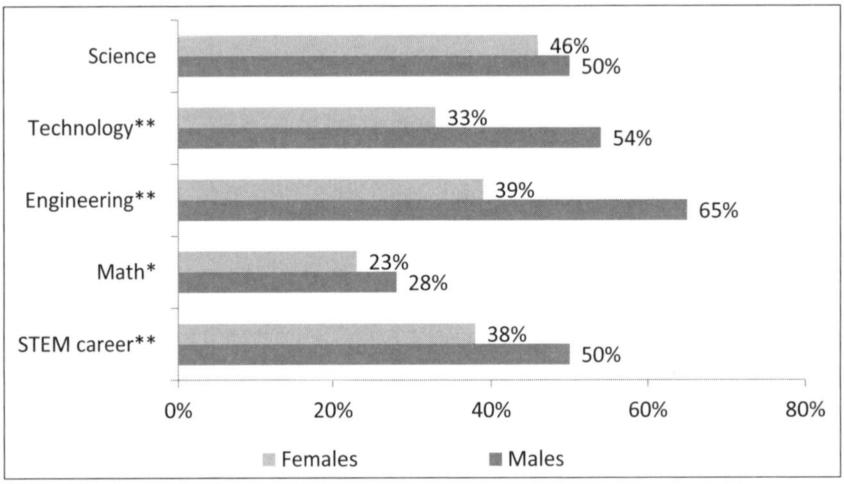


* $p < 0.05$; ** $p < 0.001$

*Source:* Regional Scale-Up Program, Student Survey

## Conclusions

Iowa's statewide STEM initiative is characterized by broad stakeholder involvement, rapid implementation, short- and long-term strategic goals, and robust evaluation. In the course of implementing the ISMP, several important process-oriented lessons were learned. Specifically, three themes emerged as best practices to ensure efficient and effective momentum.

1. *Collaboration is necessary, and coordinated collaboration is best.* When the ISMP was developed, several highly qualified and competent organizations pooled their tangible and intangible resources for a common goal. Each organization brought a unique specialty and skill set to the project and enhanced the group's ability to think creatively about ways to systematically track and monitor STEM in Iowa. An important component of this collaboration was that it occurred in a coordinated way with one organization and individual serving as the primary liaison between the Iowa STEM Advisory Council and the ISMP partners. Having a designated liaison improved the flow of communication between the Council and ISMP and among the partners involved.
2. *Alignment of evaluation methodologies with state priorities is key.* The Iowa STEM Advisory Council identified several targeted priorities on which to focus. Throughout the ISMP development process, the partners worked to clearly align the methods and data

collection instruments with the priorities and goals of the Council activities. Without such an alignment, none of the ISMP data would be relevant or useful.

3. *Start small, then add components.* During the initial development of the ISMP, the partners included a much wider array of evaluation methodologies in the project implementation plan. However, due to budget constraints, the plan was scaled back considerably to include the four components described above. Limiting the scope of the ISMP proved to be an advantage to the partners and to stakeholders. We were able to devote the necessary time and consideration to planning and initiating activities. This would not have been possible with a larger scope of work. After one full year with the ISMP in place, several methods (such as the regional Scale-Up program data collection) have been implemented, refined based on process evaluation, and initiated as systematic and routine components. In the future, additional methods such as case studies, qualitative data collection, targeted quantitative data collection, social network analysis, or asset mapping may be possible.

The data compiled, collected, and synthesized for this report come from a variety of sources. The data represent a wide range of characteristics, including periods of time, sub-populations, and data collection methods. This variation can lead to difficulty in synthesizing and interpreting the data. The purpose of this first report is to present a baseline assessment of STEM education and workforce development centered on the activities of the Iowa Governor's STEM Advisory Council. Future monitoring activities will work to refine ISMP measures, indicators, and data collection/compilation systems and to strengthen relationships with data partners in the state.

## References

Alliance for Science and Technology Research in America (ASTRA). 2011. State STEM Report Card. *http://stemconnector.org/sites/default/files/sbs/ASTRA%20STEMEd%20Iowa%202011.pdf*

Atkinson, R. D. 2012. Why the current education reform strategy won't work: To improve innovation and boost the economy, the nation needs a fundamentally new approach to education in science, technology, engineering, and mathematics. *Issues in Science and Technology* 28 (3): 29–36.

Barber, B. K., and J. A. Olsen. 2004. Assessing the transitions to middle and high school. *Journal of Adolescent Research* 19 (1): 3–30.

Battelle Technology Partnership Practice. 2011. Briefing paper: Iowa advanced manufacturing industry cluster: Prepared for Iowa department of economic development. *www.iowaeconomicdevelopment.com/industries/advancedmanufacturing*

Bybee, R. W. 2010. Advancing STEM Education: A 2020 Vision. *Technology and Engineering Teacher* 70 (1): 30–35.

Carnevale, A. P., N. Smith, and M. Melton. 2011. Georgetown University Center on Education and the Workforce. *http://cew.georgetown.edu/stem*

Change the Equation. 2011. *Vital signs: Reports on the condition of STEM learning in the U.S.* Washington, DC: Change the Equation.

Dotterer, A. M., S. M. McHale, and A. C. Crouter. 2009. The development and correlates of academic interests from childhood through adolescence. *Journal of Educational Psychology* 101 (2): 509.

Duschl, A., H. Schweingruber, H., and A. Shouse. 2007. Taking Science to School: Learning and Teaching Science in Grades K–8. Committee on Science Learning, Kindergarten through Eighth Grade; National Research Council, National Academies Press. *http://books.nap.edu/catalog.php?record_id=11625*

Eccles, J. S., C. Midgley, and T. Adler. 1984. Grade-related changes in the school environment. In *The development of achievement motivation*, ed. J. G. Nicholls, 283–331. Greenwich, CT: JAI Press.

Grey, M. 2006. New Americans, New Iowans: Welcoming immigrant and refugee newcomers. Iowa Center for Immigrant Leadership and Integration. *www.newiowans.com*

Hanushek, E., P. Peterson, L. Woessmann. 2012. Achievement growth: International and U.S. state trends in student performance. Cambridge, MA: Harvard's Program on Education Policy and Governance & Education. *www.hks.harvard.edu/pepg/PDF/Papers/PEPG12-03_CatchingUp.pdf*

Langdon, D., G. McKittrick, D. Beede, B. Khan, and M. Dorns. 2011. *STEM: Good Jobs Now and for the Future*. Washington, DC: U.S. Department of Commerce. *www.esa.doc.gov/sites/default/files/reports/documents/stemfinaljuly14.pdf*

Mayes, R., and T. R. Koballa, Jr. 2012. Exploring the science framework: Making connections in math with the Common Core State Standards. *Science and Children* 50 (4): 8–15.

President's Council of Advisors on Science and Technology (PCAST). 2010. Prepare and inspire: K–12 science, technology, engineering, and math (STEM) education for America's future. *www.whitehouse.gov/sites/default/files/microsites/ostp/pcast-stemed-report.pdf*

Sanders, M. 2009. STEM, STEM education, STEMmania. *The Technology Teacher* Dec./Jan.: 20–26.

Thomasian, J. 2011. *Building a science, technology, engineering, math agenda*. Washington, DC: NGA Center for Best Practices.

Triplett, S. 2012. Senior Account Executive – Strategic Partnerships, ACT, Inc. Personal communication.

Tsupros, N., R. Kohler, and J. Hallinen. 2009. STEM education: A project to identify the missing components. Intermediate Unit 1: Pittsburgh, PA: Center for STEM Education and Leonard Gelfand Center for Service Learning and Outreach, Carnegie Mellon University.

# Like a Scientist

## Using Problem-Based Learning to Connect Practice With Content in STEM Education

*Donna R. Sterling, Mollianne G. Logerwell, and David E. Long*
*George Mason University*
*Fairfax, VA*

### Setting

George Mason University is a major research university in Northern Virginia, minutes from Washington, D.C. Since it was founded in 1972, Mason has grown to more than 34,000 students in nearly 170 degree programs at undergraduate, masters, doctoral, and professional levels.

The College of Education and Human Development (CEHD) at Mason trains more than a quarter of the educators in Northern Virginia and is the second-highest annual producer of new teachers and school administrators for Virginia. The CEHD Graduate School of Education's degree, licensure, and certificate programs focus on teacher and counselor preparation, advanced studies for teachers and school leaders, instructional technology, and research training.

The Center for Restructuring Education in Science and Technology (CREST) in the College of Education and Human Development at George Mason University focuses on providing quality science, technology, engineering, and mathematics (STEM) education from early childhood through adulthood. CREST provides leadership in education by developing partnerships with the education and business communities to conduct teacher professional development, research, and public service. CREST

- fosters professional development of leaders in education;
- conducts research in professional development, teaching, and learning; and
- transforms education and public policy through collaboration and data-driven decision making.

Working with scientists, engineers, and educational leaders, the center involves teachers in collaborative programs that are designed to enhance their knowledge of integrated STEM teaching. Underrepresented groups such as females, minorities, and special-needs populations are part of the teaching and research focus. CREST activities include:

- Creating professional teacher development programs that enable teachers to be effective learners, facilitators of learning, and mentors;
- Conducting research on science teaching and learning including the implementation of effective teaching strategies, science concept development and assessment, collaborative action research, and the team planning and teaching process;
- Developing student-centered, integrated, hands-on, inquiry-based thematic science curricula and technology applications; and
- Assessing program effectiveness and developing performance assessment strategies and rubrics.

## Overview of Program

CREST uses science summer camps as part of its teacher professional development programs (Frazier and Sterling 2007, 2008; Sterling 2007, 2010; Sterling and Frazier 2006; Sterling and Hargrove 2012). For 15 years, preservice and inservice elementary teachers have planned and implemented a summer enrichment university-based day camp for students entering grades 5 to 7. The main purpose of the camp is to give teachers an opportunity to immediately implement effective, research-based instructional strategies in a risk-free environment. Given that elementary teachers have often had little in-depth science education training as part of the university programs of study, these safe spaces are crucial to incubate confidence amongst a cadre of supportive professional colleagues.

CREST uses Problem-Based Learning (PBL) (Delisle 1997; Greenwald 2000; Krynock and Robb 1999; Shack 1993; Stepien and Gallagher 1993) as a teaching strategy in which students and teachers experience, usually for the first time, purposeful science investigation in a meaningful context as they solve real-world problems over an extended period of time. Instead of exposing students to disconnected and decontextualized science experiences, PBL provides an opportunity for teachers to plan and implement lessons that work together to build students' science knowledge as they find possible solutions to the proposed problem. CREST's professional development programs also emphasize the use of best-practice inquiry and process-based experiences (NGSS Lead States 2013; NRC 2012) where students ask questions, develop investigations, and draw conclusions based on data collected. By experiencing firsthand what scientists do in an exciting and fun way, students improve both their science knowledge and their attitudes toward science (Frazier and Sterling 2008; Sterling, Matkins, Frazier, and Logerwell 2007).

Attending to the varied needs of teachers in training to those already in service, CREST designed professional development that meets the needs of both. For preservice teachers, the summer camp experience is integrated into their science teaching methods course. For inservice teachers it is part of a summer academy. In both formats, teachers spend approximately 30 hours first learning about PBL and inquiry-based instruction and then planning a two-week science camp for upper-elementary students. During camp, teachers work in groups of three or four to implement two days of the PBL unit the whole cohort designs. Each teaching group is responsible for students from the time they arrive on campus at 9 a.m. until they leave at 3 p.m. After camp, teachers spend approximately eight hours debriefing the experience, including analyzing videos of their camp teaching. An overarching theme of continuous improvement permeates the program (Akerson, Cullen, and Hanson 2009).

There were several features of CREST's summer camps that supported family engagement as well as college access and success. Family members were invited to attend the last day of camp, when students presented their solutions to the science problems that they had been investigating. When grant funding was available, scholarships that included registration, transportation, and lunch were offered to high-needs children to attend camp. Additionally, there was a morning session on the last day of camp where parents and other family members met with college counselors to learn about student preparedness, the college admission process, and financial aid. For many economically marginalized students, attending a college or university is often not part of their extended family's experience, while many such parents wish that it could be. These camp experiences work to provide a bridge mechanism where their student's success at a university sponsored camp is underscored by a practical campus experience where this success is validated for parents and the college going process is made clear. Each year, parents consistently report this was an invaluable experience and welcome the knowledge they have to prepare their children for a future with post-secondary education as a reality.

## Features of Instructional Program

Based on inspiration Donna Sterling experienced moving into the science education profession from practicing laboratory science, problem-based learning was made the thrust of CREST summer institutes. The logic was clear. From her experience as a scientist, as she and other scientists drew strength and inspiration from solving real-world problems using experimental methods, so should students. This was not the case in Sterling's entry into the field of science education. At the time, science education was rutted in a pedagogy concerned with concept mastery alone. Sterling set out to do her part in changing the field through exciting, highly engaging student-centered problem-based learning experiences. Problem-based learning would become the organizing principle from which other critical science education concepts (inquiry, nature of science, hands-on learning) would be operationalized.

At the core of each PBL unit was a real-world science scenario and overarching problem questions. These questions were designed to engage both a students' authentic fund of knowledge (Gonzales, Moll, and Amanti 2005), but also connect this knowledge to larger societal problems for which they could work toward practical, STEM-based solutions. Each summer a different theme was chosen by CREST staff based on perceived student and parent interest, and teachers designed the two-week camp PBL units around that theme. Table 20.1 (p. 336) details a representative sample of camp units (Frazier and Sterling 2007, 2008; Sterling 2010; Sterling and Frazier 2006).

After developing the scenario and problem question, teachers planned connected activities that motivated students to learn science content while working toward possible solutions for the problem that had been presented. For example, in the "Exploring Space" camp, each morning students received a message from NASA containing newly acquired data from the space probe (e.g., spectrographic information, soil composition, and atmospheric conditions). Students designed experiments and completed other activities to help them understand what the

**Table 20.1.** Representative Camp Problem Questions and Scenarios

| Theme | Problem Question | Abbreviated Scenario |
|---|---|---|
| **Motor Mania** | Can you make it to Virginia in time for the George Mason Grand Prix? | Students created cars capable of overcoming various "detours" (physical/mechanical challenges – e.g., too much luggage, mechanical failure) and "roadblocks" (environmental/travel challenges – e.g., excessive heat/cold, traffic) in order to win a road race. |
| **Weather Tamers** | How can buildings be made disaster-proof? | As junior FEMA scientists, students developed a model city that could withstand natural disasters such as hurricanes, tornadoes, and earthquakes. |
| **Dig It** | Who's buried at Mason? | Students determined the identity and time period of skeletal remains found at a construction site on campus. |
| **Exploring Space** | Where is the lost Patriot Probe? | As mission control specialists for NASA, students determined where a lost space probe was. |
| **Watershed Mysteries** | How can we repair and prevent further decline of our watershed? | Students determined what was causing the ducks in the area to become sick. |

data meant, narrowing down a list of possible locations to determine the location of a lost space probe. Following best practice and expectations of a fun summer camp, students were excited by the prospect of presenting the findings of their work to parents and important community members during a last day culminating project to wrap up camp activities. For the culminating project, students presented their solution and then lead their family members through some of the labs they had done during camp. The overall impact of these activities always resonated with a significant number of camper parents as they reflected upon the impact of the camp on their child's interest in science. Indicative of many parents' comments on the impact of camp on home dialogue during camp evenings, one parent remarked that during and after the camp, "My child wouldn't quit talking about the camp … about the problem they were working on … she came home from camp each day and wanted to test everything. Everything now is about science!"

The experience of seeing the campers changed during the process was not isolated to the camp parents. Camp directors and staff saw the same change as teachers and parents watched in awe. Reflecting on the campers taking ownership of the camp environment, a camp director explains:

> *I still remember that the last day, in the morning, we were prepping for everything—for the parents and visitors to come in and see the culminating project. One by one, the groups of campers decided that they were going to run it. The teachers would be there to support them … but they would be the scientists showing their parents what they did. They explained the science content behind it, what experiments they did, and how this process helped them answer the problem-based learning's overarching question. The kids ran the entire show. Parents were beaming. The campers would show their parents that to be good scientists, they have to take careful notes of their observations—they were adamant.*

## Evidence for Success

In addition to conducting high-quality professional development through the summer Elementary Science Institute, CREST has conducted ongoing research to evaluate the success of the institute and summer science camps. Both of these are used for teacher professional development and student learning to take the next steps in improving STEM educational practice. The following three summer institute experiences are indicative of the growth experienced by teachers and students, along with the lessons learned. Analysis of program outcomes informed the next steps in program development, leading to the current program research.

## Developing Inservice and Preservice Teachers

### *Medical Mysteries*

Meeting the needs of both teacher professional development and underachieving students of color, CREST professional development is designed from the ground up to integrate multiple goals. As shown in our published work (Sterling, Matkins, Frazier, and Logerwell 2007), science camps are transformative experiences for students, parents, and teachers in the urban setting.

Based on prior experience working with pre- and inservice teachers, this camp was designed to meet the needs of preservice teacher field placements, while also meeting the needs of local schools. In partnership with the school district, the camp theme of "medical mysteries" was chosen to have students work through problem based learning activities to support their science learning while also increasing their health and medical knowledge.

A disease outbreak was presented to the campers to organize their problem-solving skills. Using the fictionalized town of Masonville, students worked through problems of disease identification, building conceptual knowledge of bacteria, viruses, and epidemiology. All camp activities built on and connected to helping students investigate and solve what caused the outbreak. Student engagement was high as they cultured bacteria from their own hands during the camp, building anticipation over the three-day incubation period. Students were both in awe and giddy at the fact of being in a real, functioning science lab, while also having the scientists on hand take them and their problem-solving seriously. Other hands-on activities that facilitated the students' understanding and problem solving were epidemiological modeling experimentation where the students conducted a phenolphthalein test of liquids, which modeled disease transmission among carriers.

Pre-, mid-, and post-assessments were taken of student knowledge about life science, especially as pertaining to disease vectors and epidemiology. Student growth on the Student Life Science Content Knowledge and Skills instrument were as follows: pretest = 50% correct, posttest = 59% correct $t$ (49) = 3.54, p = 0.00. Also measured was general attitudinal change toward science and science career aspirations, using the Student Attitudes Toward Science instrument. Significant positive change in attitude toward asking more questions was observed; from pre to post, $t$ (48) = 2.35, $p$ = 0.02

Explicit attention was paid to integrate attention to working on closing achievement gaps between racial majority and minority students by focusing the camp toward underperforming minority students. In prior camps, as in this one, students from across the socioeconomic spectrum were shown to produce equally creative, high-quality results, regardless of home circumstances. As various grant funds were available, CREST explicitly looked to support students from such backgrounds as a support to leveling inequities of opportunity. As shown later in current work, this facet of institute planning remains key to the commitment of social justice in education.

## *Weather Tamers*

For several years, preservice elementary teachers planned and implemented CREST's summer science camp as the field-experience component of their graduate science methods course. As CREST summer institutes continued to develop, it became clear to CREST faculty that the integration of positive teaching strategies learned within science methods courses could be exercised within the camp structure. Hands-on, problem-based learning would excite the campers, while teachers would gain invaluable experience practicing in a non-school, lower-stakes environment. Additionally, teachers would leave with a fairly polished transportable unit plan for practice within their school.

During the camp, teacher content knowledge, teacher efficacy beliefs using the Science Teaching Efficacy Beliefs Instrument for Preservice Teachers (STEBI-B) instrument, along with reflective writing data were collected to analyze camp outcomes for teachers attending. The STEBI-B was selected to ascertain preservice teachers' beliefs in their ability to effectively teach science as well as their belief that teaching, in general, can influence student achievement in science. Based on Bandura's (1977) social learning theory and following Gibson and Dembo's (1984) general Teacher Efficacy Scale, the STEBI-B was developed by Enochs and Riggs (1990) to measure a preservice teachers' efficacy specifically for science teaching. CREST selected this instrument to ascertain preservice teachers' beliefs in their ability to effectively teach science as well as their belief that teaching, in general, can influence student achievement in science. This instrument is a 23-item Likert-type survey with five possible responses (1 = strongly agree, 5 = strongly disagree). There were two subscales, the Personal Science Teaching Efficacy Belief Scale (PSTE) and the Science Teaching Outcome Expectancy Scale (STOE). Items in the PSTE included statements like "I will find it difficult to explain to students why science experiments work" and "I understand science concepts well enough to be effective in teaching elementary science." Items in the STOE included statements like "When a student does better than usual in science, it is often because the teacher exerted a little extra effort" and "The teacher is generally responsible for the achievement of students in science." Thirteen items were positively worded, and ten items were negatively worded. Psychometric analyses have produced Cronbach's alphas of 0.92 for the PSTE and 0.77 for the STOE (Riggs 1988). Previous studies have shown relationships between the teachers' STEBI scores and their content knowledge (Cakiroglu 2000; Chang 2003; Koc 2006; Schoon and Boone 1998) as well as their students' achievement test scores (Staples 2002). Participants completed this instrument during class on the first and last day of their science methods course.

As CREST staff found out, planning for such dynamic and intensely child-centered science pedagogy in and of itself was one of the most valuable experiences that teachers reported finding valuable. Camp planning, as one camp director explained, was very exciting:

*It was a real on your feet, experiential way to learn to teach using inquiry. The teachers came to see it as a safe space where they could explore inquiry learning .... It also was helpful that they were taking their methods courses with me at night, so we could debrief what worked during the day, and continue to refine what we do the next day or week. We were forming a bond doing camp and then teaching at night. The debriefing of this practical experience was also helping the next groups teaching in camp to prepare for their turn to teach.*

In addition to teachers finding the camp planning experience valuable, the campers were often caught in moments of discovery where their delight at insight was palpable. While investigating historical maps of hurricane trajectories across the Atlantic and Caribbean, campers used their deductive reasoning skills to explain why hurricanes prior to a certain year had no points of origin within the central Atlantic, whereas those after did. During the same camp, an engineering design challenge saw campers work to solve the problem of how to design a hurricane proof home (Frazier and Sterling 2007). All camp activities related to and connected with the causes of different types of weather and how buildings could be made disaster-proof. Using materials that modeled real-world construction, students worked through simple experimental design challenges where wind was increased until their models failed. Campers were challenged to redesign for improvement with great success. Practically, the camp format within the institute offered forms of experience that teachers later realized that they acquired—which wouldn't have been as easy to acquire in a traditional field placement. The process of designing the camp content and structuring activity which authentically connected to students funds of knowledge also led to using guests and experts from the community to support the camp problem narrative. Breaking out of their comfort zone, many teachers gained new skills in working through who might support their efforts, who might be willing to work with students to increase their learning through real-world community connections, and how to go about making contacts to set these up. All of these things were unlikely to have been part of a traditional preservice experience (see Table 20.2).

## **Table 20.2.** Summer 2006: Weather Tamers and Crime Busters*

(pre = first day of class, post = last day of class)

| | Pre | Post | *t* | *p* |
|---|---|---|---|---|
| **STEBI-B** | 3.23 | 3.74 | 4.55 | 0.000 |
| **PSTE subscale** | 3.23 | 3.69 | 4.02 | 0.001 |
| **STOE subscale** | 3.22 | 3.71 | 4.45 | 0.000 |

Degrees of freedom = 20;
*Weather Tamers data is presented as one of two camps held and measured concurrently.

Weather Tamers camper science content knowledge
Pretest = 63%, posttest = 71%, $t$ (21) = 2.90, $p$ = 0.008

Crime Busters camper science content knowledge
Pretest = 56%, posttest = 65%, $t$ (22) = 5.19, $p$ = 0.000

## *Exploring Space and Watershed Mysteries*

Building off of prior CREST camp experiences, CREST's most comprehensive elementary school teaching research to date (Logerwell 2009) was conducted to explore the effects of a summer science camp teaching experience on preservice elementary teachers' science content knowledge, science teaching efficacy, and understanding of the nature of science—three variables that have been linked to increased student learning (Anderson, Greene, and Loewen 1988; Armor et al. 1976; Ashton and Webb 1986; Druva and Anderson 1983; Goldhaber and Brewer 1997; Hannum 1994; Kerley 2004; Ledford 2002; Luera, Moyer, and Everett 2005; Midgley, Feldlaufer, and Eccles 1989; Monk 1994; Moore and Esselman 1992; Muijs and Reynolds 2002; Ross 1992, 1998; Staples 2002; Watson 1991; Wenglinsky 2000).

Three cohorts of preservice elementary teachers enrolled in different sections of the same elementary science methods course taught by the same professor were recruited for the study. Table 20.3 describes the length and field placement experiences for each section.

**Table 20.3.** Watershed Mysteries Camp Groups

| Group | Number of Participants | Length of Methods Course | Field Placement Experience |
|---|---|---|---|
| Franklin* Partnership Schools (FPS) | 21 | 7 weeks | Summer science camp for 4th–6th grade students |
| Lincoln* Partnership Schools (LPS) | 15 | 3 weeks | 15 hours of classroom observation in an elementary classroom |
| Professional Development Schools (PDS) | 24 | 16 weeks | PDS internship, including teaching one science lesson |

* Pseudonym

The summer science camp experience was considered a treatment group as it was not a typical field placement experience for teachers in training. During the summer in which the study took place, there were two camps, each lasting six hours per day for two weeks. The preservice teachers were divided into teaching groups of two to three teachers; each group was responsible for planning and teaching two days for one of the camps. Half of the cohort was responsible for an "Exploring Space" camp. The preservice teachers in this group designed a problem-based unit in which the students worked to figure out the location of a lost space probe. The science content of this camp included astronomy, Earth science, and chemistry. The other half of the cohort was responsible for the "Watershed Mysteries" camp. The preservice teachers in this group designed a problem-based unit in which students worked to figure out what was causing ducks in a local pond to get sick. The content of this camp included chemistry, biology, and Earth science.

The comparison groups completed either 15 hours of observation in an elementary classroom or a yearlong professional development internship that included teaching one science lesson in their cooperating teachers' classroom. These two formats represented the two most common programmatic experiences that the university science education program had run, and were the basis of comparison for the summer institute's unusual science camp field component.

## Measurements of the Institute

A battery of instruments plus the STEBI-B addressed above were selected to assess the efficacy of the teacher professional development. Building upon the findings of past summer institutes, more in-depth measures of teacher attitudes and whether the type of field placement had a differential impact on teachers were assembled.

### *Personal Data Questionnaire (PDQ)*

The PDQ was developed to obtain demographic information about the participants. This instrument was administered in class on the first day and last day of the science methods course. Analyses were conducted to ascertain differences between the three groups on demographic variables such as age, gender, ethnicity, college major, number of science courses completed, and science teaching experience.

### *Science Content Knowledge Assessment (SCKA)*

The SCKA was developed by Logerwell (2009) for the CREST study to measure individuals' science content knowledge because no general science content tests for teachers were available at the time. This instrument is a 40-item multiple-choice test that uses questions from the Virginia Standards of Learning Earth science, biology, and chemistry high school end-of-course exams and the Harvard University Project MOSART grade 9–12 physics test. The Jefferson Lab's Virginia State Standards of Learning practice test website (*http://education.jlab.org/solquiz/index.html*) was used to randomly generate 10 questions each for the Earth science, biology, and chemistry portions of Version A. Ten questions were also randomly selected from a hardcopy of Project MOSART's grade 9–12 physics test form 741. Version B was created by selecting additional topic-matched questions from the Jefferson Lab website and Project MOSART test. Face validity of both versions were obtained by a panel of experts, which included science and science education faculty. Each assessment was also administered to a pilot group of preservice teachers ($n$ = 19) in order to identify any typographical errors and determine the time needed to complete the instrument. Analyses produced Cronbach's alphas of 0.53 for Form A and 0.86 for Form B.

This instrument was administered two times. Participants completed Version A on the first day of their science methods course and Version B on the last day of their science methods course. Analyses were done for both the whole instrument and for each content subscale. The biology subscale corresponded to questions 1–10, the chemistry subscale corresponded to questions 11–20, the Earth science subscale corresponded to questions 21–30, the physics subscale corresponded to questions 31–40.

### *Views of Nature of Science Questionnaire (VNOS)*

This instrument, comprised of eight open-ended questions, was a modified version of the 10-item VNOS-C development by Lederman, Abd-El-Khalick, Bell, and Schwartz (2002). The validity of the VNOS-C was established by Abd-El-Khalick (1998, 2001) through the use of systematic comparison of participants' NOS profiles which had been independently generated through separate analyses of corresponding questionnaire and interview transcripts. Studies that have administered the VNOS-C to preservice elementary teachers include Abd-El-Khalick (2001); Abd-El-Khalick and Lederman (2000); Cochrane (2003); Lederman, Schwartz, Abd-El-Khalick, and Bell (2001); and Schwartz, Lederman, and Crawford (2000). Because the targeted tenets of NOS were addressed by multiple items on the VNOS-C, the version used in this study eliminated two items that had considerable overlap with the remaining items in order to reduce the estimated time it takes to complete the questionnaire.

CREST selected the VNOS to measure participants' understanding of key aspects of the nature of science as identified in science education literature. The VNOS was administered during class on the first and last day of the science methods course. In order to obtain a general overview of the participants' views of the nature of science, a rubric was used to rank the responses to each question on a 5-point scale (1 = naïve view, 5 = sophisticated view) and a composite rubric score for the entire instrument was calculated. In order to explore differences among the groups' responses as well as change within each group in more depth, participants' responses were also coded for concurrence with central tenets of the nature of scientific knowledge (Lederman et al. 2002). An independent rater with NOS expertise verified the rubric scores and coding for a representative sample of participants' responses.

One-way analysis of variance tests were run to ascertain between group differences for age and number of science courses taken. For all other variables, a mixed analysis of variance (ANOVA) with time (pretest, posttest) as a within-subject factor and group (FPS, LPS, PDS) as a between-subjects factor was performed. Tukey post hoc analyses were done to explore differences between each pair of groups. Follow-up paired-samples $t$ tests were performed when there were significant within-group differences. SPSS 15.0 was used for all quantitative analysis, and an alpha level of 0.05 was used for all statistical tests.

Descriptive statistics on the participants' Science Teaching Efficacy Belief Instrument for Preservice Teachers (STEBI) data are provided in Table 20.4. Results revealed significant differences among the groups on the Personal Science Teaching Efficacy (PSTE) subscale, $F\ (2, 55) = 4.66, p = 0.01$, and overall STEBI, $F\ (2, 55) = 5.77, p = 0.005$, but not the Science Teaching Outcome Expectancy (STOE) subscale, $F\ (2, 55) = 1.99, p = 0.15$. Tukey post hoc tests showed that participants in the FPS group had significantly higher PSTE scores than the PDS group, $MD = 0.33, p = 0.01$, and significantly higher STEBI scores than both the PDS group, $MD = .25$, $p = 0.007$, and LPS group, $MD = .23, p = 0.04$.

**Table 20.4.** Science Teaching Efficacy Belief Instrument Descriptives and Analysis of Variance

| | | FPS ($n$ = 21) | | LPS ($n$ = 15) | | PDS ($n$ = 24) | | $F$ |
|---|---|---|---|---|---|---|---|---|
| | | $M$ | $SD$ | $M$ | $SD$ | $M$ | $SD$ | |
| **PSTE** | | | | | | | | |
| | Pre | 3.78 | 0.41 | 3.55 | 0.50 | 3.46 | 0.39 | 4.66* |
| | Post | 4.35 | 0.38 | 4.17 | 0.42 | 4.01 | 0.39 | |
| **STOE** | | | | | | | | |
| | Pre | 3.71 | 0.44 | 3.52 | 0.43 | 3.68 | 0.34 | 1.99 |
| | Post | 3.90 | 0.46 | 3.54 | 0.60 | 3.64 | 0.38 | |
| **STEBI** | | | | | | | | |
| | Pre | 3.75 | 0.30 | 3.54 | 0.29 | 3.55 | 0.27 | 5.77** |
| | Post | 4.16 | 0.33 | 3.90 | 0.39 | 3.85 | 0.21 | |

*$p < 0.05$; **$p < 0.01$

There were also significant within-group effects for the PSTE, Wilks' $\Lambda$ = .32, $F$ (1, 55) = 116.24, $p$ = 0.001, partial $\eta^2$ = 0.68, and STEBI, $\Lambda$ = 0.37, $F$ (1, 55) = 92.68, $p$ = 0.001, partial $\eta^2$ = 0.63. Follow-up paired-samples $t$ tests (see Table 20.5) revealed that all three groups had significant, positive changes in both their PSTE and overall STEBI from the pretest to the posttest. Additionally, the FPS group had a significant, positive change in their STOE from pretest to posttest.

**Table 20.5.** Science Teaching Efficacy Belief Instrument Paired Samples t-Tests

| | FPS ($n$ = 21) | | LPS ($n$ = 15) | | PDS ($n$ = 24) | |
|---|---|---|---|---|---|---|
| | $t$ | $p$ | $t$ | $p$ | $t$ | $p$ |
| **PSTE** | 6.79 | 0.001 | 6.49 | 0.001 | 6.17 | 0.001 |
| **STOE** | 2.07 | 0.049 | 0.34 | 0.74 | 0.48 | 0.64 |
| **STEBI** | 7.13 | 0.001 | 5.21 | 0.001 | 4.95 | 0.001 |

Related to science teaching efficacy, the analysis revealed significant differences among the groups, with the treatment group (FPS) having higher general science teaching efficacy than both comparison groups and higher personal science teaching efficacy than the PDS group. Additionally, all three groups experienced significant increases in general and personal science teaching efficacy, while only the treatment group experienced a significant increase in science teaching outcome expectancy.

Descriptive statistics on the participants' Science Content Knowledge Assessment (SCKA) data are provided in Table 20.6 (p. 344). A mixed ANOVA with time (pretest, posttest) as a within-subject factor and group (FPS, LPS, PDS) as a between subjects factor was performed. Results revealed no significant differences among the groups on the overall SCKA or any of the subscales. However, there were significant within-group effects for the chemistry subscale, Wilks' $\Lambda = 0.60$, $F(1, 57) = 37.68$, $p = 0.001$, partial $\eta^2 = 0.40$, and overall SCKA instrument, Wilks' $\Lambda = 0.88$, $F(1, 57) = 7.84$, $p = 0.007$, partial $\eta^2 = 0.12$.

**Table 20.6.** Science Content Knowledge Assessment Descriptives and Analysis of Variance

| | | FPS (*n* = 21) | | LPS (*n* = 15) | | PDS (*n* = 24) | | *F* |
|---|---|---|---|---|---|---|---|---|
| | | *M* | *SD* | *M* | *SD* | *M* | *SD* | |
| **Biology** | | | | | | | | |
| | Pre | 6.81 | 1.36 | 7.13 | 1.13 | 7.00 | 1.35 | 1.59 |
| | Post | 8.00 | 1.26 | 6.89 | 2.61 | 6.42 | 2.32 | |
| **Chemistry** | | | | | | | | |
| | Pre | 3.29 | 1.68 | 3.40 | 1.99 | 3.21 | 1.74 | .31 |
| | Post | 5.57 | 1.86 | 5.20 | 2.62 | 4.92 | 2.06 | |
| **Earth Science** | | | | | | | | |
| | Pre | 6.38 | 1.02 | 6.00 | 1.69 | 6.71 | 1.71 | 1.14 |
| | Post | 6.90 | 1.58 | 5.80 | 2.27 | 5.89 | 2.23 | |
| **Physics** | | | | | | | | |
| | Pre | 5.05 | 1.66 | 4.47 | 1.77 | 4.25 | 1.59 | 2.27 |
| | Post | 5.48 | 1.66 | 5.47 | 2.26 | 4.50 | 2.27 | |
| **SCKA** | | | | | | | | |
| | Pre | 21.52 | 3.98 | 21.00 | 3.63 | 21.17 | 4.12 | 1.40 |
| | Post | 25.95 | 4.74 | 23.33 | 8.35 | 21.71 | 7.65 | |

Follow-up paired-samples *t* tests (see Table 20.7) revealed that participants in both the LPS and PDS groups had significant positive changes on the chemistry subscale, while participants in the FPS group had significant positive changes from pretest to posttest on the instrument overall as well as the biology and chemistry subscales.

**Table 20.7.** Science Content Knowledge Assessment Paired Samples t-Tests

| | FPS (*n* = 21) | | LPS (*n* = 15) | | PDS (*n* = 24) | |
|---|---|---|---|---|---|---|
| | *t* | *p* | *t* | *p* | *t* | *p* |
| **Biology** | 3.34 | 0.001 | –0.37 | 0.72 | –1.15 | 0.26 |
| **Chemistry** | 5.67 | 0.001 | 2.16 | 0.05 | 3.81 | 0.001 |
| **Earth Science** | 1.42 | 0.17 | –0.36 | 0.72 | –1.78 | 0.09 |
| **Physics** | 0.88 | 0.39 | 1.50 | 0.16 | 0.48 | 0.64 |
| **SCKA Total** | 4.47 | 0.001 | 1.17 | 0.26 | 0.36 | 0.72 |

Related to science content knowledge, the analyses showed that there were no significant differences among the groups. However, the treatment group experienced significant improvement in overall knowledge as well as biology- and chemistry-specific knowledge, while the comparison groups only experienced significant improvement in chemistry-specific knowledge.

Descriptive statistics on the participants' Views of Nature of Science (VNOS) data are provided in Table 20.8. The average rubric score for all groups was in the "somewhat naïve view" range on both the pre- and posttest. A mixed ANOVA with time (pretest, posttest) as a within-subject factor and group (FPS, LPS, PDS) as a between-subjects factor was performed. Results revealed no significant differences among the groups on the VNOS. However, Tukey post hoc tests revealed that participants in the FPS group had significantly higher VNOS scores than participants in the LPS group, $MD = 1.82, p = 0.05$.

**Table 20.8.** Views of Nature of Science Descriptives and Analysis of Variance

| | FPS ($n$ = 21) | | LPS ($n$ = 15) | | PDS ($n$ = 24) | | $F$ |
|---|---|---|---|---|---|---|---|
| | $M$ | $SD$ | $M$ | $SD$ | $M$ | $SD$ | |
| Pre | 14.14 | 2.85 | 14.13 | 3.80 | 15.00 | 1.77 | 2.95 |
| Post | 18.57 | 2.54 | 15.50 | 2.38 | 16.09 | 2.86 | |

The results also indicated that there were significant within-group effects, Wilks' $\Lambda = 0.56, F(1, 55) = 42.90, p = 0.001$, partial $\eta^2 = 0.44$. Follow-up paired-samples $t$ tests revealed that there were significant positive changes in the VNOS rubric scores from the pretest to the posttest for participants in both the FPS, $t(20) = 8.85, p = 0.001$, and LPS, $t(13) = 2.31, p = 0.04$, groups, but not the PDS group, $t(22) = 1.80, p = 0.09$.

VNOS data were also qualitatively analyzed. Responses to each question were coded to determine their concurrence with central tenets of NOS (Lederman et al. 2002). A matrix (Krippendorff 1980; Miles and Huberman 1994) was developed to organize the data by question number, group (i.e., FPS, LPS, PDS), and time period (i.e., pre and post). All responses to question 1 were coded first, followed by all responses to question 2, and so on. During coding, data was examined phrase-by-phrase in terms of their meaning interpreted by the rater (Krippendorff 1980). Information categories, or "nodes" as termed by Gibbs (2002), were created that summarized the perceived meaning of each phrase in the context of the entire response. For example, in response to question 2, one participant said, "Scientists need to experiment because it's the only way to obtain proof." This statement was coded as "proof." Once all coding was complete, similar nodes were categorized together in order to aid in further analysis. For example, for question 2 the nodes "proof," "support," and "facts" were categorized together. Next, the total number of responses coded at each category by group and time period were determined (Krippendorff 1980). Categories for each question were then examined to determine patterns in the responses between the groups at each time period and changes within each group from pre- to posttest (Krippendorff 1980).

All three groups responded similarly on the pretest for most of the VNOS questions, with only three questions—numbers 5, 6, and 7—showing somewhat different patterns of responses.

For question 5, the three groups were similar except that a higher percentage of participants in the FPS group indicated that scientists had seen an atom with a microscope, while a higher percentage of participants in both the LPS and PDS groups noted that atomic theory was all speculation. Likewise, for question 6, all three groups responded similarly except that more members of the FPS and PDS groups said that the different theories were a result of old or lost evidence or stated that both theories were possible without giving any further explanation. For question 7, more participants in the LPS group believed that science was culturally influenced, followed by the PDS group, and then the FPS group. The pattern was reversed when looking at who said that science is a combination of universal and cultural elements, and approximately equal percentages of each group thought science was universal.

There were more differences between the groups at the posttest compared to the pretest, with only questions 1 and 3 having similar patterns of responses across all three groups. For question 2, all members of the PDS group stated that science required experimentation, while a growing number of participants in the FPS and LPS groups indicated that science can be based on observations instead of experimentation. For question 4, large percentages of both comparison groups continued to believe that laws were proven facts while theories were unproven opinions. Members of the FPS group, however, tended to describe theories as explanations for scientific phenomena. For question 5, participants in the FPS and LPS groups were approximately equally divided between saying that scientists had seen atoms with a microscope, atomic theory was based upon evidence from some sort of experiment, or they did not know what the basis for atomic theory was. More members of the PDS group, however, indicated that atomic theory was based on evidence from experiments, while fewer members stated that scientists had seen atoms with microscopes or that they did not know what evidence scientists used. For question 6, over two-thirds of the FPS group attributed the different theories to differing perspectives, while just less than half of the LPS group and just over a third of the PDS group did so. Similarly, about a quarter of each comparison group indicated that the difference was due to lost or old evidence, while only 5% of FPS did so. For question 7, a much higher percentage of the FPS group indicated that science is culturally influenced compared to the LPS and PDS groups (81%, 53%, and 46%, respectively), while a higher percentage of the LPS group stated that science is universal compared to the PDS and FPS groups (27%, 12%, and 8%, respectively). No members of the LPS group noted that science is a combination of cultural and universal elements on the posttest, while over one-fifth of the PDS group and a few members of the FPS group did. For question 8, a higher percentage of the FPS group stated that scientists use creativity throughout their investigations compared to the PDS and LPS groups (55%, 26%, and 18%, respectively), while more members of the LPS group indicated that scientists only use creativity during the planning and design stages compared to the PDS and FPS groups (70%, 55%, and 30%, respectively).

Overall, the analysis showed that all three groups had similar understanding regarding NOS on the pretest. However, on the posttest, the treatment group had demonstrated the most growth toward a more informed view of NOS, followed by the LPS group, and, finally, the PDS group. Even though each group demonstrated some improvement, there were still major misconceptions regarding NOS at the posttest, including the beliefs that science requires experimentation, laws are proven facts while theories are unproven opinions, theories become laws once they have garnered sufficient support, scientists have seen atoms with microscopes, differing scientific theories are attributable to the fact that scientists either did not observe the phenomenon first-hand or manipulated the data to support their opinion, science is devoid of cultural influence, and scientists only use creativity in select aspects of their investigations.

Student outcomes (alongside teacher growth detailed above) were also very positive. Campers completed a 36-question science content test on the first and last day of camp. The 35 multiple-choice questions were taken from released fifth- and eighth-grade Virginia *Standards of Learning* science tests, *National Assessment of Educational Progress* (NAEP) tests, and *Trends in International Mathematics and Science Study* (TIMSS) tests. There was also one short-answer question that asked students to identify and explain what they knew about two scientists. Based on paired *t*-test analysis of the pre- and posttest data, as seen in Table 20.9, students in both camps demonstrated increased content knowledge but the gains were not significant when the two camps were examined independently. However, when the data were aggregated, there were significant improvements in students' science content knowledge.

**Table 20.9.** Student Growth in Science Knowledge During Camp

| | Pre | | Post | | | |
|---|---|---|---|---|---|---|
| **Camp** | **Mean** | ***SD*** | **Mean** | ***SD*** | ***t*** | ***p*** |
| **Space Exploration** | 31.10 | 4.55 | 31.98 | 5.41 | 1.61 | 0.119 |
| **Watershed Mysteries** | 27.45 | 4.68 | 28.94 | 4.53 | 1.74 | 0.100 |
| **Combined** | 29.56 | 4.91 | 30.66 | 5.24 | 2.39 | 0.021 |

Each next step CREST has taken has been informed by prior research to inform the road ahead. From summer institutes that held camps such as Medical Mysteries, measures of student science content knowledge were clearly shown to increase. In subsequent years, camps such as Weather Tamers showed clear growth in teacher confidence to engage with reform-based science pedagogy, while also showing increases in student science knowledge. Expanding further, while continuing to sustain reform based pedagogy and increasing student science knowledge, CREST continued to push our understanding of current, critical science research focuses, such as advancing understanding of NOS, as shown in the Watershed Mysteries camp. These positive steps led CREST to its current signature project.

## Next Steps

Informed by past successes, the successor to the programs above is the Virginia Initiative for Science Teaching and Achievement (VISTA), which expands science professional development to teachers throughout Virginia. Additionally, professional development programs for leaders from school district or division science coordinators, as well as college and university science education faculty were added as instrumental to the VISTA vision. Where states, districts, and society in general were not meeting the need to provide a structure to ensure a more robust pipeline of STEM career and advanced education–ready high school graduates, VISTA would begin to meet this need.

VISTA is a large-scale validation research grant funded through the Investing in Innovation (i3) program of the U.S. Department of Education. One of VISTA's four professional development programs, the Elementary Science Institute (ESI) integrates the summer science camp model. This model is being expanded and implemented from one to four universities across Virginia. The institute includes a four-week summer program and academic year follow-up support for teams of teachers from the same school as they experiment with problem-based learning in their classrooms.

During the summer, the teachers spend the first week learning how to conduct hands-on, inquiry-based science teaching using PBL, by planning for and implementing a summer camp. The camp topic is based on identifying the lowest-scoring areas of the previous year's state science assessments, and putting authentic, problem-based learning opportunities in place for teachers and students to investigate. In the second and third weeks, teachers in small groups collaboratively teach two days of the co-planned camp unit to upper elementary students. They also observe and assist during three days of camp. When they are not teaching or observing camp, teachers work through additional professional development modules in science teaching strategies and science content. Then, in the fourth week, teachers reflect on their summer experience and plan to implement a PBL science unit in their own classrooms. Throughout the summer program, teachers have access to university science educators, scientists, engineers, math specialists, special education specialists, and ELL specialists to help them plan the summer camp as well as their academic year PBL units. Additionally, principals and school district science coordinators attend a one-day seminar in order to learn about the science teaching methods advocated by VISTA and meet with their teachers to discuss ways to support hands-on, inquiry- and PBL-based science teaching in their schools and school districts.

As research (Ladson Billings 1994; Tobin, Elemesky, and Seiler 2005) has increasingly shown that economically marginalized students have continued to have less access to high-quality science teachers and classrooms outfitted with rich science materials, the VISTA program works explicitly to begin curtailing this problem. By inviting only economically disadvantaged students—a large number of them students of color — VISTA has built into its reform strategy both a training platform for teachers, but also serving the public good by supporting all students to see science and careers in science as something to be valued and a potential future line of work.

Returning to their classrooms, teachers implement their hands-on, inquiry-based PBL science units and attend follow-up sessions that continue to strengthen their ability to implement the pedagogical strategies introduced during the summer as well as give them a chance to share their challenges and successes. Additionally, a trained instructional coach is assigned to each teacher and provides a total of 24 hours of science classroom support during the academic year.

Looking to the future, teachers are cultivated as professionals. VISTA supports teachers to attend its state-level professional association, the Virginia Association of Science Teachers (VAST) annual conference. There, they learn about other teacher's successes and present their science teaching, learning, and problem-based learning resources to a statewide audience, meeting with colleagues for further professional development.

VISTA additionally has begun a program for new secondary teachers, providing them supports of a cohort-based, best practices graduate courses over two years, the means to travel to the state science teacher's professional association meetings, and financial remuneration for their time. Prior research conducted through CREST to create a supportive network of new science teachers showed an almost 10% increase in treatment teachers' student state assessment scores for science (Sterling, Matkins, and Kitsantas 2004). Directly informed by this prior work, current VISTA research on the secondary program's efficacy has been commenced, with preliminary data indicating VISTA participating teachers show greater gains on state science assessments than veterans of many more years. VISTA additionally has begun conducting research on the two science education academies—one for school district science coordinators, one for college faculty.

All the while VISTA conducts its statewide professional development, the research above is executed by parallel teams of researchers assessing VISTA's effects. As outside evaluators, a team of researchers from Oregon State University and the University of Virginia are assessing the program's effects on student achievement. Using interview, survey, observational, and state standardized test scores, a composite picture of VISTA's successes is being illustrated. At the same time, an internal team of implementation researchers led by a team at George Mason University are assessing the social and structural factors outside VISTA's control that place obstacles in the way of robust, reform-based science education in American schools. Together, these two parallel research strands have the potential to offer one of the most comprehensive, methodologically rich pictures of science education reform in action in the contemporary United States.

## Ties to Other Reform Efforts

As the national focus on STEM has evolved, so has CREST and VISTA's focus. The "E" in STEM for engineering is now firmly a part of the *Next Generation Science Standards* (NGSS Lead States 2013). In this reform climate, we have expanded our original work in science problem-based learning (Sterling 2007, 2010), linking engineering design as a research process parallel to science investigations (NRC 2012; Sterling 2000).

*A Framework for K–12 Science Education* states that "new insights from science often catalyze the emergence of new technologies and their applications, which are developed using engineering design" (NRC 2012, p. 210). The *Framework* emphasizes the interdependence of science, technology, and engineering to harness the power of design activities to support authentic learning of science and engineering concepts. It also advocates that students during their K–12 years carry out many scientific investigations and engineering design projects. Supporting this, VISTA's 2013 summer Elementary Science Institute seamlessly integrated both rich science and engineering education principles into its problem-based learning. Teachers and students investigated issues of extractive and sustainable forms of energy, using both science content and engineering process skills to design solar-powered cars, wind turbines, and fish ladders. Students engage their creative efforts to solving critical real-world problems that not only Virginia teachers and students, but all American citizens face.

The hands-on nature of problem-based science and engineering design activities engages students, fosters higher-order thinking, and deepens understanding of how integrated science, technology, engineering, and mathematics (STEM) influences the world around them. For understanding society and the natural world, students need to understand the interdependence of science, engineering, and technology as well as the engineering design process that includes defining an engineering problem, developing potential solutions, and optimizing the design solution. By exposing students to authentic science and engineering activities, you help students develop a career readiness perspective and skills.

## References


Abd-El-Khalick, F. 1998. The influence of history of science courses on students' conceptions of the nature of science. *Dissertation Abstracts International* 59 (07): UMI No. 984244.

Abd-El-Khalick, F. 2001. Embedding nature of science instruction in preservice elementary science courses: Abandoning scientism, but.... *Journal of Science Teacher Education* 12: 215–233.

Abd-El-Khalick, F., and N. G. Lederman. 2000. The influence of history of science courses on students' views of nature of science. *Journal of Research in Science Teaching* 37: 1057–1095.

Akerson, V. L., T. A. Cullen, and D. L. Hanson. 2009. Fostering a community of practice through a professional development program to improve elementary teachers' views of nature of science and teaching practice. *Journal of Research in Science Teaching* 46: 1090–1113.

Anderson, R., M. Greene, and P. Loewen. 1988. Relationships among teachers' and students' thinking skills, sense of efficacy, and student achievement. *Alberta Journal of Educational Research* 34 (2): 148–165.

Armor, D., P. Conroy-Oseguera, M. Cox, N. King, L. McDonnell, A. Pascal, E. Pauly, and G. Zellman. 1976. *Analysis of the school preferred reading programs in selected Los Angeles minority schools. www.rand.org/pubs/reports/2005/R2007.pdf*

Ashton, P. T., and R. B. Webb. 1986. *Making a difference: Teachers' sense of efficacy and student achievement.* New York: Longman.

Bandura, A. 1977. Self-efficacy: Towards a unifying theory of behavioral change. *Psychological Review* 84: 191–215.

Cakiroglu, J. 2000. Preservice elementary teachers' self-efficacy beliefs and their conceptions of photosynthesis and inheritance. *Dissertation Abstracts International* 61 (08): UMI No. 9980981.

Chang, Y. 2003. An examination of knowledge assessment and self-efficacy ratings in teacher preparation programs in Taiwan and the United States. *Dissertation Abstracts International* 64 (06): UMI No. 3095143.

Cochrane, B. 2003. Developing preservice elementary teachers' views of the nature of science (NOS): Examining the effectiveness of intervention types. Paper presented at the Annual International Conference of the Association for the Education of Teachers of Science, St. Louis, MO.

Delisle, R. 1997. *How to use problem-based learning in the classroom*. Alexandria, VA: Association for Supervision and Curriculum Development.

Druva, C. A., and R. D. Anderson. 1983. Science teacher characteristics by teacher behavior and by student outcome: A meta-analysis of research. *Journal of Research in Science Teaching* 20: 467–479.

Enochs, L. G., and I. M. Riggs. 1990. Further development of an elementary science teaching efficacy belief instrument: A preservice elementary scale. *School Science and Mathematics* 90: 694–706.

Frazier, W. M., and D. R. Sterling. 2007. Weather tamers. *Science Scope* 30 (7): 26–31.

Frazier, W. M., and D. R. Sterling. 2008. Problem-based learning for science understanding. *Academic Exchange Quarterly* 12 (2): 111–115.

Gibbs, G. 2002. *Qualitative data analysis: Explorations with NVivo*. Philadelphia, PA: Open University Press.

Gibson, S., and M. Dembo. 1984. Teacher efficacy: A construct validation. *Journal of Educational Psychology* 76 (4): 569–582.

Goldhaber, D. D., and D. J. Brewer. 1997. Evaluating the effect of teacher degree level on educational performance. In *Developments in School Finance 1996*, ed. W. J. Fowler, 197–210. Washington, DC: National Center for Educational Statistics.

González, N., L. C. Moll, and C. Amanti. 2005. *Funds of knowledge: Theorizing practices in households, communities, and classrooms*. Mahwah, NJ: L. Erlbaum Associates.

Greenwald, N. L. 2000. Learning from problems. *The Science Teacher* 67 (4): 8–12.

Hannum, J. W. 1994. The organizational climate of middle schools, teacher efficacy, and student achievement. *Dissertation Abstracts International* 55 (12): UMI No. 9514121.

Hmelo-Silver, C. E. 2004. Problem-based learning: What and how do students learn? *Educational Psychology Review* 16: 235–266.

Kerley, D. M. 2004. Exploring connections among relational trust, teacher efficacy, and student achievement. *Dissertation Abstracts International* 65 (12): UMI No. 3158253.

Koc, I. 2006. Preservice elementary teachers' alternative conceptions of science and their self-efficacy beliefs about science teaching. *Dissertation Abstracts International* 68 (01): UMI No. 3248024.

Krippendorff, K. 1980. *Content analysis: An introduction to its methodology*. Thousand Oaks, CA: Sage.

Krynock, K., and L. Robb. 1999. Problem solved: How to coach cognition. *Educational Leadership* 50 (3): 8–12.

Ladson-Billings, G. 1994. *The dreamkeepers: Successful teachers of African American children*. San Francisco: Jossey-Bass.

Lederman, N. G., F. Abd-El-Khalick, R. L. Bell, and R. S. Schwartz. 2002. Views of nature of science questionnaire: Toward valid and meaningful assessment of learners' conceptions of nature of science. *Journal of Research in Science Teaching* 39: 497–521.

Lederman, N. G., R. Schwartz, F. Abd-El-Khalick, and R. L. Bell. 2001. Preservice teachers' understanding and teaching of nature of science: An intervention study. *Canadian Journal of Science, Mathematics, and Technology Education* 1: 135–160.

Ledford, D. M. 2002. Teachers' level of efficacy as a predictor of the academic achievement of students with disabilities. *Dissertation Abstracts International* 63 (05): UMI No. 3052726.

Logerwell, M. G. 2009. The effects of a summer science camp teaching experience on preservice elementary teachers' science teaching efficacy, science content knowledge, and understanding of the nature of science. Unpublished dissertation. UMI No. 3367054.

Luera, G. R., R. H. Moyer, and S. A. Everett. 2005. What type and level of science content knowledge of elementary education students affect their ability to construct an inquiry-based science lesson? *Journal of Elementary Science Education* 17: 12–25.

Midgley, C., H. Feldlaufer, and J. S. Eccles. 1989. Change in teacher efficacy and student self- and task-related beliefs in mathematics during the transition to junior high school. *Journal of Educational Psychology* 81: 247–258.

Miles, M. B., and A. M. Huberman. 1994. *Qualitative data analysis: An expanded Sourcebook*. 2nd ed. Thousand Oaks, CA: Sage.

Monk, D. H. 1994. Subject matter preparation of secondary mathematics and science teachers and student achievement. *Economics of Education Review* 13 (2): 125–145.

Moore, W., and M. Esselman. 1992. Teacher efficacy, power, school climate and achievement: A desegregating district's experience. Paper presented at the Annual meeting of the American Educational Research Association, San Francisco, CA.

Muijs, D., and D. Reynolds. 2002. Teachers' beliefs and behaviors: What really matters? *Journal of Classroom Interaction* 37 (2): 3–15.

NGSS Lead States. 2013. *Next Generation Science Standards: For states by states*. Washington, DC: National Academies Press. *www.nextgenscience.org/next-generation-science-standards*

National Research Council (NRC). 2012. *A framework for K–12 science education: Practices, crosscutting concepts, and core ideas.* Washington, DC: National Academies Press.

Project MOSART. 2007. *Grades 9–12 physics test form 741. www.cfa.harvard.edu/sed/mosart*

Riggs, I. M. 1988. The development of an elementary teachers' science teaching efficacy belief instrument. *Dissertation Abstracts International* 49 (12): UMI No. 8905728.

Ross, A. T. 1998. Exploring connections among teacher empowerment, teacher efficacy, transformational leadership, and student achievement. *Dissertation Abstracts International* 59 (09): UMI No. 9906642.

Ross, J. A. 1992. Teacher efficacy and the effect of coaching on student achievement. *Canadian Journal of Education* 17 (1): 51–65.

Schoon, E., and W. Boone. 1998. Self-efficacy and alternative conceptions of science of preservice elementary teachers. *Science Education* 82: 553–568.

Schwartz, R. S., N. G. Lederman, and B. Crawford. 2000. Understanding the nature of science through scientific inquiry: An explicit approach to bridging the gap. Paper presented at the Annual Meeting of the National Association for Research in Science Teaching, New Orleans, LA.

Shack, G. D. 1993. Involving students in authentic research. *Educational Leadership* 50 (7): 8–12.

Staples, K. A. 2002. The effect of a nontraditional undergraduate science course on teacher and student performance in elementary science teaching. *Dissertation Abstracts International* 63 (05): UMI No. 3051995.

Stepien, W., and S. Gallagher. (1993). Problem-based learning: As authentic as it gets, *Educational Leadership* 50 (7): 25–28.

Sterling, D. R. 2000. Science and engineering. *Science Scope* 24 (2): 24–29.

Sterling, D. R. 2007. Modeling problem-based instruction. *Science and Children* 45 (4): 50–53.

Sterling, D. R. 2010. Hurricane proof this! *Science and Children* 47 (7): 48–51.

Sterling, D. R., and W. M. Frazier. 2006. Collaboration with community partners. *The Science Teacher* 73 (4): 28–31.

Sterling, D. R., and W. M. Frazier. 2010. *New science teachers' support*. Evaluation report submitted to the National Science Foundation.

Sterling, D. R., W. Frazier, S. Roche, and M. Logerwell. 2008. *Stars and splash: Science for preservice elementary teachers*. Evaluation Report submitted to State Council of Higher Education for Virginia, Richmond.

Sterling, D. R., and D. L. Hargrove. 2012. Is your soil sick? *Science and Children* 49 (8): 51–55.

Sterling, D. R., J. J. Matkins, W. M. Frazier, and M. G. Logerwell. 2007. Science camp as a transformative experience for students, parents, and teachers in the urban setting. *School Science and Mathematics* 107 (4): 134–148.

Sterling, D. R., J. J. Matkins, and A. Kitsantas. 2004. New science teachers without teacher training: What works to keep them in the profession? [CD-ROM] *NARST Conference 2004*. Vancouver, BC, Canada: National Association for Research in Science Teaching.

Tobin, K. G., R. Elmesky, and G. Seiler. 2005. *Improving urban science education: New roles for teachers, students, and researchers*. Lanham, MD: Rowman & Littlefield.

Watson, S. 1991. A study of the effects of teacher efficacy on the academic achievement of third-grade students in selected elementary schools in South Carolina. *Dissertation Abstracts International* 53 (06): UMI No. 9230552.

Wenglinsky, H. 2000. *How teaching matters: Bringing the classroom back into discussions of teacher quality.* Princeton, NJ: Educational Testing Service.

# Integrating Technology and Engineering in a STEM Context

*Barry N. Burke*
*International Technology and Engineering Educators Association*
*Reston, VA*

*Philip A. Reed*
*Old Dominion University*
*Norfolk, VA*

*John G. Wells*
*Virginia Tech*
*Blacksburg, VA*

## Setting

Imagine students entering the classroom with an enthusiasm that cannot be contained. They come from all walks of life and with different experiences and backgrounds and are eager to engage in learning. Inspiration and innovation are on their mind. What they learned in their science and math classes is now being applied in another class they take called Technology and Engineering. Opportunity is what they see for their future. Something about connecting all the dots from all their classes propels them to change their outlook, to get involved, to get excited about school and to envision their future.

This is just what is happening in over 1,800 classrooms, with over 53,000 students in over 580 schools nationwide. Teachers in these classrooms are using a program called Engineering byDesign (EbD) to deliver Technology and Engineering in a STEM context. Schools in inner-city, urban, suburban, and rural settings are all participating in the program as "EbD-Network Schools." Network schools have agreements in place that are signed by the teacher, principal, supervisor, and superintendent. The EbD-Network has experienced an average annual growth rate of 35% since its inception in 2007 (ITEEA 2012). EbD is successful because it is hands-on, relevant to the student, and uses real-world problems as the context for teaching and learning.

Engineering byDesign is a standards-based integrative STEM education model program that was developed by the International Technology and Engineering Educators Association's STEM Center for Teaching and Learning. The vision was to take multiple sets of content standards and transform them into classroom practice that brings the technology and engineering to STEM. In its infancy, EbD focused on *Standards for Technological Literacy* (ITEEA), *National Science Education Standards* (NRC), *Benchmarks for Science Literacy* (AAAS), and *Principles & Standards for School Mathematics* (NCTM). Since late 2011, EbD has moved to work specifically with the *Common Core State Standards,* in mathematics and English Language Arts. As the *Next Generation Science Standards* (*NGSS*) were developed (NGSS Lead States 2013), EbD has worked to include science and engineering practices, crosscutting concepts, and disciplinary core ideas to ensure that students are technologically literate using *NGSS* materials and *Standards for Technological Literacy* (ITEEA 2000, 2005, 2007).

To set the stage for integrating technology and engineering in a STEM education context, the authors begin with a common understanding of not just STEM education, but Integrative STEM education. Integrative STEM education is operationally defined as "the application of technological/engineering design based pedagogical approaches to *intentionally* teach content and practices of science and mathematics education concurrently with the content and practices of technology/engineering education. Integrative STEM education is equally applicable at the natural intersections of learning within the continuum of content areas, educational environments, and academic levels" (Wells and Ernst 2012). Using the Wiggins and McTighe (1998) Understanding by Design Model, curriculum and assessments have been developed and has driven the development of focused professional learning communities.

## Overview of the Program

EbD is a standards-based model that address the four *National Science Education Standards* (*NSES*) goals (NRC 1996) in an integrative STEM context. As EbD was developed, authors from the science, technology and engineering, and mathematics community coordinated their writing efforts to address the ideals and underlying goals from each of the respective content standards. These broad overarching goals were used to ensure content richness and depth:

1. Knowing and understanding the natural and the designed world;
2. Using appropriate scientific and engineering processes to inform decision-making;
3. Engage the public in matters of technological and scientific awareness and concern;
4. Use data to inform productivity as it relates to the natural and designed worlds in today's global marketplace.

With the introduction of the *Next Generation Science Standards* (*NGSS*; NGSS Lead States 2013), the model has reworked content to not just "align" with the standards but carry on the tradition of a standards-based approach to development and implementation. The EbD program fits neatly into the advances in the *NGSS*. An example of the crosswalk between *NGSS* and *Standards for Technological Literacy* follows in Figure 21.1.

## Figure 21.1. Middle School *NGSS* Alignment (Partial)

| NEXT GENERATION SCIENCE STANDARDS | KEY: 4 = Benchmark must be covered in detail, lessons and assessments cover this content<br>3 = Benchmark is covered, but topics and lessons do not center on them<br>2 = Topics and lessons refer to previous knowledge and integrate content covered<br>1 = Topics and lessons refer to previous knowledge | Standards for Technological Literacy | Engineering byDesign™: Exploring Technology | Engineering byDesign™: Invention & Innovation | Engineering byDesign™: Technological Systems |
|---|---|---|---|---|---|
| 4 | Construct an argument supported by evidence for how increases in human population and per-capita consumption of natural resources impact Earth's systems. | 5-F | 4 | 3 | |
| MS-ETS1- | **Engineering Design** | | | | |
| | **Students who demonstrate understanding can:** | | | | |
| 1 | Define the criteria and constraints of a design problem with sufficient precision to ensure a successful solution, taking into account relevant scientific principles and potential impacts on people and the natural environment that may limit possible solutions. | 8-G | 4 | 4 | 4 |
| 2 | Evaluate competing design solutions using a systematic process to determine how well they meet the criteria and constraints of the problem. | 2-S, 8-F, 11-I | 4 | 4 | 4 |
| 3 | Analyze data from tests to determine similarities and differences among several design solutions to identify the best characteristics of each that can combined into a new solution to better meet the criteria for success. | 11-K,11-L | 4 | 4 | 4 |
| 4 | Develop a model to generate data for iterative testing a modification of a proposed object, tool, or process such that an optimal design can be achieved. | 2-T, 9-H | 4 | 4 | 4 |

The goals and organizing principles of EbD are based on STL and aligned with *NGSS*, *NSES*, and the *Common Core State Standards*. The program is organized around 10 principles and has established the goal to restore America's status as the leader in innovation, by providing a program for students that constructs learning from a very early age and culminates in a capstone experience that leads students to become the next generation of engineers, technologists, innovators, and designers (ITEEA 2012). These principles are very large concepts that identify major content organizers for the program. The 10 organizing principles are:

1. Engineering through design improves life.
2. Technology and engineering have affected, and continues to affect everyday life.
3. Technology drives invention and innovation and is a thinking and doing process.
4. Technologies are combined to make technological systems.
5. Technology creates issues and impacts that change the way people live and interact.
6. Engineering and technology are the basis for improving on the past and creating the future.
7. Technology and engineering solve problems.
8. Technology and engineering use inquiry, design, and systems thinking to produce solutions.
9. Technological and engineering design is a process used to develop solutions for human wants and needs.
10. Technological applications create the designed world.

## EbD Development: A Unique Approach

In the beginning (1998), development began on the creation of a standards-based model. It was focused on how to deliver newly developed standards—to translate them from broad statements to student learning objectives and professional development. As EbD was conceived, it was not about more math and science, but about connecting math and science to technology and engineering. Author teams of science, mathematics, and technology/engineering were brought together to develop each guide based on the standards and benchmarks in their content area to ensure STEM content. Each unit and lesson prescribes the level of coverage that authors use in developing the content into classroom instruction. The grid in Figure 21.2 shows the relationship between *Common Core State Standards, Mathematics*; *Standards for Technological Literacy*; and EbD.

**Figure 21.2.** *Common Core:* STL Responsibility Matrix Used by Curriculum and Assessment Teams

COMMON CORE STATE STANDARDS INITIATIVE
PREPARING AMERICA'S STUDENTS FOR COLLEGE & CAREER

KEY
4 = Benchmark must be covered in detail, lessons and assessments cover this content
3 = Benchmark is covered, but topics and lessons do not center on them
2 = Topics and lessons refer to previous knowledge and integrate content covered
1 = Topics and lessons refer to previous knowledge
© International Technology and Engineering Educators Associaton

| Mathematics- Middle School | | STL | Exploring Technology | Invention & Innovation | Technological Systems |
|---|---|---|---|---|---|
| **6.RP** | **Ratios and Proportional Relationships** | | | | |
| 6.RP.1 | Understand the concept of a ratio and use ratio language to describe a ratio relationship between two quantities. For example, "The ratio of wings to beaks in the bird house at the zoo was 2:1, because for every 2 wings there was 1 beak." "For every vote candidate A received, candidate C received nearly three votes." | 9-D,E | 4 | | |
| 6.RP.2 | Understand the concept of a unit rate a/b associated with a ratio a:b with b ≠ 0, and use rate language in the context of a ratio relationship. For example, "This recipe has a ratio of 3 cups of flour to 4 cups of sugar, so there is 3/4 cup of flour for each cup of sugar." "We paid $75 for 15 hamburgers, which is a rate of $5 per hamburger." | 11-C | 3 | | |
| 6.RP.3 | Use ratio and rate reasoning to solve real-world and mathematical problems, e.g., by reasoning about tables of equivalent ratios, tape diagrams, double number line diagrams, or equations | 11-E | 4 | | |
| **7. N.S.** | **The Number System** | | | | |
| 7.NS.1. | Apply and extend previous understandings of addition and subtraction to add and subtract rational numbers; represent addition and subtraction on a horizontal or vertical number line diagram. | 13-C | | 1 | |
| 7.NS.2 | Apply and extend previous understandings of multiplication and division and of fractions to multiply and divide rational numbers. | 13-E | | 4 | |
| 7.NS.3 | Solve real-world and mathematical problems involving the four operations with rational numbers. | 11-G | | 4 | |
| **8.G** | **Geometry** | | | | |
| 8.G.5 | Use informal arguments to establish facts about the angle sum and exterior angle of triangles, about the angles created when parallel lines are cut by a transversal, and the angle-angle criterion for similarity of triangles. *For example, arrange three copies of the same triangle so that the sum of the three angles appears to form a line, and give an argument in terms of transversals why this is so.* | | | | 0 |
| 8.G.6 | Explain a proof of the Pythagorean Theorem and its converse. | | | | 0 |
| 8.G.7 | Apply the Pythagorean Theorem to determine unknown side lengths in right triangles in real-world and mathematical problems in two and three dimensions. | 13-E | | | 2 |
| 8.G.8 | Apply the Pythagorean Theorem to find the distance between two points in a coordinate system. | 13-F | | | 2 |
| 8.G.9 | Know the formulas for the volumes of cones, cylinders, and spheres and use them to solve real-world and mathematical problems. | 13-G | | | 3 |
| **8.SP** | **Statistics & Probability** | | | | |
| 8.SP.1 | Construct and interpret scatter plots for bivariate measurement data to investigate patterns of association between two quantities. Describe patterns such as clustering, outliers, positive or negative association, linear association, and nonlinear association. | 13-G | | | 1 |

## EbD: A STEM Program for *All* Students

Throughout development, the focus had to be on a program that could be implemented in any school in the country; be integrative STEM; be rigorous enough to challenge the brightest; and be flexible, affordable, and accountable. Foremost in the minds of the designers, this meant that the material presented had to be for *all* students. Therefore, EbD was designed with the "little e" in mind—providing the experiences a student will need to understand how the natural world and the designed world are used to design the future (engineering, little "e" used as a verb: Teach all students to think or learn to engineer or use engineering concepts [ITEA 2006]).

There is a distinct difference between helping all students to learn about an engineering way of thinking, versus the knowledge and skills required to prepare a student whose goal is to become an Engineer (the Big "E," used as a noun: Prepare students to be engineers, career-oriented. [ITEA 2006]). Further, the developers understand that if students grasp the little "e" that the Big "E" will certainly follow. That is, they will be prepared for careers as engineers.

Throughout the building blocks (STEM for grades K–5) and the secondary courses, materials are presented in a 5-E (Bybee 1998)/ 6-E lesson plan (Burke 2014) format. This format uses extension lessons that address further development of content connections with students.

## EbD Curriculum: An Integrative Approach for Teachers

EbD materials are classroom ready, so teachers can focus on student learning, not on "how" to deliver a lesson. Valuable time can be lost if a teacher is unsure of what comes next. Moreover, if a teacher does not understand how the unit and subsequent lessons flow, vital portions of a unit may not be covered as intended or not covered at all.

EbD is now available in two versions. The StandardEdition (EbD-SE) is what can be obtained from the ITEEA store (*www.iteea.org*), runs on a CD, and can be used in any PC or Mac computer. The MediaRichEdition (EbD-MRe) is completely web-based, only available for schools in the EbD-Network, and is constantly updated with changes, resources provided by teachers, and as its name implies, is *media rich*.

Engaging teachers with a dynamic curriculum, integrated online learning community and online assessment tools that can form the basis for informing instruction required a multi-faceted approach. The MRe, being web-based, provides the platform for updating content on a daily basis when needed or for rearranging content. In 2011, an integrated approach to curriculum, professional development, and assessment was unveiled through the creation of the EbD-Portal (Figure 21.3). The Portal connects what teachers need most when they need it most: online curriculum (MRe), online learning communities, and Pre-Post assessment tools (Student Assessment and Design Challenge).

**Figure 21.3.** EbD-Portal Resources

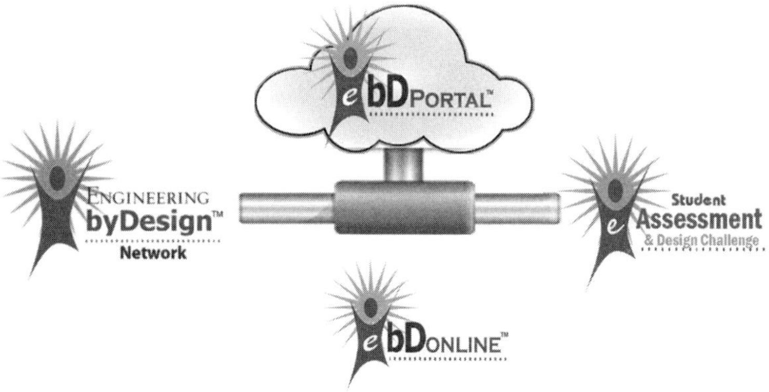

## EbD Core Program

The EbD model (Figure 21.4) consists of building blocks at grades K–5 and courses in each of the grade bands for middle school (grades 6–8) and high school (grades 9–12). Each elementary EbD-TEEMS building block consists of 20 lessons and incorporates an integrative STEM approach to delivering material that was previously presented in a traditional manner. Building blocks may be completed in a 1-week period, or implemented over a 6-week period. The building blocks are the first materials in EbD to be based on *NGSS, NSES, CCSS, STL,* and aligned to the NAE's Grand Challenges for Engineering.

**Figure 21.4.** The EbD Core Program

| CORE PROGRAM | Grade | HS Choices | Course | Duration |
|---|---|---|---|---|
| | K–2 | | EbD-TEEMS™ | 1-6 weeks |
| | 3–5 | | EbD-TEEMS™ *I³* | 1-6 weeks |
| | 6 | | *Exploring Technology* | 18 weeks |
| | 7 | | *Invention and Innovation* | 18 weeks |
| | 8 | | *Technological Systems* | 18 weeks |
| | 9 | | ***Foundations of Technology*** | 36 weeks |
| | 10–12 | HS Choices | ***Technology and Society*** | 36 weeks |
| | 10–12 | HS Choices | ***Technological Design*** | 36 weeks |
| | 11–12 | HS Choices | ***Advanced Design Applications*** * | 36 weeks |
| | 11–12 | HS Choices | ***Advanced Technological Applications*** * | 36 weeks |
| | 11–12 | | ***Engineering Design*** *(Capstone)* | 36 weeks |

The middle school program consists of three courses that explore the relationship between inquiry and design; then uses the knowledge and skills learned to invent, innovate, and then apply the engineering design processes to further develop understanding of how to combine the core areas of technology to create systems.

The high school program provides for a foundation that builds on the knowledge and skills learned in elementary and middle school to develop deeper understanding and skills around the natural and designed world. While there are six courses in the core sequence, it is anticipated that a high school would offer the Foundations course in grade 9 and Engineering Design (capstone course) in grade 12. This would leave two courses that could be chosen from the remaining four in the core as time, resources, and teacher expertise allows.

## EbD-Network of Schools

One of the challenges of a standards-based, dynamic curriculum is the ability to ensure that the materials are teacher-ready and that the infrastructure is easily updated. More important is to have a committed group of teachers that implement the materials with fidelity, use the assessment tools as they were designed, and participate in the online learning community. The EbD-Network of schools is comprised of teachers who have committed to all of these points. Figure 21.5 shows the growth in the network school program. Since 2007 the program has grown at a rate of approximately 35% per year.

Membership in the network varies. Individual schools as well as districts large and small have joined the network, providing the MRe resources to all their teachers. The network is comprised of inner-city schools; private schools; STEM academies; technical centers; and urban, rural, and suburban schools.

## Curriculum Foundations

EbD enhanced validity by actively engaging with several states involved with the requirements for Race to the Top (U.S. Department of Education 2014). Specifically, EbD focuses on the five core education reform areas. First, the nationally recognized standards upon which

**Figure 21.5.** EbD-Network School Growth

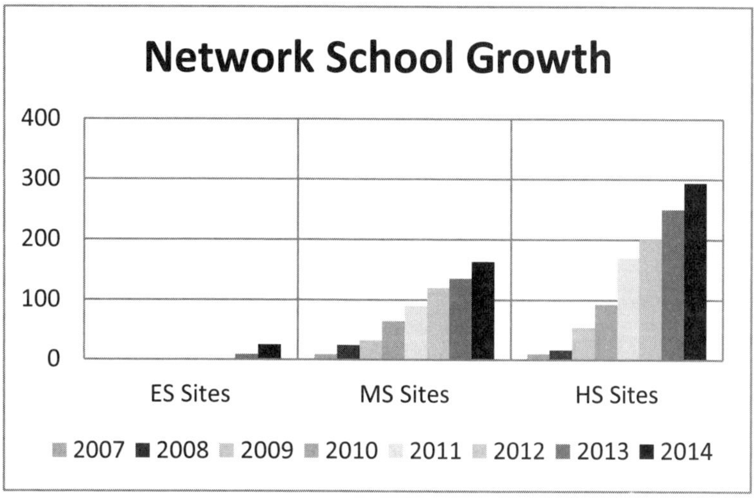


EbD curriculum and assessments are based help prepare students to succeed in college and the workplace and to compete in a global economy. Second, the system for collecting and reporting EbD assessment data measures student growth and success formatively as well as summatively, which informs teachers and principals about how they can improve instruction. Third, the STEM CTL's consortium of states developed a system that provides real-time data for teachers on student progress and the integration of assessments and curriculum as determined by Race to the Top. Opportunities for state, district, and local professional development can take place with trained Teacher Effectiveness Coaches (TECs) from the STEM⊕CTL. EbD materials are created using sound curriculum models and are coordinated and mapped to the three areas for the National Assessment of Educational Progress (NAEP) Technology and Engineering Literacy Assessment (WestEd 2009) as well as the Engineering Grand Challenges (NAE 2010). The 6E Learning byDeSIGN Model (Burke 2014) found in Table 21.1 provides students with a solid foundation for future STEM learning throughout the K–12 materials. A student-centered model, it is designed to maximize the connections between design and inquiry in STEM classrooms. Additionally, the program is built on constructivist models and creates awareness and competence over time as it builds on learned knowledge and skills.

**Table 21.1.** The ITEEA 6E Learning byDeSIGN Instructional Design Model

| | |
|---|---|
| **Engage** | The purpose of the ENGAGE phase is to pique student interest and get them personally involved in the lesson, while pre-assessing prior understanding. |
| **Explore** | The purpose of the EXPLORE phase is to provide students with the opportunity to construct their own understanding of the topic. |
| **Explain** | The purpose of the EXPLAIN phase is to provide students with an opportunity to explain and refine what they have learned so far and determine what it means. |
| **eNGINEER** | The purpose of the eNGINEER phase is to provide students with an opportunity to develop greater depth of understanding about the problem topic by applying concepts, practices, and attitudes. They use concepts learned about the natural world and apply them to the man-made (designed) world. |
| **Enrich** | The purpose of the ENRICH phase is to provide students with an opportunity to explore in more depth what they have learned and to transfer concepts to more complex problems. |
| **Evaluate** | The purpose of the EVALUATION phase is for both students and teachers to determine how much learning and understanding has taken place. |

## Professional Development

For EbD-Network schools, the online learning community is part of their "network" agreement. In addition to the online learning community, the center provides summer professional development opportunities around the country each summer. These institutes are typically a one-week professional development experience where teachers experience the content of the course. Included in this PD are the integrated STEM connections to mathematics and science so that teachers are able to return to the classroom and implement a successful integrative STEM program. There are additional PD opportunities online and at the ITEEA annual conference.

The EbD curriculum and professional development model challenges the existing silo mentality framework by presenting a viable alternative for teaching STEM education as a learner-centered integrative process (Humphreys, Post, and Ellis 1981). Furthermore, research has revealed that students engaged in integrative instruction outperform those in traditional classrooms on standardized tests (Hartzler 2000). Specific to the pedagogical connections within EbD curriculum, the integrative STEM education technological/engineering (T/E) design-based pedagogical model presented in Figure 21.6 (Wells 2008) depicts the integration of T&E design where scientific inquiry is an integral element of design. In upper-level EbD courses the interdisciplinary approach is more the norm for addressing design challenges that require discipline-specific content at varying levels of complexity in the development of a design solution. This approach helps students recognize the natural intersect between T&E design-based learning and scientific inquiry (Klein 1996; Lewis 2006). The EbD curriculum is intended to capitalize on the intersections of STEM content and practices in a manner congruent with how the brain organizes information and constructs knowledge (Bruning, Schraw, Norby, and Ronning 2004; Shoemaker 1991).

**Figure 21.6.** Integrative STEM Education T&E DBL Pedagogical Model

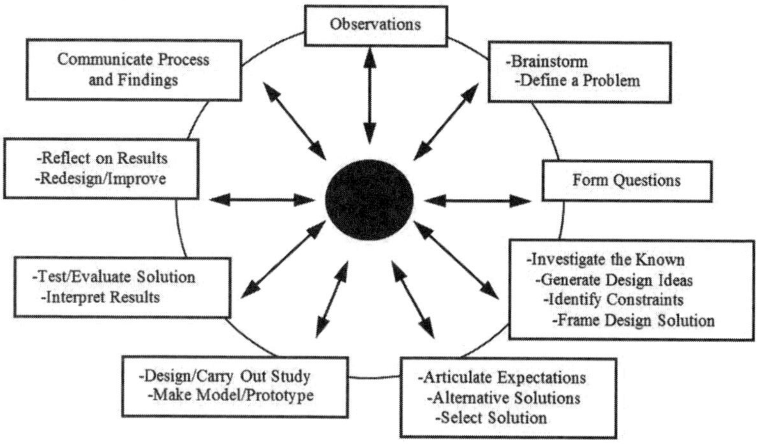


The EbD-Portal professional development model provides a unique environment based on a pedagogical commons approach (Wells 2008, 2010) whereby teachers engage in a common curriculum using a variety of appropriate instructional strategies and assessment of integrative achievement found to effectively promote STEM integration (Miller 2005; Satchwell and Loepp 2002).

## Collaborators

EbD has collaborators at all levels—from instructional design to corporate support. Eighteen states participate in the EbD Consortium of States that drive the development of materials and the EbD-Network. Schools in an additional five states also participate in the network. In Figure 21.4 (EbD Core Program), logos represent where collaborations with National Science Foundation (NSF) and the National Aeronautics and Space Administration (NASA)-funded projects that developed individual units or courses.

## Evidence for Success

### *Types of Information Collected*

Information, including demographics, is collected on network schools. For students, a pretest is used to ascertain their prior knowledge and provides the teacher with information necessary to plan instruction that is responsive to students' needs. The student pretest is intended to be both an embedded assessment and a methodology for connecting students' prior knowledge to content and skills. It is also a tool to determine grouping for collaborative learning. Formative assessments are included in the course guides and are recommended throughout instruction.

These are used to obtain information in order to adjust teaching based on the learning needs of the students.

The summative assessments are used to obtain final data about student learning gains, achievement, and instructional effectiveness. There are two summative assessment options included: a rubric to score students' solutions to the design challenge and a more traditional assessment (posttest) that reflects the standardized testing format employed by states for accountability purposes. In the current era of standards and accountability, the use of both summative assessment options is recommended. The following are findings from the Middle School courses offered by the EbD program (ITEEA 2012).

1. In the 2012–13 school year, Asian Females (14.57%) and African American Males (12.10%) reported the highest gains on EbD assessments.
2. Of the states reporting a minimum of 300 students, the three states that provided one-week professional development saw the highest student gains on the EbD assessments.
3. In the three middle school courses (Exploring Technology, Invention & Innovation, and Technological Systems), between 2009 and 2011, the student perception of the relevance of science has grown. In 2009, 66.1% of the students indicated that science was very relevant or relevant to the course and in 2011 this number increased to 75.6%. This is a growth of 13.6%.
4. Specifically, in Exploring Technology, the student perception of the relevance of science at the end of the course has grown from 29.6% in 2009 to 43.7% in 2011, a growth of almost 34%.
5. In 2011, when students began a middle school EbD course, almost 50% of them indicated that mathematics is very relevant. This is an increase of 23% from 2009 when only 27.1% of the students believed mathematics was very relevant. This may have indicated that students are seeing the value of mathematics and science when studying technology.
6. In middle school EbD courses, the percentage of students considering a career in an engineering field has increased from 7.6% in 2009 to 10.6% in 2011. While this is still a small overall percentage of the students considering engineering, the increase is notable.

## *Varied Users of the Program*

Endorsement of EbD is documented by the 18 consortium states, over 500 participating school systems reaching over 50,000 students in grades 6–12, and other organizations. The foundational document, *Standards for Technological Literacy* (ITEA 2000, 2005, 2007), went through a rigorous review cycle that included a review by the National Research Council. The foreword is by William A. Wulf, President of the National Academy of Engineering at the time of publication, and states, among other things, that: "[ITEEA] has successfully distilled an essential core of technological knowledge and skills we might wish all K–12 students to acquire." Additionally, EbD has been endorsed by the States' Career Clusters (NASDCTEc 2013) for the Science, Technology, Engineering, and Mathematics (STEM) and Information Technology (IT) clusters.

## *Outside Evaluation/Observers*

Most EbD curriculum was initially developed with support from the National Aeronautics and Space Administration (NASA) and the National Science Foundation (NSF) (see Figure 21.4). EbD staff, TECs, consortium members, and other partners are continually demonstrating in their classrooms and sharing at meetings and conferences. Presentations have included the NSTA annual conferences, NSTA STEM Forum, and NSTA Professional Development Institutes.

## *Voices of Instructors/Students*

Over the past five years, the STEM Center for Teaching and Learning has engaged teachers in the program in summer institutes where they learn the pedagogy and technical workings of the EbD materials. Professional development participants are engaged in curriculum and assessment activities so they experience the EbD materials they will use with students. Pre- and postsurveys are given at each workshop and participant comments provide insight into various aspects of the program. Some of the quotes deal with the interactive nature of the curriculum: "EbD curriculum put the E in Engaging," while others focus on the implementation model: "EbD places STEM at the fingertips of America's students."

A sixth-grade student, in an article in a local newspaper, wrote:

> *The next thing we learned about was the Engineering Design Process of input (the problem), process (how you get to your solution), output (the solution), and feedback (how well it works). We also learned about journaling and scale drawings as part of this lesson. Then, to put it all together, we had to create a solution to make a pencil that we couldn't lose. Now we are learning about transportation subsystems and working on a project to create a vehicle that can be propelled by wind across ice. This helps us apply our knowledge of control, guidance, structure, support, suspension, and propulsion as well as our knowledge of the Engineering Design Process. Tech Ed is one of my favorite subjects. If you're going to take it, look forward to it!"*

A ninth-grade student remarked the following: "I never really understood the importance of science until I took this course. When we do an activity, our teacher is always showing us how this relates to the science and math we learn. I never had a class that helped me better understand other classes [subjects]."

A STEM Supervisor had this to say about the program: "The EbD program at the middle school level is technology and engineering education with math and science embedded in the curriculum."

A postsecondary partner had this to say: "EbD provides exemplary standards-based curriculum and instructional materials for preservice technology and engineering education teachers to model and use."

An elementary EbD teacher and teacher effectiveness coach said,

> *Math and science are an integral part of the activities and challenges presented in the EbD materials. While students are designing and building, they have the opportunity to learn many*

*concepts For example, in math: measurement, money, graphing, comparing numbers, time, temperatures, weight, angles, and geometric shapes. Science concepts may include: the natural world, matter, animal shelters, weather magnets, simple machines, pneumatics, and the sun. As a teacher, how do you use the materials? Each year I align the curriculum I must teach with the activities and challenges within each Engineering by Design material. My main focus as I look through the activities is to connect them with the State science and social studies objectives. For instance, in science my students must learn about magnets. In order for them to gain a better understanding of repelling and attracting teams of students design and build a maglev train that actually works. As I watch my students participate in many of the activities in the EbD materials they are active participants who are enjoying themselves as they learn. I am a facilitator as they use their minds and hands to design and build.*

## Assessment Foundations

All assessments are based on the EbD Responsibility Matrix (see Figure 21.2), used by authors in the development of each course. The matrix is based on *Standards for Technological Literacy* (ITEA 2000, 2005, 2007) and lists all standards, benchmarks, and EbD courses. The codes listed at the top of Figure 21.2 are inserted to ensure curriculum and assessment developers are creating articulated materials that target the proper benchmarks. These codes are placed in the Responsibility Matrix to align courses and benchmarks so curriculum writers, assessment developers, and professional development providers can quickly identify content covered.

An assessment blueprint and table of specifications is developed to further help the assessment team create items that match the EbD Responsibility Matrix. A blueprint lists the STL benchmarks as well as other standards (i.e., *Common Core State Standards, Mathematics and ELA*; *NGSS*) that have been cross-walked in the curriculum and the depth of coverage. This assists the writers in determining how many assessment items need to be written for each benchmark. Processes include the annual refinement of existing items and the development of new test items to support the pre-post testing. Additionally, the assessment review team creates and updates the end-of-course design challenges. Here students work in groups to develop solutions to a design problem and then are rated on their knowledge of the design process and their entries in their engineering design journal (EDJ). Figure 21.7 shows the assessment participation rates for the past seven years.

**Figure 21.7.** Nationwide Assessment Participation

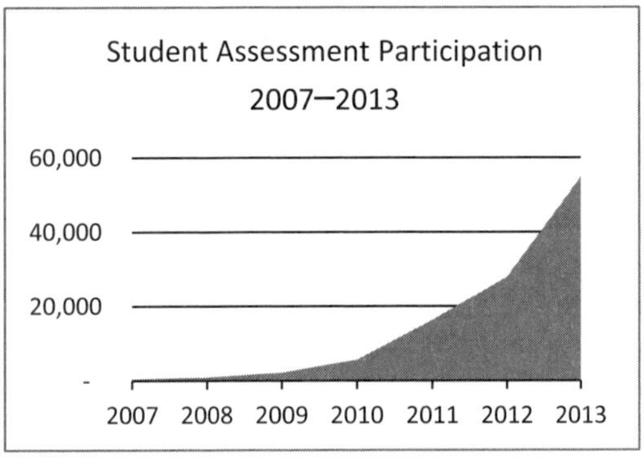


## Integrative STEM Education and EbD: What Does It Look Like?

Foundations of Technology (FoT) is the first EbD course (ninth grade) for high school students, because it builds on the knowledge and skills learned in elementary and middle school. Students develop deeper understanding and skills around the natural world and the designed world by studying key concepts such as the engineering design process. The following lesson is typical of EbD lessons for grades 6–12. Grades K–5 use a slightly different system of building blocks consisting of 20 standards-based lessons. The overview that follows is an exemplar from FoT, Unit 2, Lesson 1:

# Unit 2: Design
# Lesson 1: The Engineering Design Process
# Lesson Snapshot

**Big Idea:** The Engineering Design Process is a systematic, iterative problem-solving method that produces solutions to meet human wants and desires.

***Teacher Note:*** *Big ideas should be made explicit to students by writing them on the board and/or reading them aloud. For deeper understanding, have students write the Big Idea in their own Engineering Design Journal (EDJ), using their own words if they choose.*

**Purpose of Lesson:** Unit 2, Lesson 1 introduces students to the engineering design process and requires that they apply it.
**Lesson Duration:** Eight (8) hours.

**Activity Highlights:**
***Engagement:*** Students will watch a video entitled, "How I Harnessed the Wind," from *www.ted.com*. Students will record notes on the process used in the video to harness the wind. The teacher will lead a discussion on the process that was used by William Kamkwamba to harness the wind.
***Exploration:*** Given the steps of the Engineering Design Process on note cards (one step per card) (File 2.1.1or File 2.1.2), students will attempt to place the steps in the correct order. Students will use prior knowledge and the sequence demonstrated in the engagement example to determine the order. The teacher will give feedback and prompt students to justify their order.
***Explanation:*** The teacher presents the students with the correct sequence and delivers a presentation on the Engineering Design Process (Presentation 2.1.1). Students will record notes in their Engineering Design Journals (EDJ). A graphic organizer can be used to help students transition to the expanded Engineering Design Process (File 2.1.3). The teacher will deliver a presentation on the Pythagorean Theorem (Presentation 2.1.2), and use the Pythagorean Theorem Review (File 2.1.4) to work with students. Additional instructional resources are available in (Video 2.1.3).
***Extension:*** Students will apply the steps of the Engineering Design Process to a simple design problem (File 2.1.5). Students will document the Engineering Design process in their EDJ. Students will apply mathematical concepts related to the design challenge (File 2.1.5 and File 2.1.6).
***Teacher Note:*** *The data collected during the testing/evaluation of the design challenge will be used in Unit 2, Lesson 2. The teacher should make sure all data is recorded.*
***Evaluation:*** Student knowledge, skills, and attitudes are assessed using selected response items, brief constructed-response items, and performance rubrics for class participation, discussion, and design briefs.

For each lesson, teachers are provided with an overview that includes standards and benchmarks, learning objectives, resource material lists, required student knowledge and/or skills, and student assessment tools and/or methods (including rubrics). A lesson plan that follows the 6E model is provided for each lesson along with a file detailing recommended laboratory-classroom preparation notes. Finally, all files associated with the lesson are provided. If there is a student activity or worksheet, exemplars are provided to help teachers with the teaching and learning process. For example, the following handout is a student worksheet of the engineering design process with all of the blanks completed:

| Name: | Period: | Date: |
|---|---|---|
| Foundations of Technology<br>Unit 2 Lesson 1: The Engineering Design Process<br>File 2.1.3: Engineering Design Process Graphic Organizer | | |

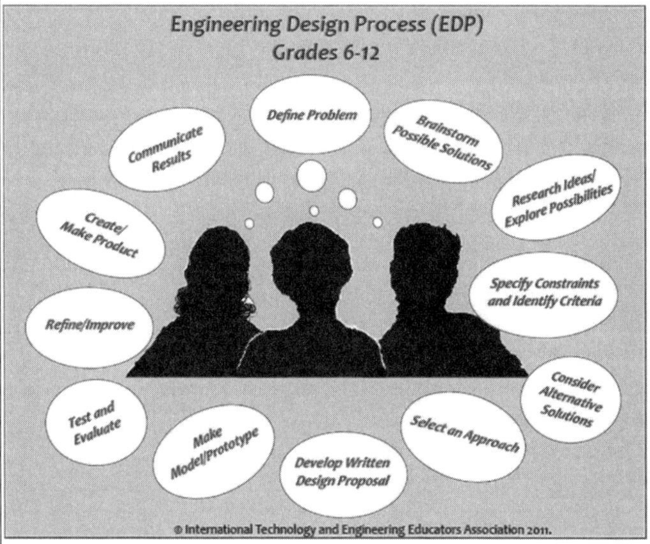

## Next Steps

In the past decade the focus on STEM education as an agenda for educational reform has brought about change not only in these four core disciplines, but in all disciplines and at all levels. This vision of teaching STEM content and practices as an integrative instructional approach has been the pedagogical premise of Technology/Engineering (T/E) Education since the early 1900s and continues today as reflected in the opening pages (pp. 6–9) of the *Standards for Technological Literacy* first published in 2000 (ITEA 2000/2005/2007). Unique to Integrative STEM Education (I-STEM ED) for Technology and Engineering Education is the use of technological and engineering design-based learning (T&E DBL) to intentionally teach content and practices of not only T/E, but science and mathematics as well (Wells 2013, p. 29). As the flagship curriculum for ITEEA, EbD was designed to be the pathway for implementing the AAAS vision and its application of the I-STEM ED approach the vehicle for bringing together traditionally silo STEM disciplines. The hallmark of this curricular approach is the use of T&E to *intentionally* teach STEM content and practices as an integrative endeavor. Critical to the sustainability of EbD will be a continuous evolution in its evaluation of the model used for achieving 21st-century integrative STEM education learners.

A particularly daunting challenge for EbD PD is developing the required level of pedagogical content knowledge (PCK) demanded of the teacher attempting to implement T&E design-based learning strategies. To evaluate the extent to which participating teachers have gained the ability to meet these demands, EbD is designing PD assessment that seeks to document the teacher learning process and ensuing changes in their pedagogical practices. Baseline information on participant characteristics is gathered through demographic data, and their propensity to fully adopt the EbD instructional model is determined using the Stages of Concern (SoC) instrument. Evaluation of the instructional strategies employed by EbD teachers will be accomplished using the Indicators of Instructional Change (IIC) instrument for pre/post lesson analysis (Wells 2007) in concert with an instructional observation protocol designed to gauge their level of PCK (Wells 2011).

## Yet to Try

As initially envisioned, EbD is a standards-based model designed to integrate technology and engineering within a STEM education context. The model is being implemented and practiced in more than 1,800 classrooms across multiple states and annually engages more than 50,000 students nationwide. A basic tenant of EbD is fostering student learning through T&E design-based learning using integrative STEM education approaches. Achieving change of this order requires sustained systematic modifications to schooling, rethinking traditional approaches to pre/inservice professional development, and a fundamental redesign of the current teacher preparation process. Recognizing such large-scale change must be done in concert with state and national initiatives, EbD has worked in concert with the *Common Core State Standards, Mathematics and ELA*, as well as the *Next Generation Science Standards* for specifically addressing the practices, concepts, and disciplinary core ideas necessary to ensure technological literacy for all learners. In collaboration with these national STEM education initiatives, EbD provides the educational infrastructure necessary for developing 21st-century educators capable of preparing today's students for tomorrow's global challenges.

## Proposed Use of the Data: EbD Assessment

Assessing the extent of student learning as a result of participating in EbD is challenging, given the very nature of integrative STEM education teaching practices and both individual and team approaches employed in T&E design-based learning activities. EbD currently follows a fairly traditional method of student assessment using pre/post EOC gain scores as a measure of changes in student content knowledge. In contrast, the T&E design challenges serve as a more progressive EOC summative assessment metric requiring alternative approaches to evaluating student comprehension as revealed in the evidence embedded in their design solutions. Together these data provide a measure of the extent to which participation in EbD is promoting STEM literacy. As a result of the Race to the Top initiatives in many states, teachers have begun to use the pre-post assessments in ways that help the teacher identify student learning gains. In 2014, Maryland and New York teachers use the pretest to identify areas of strengths and weaknesses. They are then able to modify instructional strategies to help students achieve higher gains. These gains (or losses) are used by the teachers as part of the "Standards of Learning" that translates to a portion of their teacher effectiveness—or annual teacher evaluation. Scaling this model to other states so that teachers can be more efficient and successful is a proposed upgrade to the system.

## Ties to Other Reform Efforts

In the context of global assessment metrics such as the *Programme for International Student Assessment* (PISA; OECD 1997), national assessment of student learning in the United States is evolving toward the use of open-ended, novel design-based scenarios that require learners to demonstrate understanding rather than recall. The dynamic and complex nature of T&E design-based learning places unique cognitive demands on students and requires their use of STEM practices in producing viable design solutions. To evaluate development of these higher-order cognitive skills, EbD is developing its assessment strategies to be in line not only with international tools (PISA), but national measures as well such as those found in both the NAEP 2014 Technology and Engineering Literacy Assessment (WestEd 2009) and the NAEP 2009 Science Assessment Framework (NAGB 2008). Student performance expectations correlate well with their ability to respond to a set of four cognitive demands (knowing that, knowing how, knowing why, and knowing when and where to apply knowledge) which can be assessed at the basic, proficient, and advanced levels. These cognitive demands offer a means of assessing knowledge gained along the declarative, procedural, schematic, and strategic continuum (Wells 2008, 2010). EbD is incorporating these national assessment strategies and looking to document the connections between T&E design-based instructional strategies and the cognitive domains of learning through this integrative STEM education approach.

## Questions About EbD by Others

There are traditionally three questions asked by others (and responses) with regard to the program:

1. How much does it cost for the curriculum? The equipment? The materials? The software?

   In a state that is a member of the EbD Consortium, the curriculum is free. Non-Consortium state schools may opt in by becoming part of the EbD network or purchasing the course guide from the ITEEA web store. Some small processing equipment and hand tools are required. Each course has a list that is provided as part of the course guide. Most of the materials that are used in the EbD program are ones that can be purchased locally. The costs vary by course, and are provided as part of each course guide. The software required includes an office suite (e.g, MS Office) and design software. EbD-Network schools are eligible to receive the Design Academy Suite of products from Autodesk, Inc. at no charge through a partnership agreement.

2. Professional Development: Where? When? How long? Is it required?

   Professional development is available each summer at various locations around the country. The PD Planner can be found at *www.iteea.org/PD*. Institutes are generally one week long and cost approximately $425 for the week. PD is not required, but highly recommended. All institutes are led by ITEEA-authorized teacher effectiveness coaches and include all the materials and access to the MRe version of the guides. All PD is hands-on.

3. If we are to teach STEM in our school, how do we teach engineering? We don't have an engineer in our school.

   Most schools have a technology and engineering teacher in their school. This teacher may teach design or other hands-on type of class. Some schools call it "technology education." These teachers can be a significant component to an integrative STEM program. A team of teachers—from science, mathematics and technology/engineering—can effectively deliver the STEM program such as EbD, each providing the content to make the instruction stronger.

For More Information About Engineering byDesign:

- *www.engineeringbydesign.org* (general information)
- *www.iteea.org/EbD/Resourses/EbDresources.htm* (resources and PowerPoints)
- *www.iteea.org/EbD/CATTS/cattsconsortium.htm* (consortium of states)
- *www.iteea.org/EbD/PD/index.htm* (professional development)

## References

American Association for the Advancement of Science (AAAS). 1993, 2009. *Project 2061: Benchmarks for science literacy.* Washington, DC: AAAS.

Bruning, R. H., J. G. Schraw, M. M. Norby, and R. R. Ronning. 2004. *Cognitive psychology and instruction.* Columbus, OH: Pearson.

Burke, B. N. 2014. The ITEEA 6E learning byDeSIGN model: Maximizing informed design and inquiry in the integrative STEM classroom. *Technology and Engineering Teacher* March: 32–38.

Hartzler, D. S. 2000. A meta-analysis of studies conducted on integrated curriculum programs and their effects on student achievement. Unpublished doctoral dissertation, Indiana University, Bloomington, IN.

Humphreys, A., T. Post, and A. Ellis. 1981. *Interdisciplinary methods: A thematic approach.* Santa Monica, CA: Goodyear Publishing.

International Technology Education Association (ITEA). (2000, 2005, 2007). *Standards for Technological Literacy: Content for the Study of Technology.* Reston, VA.: Author.

International Technology and Engineering Educators Association (ITEEA). 2012. *Assessment consortium report, 2012.* Reston, VA: ITEEA.

National Academy of Engineering (NAE). 2010. NAE grand challenges for engineering. *www.engineeringchallenges.org*

National Assessment Governing Board (NAGB). 2008. *Science framework for the 2009 national assessment of educational progress.* Washington, DC: NAGB.

National Council of Teachers of Mathematics (NCTM). 2000. *Principles and standards for school mathematics.* Reston, VA: NCTM.

National Governors Association Center for Best Practices and Council of Chief State School Officers (NGAC and CCSSO). 2010. *Common core state standards.* Washington, DC: NGAC and CCSSO.

NGSS Lead States 2013. *Next Generation Science Standards: For states, by states.* Washington, DC: National Academies Press. *www.nextgenscience.org/next-generation-science-standards*

Organisation for Economic Co-operation and Development (OECD). 1997. Programme for International Student Assessment (PISA). *www.oecd.org/pisa/faqoecdpisa.htm*

Satchwell, R. E., and F. Loepp. 2002. Designing and implementing an integrated mathematics, science, and technology curriculum for the middle school. *Journal of Industrial Teacher Education* 39 (3): 41–66.

U.S. Department of Education. 2014. Race to the top fund. *www2.ed.gov/programs/racetothetop/index.html*

Wells, J. G. 2007. Key design factors in durable instructional technology professional development. *Journal of Technology and Teacher Education* 15 (1): 101–118.

Wells, J. G. 2008. Potentials of integrative STEM education in technology education. Paper presented at the Mississippi Valley Technology Teacher Education Conference, St. Louis, MO.

Wells, J. G. 2010. Research on teaching and learning in science education: Potentials in technology education. In *Research in technology education*, ed. P. Reed and J. LaPorte, 145–157. Muncie, IN: Ball State University.

Wells, J. G. 2011. Beyond preordinate evaluation of professional development: Responsive assessment of enacted projects. *Journal of Research in Education* 17 (35): 219–257.

Wells, J. G. 2013. Integrative STEM Education at Virginia Tech: Graduate preparation for tomorrow's leaders. *The Technology and Engineering Teacher* 72 (5): 28–34.

Wells, J. G., and J. V. Ernst. 2012. Integrative STEM education. *www.soe.vt.edu/istemed*

WestEd. 2009. Framework and assessment and item specifications for the 2014 national assessment of educational progress (NAEP) in technology and engineering literacy. *www.nagb.org/publications/frameworks/technology/2014-technology-specification.html*

Wiggins, G., and J. McTighe. 1998. *Understanding by design.* Alexandria, VA: ASCD.

# RIP~ing Through STEM

*Robert E. Landsman*
*ANOVA Science Education Corporation*
*Honolulu, HI*

*R. Gordon Schaubhut*
*R/G Engineering*
*Glen Rock, NJ*

## Setting

Authors Robert Landsman and Gordon Schaubhut met in high school through common interests and experiences, and became good friends. College and graduate school diverted them into different career pathways—neuroscience and electrical engineering. Thirteen years later at the American Museum of Natural History and the City University of New York they combined mathematical knowledge and practices with their STEM fields to engineer the design of a scientific research procedure using a variety of technologies. Their vision to one day collaboratively share how STEM thinking is authentically executed through the fields of science and engineering with the use of mathematics and technology is realized in this chapter.

The chapter webpage, *www.anovascience.com/RIP~ingThroughSTEMwebpage.htm,* also must be accessed by the reader to fully understand this chapter. It contains the figures and tables cited in this chapter and, where indicated, expanded discussion, explanations, and supporting data and data analyses.

The Research Investigation Process (RIP) developed by research neuroscientist Robert E. Landsman is a scientific inquiry model implemented in classrooms through the assistance of ANOVA Science Education Corporation. This company furnishes RIP professional development for teachers, including initial RIP learning workshops, in-classroom mentoring and coaching, and online curriculum development support.

The RIP has always been a model for STEM, whether K–12 students were engineering and designing mountains to serve as models to conduct their investigations on the science of weathering, using oscilloscope technology to transform electrical energy into light energy to conduct mathematical analyses on the data from their investigations on electric fish communication behavior, or designing model oceans as part of the methodology or oil spill clean-up devices based on their findings for their investigations on mitigating the effects of oil spills on aquatic environments (Landsman, Evans, and Kamimura 2012; LoRusso 2012; Miller 2010; Sayres 1996).

When classrooms are "RIP~ing," they are learning about one or more of the STEM content areas through STEM while using practices characteristic of all four STEM fields. The concept of *engineering* as a crucial component of learning for any STEM (or other) content area is inherent in the RIP model. The scientific and engineering practices are used along with mathematics and technology to envision, design, and conduct investigations and conclude from the findings of those investigations with students, not teachers, making the primary contribution of doing and thinking effort in the learning experience.

This chapter describes how the RIP model was applied to a single theme in the elementary classroom (grade K) and middle school classroom (grade 7), and how for both cases the students learned about STEM content through the application of STEM practices. The RIP has been successfully implemented in a variety of grades K–12 settings with consistent teacher and student impact. Quantitative and qualitative impact data are presented from a wide range of schools with varied demographics. These data show marked positive changes in teacher and student attitudes towards teaching and learning, respectively. They also demonstrate growth in teacher and student understanding and performance related to the use of scientific inquiry and the scientific and engineering practices. Teachers attribute these changes to use of the RIP model.

## The RIP Scientific Inquiry STEM Model

The Research Investigation Process is a scientific inquiry-based critical-thinking model that leads to attainment of the four goals of science education by facilitating establishment of the *National Science Education Standards* (*NSES*) *More Emphasis* conditions necessary for achieving science education standards (Landsman 2005b; Landsman and Kamimura 2012; NRC 1996; Yager 2012; see also chapter web page subsection: "RIP Model—*More Emphasis* Conditions and the Four Goals of Science Education"). The RIP also enjoys success as a STEM model for K–12 education.

The ultimate goal of RIP-based inquiry is to have students lead the thinking and performance of actions required within all of the components of the investigation (Landsman 2005b, 2012). Motivation is one of the primary factors for the current "innovative generation" K–12 student's success in the classroom (Wagner 2012). The RIP model of scientific inquiry is designed to enhance student intrinsic motivation and foster development and application of scientific and engineering practices for thinking and learning in any content area. RIP enables and facilitates the learner to not only engage in every aspect of the investigation, but also to assume a leadership role in all aspects of the investigation (where developmentally and experientially feasible), from conception through culmination of the investigation process.

"Engaging in 'full and open' inquiry (NRC 2000) using the RIP is consistent with the scientific inquiry and scientific and engineering practices described in the *NSES* and *Next Generation Science Standards* (*NGSS*; see chapter web page subsection: "RIP Model—Full and Open Inquiry"). This approach necessitates that students engage in full and open inquiry as early as possible, depending upon developmental abilities and previous experience with the scientific and engineering practices (Landsman 2005a, 2012; Landsman and Kamimura 2012). Both the *NGSS* and *NSES* specify that students at all grade levels are to engage in complete scientific investigations (NGSS Lead States 2013; NRC 1996, 2012). It is, in part, because of the full openness of the inquiries themselves that the RIP engages students in a STEM learning experience. The investigations that are presented in the next section describe and illustrate how the RIP is an inquiry-based model for STEM learning that facilitates students engaging in "full" open inquiry, or "partially" open inquiry (especially during initial implementation in the classroom and for primary grades). We then present qualitative and quantitative evidence indicating that engaging in teaching and learning through the RIP model fosters positive attitudes toward learning STEM content, the development of 21st-century skills, and the development of abilities necessary to achieve the four goals of science education.

## Overview of the RIP Scientific Inquiry STEM Model

More than ever, there is a predominance of jobs requiring STEM-related knowledge and a set of 21st-century "survival skills" (Bybee 2013; Wagner 2008; Yahoo News 2013). U.S. students are lagging and likely falling further behind most of the developed world in STEM education (OECD 2010; Ripley 2013; Thomasian 2011; Toulmin and Groome 2007; Yahoo News 2013). Of major growing concern is that fewer students appear to be interested in STEM careers (CNN Money 2013). This may be related to the lack of 21st-century skill development (e.g., critical thinking and reasoning, attention to detail, finding new ways to approach problems, and understanding data) resulting in part from ineffective instructional strategies and shortage of curriculum emphasizing STEM in K–12 education (Ripley 2013; Thomasian 2011; Toulmin and Groome 2007; Wagner 2008, 2009; and Yager 2012). RIP is a scientific inquiry-based critical-thinking instruction and learning model that naturally incorporates STEM and 21st-century skills into K–12 education (Yager 2012).

The acronym "STEM" is widely used throughout the academic community and STEM fields, while there is no one commonly accepted definition for this term (Bybee 2013). STEM can be infused in teaching pedagogy and curriculum through an individual focus on one or more of the four included content areas, or "taught in an integrated and a cross-curricular manner, not just in 'silos' where the individual subject areas dominate and the other subjects are only afterthoughts" (Dugger 2011). Regardless of STEM definition, the RIP is an integrated and flexible STEM model for K–12 education.

The RIP invites thematic-based learning into the classroom, a concept grounded in cognitive psychology and cognitive neuroscience research findings related to episodic memory (Dere, Easton, Nadel, and Huston 2008; Tulving 1983, 1985). This type of memory involves representations of personally experienced events prompted by location or circumstance that eventually become part of semantic memory learning. It is enhanced by intensified sensory input and considered to be of unlimited capacity, effortless, and used by everyone. Early in learning, knowledge is more likely to be retained in episodic form, which is readily remembered in detail by students. The more distinctively rich the experience, or theme, surrounding the content to be learned, the more readily the memory content shifts from episodic to semantic memory and "just knowing" during recall (Herbert and Burt 2004).

Using STEM to learn by using the RIP model involves the integration of the four content areas (science, technology, engineering, and/or mathematics). Student and teacher responses indicate that using practices and content from these areas enriches their instruction and learning experiences. Memory consolidation (turning electrical and chemical input into memory) occurs in the hippocampus of the brain, one of the areas implicated in neurogenesis, and takes time. Thus, it is not surprising that studies show more retention and learning of information when it is consumed embedded in small deeper contexts. Two known reward centers in the brain that "light up" when a person is engaged in behavior that is of interest or rewarding to them are found in the hippocampus and cerebellum, both areas of the brain implicated in learning and memory! Curiosity, novelty, and expectations motivate episodic memory and fMRI studies have demonstrated neural pathways through which motivation can enhance memory formation (e.g., Adcock et al. 2006). Students can satisfy their own curiosity, learn in novel ways, and have expectations of success in answering their own questions—all motivators for learning. The RIP model also motivates learning by leveraging technology and firsthand experiences, both well-recognized learning motivators. Inclusion of the diverse STEM content found within the RIP instructional and learning process heightens the likelihood that one or more of these four subject matter areas might already appeal to a student. The combining of two or more of these STEM subjects in new and/or different ways may synergistically stimulate the motivation to learn.

The RIP model engages teachers in instructing and students in learning through the designing and conducting of extended research investigations. As a scientific inquiry-based STEM model, the RIP naturally brings the practices of scientists and engineers into the realm of the classroom. "Science and engineering practices should be thought of as both learning outcomes and instructional strategies" (Bybee 2011, p. 6). A focus on instructional practices of teachers is of central importance to successful reform covering STEM and the skills envisioned in 21st-century education (Ripley 2013; Yager 2012). "The most important in-school factor [in quality education] is the teacher" (Yahoo News 2013). Teachers are trained in the RIP inquiry model through intensive professional development lasting four to six consecutive days. This involves placing the teacher in the role of the student, allowing them to experience the same challenges and successes that their students will encounter. Once implemented in the classroom, the RIP model professional development focuses on teacher mentoring throughout the school year to strengthen and sustain inquiry instructional skills. The RIP program is flexible and adaptable so that it can be tailored directly to the standards-based curriculum for any school system and student demographic. RIP curriculum development follow-up teacher support sessions are molded to the needs and goals of the classrooms, teams/departments, grade levels, schools, and so on. In a manner similar to how they are taught and coached, the teachers then introduce the RIP model into their classrooms and partner with their students as mentors and coaches.

## RIP Model: STEM and the Scientific and Engineering Practices

The Research Investigation Process naturally invites the integration of the four content areas of STEM within each investigation. Whether intentional or not, using the RIP as a critical-thinking model and instructional framework will *always* include STEM content as we will illustrate in this chapter. First, the RIP can be used to facilitate the learning of new information in one or more of the four STEM content areas. Also, at the very least, every RIP-based research investigation will involve student use of most or all of the scientific and engineering practices described in the *NGSS* (NGSS Lead States 2013), application of mathematics for the analysis of data, and the use of technology for a variety of purposes.

The ease of incorporation of STEM concepts and skills into the learning process makes the RIP an ideal model for STEM education. RIP scientific inquiry initiated in the classroom may or may not directly focus on learning new science content; however, the investigation will always include application of scientific inquiry and the *NGSS* scientific practices. *Technology* is used as a resource and is heavily embedded throughout RIP investigations. Some research investigations require technology for the engineering design and construction of models or products. Technology in the RIP investigation also appears through the use of computers and mobile devices with internet access for gathering background information and for communication with research colleagues and experts in the research field; computer hardware and software for collection, maintenance, summarization, analysis, and presentation of data that will be used as evidence for testing a hypothesis; and the use of scientific or engineering tools (e.g., scopes for magnifying, laser measurement tools, and so on). During the proposal and final phases of the research investigation, technology is used to deliver presentations to an audience for assessment and knowledge sharing purposes. On the other hand, technology use or products may be the target of the investigation, such as studying cell phone reception in different locations. *Engineering* is represented through the designing of the method within the research investigation, including selection or development of materials used, research design, and procedure; the use of the *NGSS* engineering practices; and the designing of models or building of a product as a solution to a problem. *Mathematics* is integral to scientific inquiry (NGSS Lead States 2013). The RIP model highlights for the students the importance and use of mathematics in the decision-making process. The results component of the RIP investigation will always involve collecting, organizing, analyzing, and presenting data in some form. Computational thinking practices, a key feature of the RIP, enable students to apply algorithms using computer technology resources (when available) to organize and analyze their data. In addition, the investigation may directly target the learning of mathematical concepts. Thus, the development of mathematical skills necessary to conduct the investigation also becomes the basis for learning these skills in context, while the understanding and use of particular mathematical concepts may be a primary focus of the investigation.

The grades K and 7 ocean oil spill investigations presented in in this chapter illustrate how the RIP model seamlessly incorporates STEM skills, knowledge, and practices into the learning experience.

The eight *NGSS* scientific and engineering practices, and other essential practices used by the scientist and engineer, are incorporated into the RIP model (NGSS Lead States 2013; Bybee 2011). Equally important to understanding the scientific and engineering practices is the ability to achieve them in the classroom through investigation-based curriculum. Qualitative and quantitative evidence of skills developed toward these practices are presented in this chapter (and on the chapter web page).

## STEM Practices in the RIP Model Components

The practices engaged in by scientists and engineers in their everyday career activities as they perform investigations include the components found in the RIP model of scientific inquiry. Many important 21st-century survival skills are imbedded within these practices (Wagner 2008). This section will describe how STEM and these practices are incorporated into the RIP components: Making Observations, Asking Research Questions, Gathering Background Information, Constructing the Hypothesis, Method: Designing and Conducting the Study, Analyzing the Results, Discussion and Conclusion, and the Next Step (see Figure 22.1 on our web page for RIP model flowchart and Landsman, 2005a and 2005b for in-depth descriptions and discussion of these components). Although the components presented here are ordered for discussion's sake, professional scientists and engineers (as well as developmentally ready students) will apply them in any sequence necessary to achieve the goals of their investigation.

### *Making Observations*

The observation component of the RIP scientific inquiry model fosters development of attention to detail skills and critical-thinking abilities, important practices used by both scientists and engineers (Landsman 2006, pp. 14–16; 2012, p. 10). "Attention to detail" was one of the most important skills that former high school RIP students frequently identified as having been developed from their RIP learning experience (Landsman 2005b). RIP teacher professional development includes activities that target these important thinking characteristics. For example, observations made by teachers are analyzed by them for objective and subjective characteristics, highlighting the important role of attending to details and thinking about one's own thinking (see Making Observations Koi Activity, Landsman 2006, pp. 16–18). At the very start of their very first investigation, students begin to develop these characteristics by engaging in these same activities.

They learn the importance of attending to details and thinking about them when making their observations and to begin appreciating why this is important for any investigation including the STEM fields alone or together. The practice of using models, such as simulations of a large-scale event like an earthquake or tornado or a scale model of a solar system or automobile, in this RIP component provides students opportunities to make observations so they can "understand how things work" (NRC 1996, p. 117).

### *Asking Research Questions*

Well-thought-out RIP research questions will lead students to gaining "more information about the natural and/or designed world(s)" (NGSS Lead States 2013). The research question in the RIP model originates from the observation component and includes the *NGSS* practice, "observation of phenomena, or unexpected results, to clarify and/or seek additional information" (NGSS Lead States 2013, p. 17). Of equal importance are RIP research questions that originate from observing the results from previous investigations, as science builds upon science (Landsman 2005a). The research question defines the content (STEM and non-STEM) that will be searched for and consumed by students in the subsequent RIP component, Gathering Background Information, and throughout the entire investigation. The research question must be answerable; in other words, it generates tentative answers that can be tested through an investigation. Because it must be answerable, its formulation is based on thinking through all of the subsequent inquiry components that make up the RIP. Each of these components, as described below, incorporates the use of STEM skills and knowledge. All of these attributes of asking research questions are consistent with the *NGSS* practice of "Asking Questions and Defining Problems" (NRC 2012, p. 56; NGSS Lead States 2013; Bybee 2011).

## *Gathering Background Information*

The RIP background component includes a combination of standards-based content and information pertaining directly to the research question and the investigation that is being conducted. This information enters the RIP as input to a number of the components, dependent upon the particular investigation, and is used in different components throughout the inquiry. The Observation component of the RIP in particular will usually include information that becomes part of the background information. Background information may also be derived from previous experience.

Science and engineering are typically collaborative endeavors. The practice of "Obtaining, Evaluating, and Communicating Information" is reflected in students frequently working together in groups on their RIP investigations (NGSS Lead States 2013; Bybee 2011). This involves communicating and sharing information within and between student groups.

Scientists and engineers "gather, read, and evaluate scientific and/or technical information from multiple authoritative sources, assessing the evidence and usefulness of each source...." (NGSS Lead States 2013, p. 31). Evaluating the quality (reliability, validity, and credibility of source) of the background information obtained by students is an integral characteristic of the RIP. Teachers are trained in how to discern the quality of the information found in printed text as well as electronic media (e.g., websites, digital storage media, and so on) and transfer this information into the classroom by engaging students in relevant RIP activities.

This component requires the use of technology and technology skills involved in obtaining background resource information. It will also include STEM standards-based information to be learned that is pertinent to the specific content of the investigation topic and/or the successful implementation of each RIP component. For example, learning new mathematical concepts for data analysis (including computational thinking algorithms) that are utilized later in the results component is part of the background information necessary for students to complete their investigation. Students are expected to learn how to hand-calculate the same formulas and algorithms that are part of the software used to actually analyze their RIP investigation data. RIP characteristics and activities also foster and support the development of students' Common Core English language arts skills relevant to obtaining, evaluating, and communicating information (NGAC and CCSSO 2010).

## *Constructing the Hypothesis*

The RIP hypothesis is the student's *tentative answer* to the research question. "It is creatively constructed, based on the inter-relatedness of a number of factors such as critical thinking and reasoning, and on all the evidence available to the student and relevant to the investigation, including background information, other scientific knowledge, and personal experience"

(Landsman 2006). The RIP hypothesis format is used for multiple purposes: it (1) assesses student logic and the inclusion of appropriate and relevant background information for its justification, (2) ensures its testability, and (3) incorporates and ties together the major components of the RIP (Landsman 2005a; Landsman and Kamimura 2012). To be successful in the construction of their hypothesis, students must apply assessment to their own thinking in terms of how the three parts of a hypothesis (discussed below) fit together—each depending upon the other two. In constructing a hypothesis, students also must envision the entire investigation process, focusing on details for each component and whether they are implementable. Once these tasks are accomplished, the testability of the hypothesis is ensured.

The RIP hypothesis challenges students to think deeply and logically. Logical thinking is inherent in science and engineering, both of which involve the integration of technology and mathematics. The RIP hypothesis is always constructed through the connection of "if," "then," and "because" mathematical logic statements that serve to interconnect all of the components of the RIP model (Landsman 2005a, 2006). This in essence requires students to think through the design of the scientific investigation (part of the next RIP component: Method) in the "if" statement to include engineering models and technologies needed to perform the test of the hypothesis. Formation of this statement requires review and understanding of the background information. This information leads them to their construction of the "then" statement, which serves as the predicted outcome for the comparison stated in the "if." The "then" also requires realization and understanding of the data that will be collected and how it will be analyzed mathematically in the Results component that occurs later in the RIP model. The "if" and "then" together define the research design (i.e., 2-group independent design, dependent design, 2-group with controls; Landsman 2005a) which in turn determines the mathematical analysis and algorithms that will be needed to analyze the data collected (Landsman 2005a, 2005b). The "because" statement, also based on the background information, is the rationale for the "then" statement prediction. It requires students to learn the relevant STEM content and practices and apply them in the conducting of their investigation. The evidenced-based content of the subsequent RIP Discussion and Conclusion component will depend upon the hypothesis and its relationship to the data. The test of the hypothesis and its evaluation based on the data collected and analyzed in the investigation may lead to further investigations in the Next Step that test a new hypothesis.

### *Method: Designing and Conducting the Study*

Scientists and engineers in their everyday practices typically design and perform systematic investigations that necessitate specifying the nature of the variables involved and data that will be collected and later used as evidence to evaluate their hypothesis as a viable answer to their research question (Bybee 2011; NGSS Lead States 2013). Therefore, investigations conducted by scientists and engineers involve planning the investigation before actually carrying it out. Prior to conducting their investigations, students develop RIP storyboard proposals for planning and assessment of knowledge and ideas (Landsman 2005a, 2005b), in which they incorporate their testable hypothesis, plan methods they will use in their study, and identify the variables they will manipulate and/or measure. Scientists may use the study they design to "describe a phenomenon, or to test a theory or model for how the world works" (NGSS Lead States 2013, p. 7). Engineers might wish to investigate "how to fix or improve the functioning of a technological system or to compare different solutions to see which best solves a problem" (NGSS Lead States 2013, p. 7).

The methods involved in RIP research investigations tie together aspects from the four areas of STEM. For example,

> *It is the combination of inquiry design and data analysis that serves the scientist [and engineer] as a powerful [critical thinking] tool for making unbiased decisions about hypotheses [and problem solving]. In order for students to be critical consumers of scientific claims and their associated products, they must have some degree of understanding of the analytical processes underlying decision making in science [and engineering]"* (Landsman, 2005b, p. vi).

Technologies are represented in the RIP Method in the form of models used in investigations; analytical tools for making measurements and collecting data; and computers used for a variety of tasks such as data collection, analysis, presentation, and mathematical modeling.

## *Analyzing the Results*

The same actions that scientists and engineers use that define the *NGSS* scientific and engineering practice of "Using Mathematics and Computational Thinking" are used in the data handling and analyses for a RIP investigation (NGSS Lead States 2013; Barr, Harrison, and Conery 2011; Wing 2006). The RIP Results component involves organizing and summarizing the data that students collect and presenting them in tabular and graphical representations for analysis. This may or may not include applying mathematics in the form of statistical tests. This component aligns directly with the scientific and engineering practices as described by Bybee (2011):

> *Scientific investigations produce data that must be analyzed in order to derive meaning. Because data usually do not speak for themselves, scientists use a range of tools—including tabulation, graphical interpretation, visualization, and statistical analysis—to identify the significant features and patterns in the data. Sources of error are identified and the degree of certainty calculated.... Engineering investigations include analysis of data collected in the tests of designs. This allows comparison of different solutions and determines how well each meets specific design criteria—that is, which design best solves the problem within given constraints. Like scientists, the engineers require a range of tools to identify the major patterns and interpret the results.* (Figure 4 in Bybee, 2011)

Data collected in research investigations is inherently full of uncertainty. The realization and/or calculation (dependent on grade level and student ability) of uncertainty in the form of error is a hallmark of the RIP for all grade levels. Understanding this uncertainty is crucial for the critical-thinking and decision-making process that is incorporated in the RIP model. The calculation of mathematical uncertainty can be a daunting task for a seventh-grade student let alone a kindergartner. However, the RIP model requires students as early as kindergarten to understand that mathematical concepts, algorithms, and formulas are useful for making decisions about the data they obtain. These young students are made aware of the concept of error and that error is naturally represented as part of all data collected by humans (human error, random error, subject error, and so on), although they may be developmentally unable to calculate it mathematically. Students learn that error contributes to the uncertainty that exists when they use their data to make decisions. They also come to understand that the concept of error cannot be ignored in a scientific investigation because the data will lose meaning, which may lead to erroneous conclusions.

Once the data are organized and summarized in tables and graphs are drawn, the students use mathematical procedures and algorithms to look for patterns and differences. Their level of mathematical understanding and abilities will determine the complexity of the analyses they will be able to perform.

The two investigations described in the next section demonstrate the importance of mathematical integration and quantitative analysis in determining the results of a RIP study that is used to test a hypothesis. Application of the mathematical concepts and computational thinking applied by students in analyzing the results of their investigations mimics how scientists and engineers actually utilize mathematics to understand the data they collect, analyze, and apply to make decisions every day in their professional careers. The science and engineering practice of collecting, analyzing, and using data (mathematics and computer technology) are naturally integrated into the results component of the RIP model, which clearly ties together the four areas of STEM.

## *Discussion and Conclusion*

The Discussion and Conclusion component of the RIP model requires student engagement in a number of the scientific and engineering practices (NGSS Lead States 2013). Students construct their explanations and review and discuss their design solutions in regard to the study they have conducted. Within this component of the RIP, students weigh the evidence based on the data collected, analyses performed, and interpretation from their Results component to argue for or against their hypothesis and design solutions (*NGSS* science and engineering practice #7; NGSS Lead States 2013). Students, working in their teams with information they obtained and evaluated throughout the investigation, ask questions and communicate amongst themselves while developing their discussion and conclusions to arrive at a consensus. These practices are depicted throughout the *NGSS* (NGSS Lead States 2013).

The oral and written communications produced by students discussing and reviewing STEM content from background information includes critiques of their scientific study they have conducted. For the RIP model, students typically employ the scientific practice of constructing explanations related to uncertainty and mathematical error in their discussion of the outcomes of their investigations. Their findings frequently lead to further discussions of new or revisions to engineering and technological solutions.

## *Next Step*

The final component in the RIP model involves students and their teachers developing action plans that will include such activities as communicating their findings, replicating their study, developing a new investigation, influencing decision makers to take action, and/or designing a new or redesigning an existing technology (NGSS Lead States 2013; Landsman 2005a). Student and teacher engagement in these activities and actions reflect their interest and enthusiasm resulting from their RIP scientific inquiries.

This component includes the sharing of scientific inquiry-based instructional practices and their students' accomplishments by teachers with their colleagues and the general academic world through oral presentations and publications. Students use the scientific and engineering practices involved in planning and communicating information related to their RIP investigations to their peers through models, posters, journals, and oral presentations (NGSS Lead States 2013). "An action plan that includes the proposal for future research investigations helps students to realize and appreciate the cyclical nature of science" (Landsman 2005a, p. 289). Replication of the mathematical (e.g., statistically based) results from an investigation is crucial prior to students taking the next step of using their findings to try to influence environmental, social, and health policies. Future investigations frequently include new investigations that build on the results of a prior investigation in an attempt to further advance scientific knowledge.

## Assessment Practices Through the RIP Storyboard

Continual ongoing assessment of logical thinking, designs, and methods is a constant practice shared by scientists and engineers. The RIP model contains the practice of assessing STEM content knowledge, scientific and engineering practices, and attitudes and opinions toward learning about science and mathematics and learning through STEM (the RIP scientific inquiry model). RIP scientific investigations include the development of a plan or "storyboard proposal" that describes the details that will be included for each component of the investigation process. "It portrays through writing and drawings the individual components as well as the overall plan of the research investigation, making the students' thinking and reasoning visible" (Landsman 2006, p. 38) to the teacher and student.

Because of its visibility, the proposal serves as an assessment tool through which teachers and classmates of the student investigator can then verify the clarity, accuracy and completeness of STEM content and provide feedback to ensure logical connections between the RIP components. This critical assessment ensures that the student's inquiry is valid and will be achievable. Equally important to the learning process are the conversations that happen during the assessment process between investigator, coach, and peers that strengthen the students' critical-thinking skills. "Thus, the assessment of content is incorporated into the assessment of thinking" (Landsman 2006, p. 38).

## Qualitative Results and Analysis

This section will describe two oil spill clean-up RIP investigations conducted at the elementary (Grade K) and secondary (Grade 7) levels and how they exemplify the RIP as a STEM instruction and learning model. Both oil spill investigations tapped on elements of the students' local environments (Honolulu, Hawaii and Kirtland, New Mexico, respectively) important for human survival, contributing to the interest levels of students. In addition to the general relevance related to environmental stewardship is the fact that the kindergarten students in Hawaii (Kahala Elementary School) are surrounded by an ocean, which is a major source of their food, drinking water and recreation, as well as jobs and support for the economy. Oil is a reality in the everyday lives of the New Mexico students (Kirtland Middle School). They are surrounded by oil fields and their potential detrimental consequences to the environment, as well as the effects of the oil industry on their local economy. These connections to the students' real environments added to the interest level and made STEM relevant and important for both groups of students.

The two kindergarten classes from two consecutive years ("year 1" and "year 2") were instructed by Ms. Lori LoRusso, a 22-year veteran teacher. Both classes contained 10 boys and 10 girls, most of whom came from white collar families of multiethnic backgrounds living in a middle- to upper-class community. First-year teacher Melissa Miller's seven Grade 7 classes, totaling 133 students, were from the same school year at this Title I school. Thirteen percent of these students were considered Special Education, 10% ELL, and 77% general education.

Both teachers were new to scientific inquiry as an instructional and learning tool prior to their experience with the RIP model. Ms. LoRusso's first use of this model in the classroom was with her year 1 students described below. The investigation described for Ms. Miller and her students was conducted during the latter half of her first year of teaching. She had previously introduced the RIP into her classroom during an earlier part of the year. Both teachers attended district-level RIP professional development training sessions conducted by ANOVA Science Education Corporation (Honolulu, Hawaii) prior to the described investigations. (For extensive details regarding each of these inquiries, see LoRusso 2006, 2012, for Grade K and Miller 2010, for Grade 7.)

## *Grade K Oil Spill Research Investigation*

"Observation of the world around us is where science usually begins," states teacher Lori LoRusso (LoRusso 2012, p. 41). "As a kindergarten teacher, it is my responsibility to provide my students with opportunities for observation so they can learn to appreciate and understand the world around them…." Students engaged in making observations in the classroom environment as teacher Ms. LoRusso guided them to assess their own observations for objectivity and subjectivity. She demonstrated an oil spill by adding blue food coloring to vegetable oil while students observed. This mixture was then introduced into a container filled with water. Students observed that the water turned blue and the oil floated on top of the water. They drew their observations using crayons and described in their own words what they had observed. Their involvement in this process led the students to their own definition of an oil spill in water.

The observations students made from the RIP Background Information component including the Exxon Valdez oil spill and picture books about oil spills (*Oliver and the Oil Spill*, Chandrasekhar 1991; *Oil Spill*, Berger 1994; *Alaska's Big Spill: Can the Wilderness Heal?* Hodgson 1990) were applied to their knowledge and understanding of the concept of an oil spill leading them to fully appreciate the destructive nature of these events on the environment. This led directly to their research question.

The students naturally began to ask questions such as, "How are we going to take out the oil?" As the students expressed concern about the environmental impact on fauna and flora of oil spilling into oceans and waterways, Ms. Lo Russo asked them how they thought they could clean up the oil spill in the model she had made. This became the years 1 and 2 classes' research question: "What is the best way to clean up an oil spill?"

During their inquiry, the kindergarten students engaged in student-teacher, student-student, and general classroom discussions. The resources used and background information gathered from them became central to these discussions that took place throughout the various RIP components of the kindergarteners' investigation. For example, a wall chart containing the RIP scientific inquiry components was frequently referred to by the students to help them locate where they were in the RIP model and what they should be doing. RIP-based activities learned in the RIP workshop conducted by ANOVA Science Education were used by Ms. LoRusso to assist in interpreting and communicating complex materials to her students. The students were also read an article about and were shown pictures from the Exxon Valdez oil spill, picture story books about oil spills, and internet sites containing information about oil spills and their harmful effects on the environment. As mentioned previously, the kindergarten students were required to learn the mathematics skills that they would later need for making a decision about their hypotheses. This involved learning the concept of and how to compute a median, and making and using data tables and graphs for the year 1 and year 2 grade K students (see RIP Method component for Grade K, below). The students frequently made use of their background information within their group collaboration and discussions in the observation, research question, background, hypothesis, method, results, discussion and conclusion, and Next Step components of their investigation.

To begin thinking about their hypothesis, the kindergarten students used their observations from the oil spill simulation created by their teacher, what they had observed from the Exxon Valdez as well as background information about oil spills from picture books, and excerpts read to them by Ms. LoRusso from internet articles. The teacher used the Socratic questioning technique to direct student focus on extracting from these sources new ways scientists were using to clean up oil spills and then asked her students how they thought they could clean up the oil spill she had made. The students offered five different technologies and their rationale for why these would work. For both years' classes, the students included an eyedropper, spoon, sponge, cloth, and cup (see Figure 2 on our web page for a student-drawn picture of the tools used).

The students collaboratively developed their technology ideas into testable hypotheses with the assistance of their teacher. A vote was taken and the most popular hypothesis was selected. For year 1, the hypothesis was: "If there is an oil spill and we try to clean it up with a cup, an eye dropper, a sponge, a cloth, and a spoon, then the eyedropper will pick up the most oil because it has a small tip." For year 2, the hypothesis was "If we try to clean up an oil spill with an eyedropper, a cup, a cloth, a spoon or a sponge, then the cup will pick up more oil than the others because it can scoop up larger amounts of liquid."

Through this grade K investigation and two previous "mini-lessons" experienced by both years 1 and 2 classes, the students learned how to compose and practiced putting together the three parts of the hypothesis. Students were asked to express in their own words what they had done to build their hypothesis. This process afforded students the opportunity to peer-critique and make corrections as needed.

Students again engaged in assessment of their understanding of the hypothesis at the end of the investigation through drawings and writing about the components of the inquiry process they had used. This was followed by their oral explanation of the components of their hypothesis.

The grade K investigations for both years 1 and 2 involved planning the method of the study they would conduct to compare the five tools described in the students' hypotheses. The studies would span several class periods and the winter recess. Students working together in pairs drew the elements of the method section for their investigation. The class's plan incorporating these illustrations was constructed into a 9-foot long poster that almost filled an entire classroom wall. (See *How We Conducted Our Study to Test Our Hypothesis* section in LoRusso 2006, for specific procedural details and extracted sections of the poster.)

Students worked in five groups of four students each for both years. An oil spill was created for each group by their teacher to model an ocean spill. Students' task was to use each of the five tools to try to collect as much oil as they could into a milliliter (ml)-lined cup. After the students collected the oil with a tool, the number of lines of oil in the cup collected by that tool was recorded as the data. The line data were then placed into rows in a data table arranged with the names of each of the five tools (see Figure 3 on web page for data table).

To assist her students in the analyzing of the data, Ms. LoRusso taught them how to arrange the five groups' numbers for each tool in a sequence from lowest to highest. She then modeled how to determine the middle number (median) in the sequence, starting with three numbers and progressing to seven. Ms. LoRusso's planning for the students to work in five groups and for the cups to contain five lines facilitated the data collection and analysis for these young students.

The Results component of the kindergarten investigation involved their using and thinking about their data, including aspects of its organization and analysis. Working in their groups, the years 1 and 2 students applied the mathematics they learned from the RIP Background Component to find the median number of lines of oil collected for each of the five tools (see Figure 3 on web page).

Graphs showing the median data for each tool were then drawn by the students (see Figure 4 on web page for a student-drawn graph from the year 1 class). The year 1 students found that their eyedroppers picked up the most oil from their model oceans: "As they analyzed their graphs, the students commented: 'The cup and the spoon scooped the same amount of oil,' 'The cup collected 4 ml of oil,' and 'The cloth collected the least amount of oil' " (LoRusso 2012, p. 49). Interestingly, the year 2 students' result was quite different—the cup picked up the most oil from the model oceans. While discussing the medians in their data table, the years 1 and 2 students noticed large differences in the numbers representing the amount of oil collected for a given tool. This observation led to a later discussion about how error may have influenced their data in their discussion and conclusion section (described below).

Through class discussion, the kindergarten years 1 and 2 students argued and decided based on the evidence from their study, that their hypothesis was supported. The grade K students discussed their observed differences in data obtained by different groups for the same tool, which led to more questions and a rich discussion and conclusion section for this inquiry. They mentioned problems they encountered and developed reasons for the varying amounts of oil collected from group to group. The year 1 students "noticed that some groups collected and counted both oil and water, several students put more than one tool in the container at the same time, and other groups spilled their oil—all errors that could explain why the data differed"

(LoRusso 2012, p. 50). These young students were already beginning to recognize sources of error in their investigation and the consequences these could have on the results.

Through their oral review of the details of their investigation, the students formed additional connections to "real-life situations." For example, a year 1 student offered to the class that the cloths they had used in their investigation were different than the spongelike material that was used by the experts to soak up oil in the story that Ms. LoRusso had previously read to the class (LoRusso 2012). Both years' classes came to realize the relevance of their investigation to an actual oil spill cleanup.

Both years' grade K students were asked by Ms. LoRusso to design and draw or build a model oil spill cleaner over their Christmas break. Students demonstrated a broad range of solutions for cleaning oil spills, reflecting that engaging them in RIP-based inquiry stimulated creativity. Of the 14 grade K year 1 students who brought their models to share with the class, 12 (86%) of them used an eyedropper or eyedropper-like tool as part of the clean-up model they had designed. They had incorporated the technologies and what they had learned from their investigation into the models they engineered (see Figures 5 and 6 on web page for examples of models the students built). The same held true for the year 2 students that built their models without parental assistance: They tended to use a cup-scooping device in their models, a solution consistent with the findings for that year's oil spill investigation.

Ms. LoRusso had the students integrate language arts standards into this component of their RIP by having them write a description of their inventions and present these to the class orally.

The grade K students readily communicated the excitement they experienced through the conducting of their oil spill investigations. For example, of the 14 year 1 students' parents who responded to a questionnaire relating to the students' building of their oil spill clean-up models, 13 of the parents stated that their child had shared with them what they had learned about how best to clean up an oil spill.

Ms. Lo Russo produced her own Next Step by presenting this oil spill investigation to the world in a website page and a published chapter in a professional education book (LoRusso 2006, 2012). She shared with educators the scientific inquiry-based instructional practices she used with the students that enabled them to conduct this research investigation, as well as the entire investigation itself.

## Grade 7 Oil Spill Research Investigation

The Grade 7 students' research investigation began with their observing a short clip of the Exxon Valdez oil spill. The video focused on consequences of the oil spill to the sea environment. The students recorded their observations of the video's content and assessed them for objectivity and subjectivity. When these observations were shared as a class, many students expressed surprise at the damage that was done by the oil spill resulting from the Valdez hitting Bligh Reef. Two of the observations they made related to environmental impact were, "After many years oil still remains there" and "The oil wasn't easy to clean up." Ms. Miller wanted to get her students to "buy into" this topic as early as possible so that they would take ownership and contribute to the thinking involved in all aspects of this scientific research investigation. She questioned them, focusing on detailed aspects of sources for oil spills, consequences on the ecosystem, and methods for cleaning up oil spills.

Many student questions were stimulated by discussion of their observations and through Socratic questioning by Ms. Miller. Their observations included some of the methods for cleaning up the surface of water following a spill. Student interest in clean-up ability really peaked as news of the Gulf of Mexico BP-Deepwater Horizon oil spill began to surface, making this investigation so much more relevant. Based on the background information they had collected through their observations, the students focused on two methods for cleaning the oil off the surface of the water: absorbing and skimming. This led to their research question: "When cleaning up an oil spill, which method, skimming or absorbing, is most effective for removing the oil?"

The middle school students spent two days in the computer lab, where they did background research related to oil spills and the various technologies available for cleaning oil out of waterways. They shared together as a class and discussed their research findings. Their background research covered three topics: what an oil spill is, how oil spills impact the ecosystem, and the variety of ways oil spills can be cleaned up. They learned what an oil spill is, how a discharge of oil reaches bodies of water, and how a spill negatively affects the environment. The technologies for cleaning oil from the water that the students researched and learned about included burning methods, skimming, evaporating, absorbing, and using dispersants.

Standards-based mathematics content that the grade 7 students learned as background information for later use included developing and using tables and graphs for organizing and presenting data. They also learned and later used advanced statistical tools for analyzing the data they collected (see below for a description of the method they used). Other standards-based content topics are mentioned at the end of this section.

The grade 7 students worked collaboratively in their groups to begin constructing their hypotheses on chart paper. They presented their hypotheses to the entire class for critique of the logic within and across the three parts that compose a RIP hypothesis: the "if," "then," and "because." One class's hypothesis was selected for the investigation: If an oil spill is simulated and we compare the methods of skimming and absorbing to clean up the oil, then we will find that absorbing is a more effective method because absorbing soaks up the oil and this is the most common method used in the history of oil spill cleanups.

To test their hypothesis, the grade 7 students planned and designed both their study and their ocean oil spill models (see Figure 7 on web page for one of their oil spill models).

The students learned about the nature of variables such as independent, dependent, and subject as well as control groups and the concept of random selection and assignment of subjects to groups. They applied their knowledge gained through their previous applications of the RIP to determine the type of study (e.g., experimental, correlational, ex post facto) and design (e.g., independent, dependent, matched groups) they would be using, and the nature of the variables involved (e.g., independent and dependent, subject) [Landsman 2005a]. The students also applied their understanding of the nature of the data they would collect to help them choose how they would be analyzing their data (i.e., statistical test) [Landsman, 2005a, 2005c].

The seventh graders had to solve the engineering problem of simulating a complex environment that included designing of a shoreline that contained representation of wildlife as well as the dynamics of simulated ocean wave action. Such a model lends itself to investigation of a myriad of additional environment variables and conditions (e.g., wind, temperature, salinity, water current, color of the water, time, type of cotton used for absorbing oil, and so on). Because this was a classroom investigation conducted over many days, the extent to which these variables could be addressed in this one study had to be limited. A number of the possible variables that were not studied in this investigation were, however, addressed in the students' Discussion and Conclusion and the Next Step components of their RIP (see discussion of these variables below).

The procedure for their study began by their introducing the oil into their model oceans to produce an oil spill. (See Designing the Study and Conducting the Study sections in Miller 2010 for specific procedural details.) The students produced simulated ocean waves using air pump technology (small aquarium pumps) over a period of three minutes to spread the oil over their model ocean environments (see Figure 8 on web page for oil spill model with air pump wave simulation to disperse oil). Over a period of seven minutes, students used their skimming tool technology (spoons) or absorbing tool technology (cotton balls) to collect the oil from their models and deposit it into a standard 200- ml beaker (see Figure 9 on web page for the absorption method used with cotton balls). This oil spill cleanup procedure was performed three times for each tool by each of the five groups of students in each of Ms. Miller's seven classes. Students collected and entered the number of milliliters of oil into their data tables for each cleanup they conducted. The data for the complete investigation were then entered into the data tables.

After collecting their data into data tables, the seventh-grade students summarized them using the appropriate measures of central tendency and variability. Each group first calculated the median amount of oil they collected from their three trials for each tool (see Figure 10 on web page for one group's data table). The median data were then combined for all five of the groups into a class data table (see Figure 11 on web page for two classes' data tables). The students summarized their class data by calculating mean-medians and variability in the form of standard deviation and standard error of the mean for each of the oil spill clean-up tools. Then the students graphed their results and statistically analyzed their data using a modified *t*-test, the eyeball test (see Figure 12 on web page for the class Periods' 5 and 6 graphs; Kugler, Hagen, and Singer 2003; Landsman 2005a, 2006). The students applied their knowledge about mathematical error acquired through previous RIP investigations to statistically determine whether there were differences between the mean-medians for any class period. Six of the seven grade 7 classes found that the skimming method resulted in collecting significantly more oil from the model ocean environments than the absorbing method. Only Period 6 did not find a statistically significant difference between the mean-median amount of oil collected using the two methods, although the skimming method appeared to have resulted in higher amounts of oil being collected (refer back to Figure 12 on web page for a comparison of the Period 5 and Period 6 class data). The Period 6 error (indicated by the error bars for the two groups) was comparatively larger than that for Period 5. The amount of error in the data for Period 6 resulted in the error bars (and thus the mean-medians) overlapping.

The Grade 7 students had hypothesized, "If an oil spill is simulated and we compare the methods of skimming and absorbing to clean up the oil, then we will find that absorbing is a more effective method because absorbing soaks up the oil and this is the most common method used in the history of oil spill cleanups." Applying the results they obtained from their study, that the skimming method collected more oil from their ocean models, the students argued that their hypothesis was *not* supported by the data.

Students readily recognized and discussed the variability in collected oil data across the groups in each class. They also noticed large oil data differences within some of the groups. While carefully collecting the oil using the skimming or absorbing methods, they observed that the oil they collected was unavoidably mixed with water. During their measurement of the oil they collected from their models, which occurred 10 minutes after depositing it as a mixture of oil and water into a beaker, they observed that the oil continued to separate from the water over time. They used this observation to conceptualize and explain one likely source for the variability in their data and discussed the importance of accuracy when making measurements. In their discussion of this measurement dilemma, they concluded that in a future investigation they would extend the waiting period to 24 hours prior to collecting their measurements.

Many of the seventh graders found it difficult to accept the results of their investigation. They truly had believed that the absorbing method would prove superior in removing the oil from their model oceans. This led them to a closer look at their procedure, materials, and technologies used in this investigation and a review of their background information. One of the science concepts they had learned about was the nature of natural organic sorbents that are frequently used in real-life oil spill cleanups. As they sorted out the details, they came to the abrupt realization that the cotton balls they had used in their study were formed from processed cotton and therefore probably did not possess the oleophilic (oil attracting) and hydrophobic (water repelling) characteristics that are found in natural organic cotton.

The Grade 7 classes used their RIP Next Step to develop an action plan to address a number of the items they had critiqued in their discussion and conclusions. Together in their groups, and again as a class, the students first planned out how they would try to replicate their study—with improvements. They would eliminate the blue coloring of the water in their model oceans to make it easier to discern the separation line of water and oil in their measurement beakers. They would improve their procedure by increasing the amount of time to let the oil and water separate before measuring the amount of oil collected. Additionally, the students would reduce the influence of error by increasing the sample size by having more groups of students within each class, or by pulling the data from all of the groups in all of the classes and then analyzing it. Finally, they would replace the processed cotton balls with natural cotton to improve oil absorption and water repulsion. The second outcome of their Next Step component of the RIP was a plan of a new investigation that would compare natural unprocessed to processed cotton to see whether the natural cotton would work better than the cotton balls they had previously used to absorb the oil.

For her Next Step, Ms. Miller made her classes' investigations available to the education world through a published website article and talk (Miller 2010, 2011).

## Summary and Synthesis of Grades K and 7 Investigations

For both grade levels, the teachers used questioning to focus their students' initial observations on the environmental science-related consequences of oil spills in the ocean and technologies that are used to clean them up. Students in the grade K classes were exposed for the first time to the concept of using a simple physical engineering model to simulate an oil spill in the ocean to "understand how things work" (NRC 1996, p. 117). Both science and engineering frequently make use of models in a variety of ways (Bybee 2011).

The students from both grades observed and applied their knowledge gathered from observation and background information. This led naturally to their posing questions to the class. These questions became the basis for the classes' research questions. Their research questions specified the problem to be solved: determining scientifically the best way to clean up an ocean oil spill. Aspects of the science, technology, and engineering components of STEM were integrated in the *thinking* that led to the research questions developed by the students. This led the students to begin the process of engineering (designing a study) to test if their tentative answer(s) to the research question was (were) viable. For grade K, the students' ideas led them to specify the exact technologies/tools (spoon, sponge, cup, eyedropper, and cloth) they believed would best clean up the oil spill they observed. For grade 7, their research question led the students directly to designing engineering models for simulating an ocean oil spill. Each class's groups shared their models, and their best ideas were selected to construct the class's optimal ocean spill model solution. The cleanup methods used for grade 7, a spoon for skimming and cotton ball for absorbing oil (the technologies), were also the result of the sharing of a multitude of ideas on the best engineering methods to use for simulating skimming and absorbing.

Both of the elementary and middle school oil spill investigations targeted standards-based content in science and mathematics. Students learned about ecology, humankind's impact on the environment, and environmental stewardship as part of the *NSES/NGSS* standards-based science content: Life science: organisms and environments (grade K), populations and ecosystems (grade 7); Science and technology: abilities of technological design (grades K and 7), understanding about science and technology (grades K and 7); Science in personal and social perspectives: changes in environments (grade K), populations, resources, and environments (grade 7), risks and benefits (grade 7), and science and technology in society (grade 7) [For example, *NGSS* grades K and 7 disciplinary core ideas - ESS3.A: Natural Resources; ESS3.C: Human Impacts on Earth Systems; ETS1.A: Defining and Delimiting an Engineering Problem; and ETS1.B: Developing Possible Solutions].

Both grade levels learned and applied statistical mathematical analysis to their data as part of their investigations (NGAC and CCSSO 2010). The background information component for both inquiries emphasized STEM technology and technology skills involved in obtaining background resource information and understanding the technologies utilized in the cleanup of oil spills. The relevance of these investigations to STEM is evident in the science, technology, and societal relationships realized through the gathering of background information and understanding its content. Students learned about ecology, humankind's impact on the environment, and environmental stewardship as part of the standards-based science content. The grade K students were exposed to computer technologies through the use of websites on the internet for obtaining background information. The seventh-grade students used websites for gathering their knowledge about oil spill cleanup technologies. The introduction of mathematical concepts pertinent to both investigations also comprises standards-based content for both grade levels that is later incorporated in their RIP Method component.

For both oil spill investigations, the students developed their own hypotheses to be tested as possible answers to their research questions. The "then" requires realization and understanding of the data that will be collected and how it will be analyzed mathematically. The "because" required the kindergarten and seventh-grade students to learn the scientific information about environmental science and technologies relevant to the content and conducting of their investigation.

The hypothesis presented students with the challenge of thinking through the logic of the "if, then, because." These students began to apply assessment to their own thinking in terms of how the three parts of the hypothesis fit together, each depending upon the other two (see LoRusso 2006, 2012 for grade K and Miller 2010 for grade 7). In constructing their hypotheses, students also envisioned the entire investigation process, focusing on details for each component and whether they were implementable. Once these tasks were accomplished, the testability of their hypotheses was ensured.

The four components of STEM are also tied together in the logic of the RIP hypothesis. Logical thinking is inherent in science and engineering, both of which involve the integration of technology and mathematics. This is clearly evident in the construction of the hypotheses for both grade levels, as this component of the RIP model ties together logically all other components of the investigation.

Whether used in grades K or 7, the methods involved in these oil spill studies tie together aspects from the four areas of STEM. The *NSES* (NRC 1996) and *NGSS* (NGSS Lead States 2013) science standards-based content involved Earth science, ecology, and environmental sciences in both investigations.

Technologies were used as tools by the kindergarteners to clean up their oil spills and by the grade 7 students to engineer their oil spill models that they developed. The grade 7 engineering of a model ocean environment and oil spill involved a much higher level of sophistication in planning and design than was needed for the kindergarten oil spill investigation, in which the teacher produced the oil spill model.

For the grade K students, planning the handling and analysis of the data required gaining an understanding of the mathematical concept of the "middle" of a set of data and the ability to apply median descriptive statistic calculations. The seventh graders applied their ability to identify characteristics of their research design to their choice of the mathematical technique they would use to analyze their data. They also used their knowledge of inferential statistical concepts involved in the comparing of two means and how to calculate this for two sets of data.

These two oil spill investigations (grades K and 7) illustrate how determining the results of the RIP study that is used to test the hypothesis is dependent upon the students' integration of mathematics, including quantitative analysis, into their work. How the mathematical concepts were applied by the students in analyzing the results of these investigations imitates the use of these concepts by scientists and engineers who employ mathematics to understand and make conclusions from the data they collect. The four areas of STEM were tied together through the science and engineering practices of collecting, analyzing, and using data (mathematics and computer technology), all which were naturally integrated into the results component of these oil spill investigations.

In addition to reviewing and discussing content related to the environment, Earth science, ecology, and technology and society, both grade levels also included critiques of their scientific investigations—all part of the science content in the RIP Discussion and Conclusion component. Here, and in the subsequent Next Step, the students weighed their hypotheses and/or design solutions and presented their data- and observation-based arguments (*NGSS* science and engineering practice #7, NGSS Lead States 2013). For both grades, the students employed the scientific practice of constructing explanations related to uncertainty and mathematical error in their discussion of the outcomes of their investigations. Both grade levels identified sources of error that contributed to uncertainty and the seventh graders offered possible solutions. The seventh graders also identified that the type of cotton they used was a problem that needed to be solved for this type of investigation. This problem could readily introduce all four STEM areas through a discussion of an engineering solution such as ways to gather loose natural cotton into a useful absorbent technology or to produce a processed cotton-material with better oleophilic and hydrophobic characteristics.

At both the grades K and 7 levels, the students incorporated STEM practices and applied STEM content concepts in their Next Step RIP-component activities. The models engineered and technologies used by the kindergarten students reflected the mathematical-based evidence they had gathered through their investigation. They applied the scientific and engineering practice of evidence-based decision-making and supporting their argument with evidence generated from their own investigation. Likewise, these practices were encompassed by the plans devised by the seventh graders to (1) reengineer their models using different absorption technology and (2) redesign their scientific procedure to accommodate their critique of their own investigation.

Students are engaged in each of the *NGSS* scientific and engineering practices within most or all of the RIP components. For example, the practice of "constructing explanations (for science) and designing solutions (for engineering)" was used in both grade level oil spill investigations by (1) students generating a research question and hypothesis that included and compared different tools, and conducting their investigation to see which one would pick up the most oil as the best solution to the oil spill problem; (2) students using their observations, analysis of the results, and their discussions and conclusions to explain natural phenomena that occurred during their investigations; and (3) the grade K students building their models based on the evidence that supported the best solution to the oil spill problem, while the grade 7 students engaged in talking about and resolving how they would test to determine the best solution for the cotton absorbency problem for their Next Steps.

## Quantitative Results and Synthesis

The previous section qualitatively demonstrated through the oil spill investigation descriptions that the RIP is a STEM instructional and learning model for elementary and secondary schools. These grades K and 7 research investigations illustrate the use and integration of STEM within the context of the oil spill common theme. This section provides quantitative evidence that supports the RIP scientific inquiry STEM model as a viable solution for raising student interest in and fostering positive attitudes toward learning and using STEM. Evidence supports gains in student knowledge and understanding of scientific inquiry components and processes and non-scientific inquiry STEM content through application of the RIP model. While the RIP demands rigorous student involvement on a multitude of levels, with emphasis on thinking, evidence supports it to be intrinsically motivating for learners.

## *Attitudes Toward Science (STEM) and Learning Through RIP Scientific Inquiry*

It is clearly evident that the grade K oil spill students preferred learning through RIP STEM inquiry over traditional methods (Figures 13 and 14 on web page). Significantly more years 1 and 2 students reported that they thought they learned more *about oil spills* using the RIP process learning style compared to just reading about and discussing the topic (Figure 13 on web page). They also preferred to learn *about science* by doing the RIP than by reading books and discussing them (Figure 14 on web page).

Student data firmly support that, as early as kindergarten, learning through the RIP model corresponds with development of STEM subject preferences over other subjects by the end of the year (Figure 15 on web page). For example, after learning through the RIP, 43% of the grade K students at Anuenue School (Honolulu, Hawaii) preferred science over social studies, mathematics, and language arts by the end of the year (Figure 15 on web page). Math (35%) was the second-most favored subject at that time. The RIP requires that students begin to understand the importance of organizing and quantifying data as early as possible to be used in make decisions during the scientific inquiry process.

Data from the very early grades (K and 1) consistently indicate that using the RIP model as a learning tool leads to intrinsically motivated students who feel good about their work. By the end of the school year, for example, after learning through the RIP, the number of grades K and 1 Kaimuki Complex (Honolulu, Hawaii) students reporting feeling proud of their work more than doubled (Figure 16 on web page).

The grade 7 oil spill students' *liking to learn about science* as a subject in school increased significantly by the end of the year compared to their pre-RIP implementation levels (Figure 17 on web page). This is a consistent finding for RIP-students, holding true for challenged as well as general education students (see Figure 18 on web page).

About 25% (28) of the grade 7 oil spill students initially (pre-RIP implementation) claimed that they did not desire to learn more science in school (Figure 19 on web page). However, at the end of the year, this number had decreased to only 7% (8 students). It is important to note that this is a consistent finding for RIP-educated students (see Figure 20 on web page for an example RIP versus non-RIP student comparison). Some of the end-of-year reasons that the grade 7 oil spill students provided for their responses to this item along with how they felt about learning science compared to other subjects are shown in Table 1 on the chapter web page.

There was also substantial change in Ms. Miller's students' preferences for science over other subjects in school over the RIP implementation year (Figure 21 on web page). At the beginning of the year before RIP implementation, the majority (70%) of the grade 7 oil spill students liked science the same or less than mathematics, social studies, and language arts. In contrast, the end-of-year assessment revealed the majority (69%) of students liking science more than the other subjects. Of these, 30% (33 students) claimed science to be their favorite subject (a threefold increase). Only 2% reported science to be their least favorite subject. These students reported an increased interest in scientific events and discoveries in the world by the end of the year (Figure 22 on web page).

Students indicated that they preferred to learn by using the STEM practices (scientific and engineering practices) and the RIP scientific inquiry model over traditional classroom methods. At the end of the year, the grade 7 oil spill students reported liking to learn science through conducting investigations significantly more than before they learned and used the RIP (Figure 23 on web page). Accompanying this was a statistically significant decrease in student preference for learning science through traditional methods such as by reading textbooks, doing labs, and attending to teacher lectures (Figure 24 on web page).

## *Student Knowledge and Understanding of Scientific Inquiry Components and Processes*

The RIP is a rigorous model that is readily learnable by grade K students. RIP-trained grade K oil spill students exhibited growth in their understanding of how to plan, think through, and conduct an investigation. The grade K year 2 oil spill students demonstrated a statistically significant Post-RIP implementation increase in their understanding of scientific inquiry and how to use mathematics skills to make decisions in science (Figures 25 and 26 on web page; note that the year 1 grade K students were not administered the RIP preassessment). These findings are typical and further exhibited in studies containing non-RIP-trained control students (Figure 27 on web page).

An unexpected finding for the grade K year 2 investigation was that students at this grade level were developmentally ready to learn the mathematical concept of middle, or median. When assessed for their ability to correctly determine the middle in a set of data, 18 of the 19 students (95%) correctly identified the median. This illustrates that actively engaging students in scientific inquiry can be developmentally appropriate for learning the concept of position, such as the middle, in context that brings relevance to learning.

One of the major goals for the implementation of the RIP is to develop a working understanding of scientific inquiry in students so that it can be used as a model for their thinking and learning. The grade 7 oil spill students displayed a steep, statistically significant, increase in their knowledge and understanding of the components of scientific inquiry and the scientific inquiry process at the end of the year compared to pre RIP-implementation (see Figures 28, 29, and 30 on web page).

Critical thinking plays a large role in the construction of testable hypotheses in scientific investigations. Students must be able to recognize the logic that connects the elements of the hypothesis in order to determine if it is in fact testable. The grade 7 student ability to be able to critically examine the parts of a hypothesis, to determine whether it is complete and testable, rose significantly by the end of the year from the pre-implementation assessment level (Figure 31 on webpage).

By the end of the year, the grade 7 oil spill students displayed a steep statistically significant increase in their recognition of the importance of consideration of error (uncertainty) in interpreting data and for decision making based on their demonstrated ability to apply the concepts of error and uncertainty to evidence-based decisions (Figure 32 on web page).

Understanding the design aspects of a scientific investigation is very challenging for both teachers and students, but also very important for successful development and implementation of inquiry-based curriculum and instruction. The grade 7 oil spill students demonstrated a substantial, statistically significant, increase in their research design knowledge and abilities by the end of the year (see Figure 33 on web page).

At the end of their first year of exposure to RIP-style learning, the Grade 7 oil spill students responded that scientific inquiry encouraged them to think "a lot more than usual" while learning science (see Figure 34 on web page).

ELL and Special Education students also become proficient in understanding and using the RIP scientific inquiry STEM model. For one district's RIP implementation, grade 8 general education students started the year with a higher level of scientific inquiry knowledge and skills (more than double the performance) than ELL students. Both groups exhibited statistically significant increases in scientific inquiry knowledge and skills following RIP implementation. However, the gain in knowledge and skills exhibited by the ELL students was substantially greater than that for the general education students, resulting in almost complete elimination of the difference in the performance of the two groups by the end of the year (see Figure 35 on webpage). Likewise, grade 8 Special Education students also demonstrate statistically significant gains in their knowledge and understanding of the scientific inquiry components and process (Figure 36 on web page).

## *Learning of Non-Inquiry Science Content*

During RIP implementation in the classroom, students are also assessed for their knowledge and understanding of science content other than the inquiry components and process (i.e., non-inquiry science content). Generally, students are able to learn content related to any subject area (e.g. STEM, social studies, language arts, and so on) through conducting of RIP scientific investigations (see Figures 37, 38, and 39 on web page), and this method of learning results in higher student performance compared to other classroom instructional strategies (see Figure 40 on web page).

Evidence also supports the RIP as a highly successful model for instruction and learning at the high school level for a broad range of student demographics. Using the RIP fosters high school student confidence in learning through scientific inquiry and development of and ability to apply scientific inquiry skills (see Figures 41 and 42 on web page).

The effects of using the RIP scientific inquiry STEM model extend beyond high school. Comparison of RIP- and non-RIP-trained high school students from the Academy for the Advancement of Science and Technology (Hackensack, New Jersey) indicated statistically significant clear advantages in future science education and STEM careers. The RIP student group enjoyed more college/university acceptances, acceptances at top schools of choice, and merit scholarships compared with the non-RIP student control group (see Figure 43 on web page). RIP experience also significantly enhances the likelihood that students will pursue STEM fields in higher education and careers, and postgraduate degrees (see Figures 44 and 45, respectively, on web page). One example of the profound personal and academic impact of RIP-based learning on high school students (not measurable through student assessments) comes from the experiences of two Special Education seniors from Kaimuki High School (Honolulu, Hawaii). Both stopped smoking cigarettes and applied (and were accepted!) to college STEM programs as direct outcomes of having conducted a research investigation on the chemistry of cigarettes (Bashaw 2012).

## Sample RIP Teacher Impact and Reflections

Sample teacher impact data evaluation reports from RIP-professional development activities and experiences and classroom implementation of the RIP can be found on the chapter web page, (*www.anovascience.com/RIP~ingThroughSTEMwebpage.htm*) under the subsection "Sample Teacher RIP Assessment Evaluation Reports" (and also see reflections from the two oil spill teachers in Table 2 on web page). It is important to note the consistency of the impact of the RIP as an instructional strategy across all teacher populations measured to date. In general, following RIP training and support over the year, teachers exhibit large, statistically significant increases in: (1) knowledge and understanding of and confidence in teaching scientific inquiry; (2) data organizing and analysis skills; (3) ability to apply data techniques for decision-making in science, and so on; (4) understanding the scientific inquiry (scientific process) standards and addressing the scientific inquiry benchmarks; (5) their understanding of, confidence in, and ability to implement standards-based inquiry; and (6) confidence in using and the frequency/quality of use of formative assessment with scientific inquiry.

RIP-trained teachers attribute student success to their instruction through this model. Using one school with a highly academically-challenged student population as an example (Anuenue Elementary School in Honolulu, Hawaii), students readily accepted the RIP model into their repertoire of learning strategies and behaviors when it was introduced into the classroom (Figure 46 on web page). Anuenue teachers reported much greater positive instructional impact of the RIP on their students' ability to learn science and/or other content areas compared to their other instructional practices used in prior years (Figure 47 on web page). The teachers also reported that compared to the instructional practices they used in previous years, the use of the RIP had greater positive impact on their students' *motivation to learn science* and/or other content areas and "increased" or "greatly increased" interest to learn science (Figures 48 and 49, respectively, on web page). The RIP scientific inquiry process is rigorous and challenging to learn for both teachers and students. However, student and teacher outcomes resulting from the rigor and challenges are well worth it (see web page subsection "Sample External Evaluator Conclusions"). Willingness of students to apply the inquiry process to their learning demands confidence. One Anuenue teacher (Teacher #1) wrote in her assessment, "Because we [teachers] are more confident, our students are more confident in science—STEM" (ANOVA Science Education Corp. 2011).

RIP scientific inquiry is clearly a viable instructional and learning model for bringing STEM into K–12 education (see RIP STEM model diagram, Figure 50 on web page). The process that students use to conduct their RIP investigations targeting content in any one or a combination of the STEM fields incorporates the scientific and engineering content and practices, as well as mathematics and technology as supporting tools. In short, performing a RIP scientific inquiry equates to doing STEM! The RIP model also aligns well with the *CCSS* college and career readiness goals by initiating and supporting knowledge, habits, skills, and critical thinking that lead to successful performance in reading, written and oral communications, teamwork, and problem solving. Teacher and student evidence supports that the RIP model impacts in positive ways attitudes toward and the development of skills supporting teaching and learning that lead to successful outcomes that extend well beyond the K–12 experience. This rigorous, thematic-based approach stimulates learning, excitement, and enjoyment in the student. The K–12 evidence along with the post–high school data presented in this chapter strongly support these effects and reflect the benefit of the critical thinking and decision making STEM-based skill sets developed in students that are mandatory for 21st-century success.

## Next Steps

The following paragraphs describe some valuable actions that may enhance the future usability and effectiveness of the RIP scientific inquiry STEM model. "Too often, reforms begin with a specific action, 'such as adopt new instructional materials,' and devotes little or no attention to other actions, such as getting administrative or community support for the changes" (Bybee 2013, p. 89). Teachers who use the RIP frequently ask how they can successfully incorporate investigations into the curriculum when their classrooms are constantly interrupted by special assemblies, sports or other extracurricular events, and field trips. Informing and educating administrators about the benefits of the RIP scientific inquiry STEM model and the needs involved in creating an environment that supports this program may lead them to consider affording teachers flexibility in scheduling such events.

The RIP model supports the reform efforts of the *CCSS* by initiating and inviting rigorous thinking into teaching and learning to prepare students for success in their postsecondary education and careers. It will be interesting to observe and document the impact of school implementation of the RIP on student ability to achieve the expectations and goals of the CCSS.

Engineering and the use of technology in engineering are sorely lacking in K–12 education, especially in the lower grades. We propose that engineering classes that are developed for these grade levels incorporate and benefit from the RIP model. This curriculum design would provide students with a paradigm for the four areas of STEM—including engineering and technology content.

## References

Adcock, R. A., A. Thangavel, S. Whitfield-Gabrieli, B. Knutson, and J. D. E. Gabrieli. 2006. Reward-motivated learning: Mesolimbic activation precedes memory formation. *Neuron* 50: 507–517.

ANOVA Science Education Corp. 2011. Implementation of the Research Investigation Process at Anuenue Elementary School: Year 2 Impact Evaluation. *http://anovascience.com/pressreleases1.htm*

Barr, D., J. Harrison, and L. Conery. 2011. Computational thinking: A digital age skill for everyone. *Learning & Leading With Technology* (March/April): 20–23.

Bashaw, C. 2012. Cigarette smokers conclude "not in my body!" In *Look at us now! Making scientific practices matter in the classroom… and beyond*, ed. R.E. Landsman, 315–346. Honolulu, HI: ANOVA Science.

Berger, M. 1994. *Oil spill!* New York: HarperCollins.

Bybee, R. W. 2011. Scientific and engineering practices in K–12 classrooms: Understanding *A framework for K–12 science education*. *The Science Teacher* 78 (9): 34–40.

Bybee, R. W. 2013. *The case for STEM education: Challenges and opportunities*. Arlington, VA: NSTA Press.

Chandrasekhar, A. 1991. *Oliver and the oil spill*. Kansas City, MO: Landmark Editions.

CNN Money. 2013 New survey shows teens losing interest in STEM careers while U.S. projects significant growth in field. *www.money.cnn.com/news/newsfeeds/articles/prnewswire/MM73366.htm*

Colburn, A. 2000. An inquiry primer. *Science Scope* 23 (6): 42–44.

Dere, E., A. Easton, L. Nadel, and J. P. Huston. 2008. *Handbook of episodic memory*. Amsterdam: Elsevier.

Dugger, W. E. 2011. STEM: Some basic definitions. International Technology and Engineering Educators Association. *www.iteea.org/Resources/PressRoom/STEMDefinition.pdf*

Herbert, D. M. B., and J. S. Burt. 2004. What do students remember? Episodic memory and the development of schematization. *Applied Cognitive Psychology* 18: 77–88.

Hodgson, B. 1990. Alaska's big spill: Can the wilderness heal? *National Geographic* 177 (1): 5–43.

Kugler, C., J. Hagen, and F. Singer. 2003. Teaching statistical thinking. *Journal of College Science Teaching* 32 (7): 434–439.

Landsman, R. E. 2005a. *RIP~ing through scientific inquiry: Critical thinking and effective decision making skills for middle school and high school science education*. Honolulu, HI: ANOVA Science Publishing.

Landsman, R. E. 2005b. RIP~ing away barriers to science education: Inquiry through the *Research Investigation Process*. In *Exemplary science in grades 9–12: Standards-based success stories*, ed. R. E. Yager, 51–71. Arlington, VA: NSTA Press.

Landsman, R. E. 2005c. *Data analysis and decision making in scientific inquiry: A statistical approach for middle school and high school science education*. Honolulu, HI: ANOVA Science Publishing.

Landsman, R. E. 2006. *Look at me now! Motivate young minds to think & learn through scientific inquiry*. Honolulu, HI: ANOVA Science Publishing.

Landsman, R. E., ed. 2012. *Look at us now! Making scientific practices matter in the classroom… and beyond using the research investigation process (RIP)*. Honolulu, HI: ANOVA Science Publishing.

Landsman, R. E., J. Evans, and and C. Kamimura. 2012. Bridging the digital divide with virtual inquiry. Paper presented at the National Science Teachers Association's Annual Conference on Science Education, Indianapolis, IN.

Landsman, R. E., and I. H. Kamimura. 2012. Supporting the next generation of scientific practices with *more emphasis* on learners and their thinking. In *Look at us now! Making scientific practices matter in the classroom … and beyond*, ed. R. E. Landsman, 1–39. Honolulu, HI: ANOVA Science Publishing.

LoRusso, L. 2006. Spilling of old knowledge onto new ways of learning: Kindergarten students test their hypotheses about cleaning up an oil spill. *www.anovascience.com/STUDENTSCIENTIFICINQUIRY-FEATURE2.htm*

LoRusso, L. 2012. Spilling of old knowledge onto new ways of learning. *Look at us now! Making scientific practices matter in the classroom … and beyond*, ed. R. E. Landsman, 41–66. Honolulu, HI: ANOVA Science Publishing.

Martin-Hansen, L. 2002. Defining inquiry: Exploring the many types of inquiry in the science classroom. *The Science Teacher* 69 (2): 34–37.

Miller, M. 2010. Ecosystem and oil spill: Where scientific inquiry in the classroom meets reality. *www.anovascience.com/STUDENTSCIENTIFICINQUIRY FEATURE4.htm*

Miller, M. 2011. Whorls, loops, and arches: Connecting fingerprints to inheritance through scientific inquiry! Paper presented at the NMSTA and EEANM Soar to Greater Heights: Connecting Earth and Education conference, Farmington, NM.

National Governors Association Center for Best Practices and Council of Chief State School Officers (NGAC and CCSSO). 2010. *Common core state standards*. Washington, DC: NGAC and CCSSO.

National Research Council (NRC). 1996. *National science education standards*. Washington, DC: National Academies Press.

National Research Council (NRC). 2000. *Inquiry and the national science education standards: A guide for teaching and learning*. Washington, DC: National Academies Press.

National Research Council (NRC). 2012. *A framework for K–12 science education: Practices, crosscutting concepts, and core ideas*. Washington, DC: National Academies Press.

NGSS Lead States. 2013. *Next Generation Science Standards: For states by states*. Washington, DC: National Academies Press. *www.nextgenscience.org/next-generation-science-standards*

OECD. 2010. PISA 2009 results: Executive summary. *www.oecd.org/pisa/pisaproducts/46619703.pdf*

Pergolizzi, R. 2008. RIP~ing @ Science in Kaimuki complex: Math and science partnership project annual narrative report for 2007–2008. Honolulu, HI. *http://kaimukirip.k12.hi.us*

Ripley, A. 2013. *The smartest kids in the world and how they got that way*. New York: Simon & Schuster.

Sayres, R. 1996. Effects of testosterone on the electric organ discharge waveform in electric catfish, *Malapterurus electricus*. *NCSSSMST Journal* 2 (1): 24–27.

Thomasian, J. 2011. *Building a STEM education agenda*. Washington, DC: National Governors Association. *www.nga.org/files/live/sites/NGA/files/pdf/1112STEMGUIDE.PDF*

Toulmin, C. N., and M. Groome. 2007. *Building a science, technology, engineering and math agenda.* Washington, DC: National Governors Association.

Tulving, E. 1983. *Elements of episodic memory*. New York: Oxford University Press.

Tulving, E. 1985. Memory and consciousness. *Canadian Psychology* 26: 1–12.

Wagner, T. 2008. *The global achievement gap*. New York: Basic Books.

Wagner, T. 2009. Seven skills students need for the future. *www.youtube.com/watch?v=NS2PqTTxFFc.*

Wagner, T. 2012. *Creating innovators*. New York: Scribner.

Wang, J. 2009. *Math and science partnership project annual report: Year 3 external evaluation findings report*. Honolulu, HI: Hawaii Department of Education.

Wing, J. 2006. Computational thinking. *Communications of the ACM* 49 (3): 33–35.

Yahoo News. 2013. The world's best schools. *www.news.yahoo.com/video/world-best-schools-004500495.html*

Yager, R. E. 2012. More emphasis on the how vs. the what. In *Look at us now! Making scientific practices matter in the classroom and beyond*, ed. R. E. Landsman, 41–66. Honolulu: ANOVA Science Publishing.

# Developing STEM Site-Based Teacher and Administrator Leadership

*Jo Anne Vasquez*
*Helios Education Foundation*
*Phoenix, AZ*

*"There are no 'leader-proof' reforms—and no effective reforms without good leadership."*
A Bridge to School Reform, The Wallace Foundation (2007).

## Setting

Helios Education Foundation is dedicated to creating opportunities for individuals in Arizona and Florida to succeed in postsecondary education. To achieve that goal, the Foundation invests resources across the education continuum to advance student preparedness and to foster college-going cultures in Arizona and Florida. The Foundation focuses its investments in three key priority areas: early grade success, college and career readiness, and postsecondary completion. Ultimately, the Foundation's goal is to build sustainable education systems to ensure every student graduates college career ready and completes a high-quality postsecondary certificate or degree.

Recognizing that strong, knowledgeable teacher leaders and administrators can become the agents for change within their school and district, the Helios Education Foundation is an engaged thought partner and funder of a number of initiatives focused on school leadership as part of its investment strategy across the transition years.

These initiatives improve academic rigor and relevance with an emphasis on STEM education by creating and sustaining highly skilled teachers and effective leaders as well as embedding college-going culture supported by actively engaged families and communities.

## Overview of the Programs

Rodger Bybee, executive director (retired), Biological Science Curriculum Study, has reported that "Writing … about the reform of education and reforming education are two very different activities. The former requires that a small group agree on a set of ideas and express those ideas clearly and with adequate justification. The latter requires that millions of school personnel in thousands of autonomous school districts change." (Bybee 1995). As hard as it is, defining STEM is the easy part; implementing STEM education on a large scale is the challenge. The following two projects are focused on this challenge!

The Helios Education Foundation and its funded partners define STEM education as an interdisciplinary approach to learning that removes the traditional barriers separating the four disciplines of science, technology, engineering, and mathematics and integrates them into real-world, rigorous, and relevant learning experiences for students.

## Program #1

STEM in the Middle (SIM) is a three-year project that began in the spring of 2011 to develop the mathematics and science content, assessment, and pedagogical knowledge of grades 5–8 middle school teachers of mathematics and science. The goal was to give these teachers the skills necessary to become STEM education experts in their schools and districts by providing them with deeper understanding of key concepts, skills, and reasoning methods of middle school mathematics and the sciences, strategies for designing integrated projects, implementing project-driven learning programs, differentiating instruction, and working with adult learners to enhance the STEM education of students. Teachers learned to develop proposals to fund needed materials and other resources to enhance the implementation of integrated projects.

STEM in the Middle (SIM), under the leadership of Dr. Carole Greenes, associate vice provost for STEM Education, director of the PRIME Center, and professor of mathematics education at Arizona State University, has other targeted goals beyond developing STEM Teacher leaders. These include:

1. *ClubSTEM* engages 60 grades 5–8 students in STEM explorations for three hours each Saturday for seven Saturdays per semester. They employ a Scientific Village approach, in which students engage in long-term integrated projects designed and led by scientists and mentored by high school students and college undergraduates. ClubSTEM students are actively engaged in high interest activities such as dissecting a kidney to explore the filtration system.

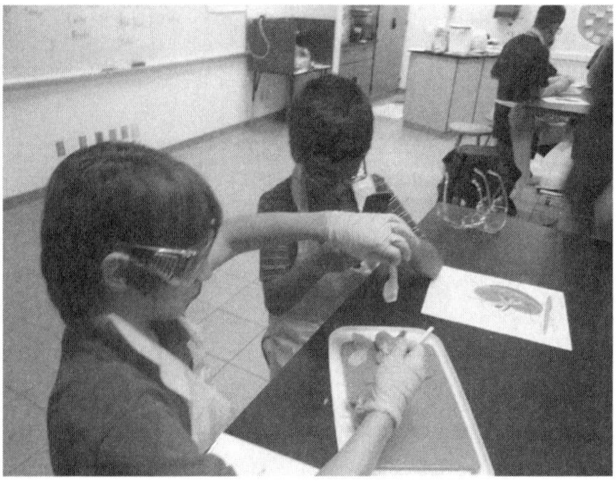

The seventh session is the Showcase Open House where students present their projects to families and the community. Topics of past STEM villages include: Anatomy and Physiology, Forensic Science, Creative Photography, Computer-Designed 3-D Structures, SumoBots, Computer Game Design-Programming and Illustration, Flight/Aviation, and Rube Goldberg Constructions.

2. Key to the success of the ClubSTEM is the mentoring program that engages high school students interested in entering STEM university programs as well as undergraduate college STEM majors to assist scientist leaders of ClubSTEM with integrated project implementation and to serve as role models for students. In the picture below, the ClubSTEM students are engaged in a Forensic Science Scientific Village, where they prepare a cast of a footprint as they investigate impression evidence left at a simulated crime scene.

   Students who apply to serve as mentors are interviewed by project staff and the senior mentor and are trained by the scientists and project staff. Depending on the projects, 10–17 mentors are appointed each semester.

3. The *MATHgazine Junior* is a four-page online magazine available to teachers, students, and families of students. It is designed to develop readers' mathematical and scientific reasoning talents. Readers solve problems and send them in to the PR"I"ME Center staff for solution verification. At the end of the academic year, the solver with the highest score receives a STEMatician certificate and is highlighted in the first issue of the subsequent academic year.

## Methodology

This initiative has many parts to their "Village" approach. The purpose in this chapter will focus only on the STEM Teacher Leader development, which is core to the SIM strategy. Following is an overview of the Saturday ConCourses for Teachers:

- Teachers meet semi-monthly for eight Saturdays during the academic year and for one week (5 days) in the summer, four hours per session. These Saturdays and summer days are spent in a combination CONFerence (STEM experts presenting key concepts and strategies) and COURSE (in-depth study of those key concepts and strategies, their implementation and their assessment). Teachers organize in Concept Study Groups (i.e., communities of practice) led by ASU faculty. In each group, teachers explore a concept in grades 5–8 mathematics or science curricula, observe how the ideas grow in complexity across the domains, and then decide how to introduce, maintain, enrich, and assess an integrated STEM project.
- In the weeks between Saturday ConCourses, teachers use their instructional materials and strategies with their own students in their own schools. They bring evidence (e.g., videos, documentation of interviews, observations, student papers, models, and computer programs) to include their implementation for feedback from the other SIM teachers and staff.
- In the final summer and academic year the teachers learn to conduct need assessments for professional development in their schools, analyzing and interpreting the results, and designing and conducting a professional development program for their staff with mentorship from SIM project leaders. With mentorship, teachers gain confidence in their abilities to provide professional development to teachers for the different grade levels.

This is all set up as a learning cycle effect that can be repeated over and over, first the teachers are the students and finally they are the teachers of other teachers.

The evaluation of the three-year SIM Project includes both quantitative (descriptive and inferential statistics) and qualitative (descriptive results of surveys, interviews, and focus groups) method. This report covers results after two years.

The methodology was designed specifically to address the overarching goals of SIM to: (1) Promote grades 5–8 students' curiosity and success with the study of mathematics and the sciences by developing their algebraic reasoning and problem-solving abilities, engaging them in activities and projects like those carried out by STEM professionals, and communicating regularly with scientists; and (2) Develop middle school teachers' knowledge of key concepts of mathematics and the sciences, how student understanding of those concepts becomes more robust, instructional strategies that promote perseverance behaviors, techniques for designing and conducting integrated projects, and methods for assessing and assisting students with varying talent.

## *Data Collection Methods and Types of Data Collected*

1. *Demographic data.* Basic information about student and teacher characteristics was gathered during the application process. Demographics include grade level, gender, self-identified ethnicity and race, school and district information for both students and teachers. Additional information on subject areas taught was collected for each teacher.

2. *Baseline assessment of algebraic and scientific reasoning.* During year 1 a validated 7-item Algebra Test of Linearity and a 15-item Test of Scientific Reasoning were adminis-

tered to all SIM teacher participants and the students in their classes, as well as all grades 5–8 students in ClubSTEM scientific villages. In year 3, SIM teachers and SIM students are to be assessed again.

3. *Participant satisfaction surveys*. Participant satisfaction surveys were administered by the external evaluator at the end of the spring session for ClubSTEM students and SIM teachers.
4. *Village leader and mentor survey*. Project staff support surveys regarding the perceptions of student growth and development as a result of student participation in ClubSTEM were administered yearly.
5. *Focus group interviews*. Random samples of students and teachers were interviewed at the start of the fall programs.
6. *Pre- and Post-village-specific knowledge assessments*. Assessments, focusing on math and science constructs and village content, were administered to all students in ClubSTEM as pre- and post-assessments in their scientific villages.

## Data Analysis

Results are included as Goals and Outcomes. Extensive amounts of data have been collected on students, including both mathematics and science previllage and postvillage assessments.

*Goal One: Promote grades 5–8 students curiosity and successes with the study of mathematics and the sciences by developing their algebraic reasoning and problem-solving abilities, engaging them in activities and projects, and communicating regularly with scientists.*

Five students were randomly selected to be part of a focus group before the end of the fall 2011 and 2012 Showcase for Open Houses for families and friends. When asked about mathematics and science in their home schools, several comments provided insight into how different Club STEM is when compared to regular classroom teaching.

- "Here we get to build and create stuff. And the college kids (mentors) think we are smart."
- "I like how the mentors help us, but don't tell us. We get to find out answers by trying our ideas and then they help us put it together."
- "I'm doing better in school and I like coming here on the weekends. Meeting other kids who don't go to my school and who like science helps."
- "I wish my teachers at my school did classes like this. I really liked my photography village. Our teacher is really smart and we create art and science at the same time."

*Goal Two: Develop middle school teacher knowledge of key concepts of STEM, how student understanding of those concepts becomes more robust, instructional strategies that promote perseverance behaviors, techniques for designing and conducting integrated projects, and methods for assessing and assisting students with varying talents.*

Here the teachers are working together in teams to check the temperature of a chemical reaction that releases energy in the form of heat.

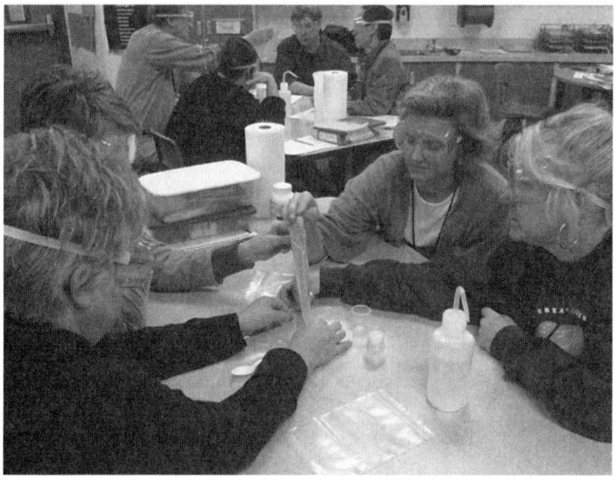

Essential to teachers' success in their own classroom is the understanding of the content by actually doing the activities and being in the role of the learner they are able to gain insight in to how their own students develop conceptual understanding.

## *Teacher Outcome 1: Increased Knowledge of Concepts/Skills and Related Fields*

Two content assessments were undertaken for teachers during each of the Summer ConCourse programs. Changes in performance from pre- to posttest were statistically ($p < .001$) significant, indicating a dramatic increase in comprehension based on the module explorations.

Feedback from interviews and surveys conducted during the school year and summer program revealed powerful quantitative and qualitative results regarding teacher insights into their own skills, abilities, and behaviors. When asked how they have grown as teachers as a result of participation in SIM, all teachers interviewed commented that they had (1) grown in their understanding of STEM, and (2) were more aware of student challenges about those same "Concept" mastering. Sixty-seven percent of those interviewed felt that they were more comfortable with offering challenging problems and a variety of problems that will "make them (teachers and students) work harder."

- "SIM has improved MY problem solving skills and that benefits my students."
- "I've gained confidence in my abilities of problem solving which will make me more likely to do these projects with kids."
- "I don't freeze or call the math teacher for help in physics. I work through math problems in science with greater ease."

- "I think I'm more open-minded when it comes to my teaching/learning. I'm also more excited about teaching science!"
- "I now spend more time on problems and challenge students to find different methods to solve the same problem."

## *Teacher Outcome 2: Increased Knowledge of Ways to Address Student Concerns and Talents*

Giving teachers experiences of actually struggling with their own understanding is a powerful way to help them put themselves in the place of their students. In the picture below the teachers are exploring the balance of forces by trying to figure out where to stand on a beam whose ends rest on scales.

The task is laid out for them but discovering the solution is up to the group. It also points out to teachers that for activities to be meaningful for all students they must build in enough time for students to go through the processes of learning.

Feedback on year-end surveys indicated that 96% of teachers felt that the educational materials provided during SIM were "excellent." Interviews revealed that teachers were more aware of student abilities and capabilities as a result of the project.

- "SIM has reinforced the importance of hands-on teaching and problems need to be relevant to all students (e.g., food, shopping, and design)."
- "I consider how each student may be processing the math/science problems now. I am more patient."
- "I learned that students benefit from struggle, and I have incorporated this philosophy with my students. I learned that when math is presented in real life contexts, it is more palatable and effective for students."
- "I believe that all students are capable of understanding and participating."
- "I feel that my students can express themselves more, which has shown me that they are more capable."
- "I have come to understand that it is not completely the fault of students for not understanding math 100%—much of the challenge lies in the material and how it's presented."

- "There are many resources available outside of textbooks to supplement instruction (like the Macy's coupons) and the importance of proportions and rates in math."
- "We are starting a STEM Summer Camp this year!"

### *Interviews Conducted With Parents During the Showcase Event Each Semester*

Speaking directly, one-on-one, with parents provided great opportunities to address their perceptions of the program, progress their children had made, and what additional supports they felt were needed to help their children proceed into STEM subjects in high school.

- "Our son is more focused and is doing better in math at his school. He was a good student and has improved. We are pleased that he is more focused to prepare for high school."
- "We were surprised at his willingness to share what he learned at STEM (in the Middle) with teachers and students at his school. Socially, he has become much more outgoing."
- "This project helps us to reinforce the idea that college is very realistic for our child. He has always liked hands-on science and he sees that he can go far. The teachers are very supportive of him here."
- "My child's appreciation of technology and not taking it for granted has been helpful. And seeing how he can use different technology tools and the 'why' of it."
- "I absolutely appreciate having him learn different aspects of science. This project-based learning has helped him greatly. He loves science and I think this is motivating him to continue science studies."

## Program #2

The Pinal County Education Service Agency's Next Generation STEM Leaders (NGSL) project is a rural countywide initiative covering a section of southeastern Arizona approximately the size of the state of Connecticut. It involves nine school districts, ranging in size from 250 to over 8,000 students. Beginning in 2010 the ESA has now worked with 36 teachers of math and science in grades 5–8 to enhance their content knowledge, develop effective pedagogy, and to help the teachers develop strategies to align effective STEM teaching and learning-based data-driven decision making.

The overarching goal of the project is that by 2014 all Pinal County districts and charter schools will have STEM teachers who have a solid understanding of STEM teaching and learning; are knowledgeable in technology application in the classroom; demonstrate knowledge of content, pedagogy, and problem-solving skills in STEM curricula; and become skilled at unwrapping academic standards for teacher instructional use. These STEM teacher leaders will become the school-based systemic change agents that will expand the STEM learning and maintain program sustainability in the coming years. In addition, The STEM teacher leaders will be able to increase student and parent exposure to STEM higher education programs and provide avenues for support for those students to who aspire at a postsecondary education.

## Methodology

The NGSL program focuses on a several activities:

- Professional development branched into three areas of concentration: (1) science and math content knowledge, (2) STEM pedagogy, and (3) leadership development. Content knowledge was developed and deepened through college-level courses each spring semester at Central Arizona College. Pedagogy and leadership were addressed through 24 hours of professional development offered throughout the school year and an intensive four-day Summer Institute each year. Through these workshops participants worked collaboratively to learn and develop STEM lessons using the Project-Based Lesson (PBL) planning model developed by the Buck Institute.
- At the beginning of the project, the majority of NGSL teachers could be characterized as being at the "entry level" of integrating technology into lessons and activities. They also had no developed means of communicating between their far-flung schools, using the internet. The project provided a tablet for each teacher, along with training on using technology and developing teacher websites.
- Outreach to students and their families to educate them on the opportunity and possibility of postsecondary STEM education was provided by "college information" nights throughout the county, family nights at Central Arizona College and, an annual Student Field Day at the college for the NGSL teachers and their students.

Upon the conclusion of the first year, the team leaders of the NGSL program reflected upon the NGSL Teacher Leaders progress and growth. It became evident there was an opportunity to increase the expansion of the STEM knowledge and learning beyond the initial core teachers by including administrators in the project. In order to facilitate and bridge the STEM teaching and learning at the district level, administrators needed to become STEM Leaders and have the opportunity to fully support the efforts of the STEM Teacher leaders through a deeper understanding of STEM Teaching and Learning. They would also become more supportive of release time for their teachers to participate and more open to the resource requests to support STEM education on their campuses.

## Data Collection for Next Generation STEM Administrators' Program

The Pinal County STEM Administrators participated in a Leadership Roundtable as well as completed interviews. The analysis of the results from this first meeting and the survey indicated needs in the areas of data-driven decision making, school collaboration networks, and technology integration. From this information, A21CE Administrators for 21st-Century Education quarterly networking and professional development meetings were scheduled to address these needs with support activities. The pyramid in Figure 23.1 (p. 410) points out the overarching goals set for the administrators: develop school collaboration networks, technology integration, and data-driven decision making.

Overall, when asked how the Pinal County Education Service Agency could best help the administrators meet their district goals and objectives, the superintendents focused on the need to develop leadership capacity in their districts. Within that overall need, several areas of concern were identified with more specific issues (e.g., dealing with budget cuts) to broader initiatives (e.g., improve parent/community relationships). In the two telephone surveys leading change through STEM implementation was mentioned fairly regularly, but cannot be considered a high

**Figure 23.1.** Administrator STEM Pyramid

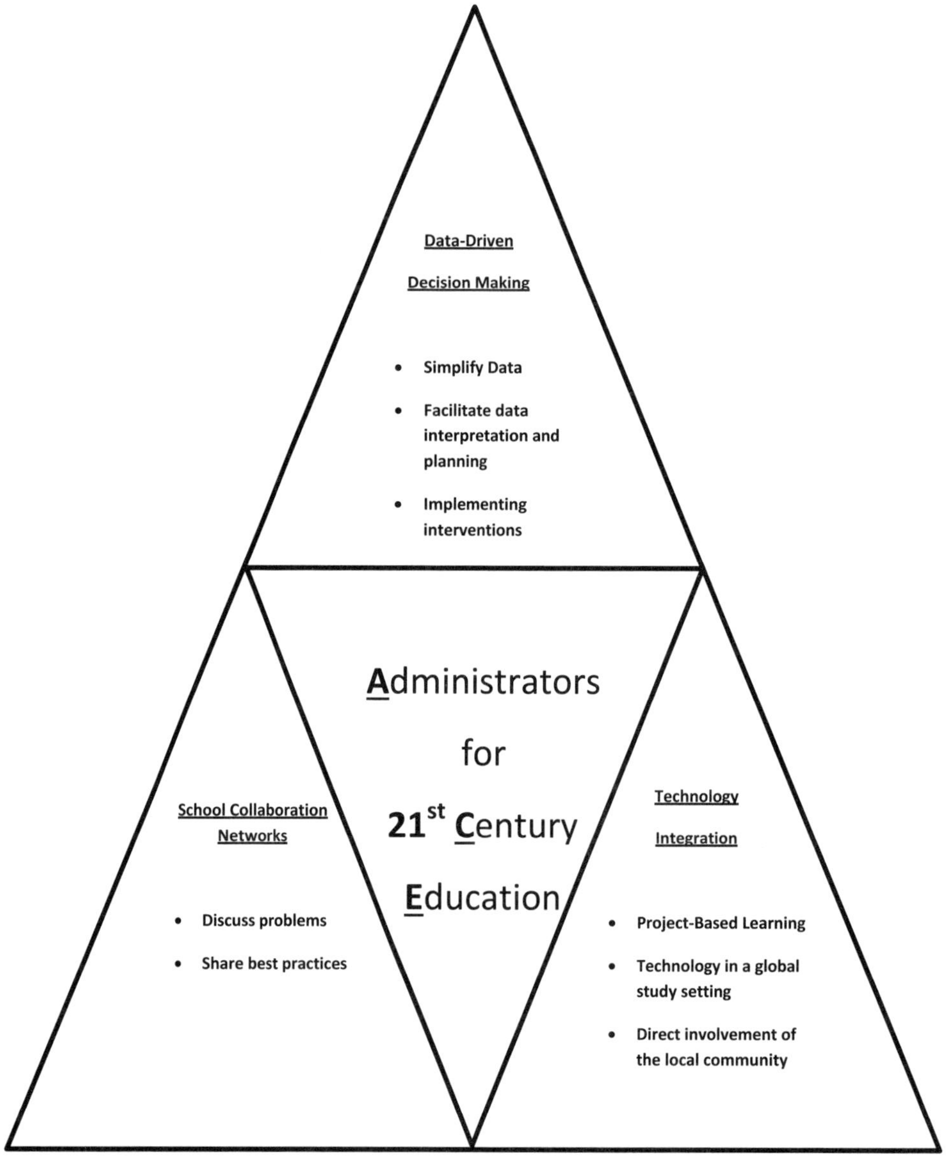
Data-Driven
Decision Making
Simplify Data
Facilitate data interpretation and planning
Implementing interventions
Administrators
for
21st Century
Education
School Collaboration Networks
Discuss problems
Share best practices
Technology
Integration
Project-Based Learning
Technology in a global study setting
Direct involvement of the local community

priority. Nevertheless, the needs most often identified by the respondents corresponded well to the goals of the NGSA project and the 21st-Century Teaching and Learning initiative in Pinal County. They can be summarized in three primary areas of need as follows:

1. Consistently high in priority across all three samples was the need for data literacy among school administrators. There is concern that administrators cannot lead teachers in data-driven decision making unless and until the administrators themselves understand how to simplify data, analyze data in a timely manner, and use data to drive adaptive change in classroom practice. Specific interest in data literacy for the administrators ranged from analyzing dips in AYP to understanding Galileo and Dibels results. In terms of leadership development, the administrators expressed a need for skill in coaching teachers on how to better utilize data in their teaching. One administrator recommended learning to lead "test talks" with teachers in a collaborative setting of administrators and teachers. Several respondents asked for support in helping teachers implement response to intervention (RTI), which is dependent upon frequent progress measurement and research-based interventions.
2. The second aspect of leadership capacity building noted by the administrators was the need to organize, manage, and sustain collaborative networks among the administrators and teachers at school sites, and across school sites within the district. Providing these opportunities for administrators to work together fosters their understanding for the need to have ongoing collaborative work throughout their schools. They may know this at one level but do not necessarily understand how effective this can be unless they themselves have experienced it. These collaborative activities also provide administrators with opportunities to experience the need in which their own teachers can feel empowered to embrace and lead change. These collaborative networks are seen as important for successful use of data in decision-making; ongoing learning among the staff; successful implementation of many new initiatives, including common core, RTI, and Beyond Textbooks; and nonthreatening, productive teacher evaluations. They also see the need for collaboration and information exchange among administrators through the district and county, however, they also do not understand how to develop and maintain a network with this capacity.
3. Throughout the surveys, the administrators noted a need to develop a better understanding of how to integrate technology in the classrooms. At present, it appears that their identified need for technology integration involves the practical aspects of utilizing specific technologies in the classrooms (e.g., interactive whiteboards or tablets). There were numerous requests for more staff development on the use of specific technologies. However, from the standpoint of leadership in the schools, it was determined that the administrators need a better understanding on how to lead their teachers from seeing technology integration as a technology-driven enterprise to a focus on technology-enabled learning. That is, a cognitive shift from functionality of isolated technologies,

to seeing that 21st-century learning uses technology to expand the classroom into the global collaborative educational network.

## Conclusions

Based on the needs assessment, it is recommended that professional development for the NGSA project addresses the following:

1. Adoption of STEM education and 21st-century teaching and learning will only succeed if there is a demonstrable need and the project-based learning model being used by the NGSL project can present demonstrable evidence of improvements in student performance. To understand the evidence, administrators will need a high degree of data literacy. Therefore, it is recommended that professional development focus on data literacy with specific exemplars from PBL assessments, AIMS, Galileo and Dibels. Data literacy should include how to simplify data, lead teachers in data interpretation and planning, and implementing and evaluating interventions in a timely manner.
2. Successful adoption of STEM education and 21st-century teaching is highly dependent upon developing a strong communication and collaboration network. Administrators will need the skills to develop and maintain networks among themselves and among their teachers. Therefore, it is recommended that professional development focus on how to organize, manage and sustain collaborative networks that empower teachers and administrators alike, provide opportunities to discuss problems, and share best practices.
3. Of the three areas, this will probably be the least intuitive for the administrators. Their responses indicate that they perceive technology integration as learning how to use specific technologies (e.g., tablets) in an individual classroom. On the other hand, technology in the 21st-century classroom *connects* students and teachers to a global learning community and promotes self-directed learning beyond the school walls. Therefore, it is recommended that professional development on leading technology integration be explored by engaging administrators in a study of problem-based learning using technology in a global setting, including direct involvement of the local community.

## Next Generation STEM Leaders Project (NGSL)

Data collection has been an ongoing activity throughout the lifetime of the NGSL and NGSA Projects. Both quantitative and qualitative data are utilized to serve a variety of purposes. While there are a variety of measures for the administrators participating in the project, the focus here is on the teacher leaders. A variety of surveys are used to inform program developers of the teacher leaders' perceived progress in understanding and implementing change and to assess ongoing professional development needs. Additional measures are used to assess short-term outcomes and the evaluation of intermediate outcomes, and long-term program impact along with its potential utility for replication in other environments.

## Data Collection Methods

1. Demographic Data. The 32 participants in the NGSL Teacher Leader cohort represent a sample of Pinal County math and science teachers in grades 5–8. Demographic data are used to track their impact on their schools and districts. These include their educational background, teaching experiences, recent professional development in math/science content, pedagogy, or curriculum, and the schools and number of students with whom they have direct contact.

2. Participant Satisfaction Surveys. ESA-developed participant satisfaction surveys were administered at the end of every professional development session to ensure quality control and assess their immediate responses to content and its applicability to the classroom.

3. Pre/Post Attitudes, Actions, and Beliefs Survey. The 31-item survey was constructed by adapting items from the TIMMS 2007 Teacher Questionnaires for eighth-grade mathematics and science and the SWEPTS Pre-Program Questionnaire. It was designed to assess participants' perceptions of themselves as teachers, their emphasis on specific teaching goals and objectives, and the teaching and learning activities used in their classrooms. Participants completed pre/post surveys yearly to inform project organizers of potential programming needs.

4. Pre/Post Leadership Confidence Survey. In year 2, NGSL organizers targeted leadership development to start preparing participants for their roles as teacher leaders in their schools and to track their self-reported growth in competency as an effective teacher. This is a 29-item survey, modeled after the seven domains of the 2010 Teacher Leader Model Standards. Perceived changes in year two were assessed by comparing pre- and postsurvey responses. Participants will complete a final survey at the end of year three.

5. NGSL Technology Use Survey. To better understand the role and purpose of technology use in the NGSL teacher leader classrooms, they completed a pre/post retrospective questionnaire each year to examine use of technology in the classrooms, their comfort level in using technology, and the level of support provided by their schools.

6. Teacher content knowledge measure: Diagnostic Teacher Assessments in Math and Science (DTAMS). To provide pre/post measures of content knowledge development in conjunction with their college coursework in mathematics and science, content-specific assessments are administered each year during the program. Results are analyzed statistically overall, and by knowledge type and content subcategories.

7. Teacher teaching skill measure: Reformed Teacher Observation Protocol (RTOP). The RTOP observational instrument is a widely accepted criterion-referenced assessment of reformed classroom activities. For the NGSL project, each teacher was videotaped while conducting a lesson in the home classroom (approximately one-hour sample each session). A trained external evaluator scored each lesson and conducted statistical tests to

assess pre/post change. Pre/post RTOPs were conducted in years 1 and 2. In year 3, each teacher will select a sample of classroom teaching for group analysis.

8. NGSL Teacher Leader Interviews. Telephone interviews were conducted in year 3 to assess participant perceptions of the overall program, their classroom applications, student responses, and administrator support.

9. Parent Survey. To examine parent attitudes and beliefs about the importance of STEM education for their children and their involvement in and support of their children's preparation for postsecondary STEM education and STEM careers. A 25-item online survey was developed in year 3. These results can be used for ongoing program activities and by the regional institutions of higher education.

10. Student Survey. In year 3, all NGSL teacher leaders' students were invited to complete a 30-item online survey to assess their attitudes about STEM disciplines as future education and career options and their perception of the others families' attitudes. As with the parent survey, these data are being used to inform future project activities as well as inform high schools and colleges in the county.

## *Data Analysis*

A fundamental goal of the NGSL project was to develop participants' STEM content knowledge and pedagogy skills. In each year of the project, both science and math cohorts have shown statistically significant gains on the DTAMS content knowledge assessments. Participants also have improved their ability to implement reformed teaching methods and activities in their classrooms, as evidenced by statistically significant gains in their RTOP scores. In addition, survey responses have shown increased willingness to incorporate in a wider variety of teaching methods and technologies in their teaching, as well as increased confidence as teacher leaders. They are making presentations on STEM education in their schools, districts, and communities. Some additional specific achievements include:

1. When the project proposal was developed, the 21st-Century Standards-Based Teaching and Learning Model (2008) for Pinal County focused solely on STEM education as shown in the Figure 23.2. Throughout the first two years of the project, it became increasingly evident to both organizers and participants that their vision of STEM education required a better framework and language to communicate that STEM was more than a program to emphasize science and mathematics as academic content areas. In their view, STEM Education was the motivation and catalyst for a paradigm shift in teaching and learning in Pinal County schools. Figure 23.3 (p. 416) reflects the evolution of this model, which is now the current vision for 21st-century teaching.

2. To support this evolving model, there was a focus on developing an identity for the group as STEM teacher leaders who were confident in their understanding of STEM education. They were individually able to create Project-Based Learning (PBL) lessons and units that were research-based, referenced cross-disciplinary standards, and integrated technology.

**Figure 23.2.** Beginning Vision Focus

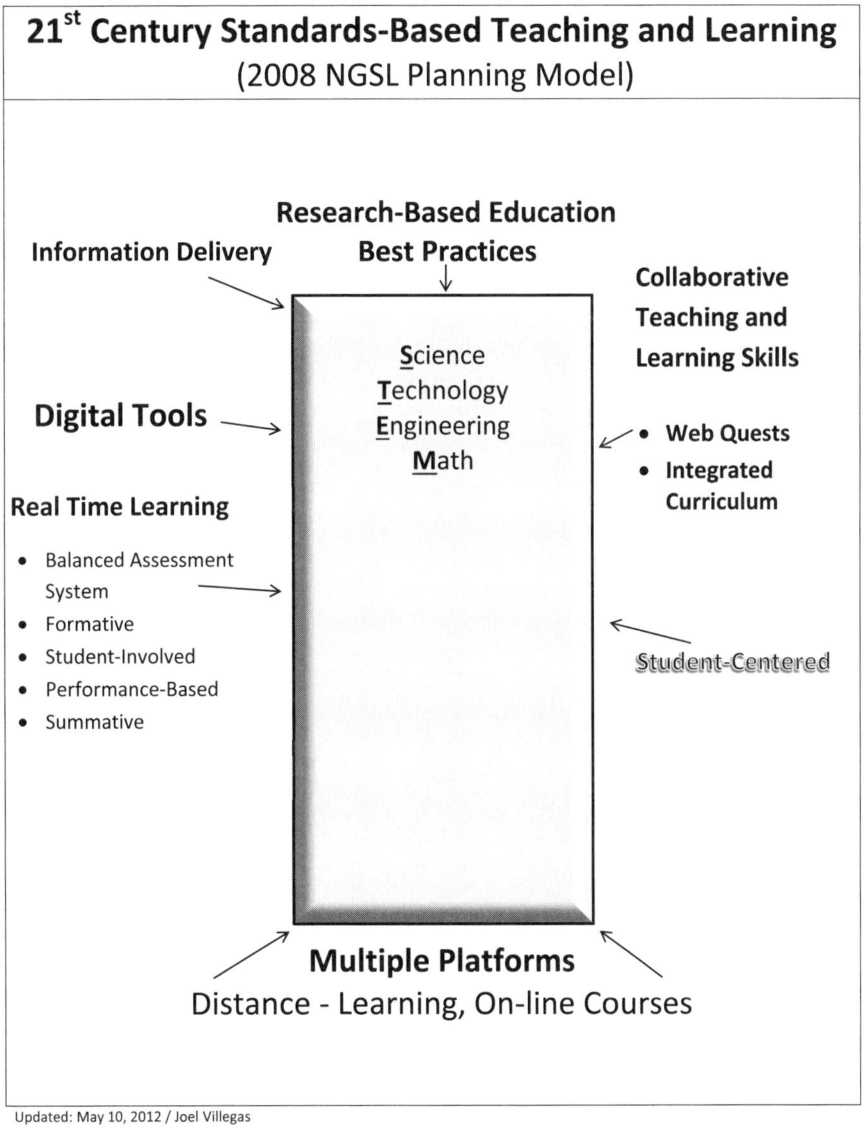
21st Century Standards-Based Teaching and Learning
(2008 NGSL Planning Model)
Research-Based Education
Best Practices
Information Delivery
Collaborative
Teaching and
Learning Skills
Science
Technology
Engineering
Math
Digital Tools
Web Quests
Integrated
Curriculum
Real Time Learning
Balanced Assessment
System
Formative
Student-Involved
Performance-Based
Summative
Student-Centered
Multiple Platforms
Distance - Learning, On-line Courses
Updated: May 10, 2012 / Joel Villegas

**Figure 23.3.** Evolution of NGSL Vision

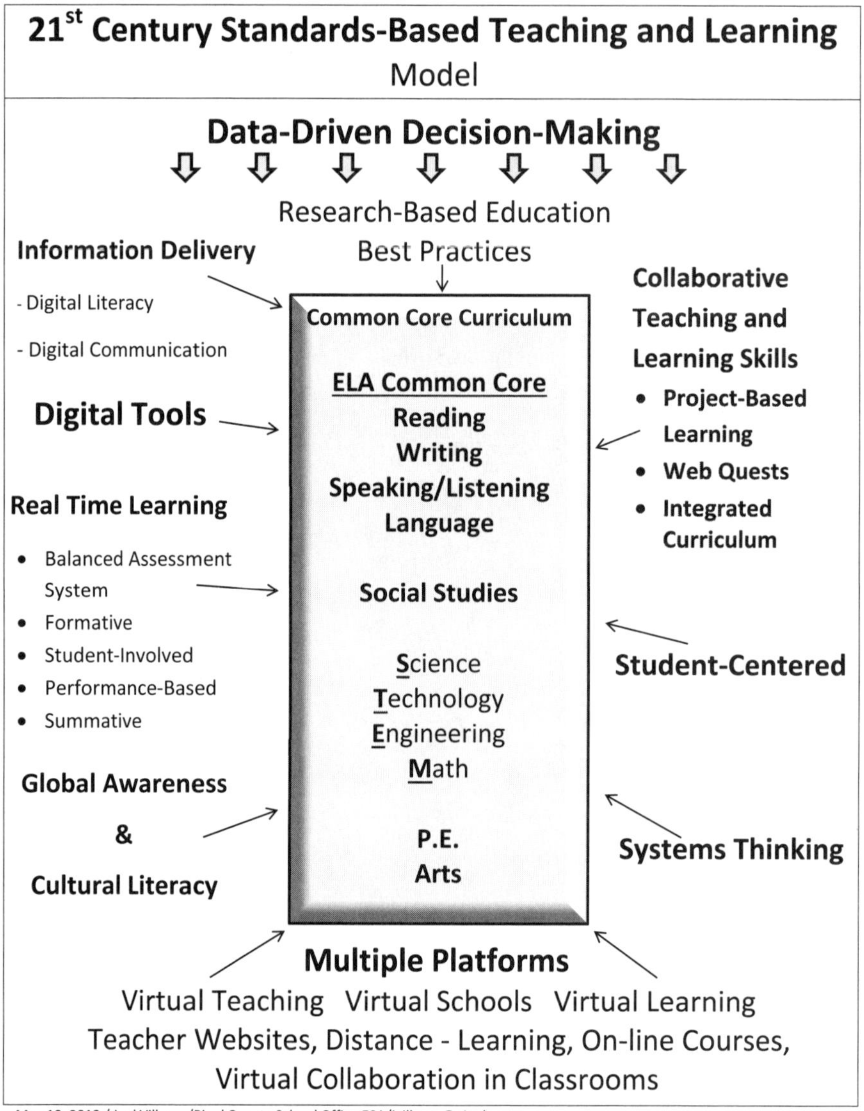


May 10, 2012 / Joel Villegas/Pinal County School Office ESA/jvillegas@pinalesa.org

They had developed a rubric for quality STEM PBLs. Throughout year 2 they continued to clarify and define their lessons and collaboratively developed STEM PBLs as a model for cross-discipline collaboration at their home schools.

3. Every teacher participant also developed aspects of each of the following characteristics or attributes of 21st-century teaching and learning. Much of the evidence is available on the NGSL website. Currently, NGSL teacher leaders are moving toward 21st-century teaching and learning and grantees plan to assess where potential participant schools currently are on the STEM Immersion Matrix (available on the Arizona Science Foundation website, *www.sfaz.org/stem*).

   - Collaborative Teaching and Learning Skills
     - » Project-based Learning
     - » Web Quests
     - » Integrated Curriculum
     - » Student-Centered
     - » Systems Thinking
   - Multiple Platforms
     - » Virtual Teaching
     - » Virtual Schools
     - » Virtual Learning
     - » Teacher Websites
     - » Distance Learning
     - » Online Courses
     - » Virtual Collaboration in Classrooms
   - Global Awareness and Cultural Literacy
   - Real-Time Learning
   - Balanced Assessment System
     - » Formative
     - » Student Involved
     - » Performance-based
     - » Summative

The purpose of Data-Driven Decision Making (D3M) in the current HELIOS grant is to establish a systematic approach to collecting, organizing, and analyzing student learning data that allows educators to monitor academic growth that occurs as a result of initiatives taken during the project. To model this process, project directors used the ongoing data collection for the evaluation of the project to monitor participant progress and to adjust their professional development. Each professional development session involved some examination of their data.

To help participants transfer this learning to the task of monitoring their students' progress, the project coordinators decided to establish benchmark assessments that assessed student learning on project-based learning. Math teachers had previously used tests such as Saxon Math. Science teachers had previously used assessments from texts or common grade-level assessments created by teaching staff. This project provided professional development to math and science teachers on a Balanced Assessment System that included student-involved, performance-based, formative, and summative assessments.

In the second year, the Summer Academy focused on creating a Balanced Assessment System. The result was better PBLs, and the assessments posted on the website reflected a deeper understanding of Balanced Assessment. Examples of PBLs are listed on the NGSL website under Project-based Learning tab at *https://sites.google.com/site/ngslarea/project-based-learning*.

The project developed a digital communication system where the information is delivered in a true 21st-century model. As an example, the NGSL developed a group website where each participant was required to design and maintain a professional website which contained their communication system with students, parents, and administrators. As a resource for administrators, specific to all Pinal County teachers, information delivery is provided through access to the website, including professional development materials, college course materials, teacher-developed materials, and student projects. Digital literacy has been facilitated by electronic communication with parents in the form of electronic flyers, emails, and surveys. In addition, tablets are used in the classroom to facilitate both communication and collaboration among students and teachers, as well as between classrooms and schools.

## Conclusions

1. *Professional development must be agile and embrace participant diversity.* Like most programs, teacher participants never started with the same levels of knowledge, skills, and abilities, or willingness to embrace change, or progressed at the same rate. In addition, over time, teachers replacing participants lost to attrition had to be absorbed into ongoing collaborations. Teachers changed schools and districts, disrupting on-site collaborations and leaving participating administrators short a STEM teacher leader. The project organizers discovered that all multiyear programming must be conceived as an open system, subject to a wide variety of perturbations. The nature and intent of programming should be resilient under change. They developed an ongoing data collection system to assess needs and agile programming that could address learners' needs wherever they were in their personal development. This agility allowed the project to embrace a newer, more comprehensive model of 21st-century teaching and learning that better reflected their evolving identity, and to incorporate new technologies in weeks/months, rather than years.
2. *Use the product to teach the process.* Similarly, project organizers discovered that their participants often got mired in the process of developing and implementing cross-discipline project-based STEM lessons. Developmentally, this occurred at different stages of the process for different teachers. As programming evolved, they found a more effective professional development model would be to start with the product and use it to discover the underlying process. That is, starting with the STEM Immersion Matrix or a PBL, participants should use the product, collect data, assess, and reflect on current strengths and needs, then modify and cycle through the process

again. This professional development model will be developed more fully in year three as the teacher leaders train more of their colleagues at their schools. This model will be supported by the development of a STEM lesson rubric.

3. *Align STEM education with district initiatives.* Initially, there was little thought given to how the standards-focused STEM lessons aligned to their district's curriculum maps and pacing guides. Administrators also did not have the adequate background education on STEM education, so they were reluctant to integrate STEM into their regular education. They were considering implementation after AIMS testing, so as not to interfere with standards-based instruction. After learning this, project leaders realized that the student achievement measurement was inadequate. Project leaders made a decision to involve administrators in the project guidance, including development of an administrator cohort. A decision was made for teacher leaders to use the "backward design model" so that PBLs were aligned to the *Common Core State Standards, Mathematics* as well as district planning guides. The administrator cohort assisted the project leaders to set an altered direction for the grant in the sense that the original goals for the project were maintained, but so the grant also met the administrator goals for standards-based instruction and increased student achievement in math and science.

4. *Retool the infrastructure of the virtual learning community.* The NGSL project has developed a successful virtual learning community by providing training and support to build professional websites for all teacher participants, the project, and administrators. To date, this infrastructure has been used as a repository for sharing information, posting PBLs and other materials. In year 3, they envision using this infrastructure for more direct instruction for the participants. Virtual professional development will assume a greater role for sharing, analysis of participants' teaching activities, reflection, and so on.

5. *Remove roadblocks to student technology use.* Initially, many of the participating rural school teachers needed basic technology tools such as document cameras. In year 2, each participant was provided with a tablet for use in the classroom. Additional sets of tablets also were made available for loaning to whole classrooms using a simple online reservation tool at the NGSL website. As the teacher participants' technology infusion skills increased, district and school policies on privacy and internet use obstructed inter-school student collaborations as part of the learning process. These policies still hamper this virtual platform. To keep up with future projects, gatekeeper's policies will need to be reviewed and revised to better serve student learning in this virtual world we now all live in.

6. *Increase student and parent exposure to higher education programs.* Initially the project directors thought that students attending isolated events were sufficient to increase student and parent exposure to higher education programs. This had limited success for a couple of reasons. First, it was difficult to work around district privacy policies that prevented regular correspondence with students and families. When NGSL organizers started a web form for parents to sign up to receive information, over 350 parents signed

up. Second, the attrition of contributing staff at the partnering college reduced participation by new staff that had different priorities. Finally, they realized that in order to get students to consider STEM post-secondary education and careers, the county needed to develop a pathway for these students beginning in elementary school.

7. *Develop a resilient evaluation plan.* The evaluation plan and data collection for the NGSL project were derived from a logic model developed at the outset of the project. Each year the logic model and evaluation plan were revised to better represent the evolving program. This also meant revision of some of the measures, with some being modified throughout the course of the project (e.g., surveys) and others discarded as they became less relevant to meaningful evaluation (e.g., RTOP in year 3). It is an ongoing challenge with program design to know in advance how to track measurements that are meaningful, consistent, and allow change in the program.

## Evidence of Success From the Two Initiatives

"Teachers are leaders when they function in professional communities to affect student learning, contribute to school improvements, inspire excellence in practice, and empower stakeholders to participate in educational improvements" (Childs-Bowen, Moller, and Scrivner 2000, p. 28).

In most professions, as the practitioner gains experience, he or she has the opportunity to exercise greater responsibility and assume more significant challenges. This is not true of teaching as it can be a flat profession. The 20-year veteran's responsibilities are essentially the same as those of the newly licensed novice. Therefore, in many settings, the only way for a teacher to extend his or her influence is to become an administrator. This desire for greater responsibility, if left unfulfilled, can lead to frustration and even in some cases dropping out of the profession all together. As shown by these two initiatives, developing strong site-based math and science leadership can provide for effective changes not only in the classrooms but within the system in which they are working. The research shows these teachers will have greater buy-in in school policy and curricular decisions, increased student achievements, and will become better communicators to parents and the whole community.

Both STEM in the Middle and the Next Generation of STEM Leaders incorporated the same approach to developing the skills and knowledge of math and science middle school teams. Their design approach, although differing in some aspects, both have shown success because they encompass the same basic key elements. They include:

- Develop STEM content and pedagogical knowledge,
- Provide a variety of strategies for implementing STEM in their own classrooms,
- Create standard-based STEM instruction materials aligned with assessment,
- Nurture and develop presentation skills in order to work with other teachers and administrators, and
- Provide and maintain connectivity between all the participants both web-based and in Professional Learning Communities.

As the quote from Rodger Bybee at the beginning of this chapter says, "reform of education and reforming education are two very different activities." In order for us to move beyond the STEM slogan and into classroom practice, the first step in making this a reality is to work on effectively changing teacher practices. Both of these initiatives have shown that working with a concentrated set of math and science teachers over a sustained period of time will produce change in their practices. The difficulty in this is trying to take this to scale. Both of these projects are now moving into their third year of implementation. The evaluation goal for Year 3 is to start to measure their influence upon their system and begin to track student achievements. This will be especially true within the Next Generation grant as most of these teachers are in the grades 6–8 span at a school site. This will be the biggest push from Helios within the evaluation for both of these projects. Until the final results are in, seeing the change in teacher practices, and their confidence, and attitudes toward teaching is considered all payback for our investments.

## References

Bybee, R. W. 1995. Science curriculum reform in the United States. In *Redesigning the science curriculum*, ed. R. W. Bybee and J. D. McInerney, 12–20. Colorado Springs, CO: BSCS.

Childs-Bowen, D., G. Moller, and J. Scrivner. 2000. Principals: Leaders of leaders. *National Association of Secondary School Principals (NASSP) Bulletin* 84 (616): 27–34.

Darling-Hammond, L., M. LaPointe, D. Meyerson, M. Orr, and C. Cohen. 2007. *Preparing school leaders for a changing world: Lessons from exemplary programs.* Stanford, CA: Stanford Educational Leadership Institute.

DeVita, C. M., R. L. Colvin, L. Darling-Hammond, and K. Haycock. 2007. *Education leadership: A bridge to school reform*. New York: Wallace Foundation.

# Integrating Science and Engineering in the Elementary Classroom

*Christine M. Cunningham and Cynthia Berger*
*Museum of Science*
*Boston, MA*

## Setting

Engineering is Elementary (EiE) is a project of the Museum of Science, Boston. The project addresses America's pressing need for effective STEM education by developing and disseminating K–8 curricula. EiE also develops and facilitates teacher professional development workshops to help educators implement these curricula. This chapter will focus on the project's flagship curriculum, Engineering is Elementary. The idea of developing an engineering curriculum for elementary-school-age students was a natural outgrowth of Dr. Christine Cunningham's experiences leading professional development for K–12 teachers to help them integrate engineering with the science, math, or technology they were already teaching. Contradicting the prevailing wisdom of the time that very young children cannot learn to engineer, Dr. Cunningham concluded that engineering education should—and could—start at a very young age. She and her team began developing the EiE curriculum in 2003.

The development of the EiE curriculum was also a natural match for the comprehensive strategy already in place at the Museum to foster technological literacy in schools nationwide. To this end, the Museum established the National Center for Technological Literacy (NCTL), which includes among its stated goals (1) integrating engineering as a new discipline in schools via standards-based K–12 curricular reform, (2) advancing technological literacy by designing K–12 engineering materials, and (3) offering educators professional development in the teaching of engineering. The EiE project is housed within NCTL.

The first phases of EiE curriculum development were supported by the Intel Foundation and subsequently by the National Science Foundation. The development team followed a rigorous process, investing more than 3,000 hours in each of the 20 EiE curriculum units and pilot testing each unit over a two-year period, first in Massachusetts classrooms and then in other classrooms nationwide. The curriculum was completed in 2011, with additional support for curriculum resources and EiE teacher professional development coming from Raytheon, S. D. Bechtel Jr. Foundation, The Liberty Mutual Foundation Inc., The Cargill Foundation, Cognizant, National

Institute of Standards and Technology, Cisco Foundation, and the Google Community Grants Fund of the Tides Foundation. Since its release, the curriculum has been widely disseminated. It is used in every state in the nation, and as of April 2014 had reached more than 65,000 teachers and 4.8 million students.

The core EiE team currently consists of about three-dozen staff members who focus on curriculum development, professional development, and research and evaluation. EiE relies on a national network of more than 50 collaborators to offer additional EiE professional development workshops, helping to support the teaching of elementary engineering in schools in their regions.

This chapter will lay out the rationale behind the design of the EiE curriculum. We will also present evidence, drawn both from EiE studies and from external evaluations, to demonstrate how the EiE curriculum supports exemplary science learning and how EiE PD workshops support exemplary science teaching.

## Overview of the EiE Curriculum

Engineering is Elementary (EiE) has been flexibly designed for use by students in grades 1–5. Teachers can choose from 20 units that focus on different fields of engineering—from traditional fields like electrical and mechanical engineering to less well-known specialties such as bioengineering and acoustical engineering. Each unit consists of a series of hands-on, project-based lessons that culminate in an engineering design challenge; students use a five-step engineering design process to explore a problem and engineer solutions. EiE is more than an engineering curriculum, however. Each unit is designed to integrate with a science topic commonly taught in the elementary grades and to support science learning. The units are also designed to support learning in other subject areas, including literacy, social studies, and math.

Though development of EiE started before the new *Next Generation Science Standards* (*NGSS*) were developed and released, the curriculum anticipated and addresses the goals set forward in *NGSS*. During the development process, EiE was expressly designed to meet the STEM objectives outlined in the National Science Education Standards (NSES) released by the National Research Council in 1996. The teacher guide for each EiE unit includes detailed tables that enable educators to map unit content with both NSES and *NGSS*. Educators can also cross reference the commonly used science curricula that EiE was designed to supplement (FOSS, STC, and GEMS).

NSES divides science topics into three categories: physical science, life science, and Earth science. Using this framework, EiE curriculum designers identified the science topics most commonly taught in grades 1 through 5 in each category, then developed an engineering design challenge for each unit that integrated with one of these science topics (Table 24.1). Teachers who are considering adopting the EiE curriculum can use an interactive application on the project website to search for the science topics they teach (or plan to teach) and find the EiE unit that integrates with that topic.

NSES differed from previous frameworks in how it moved the emphasis in K–12 education away from the mastery of lists—facts and vocabulary words—and toward the development of critical-thinking skills. The vision is that students should learn science by actively engaging in inquiry—by asking questions, organizing ideas, exploring and evaluating information, and

**Table 24.1.** Engineering is Elementary Curriculum Units and Corresponding Science Topics

| | SCIENCE TOPIC | UNIT TITLE | ENGINEERING FIELD | STORY SETTING |
|---|---|---|---|---|
| EARTH SCIENCE | Water | **Water, Water Everywhere:** Designing Water Filters | Environmental | India |
| | Air & Weather | **Catching the Wind:** Designing Windmills | Mechanical | Denmark |
| | Earth Materials | **A Sticky Situation:** Designing Walls | Materials | China |
| | Landforms | **A Stick in the Mud:** Evaluating a Landscape | Geotechnical | Nepal |
| | Astronomy | **A Long Way Down:** Designing Parachutes | Aerospace | Brazil |
| | Rocks | **Solid as a Rock:** Replicating an Artifact | Materials | Russia |
| LIFE SCIENCE | Insects/Plants | **The Best of Bugs:** Designing Hand Pollinators | Agricultural | Dominican Republic |
| | Organisms/Basic Needs | **Just Passing Through:** Designing Model Membranes | Bioengineering | El Salvador |
| | Plants | **Thinking Inside the Box:** Designing Plant Packages | Package | Jordan |
| | Ecosystems | **A Slick Solution:** Cleaning an Oil Spill | Environmental | USA |
| | Human Body | **No Bones About It:** Designing Knee Braces | Biomedical | Germany |
| PHYSICAL SCIENCE | Simple Machines | **Marvelous Machines:** Making Work Easier | Industrial | USA |
| | Balance & Forces | **To Get to the Other Side:** Designing Bridges | Civil | USA |
| | Sound | **Sounds Like Fun:** Seeing Animal Sounds | Acoustical | Ghana |
| | Electricity | **An Alarming Idea:** Designing Alarm Circuits | Electrical | Australia |
| | Solids & Liquids | **A Work in Process:** Improving a Play Dough Process | Chemical | Canada |
| | Magnetism | **The Attraction is Obvious:** Designing Maglev Systems | Transportation | Japan |
| | Energy | **Now You're Cooking:** Designing Solar Ovens | Green | Botswana |
| | Floating & Sinking | **Taking the Plunge:** Designing Submersibles | Ocean | Greece |
| | Light | **Lighten Up:** Designing Lighting Systems | Optical | Egypt |

solving problems. The EiE curriculum aligns with this vision, using guided inquiry to engage students in a series of scaffolded lessons. A corollary to the NSES goal of "engaging students in inquiry" is the goal of accommodating the inquiry process by planning classroom activities and investigations that can stretch over extended periods of time instead of being confined to a single class period. EiE specifically helps teachers meet this goal with a series of lessons that build over 6–12 class periods.

To see how EiE engages students in inquiry—and also supports the process by allowing for extended investigations—let's step through the lessons that make up an EiE unit. Every unit includes a preparatory lesson that helps to address the misconceptions young children often hold around the terms "engineering" and "technology." The prep lesson engages students in an activity called "Technology in a Bag" where the class divides into small groups and each group gets a "mystery bag" with an object inside. As children examine the object and discuss what materials it's made from, whether it's natural or human-made, and who made it, they build understanding of what technology is and what engineers do. The prep lesson sets the stage for the four lessons that follow.

Lesson 1 is built around a storybook that introduces the premise for the engineering design challenge still to come. We'll use the EiE unit "Lighten Up: Designing Lighting Systems" as an example. This unit focuses on the field of optical engineering, and the Lesson 1 storybook is about an Egyptian boy, Zane, who needs a way to illuminate the stage for a school performance during an electrical brownout. Lesson materials include a series of questions teachers can use to promote student reflection before, during, and after the story. Lesson 1 may occupy two to three class periods.

Lesson 2 is designed to give students an overview of a particular field of engineering; students learn about the kinds of work an engineer in this field does. The activities in Lesson 2, which take one or two class periods, explicitly build on what students have learned in the storybook lesson. For example, students will have read about Zane's brother, who is an optical engineer; now they learn that optical engineers design devices to make light do something useful. Then they systematically explore the properties of light, testing how light can be reflected, transmitted, or absorbed using an ordinary flashlight and a set of commonplace materials such as foil, tissue paper, and black construction paper.

EiE's curriculum developers chose the activities in Lesson 3 to illuminate the linkages between science, mathematics, and engineering—and in particular to explore the science concepts that integrate with the unit. In our example, the Lighten Up unit, students investigate how light travels. They collect sets of measurements of light intensity and angle of reflection—again, using simple materials such as mirrors, flashlights, and protractors—then analyze their data. The activities in Lesson 3 may take one to two class periods.

Finally, in Lesson 4 students engage in an engineering design challenge. Using an engineering design process based on the processes used by practicing engineers but modified by EiE to be suitable for young children, the students imagine, design, and build a technology that solves a problem; then, they evaluate and improve on their solution. In this Lighten Up unit, the problem to solve is "how to illuminate the hieroglyphics inside an Egyptian tomb." Working in small groups over the course of one to three class periods, students use what they've learned in the previous lessons about the properties of light to light up a shoebox-size model tomb. (It's worth noting that in addition to building on the earlier lessons, this lesson also connects to the social studies content in the storybook, which describes how archaeologists in the early 20th century illuminated Egyptian tombs by reflecting light inside, using mirrors—an alternative to using smoky kerosene lamps that could harm the painted tomb interiors.)

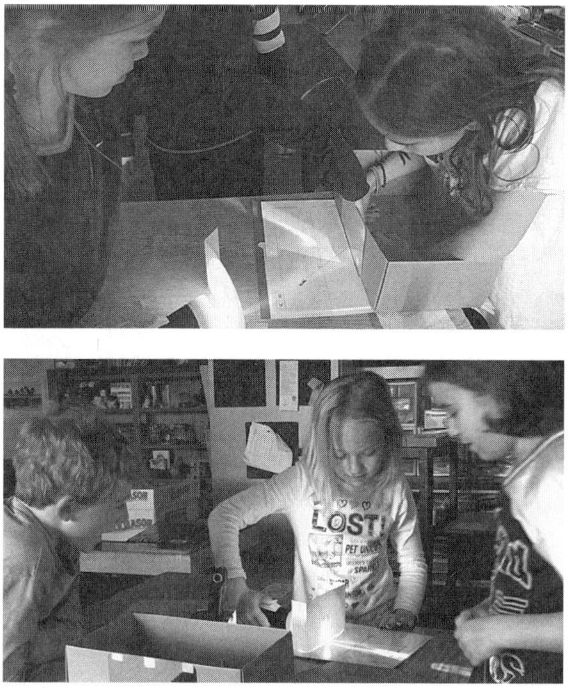


In the "Lighten Up" unit students investigate how light travels. They collect sets of measurements of light intensity and angle of reflection—again, using simple materials such as mirrors, flashlights, and protractors—then analyze their data.

An important corollary to the *NSES* goal of "engaging students in inquiry" is the goal of having students investigate real-world questions—questions that all students will find interesting, important, and relevant to their own lives. *NSES* also looks to move the emphasis of instruction away from studying a science subject in the abstract, for its own sake, and toward learning subjects in context—for example, from an historical or social perspective.

Each EiE unit expressly supports these goals. Every storybook and every engineering design challenge is based on real-world events and relevant questions. A good example is the EiE unit Now You're Cooking: Designing Solar Ovens, which focuses on green energy engineering and integrates with the teaching of physical science concepts related to energy, such as conductors and insulators. The storybook protagonist is a girl in Botswana who wants to improve the design of her family's solar oven; this story is based on real events in Botswana today, where cooking fires contribute to air pollution and deforestation, and where the government and nonprofit groups are working actively to improve the design of solar ovens and promote their use.

Another striking example of a real-world engineering design challenge is the EiE unit A Stick in the Mud: Evaluating a Landscape, which focuses on geotechnical engineering and integrates with earth science subjects such as landforms and erosion. The storybook introduces a boy in a riverside village in Nepal, where leaders are considering replacing the stepping-stones across the river with a safer gondola-like bridge called a TarPul. The boy uses what he learns about geotechnical engineering from a visiting engineer to suggest a safe site for the bridge—and to help his father understand this solution, which at first seems counter to cultural values because the location of the bridge must move. The storybook was inspired by a real nonprofit organization that installs TarPul bridges in Nepalese villages; students in one fourth-grade class in Denver were so moved by the story, they voted to donate the proceeds from their annual fundraiser to this nonprofit.

In addition to reflecting real-world events, EiE's engineering design challenges have been chosen so that each one can be easily scaled to the needs of all students. *NSES* specifically lays out this expectation—that *all* students can develop the knowledge and skills the standards describe. Engineering design challenges typically have, not one, but many different solutions, so they naturally allow students of varying academic and physical abilities to develop their own solutions. Both internal and external assessments demonstrate that the EiE curriculum supports science and engineering learning for all students, regardless of such variables as gender, cultural or ethnic background, socioeconomic status, or ability. Preliminary research suggests that EiE helps engender classroom equity, as students who may not be considered conventionally "smart" (getting good grades) may be regarded by their peers as "smart engineers."

In allowing for multiple solutions, engineering design challenges make a dramatic contrast to "cookbook" science activities that direct students toward a single correct outcome or answer. Another contrasting feature is that with engineering design challenges, there's no stigma attached to failure. Students are expected to evaluate and improve their designs as part of the overall engineering design process; if the first design fails, students use what they've learned to try again. This aspect of the EiE curriculum supports the *NSES* goal of reassigning emphasis from "getting the answer" to "using evidence to develop or revise an explanation."

EiE lessons often engage students in learning by having them work in groups—pairs or small teams of three to four students. This design was a conscious decision in line with the *NSES* goal to move away from a traditional mode of classroom discourse, where students communicate their ideas and conclusions to the teacher, and toward a framework where students communicate their ideas and work to their classmates. EiE videographers have collected footage showing students working on EiE units; in lesson after lesson you can see children as young as six working as high-functioning teams: collaborating, exercising creativity as they brainstorm, making arguments from evidence, negotiating, and compromising.

*NSES* also has the explicit goal of supporting career awareness, and EiE was designed to meet this goal through the storytelling in Lesson 1. Every EiE storybook features a character (a relative, friend, or neighbor of the child protagonist) who is an engineer and serves as a mentor. The mentor affirms the child's ability to solve his or her problem through engineering—and this message is conveyed to students as they read the story, so that they start to see themselves as future engineers. Lesson 2 also supports career awareness as students learn more about what engineers do and the real-world problems they solve. Both lessons communicate that engineering is a helping profession—that engineers make life better for other people—a message that's particularly appealing to groups currently underrepresented in engineering careers, including women and minorities. Finally, Lesson 4 builds students' engineering efficacy; after they have designed solutions to the engineering problems, many students proclaim that they want to continue to "be engineers."

We've focused on how *NSES* set goals for how teachers teach, but *NSES* also laid out goals for how teachers learn. For example, *NSES* expected that administrators would provide professional development for teachers; effective professional development is particularly important for elementary teachers, who often feel inadequately prepared to teach STEM subjects. Recognizing this need, the EiE project developed and now offers several different professional development workshops. These workshops reflect the following principles:

- Formative assessments are a good place to start. EiE workshop facilitators conduct brief, informal assessments to learn what participants already know about the subject.
- Foundational knowledge is essential. Teachers start by building their knowledge of technology, of engineering as a profession, and of the Engineering Design Process.
- Activities proceed through group work and discussion.
- Hands-on, learner-driven learning is highly effective. Workshop facilitators do very little lecture-style teaching, focusing instead on learner-driven experiences.
- Workshop participants have a change to "reflect as learners," considering the activities they complete from the perspective of student rather than teacher.
- When teachers experience the lessons from a student's point of view, they better understand where students may require extra support.
- Participants also have a chance to reflect as educators, which helps prepare them to address student misconceptions and practice effective classroom management.
- Workshop facilitators demonstrate effective pedagogical strategies, for example, using open-ended questions and encouraging participants to give evidence and explain the rationale behind their ideas.

EiE educators have facilitated hundreds of professional development workshops, both at our Boston headquarters and in locations across the country. At the end of each workshop, participants complete a survey we developed in consultation with Horizon Research; this survey has provided us with information about participants' perceptions of the workshop components, their understanding of key concepts, and how prepared they are to implement engineering instruction in their classrooms.

Our research shows that that EiE's approach to PD is highly effective. After the workshop experience, teachers report greatly increased confidence in their ability to teach the engineering units, improved understanding of engineering content, and better awareness of how science, technology, and engineering interrelate. After an EiE workshop, teachers are also more likely to agree with the assertion, "Engineering instruction CAN be implemented in an elementary classroom."

## Major Features of the Instructional Program Needed for Success With STEM

For much of the 20th century, educational reform policies in the United States focused on the amount and distribution of educational resources and the structural features of American education. In recent decades, educational reform has focused more on the role of instructional content in learning. Most nations have national standards for instructional content; America, however, leaves curriculum decisions to states or school districts. This patchwork of educational standards has had a negative impact on the quality of instruction, especially in STEM subjects.

One recent reform that aims to present clear and united goals for student learning is the *Common Core State Standards* initiative, which lays out standards in English language arts and mathematics that states can adopt voluntarily. To date 45 states, the District of Columbia, four territories and the Department of Defense Education Activity have adopted the *Common Core State Standards*.

Some of the philosophy underlying the *Common Core State Standards* is similar to that of the new *NGSS*. Both the *Common Core* math standards and *NGSS* stress procedural skill and conceptual understanding; both sets of standards call on students to apply mathematical thinking to real world challenges; and both make career readiness a goal. Like *NGSS*, the *Common Core State Standards, English Language Arts* value small-group work. Computer-based assessments designed for the *Common Core State Standards* connect to some of the *NGSS* objectives as well—they measure complex, higher-order thinking and problem-solving skills, include open-ended questions, and emphasize the understanding of ideas, not just knowing the answer. Although there are no *Common Core State Standards* for engineering, our team members are currently working to connect *Common Core* expectations with EiE—for example, we now offer a workshop that helps educators meet *Common Core* math standards through engineering activities.

## How Does EiE Exemplify the *Next Generation Science Standards*?

It's been more than 15 years since the release of *NSES*. Since then, our understanding of how students learn has advanced, leading to further reform and the new vision of science teaching presented in the *NGSS*. One way these new standards differ significantly from *NSES* is in specifying that both science *and* engineering should be integrated for students in grades K–12. *NGSS* expects that students will actively engage in both science *and* engineering practices while applying so-called "crosscutting concepts" to improve their understanding of the core ideas in both of these fields. The ultimate vision behind *NGSS* is that students will understand how STEM subjects are relevant to their everyday lives, so that as adults they will be prepared to make informed decisions in a participatory democracy—and also be both interested in and prepared for STEM careers.

The specific engineering practices that students are expected to engage in under *NGSS* include the following:

- Ask questions
- Define problems
- Develop and use models
- Plan and carry out investigations
- Analyze and interpret results
- Use mathematical and computational thinking
- Construct explanations (science) and design solutions (engineering)
- Engage in argument from evidence
- Obtain, evaluate, and communicate information

When students use the EiE curriculum, they use all of these practices as they engage with the engineering design challenge in each unit. To see how this works, we'll use the EiE unit called "The Best of Bugs: Designing Hand Pollinators" as an example. The focus of the unit is agricultural engineering, and the engineering design challenge is to build a handheld device (i.e., a technology) that can be used to successfully pollinate a particular kind of flower.

Students start by **defining the problem.** The storybook for Lesson 1, *Mariana Becomes a Butterfly*, sets the scene. Students read about a girl in the Dominican Republic who receives a gift that she can grow in her garden—a berry plant that is native to Hawaii. The problem is, the plant doesn't actually produce any berries in its new home.

This story reinforces what students understand about the science content in the unit, including pollination as a system function, the challenges of maintaining balance in an ecosystem, the concept of integrated pest management, and how a native ecosystem may be affected by introduced species. At the same time, the story defines a problem: How do you get a plant to set seed when it has no natural pollinators?

Having defined the problem, students start **asking questions** related to the engineering design challenge, which is to design a so-called "hand pollinator"— a technology that a farmer could, in the absence of an effective insect pollinator, use to pollinate a crop plant. To address this challenge, students have to **develop and use models**—including model flowers (made from test tubes and petri dishes), model pollen (represented by baking soda), and model hand pollinators (devices constructed from materials such as pipe cleaners, tape, or pompoms).

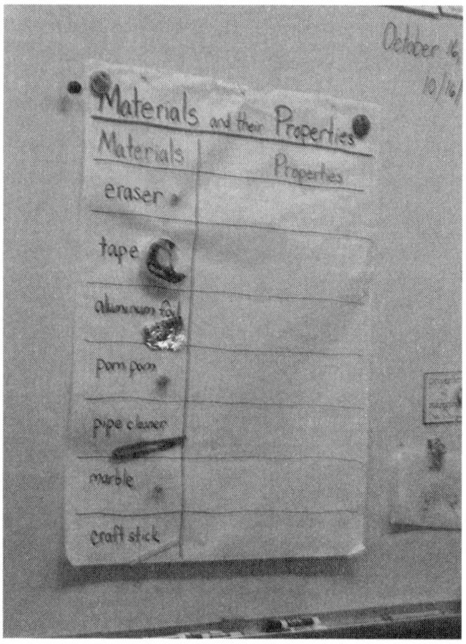


Before they can engineer a hand pollinator, though, students must **plan and carry out** a structured, controlled **investigation** that helps them understand more about the scientific and engineering principles behind the challenge. Specifically they need to investigate which materials will work best to pick up and drop off pollen.

The first step is to look at the building materials that are available. The class works together to describe the properties of each material provided (is it smooth or rough? heavy or light? and so on). Students predict which properties will be important for designing an effective hand pollinator and which will not.

Next, students work in small groups to conduct a series of controlled experiments, testing how well each of the materials picks up and deposits the "pollen." This is where modeling comes in. Students look at a real flower, which has male (stamen) and female (pistil) parts, to learn where pollen is located. Then they brainstorm how to create a **model** flower they can use to perform their tests and measure outcomes. The class tests each of the materials independently to determine how well each picks up and drops off pollen. Then all the groups share their data with the class, and as a class, the students **analyze and interpret** the results and discuss which materials worked best for the task.

At this point students often make connections between their models—the fuzzy or smooth materials they have tested—which work best (the fuzzy ones) and how this mirrors nature: the mass of fine hairs on a bee's hind legs collect and distribute pollen. Students also discuss discrepant events. Why did different groups get different results? Did they use different test methods?

With their testing complete, students continue the engineering design process as they work in pairs to design, build, and test a hand pollinator. Until now, students have worked with simple model flowers where the pollen is easy to reach; for the design challenge, they must create a hand pollinator that will work with a new model flower resembling an orchid or jack-in-the pulpit—flowers that have particularly hard-to-reach pollen. (Figure 24.1 shows four different kinds of flowers, and the models that stand in for each kind of flower. For example, the Dutchman's pipe, which has a tube-shaped flower, is modeled by a test tube.) These models introduce a new parameter into the design of the hand pollinator—not only must the tip of the device be able to pick up and drop off pollen, but the handle must be designed so the tip can reach the pollen.

Working independently, each child sketches two to four **models** of hand pollinators, and then shares these sketches with a partner. The partners must agree which design to build, a step that requires students to **construct explanations or arguments** about why they think one design is particularly strong. With a design selected, the partners **plan** in greater detail. What materials will the device be made from? What will it look like? They draw a more refined model of their design and identify the materials they will use to build it.

## Figure 24.1. The Four Flowers and Their Models

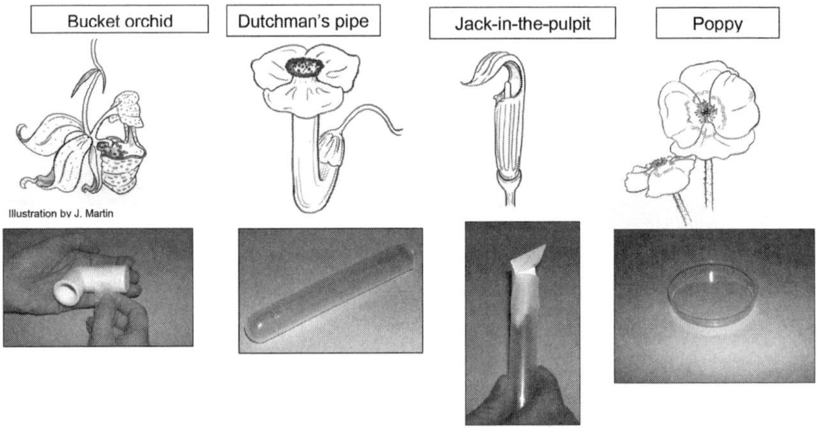


The next step is to **carry out an investigation**—to gather data about how the design performs. The partners **analyze and interpret data**—was the hand pollinator able to pick up pollen? To drop off pollen? How do they know? They identify the parts of the pollinator (their **designed solution**) that work well and discuss what they want to change, and why. As they think about ways to improve their design, students must **engage in argument from evidence** and **evaluate and communicate information.**

Finally, students redesign their pollinators, build the new devices, and once again, test—then reflect on whether or not the redesigned pollinators actually improve on the initial designs, and the evidence for their conclusions. Each pair of students shares their sketch and their final product with the class, again exercising their communication skills as they share the reasons for choosing certain materials, show what changes they made to improve the design, demonstrate how their hand pollinator works, and explain why their design worked with their model flower.

All 20 of the EiE units were designed with science and engineering practices in mind. But another *NGSS* goal is for students to understand the core ideas in a range of science and engineering fields. EiE units also address this expectation. Each unit focuses on the core ideas in a particular field of engineering, integrating these ideas with one or more science topics commonly taught at the elementary level. As an example, Table 24.2 shows the core ideas in Life Sciences (LS) and Physical Sciences (PS) that the Best of Bugs unit addresses.

**Table 24.2.** *NGSS* Disciplinary Core Ideas Progression Present in Best of Bugs Unit

| Number | Disciplinary Core Idea | Sub-idea |
|---|---|---|
| **LS1.A** | Structure and function | All organisms have external parts that they use to perform daily functions. |
| **LS2.A** | Interdependent relationships in ecosystems | Plants depend on water and light to grow. Plants also depend on animals for pollination and/or to move their seeds around. |
| **PS1.A** | Structure of matter | Matter exists as different substances that have observable different properties. Different properties are suited to different purposes. Objects can be built up from smaller parts. |

Table 24.3 (p. 434) shows how students master *NGSS* performance standards related to the core ideas listed above as they complete the *Best of Bugs* unit.

## *Who Are the Collaborators?*

As part of the curriculum development process that began in 2003–2004, EiE staff developed relationships with a number of different collaborators who have helped us to improve, evaluate, and disseminate the curriculum. One group of collaborators is comprised of the hundreds of elementary school teachers who volunteered to field test the units in their classrooms, providing valuable feedback on curriculum content and design. Teachers and teacher educators also volunteered to field test the professional development workshops that EiE developed to support curriculum implementation. To amplify our ability to provide teacher professional development outside of our home base in Massachusetts, EiE has developed collaborative relationships with dozens of organizations and individuals across the country that are qualified to deliver EiE PD workshops. Finally, EiE staff collaborated with university faculty in educational research and in curriculum evaluation and assessment.

**Table 24.3.** *NGSS* Performance Expectations Consistent With Best of Bugs Unit

| Number | Disciplinary Core Idea | Performance Expectation |
|---|---|---|
| **2-PS1-1** | Matter and Its Interactions | Plan and conduct an investigation to describe and classify different kinds of materials by their observable properties. |
| **2-PS1-2** | Matter and Its Interactions | Analyze data obtained from testing different materials to determine which materials have the properties that are best suited for an intended purpose. |
| **2-LS2-2** | Ecosystems: Interactions, Energy, and Dynamics | Develop a simple model that mimics the function of an animal in dispersing seeds or pollinating plants. |
| **K-2-ETS1-1** | Engineering Design | Ask questions, make observations, and gather information about a situation people want to change to define a simple problem that can be solved through the development of a new or improved object or tool. |
| **K-2-ETS1-2** | Engineering Design | Develop a simple sketch, drawing, or physical model to illustrate how the shape of an object helps it function as needed to solve a given problem. |
| **K-2-ETS1-3** | Engineering Design | Analyze data from tests of two objects designed to solve the same problem to compare the strengths and weaknesses of how each performs. |

## Evidence for Success With Students

### *Varied Users of the Program*

As of 2013, the EiE curriculum has been used at thousands of schools nationwide, by more than 65,000 teachers and more than 4.8 million children. Our curriculum is used in individual classrooms and schoolwide implementations; in districtwide implementations (for example, Minneapolis, Minnesota; Harford County, Maryland; and Orange County, North Carolina); and in one statewide implementation—Delaware—that involves 19 school districts, 22 charter schools, and nearly 50,000 students.

The EiE curriculum has been adopted by schools in a variety of settings, including urban, suburban, and rural areas. Most often EiE is adopted by public elementary schools, but it is also used by private schools and some middle schools. Homeschooling parents and out-of-school time programs also use EiE. Finally, a number of college- and university-based teacher preservice programs use EiE to help their students understand how to integrate engineering with science instruction.

EiE's research team has been conducting assessments and evaluations of the curriculum since 2005. In addition, three external organizations—Davis Square Research Associates of Somerville, Massachusetts; Campbell-Kibler Associates of Groton, Massachusetts; and Horizon Research of

Chapel Hill, North Carolina—have evaluated various aspects of the EiE curriculum. Links to reports and publications that present the results of these evaluations are available on the EiE website (*www.eie.org*). Building on this foundation of careful evaluation, EiE staff are currently (as of 2014) engaged in an NSF-funded "gold-standard" efficacy study of the curriculum; this research will be complete and the results released in 2016.

All EiE evaluations have collected information on how diverse student populations—including racial and ethnic minorities, English as a Second Language students, and low-income students—respond to the curriculum. Evaluations of the EiE curriculum have addressed the following questions:

- What do students understand about technology before engaging with the EiE curriculum? How does their understanding of technology change after they engage with the curriculum?
- What do students understand about engineering before they have experienced classroom instruction in this field? How does their understanding of engineering change after they engage with the EiE curriculum?
- Do students' attitudes toward science and engineering change after they engage with EiE?
  - Do students see science as a possible career path? Does this attitude change after engaging with EiE?
  - Do students see engineering as a possible career path? Does this attitude change after engaging with EiE?
  - Does experience with EiE influence students' interest in science and engineering?
- How does EiE affect students from varied demographic groups?
- Can EiE help close the achievement gap?
- Do teachers who have completed an EiE PD workshop teach the curriculum more effectively than teachers who have not been through a workshop?

## *Evidence Gathered*

Although Americans are surrounded by technologies (the products of engineering) in their daily lives, many people—and in particular many elementary students—don't have a clear understanding of what the term "technology" means or what constitutes engineering. When the EiE curriculum was under development almost a decade ago, few researchers had systematically explored student understanding and conceptions (or misconceptions) of technology and engineering. Scholars had no consensus of the concepts and process understandings that comprise technological literacy.

To address this information gap, the EiE research team has developed and employed the following evaluation and assessment tools over the past nine years:

- "Draw an Engineer" test
- "What is Engineering?" instrument
- "What is Technology" instrument
- "Engineering Interests" survey

- "Engineering Attitudes" survey
- Unit-specific assessments

The "Draw an Engineer" test was one of the earliest diagnostics EiE created. It's based on the "Draw a Scientist Test" developed by Chambers (1983) and was initially administered to more than 500 students in grades K–12. The test asks, "What is an engineer?" and students answer by drawing a picture of an engineer at work, and by describing, in writing, the picture they've drawn. This survey provided evidence that many students have a limited and/or incorrect understanding of what engineers do.

We used this information to develop a pair of survey instruments to more systematically assess how students' understanding of engineering and technology grows when they engage with EiE. The "What is Engineering?" instrument asks students to examine 16 images and descriptions of people doing different kinds of work; students are asked to circle the pictures of engineers and to complete the phrase: "An engineer is a person who…." The corresponding "What is Technology?" instrument shows 16 images and descriptions of everyday items (for example, shoes, a bridge, a tree, a book, and a lightning bolt). Students circle the items that they identify as technologies.

EiE's research team has repeatedly reviewed and improved the "What is Engineering?" and "What is Technology?" instruments and has administered them to more than 6,000 students in grades 1–5. To measure growth in understanding, the instruments are administered to students before and after they learn with the EiE curriculum. Before engaging with EiE, children typically identify engineering as being the work of construction workers and auto mechanics; or if they do understand what an engineer does, it's within the field of civil engineering, in particular building construction. Students strongly identify the term "technology" with devices that require electricity and only rarely connect everyday human-made objects such as shoes or books with technology. After engaging with EiE, students show dramatically improved understanding of technology and engineering; they are better able to identify technologies and to correctly discriminate between engineering work and non-engineering work.

## *Engineering Attitudes Survey*

EiE's Engineering Attitudes Survey for elementary students was adapted from a survey originally developed for middle school students. It consists of 20 statements about science and engineering as careers and as enterprises that affect society as a whole; students can agree/disagree on a scale of 0 to 5. Survey items include statements such as "I would enjoy being an engineer when I grow up," "I would like a job where I can invent things," and "Scientists help make people's lives better."

One EiE study reports the results after this survey was administered to 1,056 racially and socioeconomically diverse fourth- and fifth-grade students from classrooms in California, Florida, Massachusetts, New Hampshire, and Rhode Island. Some of these students were assigned to a test group and used the EiE curriculum along with a science curriculum over the course of the school year. A control group of students used the same science curriculum as the test group but did not use EiE. Students completed the survey at the beginning of the school year, and then again at the end of the year, after the test group had used EiE.

The test students who had experienced the EiE curriculum were significantly more likely than control students to agree that they would enjoy being an engineer. EiE students were also significantly more likely to agree that they were interested in—and felt comfortable with—engineering jobs and skills, and to agree that scientists and engineers make people's lives better.

Another EiE study that looked for demographic differences in students' attitudes towards engineering using data collected from 3,950 students in grades 3–5 provides evidence that EiE works well for all students. Students in this study were identified by race (White, Asian, Black, Hispanic, and other), by gender, and by whether or not they participated in reduced-price lunch or limited English proficiency programs, and completed the survey before and after using the EiE curriculum.

## *Voices of Instructors/Students Targeted*

The central mission of EiE is to foster engineering and technological literacy for *all* students. With this mission in mind, EiE chose the central characters for the 20 EiE storybooks with the goal that every student would be able to identify with at least one character. The storybook children are boys and girls of different races and from different cultures. They enjoy a variety of hobbies, have varied personalities, live in a number of different family structures, and some have disabilities (including a child in a wheelchair, a child who is blind, and a child with Down syndrome).

Our aim was to show that everyone can engineer, and it's common in an EiE classroom to hear a child exclaim that "[name of storybook character] looks like me," or say that a particular protagonist is his or her favorite. Many teachers have told us that an immigrant child from, for example, India or the Dominican Republic became the focus of positive attention from classmates because the story was set in their country of origin. One teacher told us about a class that had a long and sensitive discussion about Down syndrome after reading the storybook featuring this character, because a child in the class had a sibling with this condition. Another teacher told us this story:

> *When I first started the unit, before the extra books arrived, I used Paulo's Parachute Mission as a read-aloud, and we had great discussions on many levels from the problems Paulo faced [he moved to a new town where he had to make new friends].... The kids loved this story and almost spoke as if Paulo was their own personal friend.... It was really cool.*

Besides representing a range of real-world people, EiE storybooks also anchor the unit activities in the real world. Students often evoke the storybook characters or settings as they work on their solutions. In one second-grade class, for example, students who were working to choose a site for the Nepalese gondola-like bridge called a TarPul (described earlier) referred to the context set by the story to justify their design criteria: The boy in the story is concerned that the bridge should be located near his home because his grandmother has health problems and sometimes needs to be transported across the river to the clinic. In the dialogue that follows, the children (assigned pseudonyms here) have used small weights to model the number of people the bridge could carry:

| | |
|---|---|
| *Teacher:* | How many people could you put in your TarPul to cross the river? |
| *Sam:* | Two... |
| *Ellie:* | Three people... |
| *Teacher:* | How about the rest of you? Jenny? |
| *Jenny:* | ...four people... |
| *Observer:* | What would be the safest number? |
| *Alex:* | Two people. |
| *Teacher:* | Why would two be the safest? |
| *Ellie:* | Because if you were with your family, you might need to be with someone else. |
| *Jenny:* | If someone can't move, they can't pull themselves across. |
| *Ellie:* | And if you're a little kid with your mother, or if you're sick, you need someone to help you get across. |

The children are able to draw on the sick grandparent scenario in the storybook to conclude that a safe bridge design should accommodate at least two people crossing together.

Other comments from teachers show us that EiE really does help children learn that failure is a necessary part of the engineering process. Testing a design to failure can be fun for students—and highly engaging, as one teacher told us, "When we tested our earthquake models, the kids were delighted to be able to shake the daylights out of the structures, and they were amazed that they remained intact." When failure is set in a positive framework, children embrace the chance to hone their thinking and improve their solutions, as the following remarks from teachers show:

- Several students kept bringing in play dough that they made at home. They wanted to improve until they perfected it. When they brought it in, many wanted to share the improvements they made and why they felt it was successful. —*Grade 4 Teacher*
- I assigned an extra credit to try and improve the [play dough] recipe with further research at home. All 21 students participated! Parents communicated that the local super market was doing a booming business with packages of flour! —*Grade 3 Teacher*
- The majority of my students have continued to test materials and build parachutes during their free time (recess!) —*Grade 3 Teacher*

Communication skills are a common thread connecting the engineering practices specified in *NGSS*. Teachers often tell us that they see significant improvements in their students' ability to collaborate, work in teams, and share information, as these anecdotes illustrate:

- Three students, one from Jamaica, one from Sri Lanka and the third, a Bosnian, who never do anything together, were teamed up for this project. [The design challenge was to create a moist habitat for a frog by engineering a permeable membrane.] I marveled at their willingness to accept and test each other's ideas and analysis without conflict. Their need to test their work and "save their frog" really helped to focus their efforts in a very positive way. To observe them working was a delight, the different perspectives each brought to the task were blended to engineer a very successful membrane. *–Grade 5 teacher*

- As the teacher I got to observe over 90 students being engineers. The best thing for me was how well students got along with other students that they would not usually work with. I saw new friendships build as they worked in teams to make their parachutes. As problems came up, the teams learned that they had to work them out with each other using nice words. Some of the lower students became the leaders of the group when it was hands-on time. *—Grade 3 Teacher*

EiE activities have been specifically designed to appeal to and be accessible to all children. In addition to evaluations that provide evidence the curriculum has met this goal, we share these anecdotes. The first is from an elementary education specialist:

> *Students are placed in the room based on their IEPs [Individualized Education Plan]. Some are role models. Some children have behavioral challenges. But—these [EiE] lessons really kept them focused and engaged. They really anticipated each next step, and I feel like it was an outlet for their energy.*

Meanwhile a very skilled kindergarten teacher told us this:

> *This year I had three children with physician diagnosed ADHD. The entire year, they were constantly out of their seats moving around every 5–10 minutes until we started our EiE unit. I was shocked to observe that while they were engaged in the engineering challenges, all of them were focused and on-task for up to 40-minute stretches. It was remarkable!*

And a fourth-grade teacher told us this compelling story:

> *I worked with a very challenging student population; many of my students had been exposed to trauma.... This was often manifested through a lack of motivation ... as well as outbursts of anger.... That all changed when my class began working on the Engineering is Elementary chemical engineering unit, in which students design a process to make play dough.... My students became so motivated that almost half of the class brought in play dough samples and processes that they had developed at home.... We had a comparison of using wheat flour versus corn flour, as many of my Mexican and Central American students had access to corn flour at home rather than wheat, which brought in a cultural dimension ... that gave them pride in their identities. By using only materials that even my students of most limited resources had at home, the unit made the concepts accessible to all students. Several students ended the unit saying that they wanted to become engineers as adults, because that was something that they knew that they had the ability to do.*

## Next Steps

EiE has a number of new initiatives underway to build on the foundation already established.

1. *Developing curricula for out-of-school time programs.* EiE is applying the methods and philosophies laid out in this chapter to develop two out-of-school time engineering curricula. "Engineering Adventures" is designed for children in grades 3 – 5; "Engineering Everywhere" is for middle-school-aged children. The units for both curricula are available as free downloads from the EiE website; development of additional units continues.
2. *Conducting additional evaluation of the EiE curriculum.* In 2013 EiE began a four-year, NSF-funded, gold-standard efficacy test of the EiE curriculum. "Exploring the Efficacy of Elementary Engineering" (E4) involves field tests in hundreds of classrooms in four states.
3. *Expanding our professional development options.* EiE has recently released two new workshops. "Improving Your EiE Practice" is designed for teachers who already use the curriculum and want to build their skills; "Linking the E & M in STEM" helps teachers integrate EiE with Common Core State Standards in math.
4. *Establishing EiE "Hub Sites."* Support from the Raytheon Company and Cognizant has enabled us to establish hub sites for professional development. We are building partnerships with organizations that offer teacher professional development in Arizona, Alabama, Connecticut, Texas, New York City, and the Washington, DC, area so that schools in these regions have convenient access to EiE PD.
5. *Expanding our network of collaborators.* In addition to the formal Hub Sites, we are also building relationships with professional development providers in other locations. This improves our ability to support schools that implement EiE anywhere in the nation.

## References

Carson, R., and P. Campbell. 2007. *Museum of science: Engineering is elementary; impact on teachers with and without training*. Groton, MA: Campbell-Kibler Associates.

Cunningham, C. M. 2008. Elementary teacher professional development in engineering: Lessons learned from engineering is elementary. Paper presented at the National Academy of Engineering, Washington, DC.

Cunningham, C. M., and C. P. Lachapelle. In press. Designing engineering experiences to engage all students. In *Engineering in precollege settings: Research in practice*, ed. J. Strobel, S. Purzer, and M. Cardella. Rotterdam: Sense Publishers.

Cunningham, C. M., and C. D. Schunn. In review. Engineering practices. In *Supporting next generation scientific and engineering practices in K–12 classrooms, ed.* C. Schwarz, C. Passmore, and B. Reiser. NSTA Press.

Jocz, J., and C. Lachapelle. 2012. *The impact of engineering is elementary (EiE) on students'conceptions of technology.* Boston, MA: Museum of Science.

Lachapelle, C. P., J. D. Hertel, J. Jocz, and C. M. Cunningham. 2013. Measuring students' naïve conceptions about technology. Presented at the NARST Annual International Conference, Rio Grande, PR.

Sargianis, K., S. Yang, C. M. Cunningham. 2012. Effective engineering professional development for elementary educators. Presented at the American Society for Engineering Education Annual Conference, San Antonio, TX.

# Endword: STEM Successes and Continuing Challenges

*Herbert Brunkhorst*
*California State University, San Bernardino*
*San Bernardino, CA*

This NSTA Exemplary Science Programs (ESP) monograph illustrates a variety of programs that individuals, teams, districts, institutions, foundations, and institutions of higher education have used over the past decade to encourage science, technology, engineering, and mathematics (STEM) in response to the *National Science Education Standards* (*NSES*), *Common Core State Standards* (*CCSS*), *A Framework for K–12 Science Education*, *Next Generation Science Standards* (*NGSS*), and *21st Century Skills*. This monograph provides readers with exciting demonstrations of how STEM education has been embraced as a means of providing students and teachers with models of STEM teaching and learning that are authentic, contextual, and problem-based.

The chapters describe 24 exciting programs and approaches to various forms of integrated STEM education. The programs range from traditional classroom settings, charter/magnet schools, informal education settings, after-school and summer activities, as well as university projects. The monograph also includes suggestions for developing new statewide STEM initiatives. The Iowa model is particularly detailed enough to provide real insights into how states can address a comprehensive approach to K–16 STEM education.

Authors of each program provide settings, overviews of their programs, as well as responses to various standards as they share the major features of their efforts. Each set of authors provides evidence for success and describes the next steps for improving their programs. Some of the program descriptions are more robust than others. Although one group of authors point out that STEM education has been around since the late 19th century, the programs described in this monograph go far beyond the reform ideas during earlier times. Though a lack of consensus and specific definitions for STEM education remain, readers now have a wide variety of descriptions for STEM education, including evidences of their successes. Though there is a wide variety of environmental emphasis, engineering design emphasis, and mathematics emphasis, each of the programs use some common elements that resulted in their selection by the NSTA ESP Board of Reviewers. Each program is attentive to the *NSES*, *NGSS*, *Common Core State Standards* and *21st Century Skills*. The use of inquiry in real-life situations and an emphasis on problem-based learning are both important elements to encourage others to use these principles in the teaching and learning of STEM ideas. The chapters also show evidence of student engagement and learning in STEM contexts. A broad variety of strategies have been used to report student assessment of learning with STEM efforts and skills.

Several of the programs describe students planning and carrying out problem-based investigations, analyzing and interpreting data, using mathematics and computational thinking, and engaging in arguments regarding evidences of success. Other chapter authors describe how digital technologies enhance the teaching and learning of STEM concepts, including the use of virtual environments to extend STEM classrooms. All of the programs illustrate a high degree of collaboration—be it with students, teachers, administrators, university faculty, business/industry representatives, or institutions such as the American Museum of Natural History.

Several of the programs describe the importance of teacher professional development and provide a variety of excellent models for intensive and extended opportunities for both teachers and administrators. These include peer coaching, summer academies, and apprenticeships. Several of the models immerse teachers in the same kinds of inquiry and problem-based learning experiences and offer specific plans for engaging their own students in STEM practices.

The evidence of the impact of STEM education approaches on students, teachers, administrators, and parents include a wide variety of assessment tools—both quantitative and qualitative. Many of the programs make extensive use of pre/posttests, especially regarding STEM content. Many of the programs have tools such as open-ended responses and foci for scientific and engineering designs. Other forms of assessment include investigations, portfolios, video recordings, projects, a variety of attitude and career survey instruments, and opportunities for students to share their findings and projects with adults. In a vast majority of the programs, the evidence illustrates that the instructional approaches advocated by the NRC Science *Framework*, *Common Core State Standards*, *Next Generation Science Standards*, and *21st Century Skills* for improving education are real reforms that have significant impacts. Students display increased interest in STEM fields, greater motivation and achievement in STEM subjects, greater awareness of career opportunities in STEM areas, and increased workforce skills, including college readiness.

The programs targeting underserved students show similar gains for such students. One particular exemplary middle school engineering program serving a diverse population of students, with 73% qualifying for free-or-reduced lunch, describes exceptional assessment procedures including product fairs, runway shows, portfolio reviews with outsiders, panel reviews, service learning projects, engineering fairs, and community presentations. These formative assessments can help teachers to improve instruction continually and to develop students into self-directed learners.

In the programs emphasizing teacher professional development, evidence indicates similar gains in teacher understanding of STEM efforts, greater comfort with STEM-related pedagogical approaches, and more positive attitudes toward the new standards and STEM integration. Programs that include parents in their data collection report that parents often express their approval and pleasure of seeing their students more focused, motivated, and having a better understanding of the STEM reforms as indicators of student learning. The programs involving digital-based STEM teaching and learning (cyber learning), provide evidence that indicate similar significant gains in achievement. Some include unique assessment techniques like automated screen capture.

Each of the programs provides new exemplary ideas for the teaching and learning of STEM skills. These programs are well situated to continue to add to our understanding of student learning, teacher professional development, curricula, and teaching strategies, including digital learning, for suggesting ways to best achieve exciting K–16 STEM education. The exemplary programs demonstrate that STEM education can work at all age levels and in a variety of contexts within various educational systems.

It is hoped that this monograph will provide roadmaps and directions to those wishing to initiate STEM education programs in their schools, districts, and regions. The exemplars can jumpstart those who have wanted to begin such programs and learning about some new steps to take. Those wanting to start programs are encouraged to contact the authors who have taken the plunge. It is this kind of systemic collaboration that will improve STEM education and change education generally with students in the driver's seat.

It is important to emphasize that science is defined as "human exploration of the natural universe and seeking explanations of the objects and events encountered." Technology provides ways for all to understand our world. Engineering is the application of science-mathematics principles. Mathematics focuses on measurement tasks and techniques. All STEM features are needed to provide real learning and understanding in all educational settings. We want it to take less than 75 years to succeed (the time table suggested by Project 2061) to accomplish major educational changes. We hope the programs described in this monograph provide ample evidence for meeting the successes and emphasis of STEM efforts for all—and soon!

# Contributors

**Leonard Annetta,** coauthor of *Mission Biotech: Using Technology to Support Learner Engagement in STEM*, is a professor of science education at George Mason University, Fairfax, Virginia.

**Cynthia Berger,** coauthor of *Integrating Science and Engineering in the Elementary Classroom*, is the outreach program manager for Engineering is Elementary at the Museum of Science, Boston, Massachusetts.

**Angelette M. Brown,** author of *Middle School Engineering Education*, is a middle school engineering teacher at Columbia Academy, Columbia Heights, Minnesota.

**Barry N. Burke,** author of *Integrating Technology and Engineering in a STEM Context*, is director of STEM Center for Teaching and Learning at the International Technology and Engineering Educators Association, Reston, Virginia.

**Todd Campbell,** coauthor of *STEM Education in the Science Classroom: A Critical Examination of Mathematics Manifest in Science Teaching and Learning*, is an associate professor of science education at the University of Connecticut, Connecticut.

**Jie Chao,** coauthor of *Mixed-Reality Labs: Combining Sensors and Simulations to Improve STEM Education*, is an education researcher at the University of Virginia, Charlottesville, Virginia.

**Jennifer L. Chiu,** coauthor of *Mixed-Reality Labs: Combining Sensors and Simulations to Improve STEM Education*, is an assistant professor at the University of Virginia, Charlottesville, Virginia.

**Matthew Cieslik,** author of *Introducing STEM to Middle School Students: A World of Excitement and Inquiry*, is a teacher at the Rosa International Middle School, Cherry Hill, New Jersey.

**Renee M. Clary,** author of *Integrating Interdisciplinary STEM Approaches for Meaningful Student Learning*, is an associate professor of Geosciences at Mississippi State University, Mississippi State, Mississippi.

**Disa Lubker Cornish,** coauthor of *A State STEM Initiative Takes Root, Blossoms*, is an assistant professor at the University of Northern Iowa School of Health, Physical Education, and Leisure Services, Cedar Falls, Iowa.

**Alicia A. Cotabish,** coauthor of *STEM Starters: An Effective Model for Elementary Teachers and Students*, is an assistant professor of teaching and learning at the University of Central Arkansas, Conway, Arkansas.

**Christine M. Cunningham,** author of *Integrating Science and Engineering in the Elementary Classroom*, is the founder and director of Engineering is Elementary and a vice president at the Museum of Science, Boston, Massachusetts.

**Debbie Dailey,** coauthor of *STEM Starters: An Effective Model for Elementary Teachers and Students*, is an assistant professor of teaching and learning at the University of Central Arkansas, Conway, Arkansas.

**Jennifer Lynne Eastwood,** coauthor of *Mission Biotech: Using Technology to Support Learner Engagement in STEM*, is an assistant professor at Oakland University William Beaumont School of Medicine, Rochester, Michigan.

**Fred Estes,** author of "Rolly Pollies, Bubbles, and Wheelies: Inquiry STEM in the Early Childhood Classroom," is a teacher and LS science coordinator and science specialist at The Nueva School, Hillsborough, California.

**Allan Feldman,** coauthor of *STEMRAYS: After-School STEM Research Clubs*, is a professor of science education at the University of South Florida, Tampa, Florida.

**Margaret M. French,** coauthor of *STEM Education in the Science Classroom: A Critical Examination of Mathematics Manifest in Science Teaching and Learning*, is a doctoral student at the University of Massachusetts Dartmouth, Dartmouth, Massachusetts.

**Cheryl Christine Frye,** coauthor of *STEM Challenges and Academic Successes*, is the district STEM coach and teacher at Menifee Valley Middle School, Menifee, California.

**Jyoti Gopal,** coauthor of *Science in Our Backyard: How a School Is Turning Its Grounds Into a Living Lab*, is a kindergarten teacher at Riverdale Country School, Bronx, New York.

**S. Selcen Guzey,** coauthor of *Learning Genetics at the Nexus of Science, Technology, Engineering, and Mathematics (STEM)*, is a research associate at the STEM Education Center, University of Minnesota, St. Paul, Minnesota.

**Tony Hall,** coauthor of *STEM Starters: An Effective Model for Elementary Teachers and Students*, is an associate professor at the University of Arkansas at Little Rock, Little Rock, Arkansas.

**Mary Hanson,** author of *Urban STEM: Watch It Grow!*, is a science teacher at Humboldt High School in St. Paul, Minnesota.

**Margie Hawkins,** author of *STEM Education in the Middle School Classroom*, is a science teacher at Winfree Bryant Middle School in Lebanon, Tennessee.

**Erin O. Heiden,** coauthor of *A State STEM Initiative Takes Root, Blossoms*, is a senior research scientist at the University of Northern Iowa Center for Social and Behavioral Research in Cedar Falls, Iowa.

**Gail D. Hughes,** coauthor of *STEM Starters: An Effective Model for Elementary Teachers and Students*, is a professor at the University of Arkansas at Little Rock, Little Rock, Arkansas.

**Melinda Jodoin,** coauthor of *STEM Challenges and Academic Successes*, is a teacher in the Menifee Union School District in Menifee, California.

**Karen E. Johnson,** author of *Promoting STEM Practices for All Students*, is a K–12 science coordinator at Adams 12 Five Star Schools in Thornton, Colorado.

**Mari Kemis,** coauthor of *A State STEM Initiative Takes Root, Blossoms*, is an assistant director of the Research Institute for Studies in Education at Iowa State University in Ames, Iowa.

**Angela M. Kohnen,** coauthor of *STEM Literacy Through Science Journalism: Driving and Communicating Along the Information Highway*, is an assistant professor at Missouri State University in Springfield, Missouri.

**Terri Ladd,** coauthor of *STEM Challenges and Academic Successes*, is a teacher at Menifee Valley Middle School in Menifee, California.

**Robert E. Landsman,** coauthor of *RIP ~ing Through STEM*, is president of ANOVA Science Education Corporation in Honolulu, Hawaii.

**Elizabeth Marie Lehman,** coauthor of *Building TECHspertise: Enhancing STEM Teaching and Learning With Technology*, is a curriculum developer at the Center for Elementary Mathematics and Science Education at the University of Chicago, Chicago, Illinois.

**Deborah Arron Leslie,** coauthor of *Building TECHspertise: Enhancing STEM Teaching and Learning With Technology*, is senior curriculum developer at the Center for Elementary Mathematics and Science Education at the University of Chicago, Chicago, Illinois.

**Mollianne G. Logerwell,** coauthor of *Like a Scientist: Using Problem-Based Learning to Connect Practice With Content in STEM Education*, is director of science education for the Virginia Initiative for Science Teaching and Achievement at George Mason University, Fairfax, Virginia.

**David E. Long,** coauthor of *Like a Scientist: Using Problem-Based Learning to Connect Practice With Content in STEM Education*, is director of internal research for the Virginia Initiative for Science Teaching and Achievement at George Mason University, Fairfax, Virginia.

**Tamara J. Moore,** coauthor of *Middle School Engineering Education* and *Learning Genetics at the Nexus of Science, Technology, Engineering, and Mathematics (STEM)*, is an associate professor of engineering education at Purdue University, West Lafayette, Indiana.

**Shelly A. Muñoz,** coauthor of *STEM Challenges and Academic Successes*, is a science teacher at Menifee Valley Middle School in Menifee, California.

**Alan Newman,** coauthor of *STEM Literacy Through Science Journalism: Driving and Communicating Along the Information Highway*, is a research professor at the University of Missouri–St. Louis, St. Louis, Missouri.

**Edward A. Pan,** coauthor of *Mixed-Reality Labs: Combining Sensors and Simulations to Improve STEM Education*, is an education researcher at the University of Virginia in Charlottesville, Virginia.

**Ella Pastor,** coauthor of *Science in Our Backyard: How a School Is Turning Its Grounds Into a Living Lab*, is an environmental education facilitator at the Riverdale Country School in Bronx, New York.

**Taylor Joanelle Predmore,** coauthor of *STEM Challenges and Academic Successes*, is a statistician at Menifee Valley Middle School in Menifee, California.

**Phillip A. Reed,** coauthor of *Integrating Technology and Engineering in a STEM Context*, is an associate professor at Old Dominion University in Norfolk, Virginia.

**Ann Robinson,** coauthor of *STEM Starters: An Effective Model for Elementary Teachers and Students*, is a professor and director of the Jodie Mahoney Center at the University of Arkansas at Little Rock, Little Rock, Arkansas.

**Gillian H. Roehrig,** coauthor of *Middle School Engineering Education* and *Learning Genetics at the Nexus of Science, Technology, Engineering, and Mathematics (STEM)*, is an associate professor of science education at the University of Minnesota, St. Paul, Minnesota.

**William L. Romine,** coauthor of *Mission Biotech: Using Technology to Support Learner Engagement in STEM*, is an assistant professor of biology at Wright State University in Dayton, Ohio.

**Troy D. Sadler,** coauthor of *Mission Biotech: Using Technology to Support Learner Engagement in STEM*, is a professor of science education at the University of Missouri, Columbia, Missouri.

**Wendy Saul,** coauthor of *STEM Literacy Through Science Journalism: Driving and Communicating Along the Information Highway*, is Allen B. and Helen S. Shopmaker Professor of Education and International Studies at the University of Missouri–St. Louis, in St. Louis, Missouri.

**R. Gordon Schaubhut,** coauthor of *RIP ~ing Through STEM*, is a senior engineering manager at R/G Engineering, Inc., in Glen Rock, New Jersey.

**Donna R. Sterling,** coauthor of *Like a Scientists: Using Problem-Based Learning to Connect Practice With Content in STEM Education*, is a professor of science education and director of the Virginia Initiative for Science Teaching and Achievement at George Mason University in Fairfax, Virginia.

**Morton M. Sternheim,** coauthor of *STEMRAYS: After-School STEM Research Clubs*, is a professor of physics emeritus and director of the STEM Education Institute at the University of Massachusetts in Amherst, Massachusetts.

**Jo Anne Vasquez,** author of *Developing STEM Site-Based Teacher and Administrator Leadership*, is vice president of educational practices at the Helios Education Foundation in Phoenix, Arizona.

**Robert H. Voelkel, Jr.,** coauthor of *STEM Challenges and Academic Successes*, is an administrator at the Menifee Valley Middle School in Menifee, California.

**Lisa Waller,** coauthor of *STEM Challenges and Academic Successes*, is a teacher in the Menifee Union School District in Menifee, California.

**James H. Wandersee** (deceased), coauthor of *Integrating Interdisciplinary STEM Approaches for Meaningful Student Learning*, was an endowed professor emeritus in biology education at the Louisiana State University in Baton Rouge, Louisiana.

**Jeffrey D. Weld,** coauthor of *A State STEM Initiative Takes Root, Blossoms*, is executive director of the governor's STEM council and associate professor of biology at the University of Northern Iowa, Cedar Falls, Iowa.

**John G. Wells,** coauthor of *Integrating Technology and Engineering in a STEM Context*, is an associate professor at Virginia Polytechnic Institute and State University in Blacksburg, Virginia.

**David J. Welty,** coauthor of *STEM Education in the Science Classroom: A Critical Examination of Mathematics Manifest in Science Teaching and Learning*, is a teaching and learning supervisor at Fairhaven High School in Fairhaven, Massachusetts.

**Stephen B. Witzig,** coauthor of *STEM Education in the Science Classroom: A Critical Examination of Mathematics Manifest in Science Teaching and Learning*, is an assistant professor of science education at the University of Massachusetts Dartmouth, in Dartmouth, Massachusetts.

**Yael Wyner,** author of *Ecology Disputed: A Model Approach to STEM Education That Brings Ecology, Daily Life Impact, and Scientific Evidence Together in Secondary School Science Classrooms*, is an assistant professor of secondary education at City College of New York, part of City University of New York, New York City..

# Index

*Page numbers printed in **boldface** type refer to tables or figures.*